Hydronic Heating
Systems and Applications

Donald L. Steeby
Professor of Heating, Air Conditioning, and Refrigeration
Grand Rapids Community College
Grand Rapids, MI

Publisher
The Goodheart-Willcox Company, Inc.
Tinley Park, IL
www.g-w.com

Copyright © 2022
by
The Goodheart-Willcox Company, Inc.

All rights reserved. No part of this work may be reproduced, stored, or transmitted in any form or by any electronic or mechanical means, including information storage and retrieval systems, without the prior written permission of
The Goodheart-Willcox Company, Inc.

Manufactured in Canada.

Library of Congress Control Number: 2020943955

ISBN 978-1-64564-652-5

1 2 3 4 5 6 7 8 9 – 22 – 25 24 23 22 21 20

The Goodheart-Willcox Company, Inc. Brand Disclaimer: Brand names, company names, and illustrations for products and services included in this text are provided for educational purposes only and do not represent or imply endorsement or recommendation by the author or the publisher.

The Goodheart-Willcox Company, Inc. Safety Notice: The reader is expressly advised to carefully read, understand, and apply all safety precautions and warnings described in this book or that might also be indicated in undertaking the activities and exercises described herein to minimize risk of personal injury or injury to others. Common sense and good judgment should also be exercised and applied to help avoid all potential hazards. The reader should always refer to the appropriate manufacturer's technical information, directions, and recommendations; then proceed with care to follow specific equipment operating instructions. The reader should understand these notices and cautions are not exhaustive.

The publisher makes no warranty or representation whatsoever, either expressed or implied, including but not limited to equipment, procedures, and applications described or referred to herein, their quality, performance, merchantability, or fitness for a particular purpose. The publisher assumes no responsibility for any changes, errors, or omissions in this book. The publisher specifically disclaims any liability whatsoever, including any direct, indirect, incidental, consequential, special, or exemplary damages resulting, in whole or in part, from the reader's use or reliance upon the information, instructions, procedures, warnings, cautions, applications, or other matter contained in this book. The publisher assumes no responsibility for the activities of the reader.

The Goodheart-Willcox Company, Inc. Internet Disclaimer: The Internet resources and listings in this Goodheart-Willcox Publisher product are provided solely as a convenience to you. These resources and listings were reviewed at the time of publication to provide you with accurate, safe, and appropriate information. Goodheart-Willcox Publisher has no control over the referenced websites and, due to the dynamic nature of the Internet, is not responsible or liable for the content, products, or performance of links to other websites or resources. Goodheart-Willcox Publisher makes no representation, either expressed or implied, regarding the content of these websites, and such references do not constitute an endorsement or recommendation of the information or content presented. It is your responsibility to take all protective measures to guard against inappropriate content, viruses, or other destructive elements.

Front cover images: nikkytok/Shutterstock.com; Olga_Ionina/Shutterstock.com; Alexxxey/Shutterstock.com

Preface

Because hydronic systems for HVAC applications have come to be increasingly complex, there is a need for a textbook to deliver a comprehensive perspective covering the theory, installation, and service of today's boiler and hydronic systems. *Hydronic Heating: Systems and Applications* discusses these topics and includes areas of hydronic heating systems not currently addressed in other textbooks. Emphasis has been placed on the fundamentals of heating using hydronics as well as understanding system components and their applications. Students using this book will learn the theory and application of heating and boiler functionality and develop an understanding for the service and repair techniques of hydronic systems and controls. They will then practice understanding troubleshooting essentials, which will prepare them for on-the-job practical applications.

This textbook includes colorful and detailed illustrations and helpful technical tips based upon real-life applications. Because boilers and hydronics are used extensively for residential and commercial applications, this textbook is designed for use at colleges and universities, career schools, skill centers, labor and apprenticeship programs, and mechanical contractors as a tool for training new and skilled personnel.

It is the intent of this book to bridge the gap between high level, engineering-driven books, the more simplistic books targeted at the introductory level heating and air conditioning technician, and the "do-it-yourselfer."

About the Author

Don Steeby is a professor of heating, air conditioning, and refrigeration at Grand Rapids Community College (GRCC) where he has taught since 2007. Specifically, he teaches basic steam and hot water boiler applications, as well as training in applicable hydronic and radiant heating systems. Along with his accredited classes, Mr. Steeby trains area HVAC contractors and local manufacturers in steam and hot water boiler applications. Mr. Steeby received his Bachelor of Science degree in HVACR Applied Technology and his Master's degree in Career & Technical Education from Ferris State University.

Previously, Mr. Steeby was an adjunct professor for Ferris State University in construction management. Mr. Steeby has served on the Advisory Committee for HVAC curriculum at Ferris State University and is currently a member of the HVACR Advisory Board for Goodheart-Willcox Publishing. He is also a published author of *Alternative Energy—Sources and Systems*.

Mr. Steeby has held many additional positions in the HVAC field since 1985, including working in the industry for installation, service, sales, and project management. He worked for a manufacturer of HVAC equipment in West Michigan for over five years. He was also employed by Honeywell for 11 years. While at Honeywell, Mr. Steeby acquired extensive experience in the designing and installation of controls for hydronic and hot water systems.

Mr. Steeby holds a Mechanical Contractor's License with the state of Michigan. He has several nationally recognized certifications, including Installation and Service of Geothermal Heat Pumps, Gas and Oil Furnace Installation & Service, and Commercial Refrigeration Installation & Service with North American Technician Excellence (NATE). He is also a Certified Trainer and Installer of Geothermal Heat Pumps through the International Ground Source Heat Pump Association (IGSHPA).

Reviewers

The author and publisher wish to thank the following industry and teaching professionals for their valuable input into the development of *Hydronic Heating*:

Terry Carmouche
River Parish Community College Reserve Campus
Reserve, LA

Bob Clark
College of DuPage
Glen Ellyn, IL

Edward J. Eheander
Porter and Chester Institute
Chicopee, MA

William M. Ginocchio
Lincoln Technical Institute
Mahwah, NJ

Steven Gutsch
Chippewa Valley Technical College
Eau Claire, WI

Samuel A. Heath
Southern Technical College
Brandon, FL

Charles William (Bill) Ledford
Pikes Peak Community College & IntelliTec College
Colorado Springs, CO

J H Meeker
Northern Virginia Community College
Woodbridge, VA

Scott Moore
Fleming College
Peterborough, Ontario

Trevor Root
Algonquin College
Ottawa, Ontario

Anthony Saccavino
Porter and Chester Institute
Stratford, CT

Michael Schofield
J. Oliva Huot Technical Center
Laconia, NH

Terrence Scott
Thomas Nelson Community College
Hampton, VA

Frank J. Sloan
Lincoln Technical Institute
New Britain, CT

Ronald Stevenson
Grand Rapids Community College
Grand Rapids, MI

Richard C. Taylor
Pennsylvania College of Technology
Williamsport, PA

Acknowledgments

The author and publisher would like to thank the following companies, organizations, and individuals for their contribution of resource material, images, or other support in the development of *Hydronic Heating*.

- Air Conditioning Contractors of America (ACCA)
- American Gas Association
- Bacharach
- Bosch Thermotechnology
- Caleffi North America, Inc.
- Central Boiler, Inc.
- City of Holland
- Danfoss
- Dwyer Instruments, Inc.
- Emerson Climate Technologies
- Fenwal Controls
- Field Controls LLC
- Functional Devices, Inc.
- Honeywell, Inc.
- Ideal Industries
- International Code Council
- Master Lock
- Metraflex
- Michael Helsel
- Modine Manufacturing
- Myson
- Mueller Industries, Inc.
- National Museum of American History
- National Renewable Energy Laboratory (NREL)
- NIBCO Inc.
- Rheem Manufacturing Company
- Rinnai
- R.W. Beckett Corporation
- Sealed Unit Parts Co., Inc
- SlantFin (slantfin.com)
- Solar Panels Plus
- Solar Pathfinder
- Sterling
- Superior Refrigeration Products
- Taco Comfort Solutions
- TEC (The Energy Conservatory)
- Tekmar
- Uline
- Ultra-Fin Radiant Floor Heating System
- Uniweld Products, Inc.
- Uponor, Inc.
- U.S. Department of Energy
- Viega, LLC
- Watts
- Webstone
- Xylem, Inc.
- York International Corp.

TOOLS FOR STUDENT AND INSTRUCTOR SUCCESS

Student Tools

Student Text

Hydronic Heating introduces modern hydronic heating systems written specifically for students preparing to become entry-level technicians. This text delivers key information on sizing, installation, service, and troubleshooting required for a successful career in the HVACR field. Students will learn and build on the fundamentals of hydronic heating system design and boiler operation before being introduced to topics such as tankless heating systems, solar thermal storage applications, and outdoor wood burners. In-text features are included in each chapter to provide real-world hydronic heating content. At the end of each chapter, practice questions allow students to recall and retain the concepts covered.

Student Lab Workbook

The lab workbook that accompanies *Hydronic Heating* combines review activities and practice applications that relate to the content of the textbook chapters. Questions designed to reinforce the textbook content help students review their understanding of the terms, concepts, theories, and procedures presented in each chapter. Hands-on lab activities are included to provide an opportunity for students to apply and extend knowledge gained from the textbook chapters.

Instructor Tools

LMS Integration

Integrate Goodheart-Willcox content within your Learning Management System for a seamless user experience for both you and your students. LMS-ready content in Common Cartridge® format facilitates single sign-on integration and gives you control of student enrollment and data. With a Common Cartridge integration, you can access the LMS features and tools you are accustomed to using and G-W course resources in one convenient location—your LMS.

G-W Common Cartridge provides a complete learning package for you and your students. The included digital resources help your students remain engaged and learn effectively:

- **eBook content.** G-W Common Cartridge includes the textbook content in an online, reflowable format. The eBook is interactive, with highlighting, magnification, note-taking, and text-to-speech features.
- **Lab Workbook content.** Students can have access to a digital version of the Lab Workbook.
- **Drill and Practice.** Learning new vocabulary is critical to student success. These vocabulary activities, which are provided for all key terms in each chapter, provide an active, engaging, and effective way for students to learn the required terminology.

When you incorporate G-W content into your courses via Common Cartridge, you have the flexibility to customize and structure the content to meet the educational needs of your students. You may also choose to add your own content to the course.

For instructors, the Common Cartridge includes the Online Instructor Resources. QTI® question banks are available within the Online Instructor Resources for import into your LMS. These prebuilt assessments help you measure student knowledge and track results in your LMS gradebook. Questions and tests can be customized to meet your assessment needs.

Online Instructor Resources (OIR)

Online Instructor Resources provide all the support needed to make preparation and classroom instruction easier than ever. Available in one accessible location, the OIR includes Instructor Resources, Instructor's Presentations for PowerPoint®, and Assessment Software with Question Banks. The OIR is available as a subscription and can be accessed at school, at home, or on the go.

Instructor Resources One resource provides instructors with time-saving preparation tools such as answer keys, editable lesson plans, and other teaching aids.

Instructor's Presentations for PowerPoint® These fully customizable, richly illustrated slides help you teach and visually reinforce the key concepts from each chapter.

ExamView® Assessment Suite Administer and manage assessments to meet your classroom needs. The ExamView® Assessment Suite allows you to quickly and easily create, administer, and score paper and online tests. Included in the assessment suite are the ExamView® Test Generator, ExamView® Test Manager, and ExamView® Test Player. G-W test banks are installed simultaneously with the software. Using ExamView simplifies the process of creating, managing, administering, and grading tests. You can have the software generate a test for you with randomly selected questions. You may also choose specific questions from the question banks and, if you wish, add your own questions to create customized tests to meet your classroom needs.

G-W Integrated Learning Solution

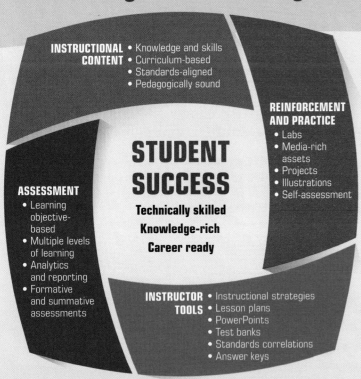

The G-W Integrated Learning Solution offers easy-to-use resources that help students and instructors achieve success.

▶ **EXPERT AUTHORS**
▶ **TRUSTED REVIEWERS**
▶ **100 YEARS OF EXPERIENCE**

EMPLOYABILITY SKILLS · TECHNICAL SKILLS · ACADEMIC KNOWLEDGE · INDUSTRY RECOGNIZED STANDARDS

Features of the Textbook

The instructional design of this textbook includes student-focused learning tools to help you succeed. This visual guide highlights these features.

Chapter Opening Materials

Each chapter opener contains a chapter outline, a list of learning objectives, and a list of technical terms. The **Chapter Outline** provides a preview of the topics that will be covered in the chapter. **Objectives** clearly identify the knowledge and skills to be obtained when the chapter is completed. **Technical Terms** list the key words to be learned in the chapter. The **Introduction** provides an overview to the chapter content.

Additional Features

Additional features are used throughout the body of each chapter to further learning and knowledge. **Safety First** features alert you to potentially dangerous materials and practices. **Code Notes** point out specific items from typical HVACR codes. **Procedures** are highlighted throughout the textbook to provide clear instructions for hands-on service activities. **Tech Tips** provide hands-on guidance that is especially applicable for on-the-job situations. **Green Tips** highlights key items related to sustainability, energy efficiency, and environmental issues. **Timeout for Math** provides key mathematical practice and explanations to on-the-job scenarios. **Did You Know?** features offer interesting facts and tidbits on topics and concepts that provide a bigger picture.

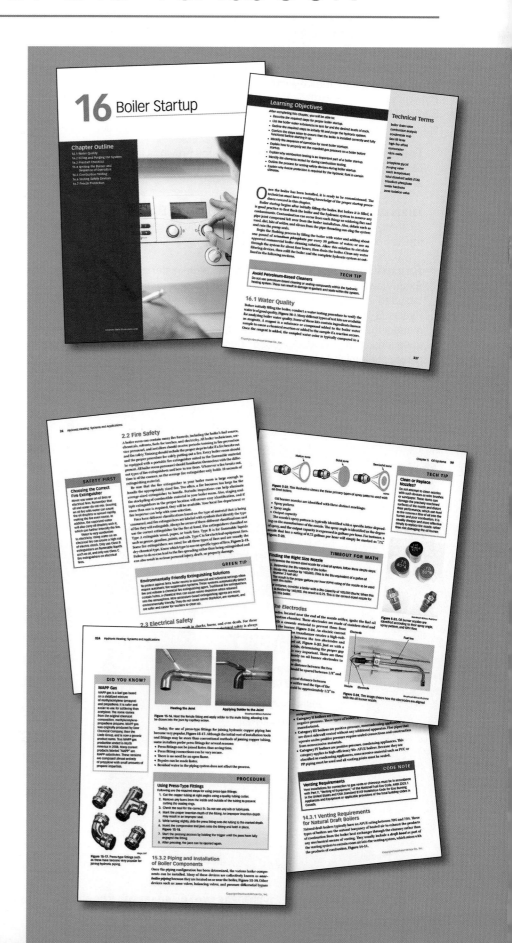

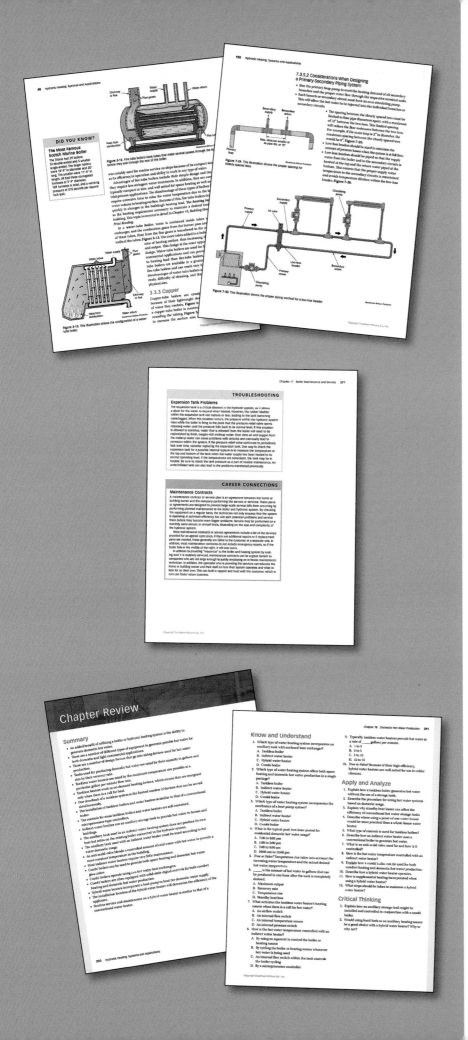

Illustrations

Illustrations have been designed to clearly and simply communicate the specific topic. Illustrations of hydronic heating systems and photographic images are provided to support learning and comprehension for students to successfully enter the field.

Expanding Your Learning

Career Connections explore the professional HVACR field and help you understand what you can expect in the workplace. **Troubleshooting** tips help you develop critical thinking, diagnostic and troubleshooting skills needed in the workplace today.

End-of-Chapter Content

End-of-chapter material provides an opportunity for review and application of concepts. A concise **Summary** provides an additional review tool and reinforces key learning objectives. This helps you focus on important concepts presented in the text. **Know and Understand** questions enable you to demonstrate knowledge, identification, and comprehension of chapter material. **Apply and Analyze** questions extend learning and help you analyze and apply knowledge. **Critical Thinking** questions develop higher-order thinking and problem solving, personal, and workplace skills.

ix

Brief Contents

1. Human Comfort and Heat Transfer 2
2. Safety 20
3. Boilers 40
4. Gas Burners and Ignition Systems 58
5. Oil Systems 84
6. Boiler Fittings and Air Removal Devices 110
7. Hydronic Piping Systems 130
8. Boiler Control and Safety Devices 154
9. Valves 172
10. Circulating Pumps 190
11. Terminal Devices 212
12. Radiant Heating Systems .236
13. Building Heating Loads and Print Reading 270
14. Boiler System Design Considerations 294
15. Boiler Installation 312
16. Boiler Startup 336
17. Boiler Maintenance and Service 348
18. Domestic Hot Water Production 374
19. Solar Thermal Storage 392
20. Outdoor Wood Boilers 422

Contents

Chapter 1
Human Comfort and Heat Transfer ... 2
1.1 Hydronic Heating 4
1.2 Basic Hydronic Configurations 4
1.3 Modern Hydronic System Configurations 7
1.4 Heating and Human Comfort 9
1.5 Benefits of Using Hydronic Heating 12
1.6 Principles of Heat Transfer 13
1.7 Heat Transfer through Piping 14

Chapter 2
Safety ... 20
2.1 Personal Protective Equipment (PPE) 22
2.2 Fire Safety 26
2.3 Electrical Safety 26
2.4 Lockout/tagout 29
2.5 Safety with Tools and Equipment 31
2.6 Working around Pressurized Gases 32
2.7 Chemical Hazards 33
2.8 Working in a Confined Space 35
2.9 Safety in the Boiler Room 35

Chapter 3
Boilers ... 40
3.1 Operating Pressures 41
3.2 Types of Fuel 43
3.3 Construction Materials and Boiler Efficiency . 45
3.4 Condensing Boilers 49
3.5 Modulating Boilers 51
3.6 Electric Boilers 53
3.7 Low NO_X Boilers 54

Chapter 4
Gas Burners and Ignition Systems ... 58
4.1 Principles of Combustion 59
4.2 The Gas Train 60
4.3 Gas Control Valves 64
4.4 Types of Gas-Fired Burners 68
4.5 Ignition Systems 70
4.6 Safety Devices for Burner Ignition Controls .. 78

Chapter 5
Oil Systems ... 84
5.1 Fuel Oil Characteristics 85
5.2 Fuel Oil Combustion 88
5.3 Fuel Oil Piping Systems 90
5.4 Oil Burner Components 94
5.5 Fuel Oil Boiler Venting 105
5.6 Combustion Analysis 105

Chapter 6
Boiler Fittings and Air Removal Devices ... 110
6.1 Pipe Fittings 111
6.2 Pressure and Temperature Gauges 116
6.3 Expansion Tanks 116
6.4 Air Removal Devices 122

Chapter 7
Hydronic Piping Systems ... 130
7.1 The Physics of Water 131
7.2 Steps for Proper Pipe Sizing and System Layout 134
7.3 Piping Arrangements 136

Chapter 8
Boiler Control and Safety Devices ... 154
8.1 Pressure Relief Valves 155
8.2 Low-Water Cutoff 157
8.3 Flow Switch 159
8.4 Backflow Preventer 160
8.5 Aquastat Relay 162
8.6 Outdoor Rest Controller 165
8.7 Electronic Controls 166

Chapter 9
Valves ... 172
9.1 Isolation Valves 173
9.2 Flow Control Valves 174
9.3 Temperature Control Valves 180
9.4 Specialty Valves 185

Chapter 10
Circulating Pumps ... 190
10.1 How a Circulator Works 192
10.2 Types of Circulating Pumps 193
10.3 Circulator Installation and Placement 196
10.4 Circulating Pump Performance 199
10.5 Pump Sizing and Selection 202
10.6 Service and Repair 206

Chapter 11
Terminal Devices 212
- 11.1 Choosing a Piping Configuration 213
- 11.2 Finned-Tube Baseboard Units 215
- 11.3 Radiators 222
- 11.4 Fan Coil Units 225
- 11.5 Unit Heaters 229
- 11.6 Other Types of Terminal Devices 230

Chapter 12
Radiant Heating Systems 236
- 12.1 Principles of Using Radiant Heating 237
- 12.2 Heating Sources for Radiant Systems 240
- 12.3 Radiant Heat Piping Systems 244
- 12.4 Radiant Floor Piping Configurations 246
- 12.5 Radiant Wall and Ceiling Panels 253
- 12.6 Designing Radiant Heating Systems 254
- 12.7 Radiant Heating Controls 261
- 12.8 Radiant Heating for Snow and Ice Melt Systems 263

Chapter 13
Building Heating Loads and Print Reading 270
- 13.1 Performing Building Heat Load Calculations . . . 272
- 13.2 Heat Loss through Conduction 272
- 13.3 Heat Loss through Infiltration 280
- 13.4 Print Reading 283
- 13.5 Types of Blueprints 284

Chapter 14
Boiler System Design Considerations 294
- 14.1 Boiler Selection 295
- 14.2 Combustion and Ventilation Air Requirements 299
- 14.3 Boiler Venting 303
- 14.4 Gas Piping 306

Chapter 15
Boiler Installation 312
- 15.1 Boiler Installation 313
- 15.2 Boiler Preparation 315
- 15.3 Hydronic Piping 320
- 15.4 Gas Pipe Installation 327
- 15.5 Condensate Disposal 328
- 15.6 Field Wiring 329
- 15.7 Cascade Boiler Operation 332

Chapter 16
Boiler Startup 336
- 16.1 Water Quality 337
- 16.2 Filling and Purging the System 338
- 16.3 Prestart Checklist 340
- 16.4 Igniting the Burner and Sequence of Operation 341
- 16.5 Combustion Testing 343
- 16.6 Testing Safety Devices 344
- 16.7 Freeze Protection 345

Chapter 17
Boiler Maintenance and Service . . . 348
- 17.1 Boiler Maintenance 349
- 17.2 Boiler and Hydronic System Service 355

Chapter 18
Domestic Hot Water Production . . . 374
- 18.1 Types of Equipment Used for Domestic Hot Water Production 375
- 18.2 Sizing for Domestic Hot Water Production . . . 376
- 18.3 Tankless Boilers 378
- 18.4 Indirect Water Heaters 381
- 18.5 Combi Boilers 384
- 18.6 Hybrid Water Heaters 387

Chapter 19
Solar Thermal Storage 392
- 19.1 Passive Storage Systems 394
- 19.2 Active Storage Systems 396
- 19.3 Solar Collectors 401
- 19.4 Application Selection for Solar Thermal Storage 403
- 19.5 Thermal Storage System Installation 404
- 19.6 System Piping 409
- 19.7 Control Strategies for Solar Thermal Systems 412
- 19.8 Filling and Starting up the Systems 414
- 19.9 Additional Applications for Solar Thermal Storage 416

Chapter 20
Outdoor Wood Boilers 422
- 20.1 Sizing the Boiler 423
- 20.2 Installation of Outdoor Boilers 425
- 20.3 Additional Applications 430
- 20.4 Boiler Maintenance 431

Glossary 436
Index 449

Feature Contents

CODE NOTE

Shut It Down!...............................29
Lockout/tagout..............................30
PP versus PVC Pipe..........................51
Fuel Oil and Piping Rules...................92
Pipe Reducers versus Bushings..............113
PVC versus Polypropylene...................116
Pressure Relief Valves and Discharge Tubes.......156
Low-Water Cutoff Requirements..............158
Backflow Prevention........................161
Outdoor Airflow for Ventilation............231
Mechanical Code Standards for Radiant Piping
    Materials..............................246
Proper Combustion and Ventilation Air
    Requirements...........................299
Combustion Air Venting Installation........302
Venting Requirements.......................304
Installation in Garages....................318
Reducing Bushings..........................328
Condensate Piping Compliance...............329
Meeting Code Requirements..................355
Investigate Local Building Codes...........405
Installation of Solar Thermal Storage Systems.....407
Following Electrical Codes.................408

PROCEDURE

Lifting Heavy Objects.......................24
Steps for Lockout/tagout....................30
Lighting the Pilot..........................72
Diagnosing the Hot Surface Igniter..........76
Measuring the Signal Strength of the Flame Rod....79
Converting a One-Pipe System to a Two-Pipe
    System.................................93
Adjusting the Air Bands.....................96
Recharging the Diaphragm-Type Expansion
    Tank with Air.........................120
Hydronic System Piping Design..............134
Installing a Differential Pressure Bypass Valve.....185
Calculating Pump Size......................203
Sizing a Finned-Tube Heater................218
Soldering Copper Tubing....................322
Using Press-Type Fittings..................324
Wiring a Cascade Boiler System.............333

TIMEOUT FOR MATH

Order of Operations..........................8
Finding the Right Size Nozzle...............99
Calculating GPM............................122
Calculating Pressure Requirements..........133
Calculating Total Loop Length..............259
Calculating Heat Loss......................278
Confined or Unconfined?....................299
Recovery Rates: Example 1..................377
Recovery Rates: Example 2..................378

TROUBLESHOOTING

Sensible Heat Formula........................7
Checking the Thermopile.....................72
Diagnosing a Defective Thermocouple.........73
Do Not Be a Parts Changer...................97
Diagnosing a Defective Control Board.......168
Freeing a Struck TRV.......................184
Repairing Radiant Tubing Leaks.............249
Diagnosing Sizing Issues...................357
Diagnosing Installation Issues.............358
Line-Voltage Issues........................358
Low-Voltage Issues.........................359
The Hopscotch Method of Troubleshooting....360
Burner Circuits: Standing Pilots...........360
Burner Circuits: Intermittent Ignition Systems.....361
Spark Ignition Systems.....................361
Hot Surface Ignition Systems...............362
Checking the Flame Rod Signal..............362
Checking Venting Problems..................363
Replacing the Flue Liner...................364
Checking Problems with Combustion Air......365
Circulating Pump Issues....................366
Problems with Air in the System............367
Expansion Tank Problems....................371
Descaling a Heat Exchanger.................387

1 Human Comfort and Heat Transfer

Chapter Outline
1.1 Hydronic Heating
1.2 Basic Hydronic Configurations
1.3 Modern Hydronic System Configurations
1.4 Heating and Human Comfort
1.5 Benefits of Using Hydronic Heating
1.6 Principles of Heat Transfer
1.7 Heat Transfer through Piping

DiKiYaqua/Shutterstock.com

Learning Objectives

After completing this chapter, you will be able to:
- Demonstrate how heat energy travels from warm to cold.
- Explain the evolution of hydronic heating systems.
- Describe the functionality of a modern hydronic heating system.
- Discuss why heating is necessary to maintain human comfort.
- Describe the benefits of using hydronic heating over other conventional sources.
- Describe how heat is transferred by conduction, convection, and radiation.
- Explain the process of heat transfer through both air and water.
- Differentiate between specific heat, sensible heat, and latent heat.
- Demonstrate how the sensible heat formula is used to calculate Btu capacities.

Technical Terms

American Society of Heating, Refrigerating and Air-Conditioning Engineers (ASHRAE)
Annual Fuel Utilization Efficiency (AFUE)
boiler
comfort zone
conduction
convection
enthalpy
heat
heat transfer
hydronic heating system
latent heat
order of operations
piping
psychrometric chart
radiation
relative humidity
sensible heat
sensible heat formula
set point
specific heat
stratification
temperature difference
temperature droop
terminal device
thermostat

Since the beginning of recorded time, humans have known that they must learn to stay warm in order to survive. With the discovery of fire, people soon realized how to harness the power of combustion as a method of effective heat transfer, **Figure 1-1**. As they became domesticated, they learned how to heat their domiciles and found new ways to make fuel sources more efficient.

As heating sources evolved, people applied basic physics to improve various methods of keeping warm. For example, the principle of convective currents—that hot air rises and cold air falls—was utilized to power boilers and circulating pumps even before electricity. This principle has been applied to older gravity-type heating systems, and it is the basis of modern terminal units and their placement in hydronic heating systems, **Figure 1-2**.

Esteban De Armas/Shutterstock.com

Figure 1-1. The discovery of fire allowed individuals to use it as a basic means to stay warm.

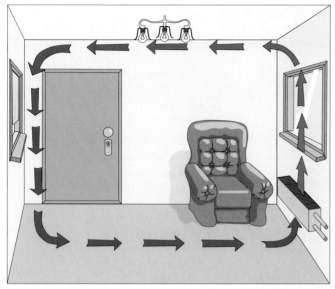

Figure 1-2. An example of convective heating currents used to heat a home.

In the beginning, various readily available fuels, such as wood, coal, and other types of biomass, were used as heat sources. These early sources of heating fuels gave way to modern types of fuels, such as natural gas, propane, and fuel oil, as well as renewable sources, such as solar and geothermal. Early cast-iron boilers using a standing pilot as a fuel ignition system evolved into condensing boilers with electronic fuel ignition systems, **Figure 1-3**. Today there are numerous heating systems available for keeping our homes and buildings warm—namely hydronic heating.

1.1 Hydronic Heating

Hydronic heating systems circulate conditioned water to an occupied space to heat the area. By definition, *heat* is energy added to a substance that causes that substance to rise in temperature. Heat is transferred through fluid by a difference in temperature, or the intensity of the heat within a material, in order to warm the water.

The fundamentals of hydronic heating can be better understood with two basic principles of heating and *heat transfer* (the movement of thermal energy from one substance to another, creating a difference in temperature):

1. Matter (or energy) cannot be created or destroyed. It can only be transferred from one form—solid, liquid, or gas—to another.
2. Heat always moves from a warmer temperature to a cooler temperature, or simply put, heat moves from warm to cold. This is because as the atoms that make up a substance are warmed, they move faster. Conversely, colder atoms move slower. When faster-moving warmer atoms come in contact with slower-moving colder atoms, they transfer their kinetic energy.

These principles allow us to apply several different methods and techniques to accomplish heat transfer in modern hydronic heating applications.

1.2 Basic Hydronic Configurations

There are many hydronic system designs or configurations used in residential and commercial structures. A basic hydronic heating system, **Figure 1-4**, is comprised of the following three components:

- Boiler
- Terminal device
- Piping

Boilers are pressurized vessels containing water or other types of fluids, such as glycol. The fluid is heated to a certain temperature by burning fossil fuels or other types of combustible materials. The actual term *boiler* can be somewhat misleading because hot water boilers used for residential and commercial applications do not actually heat the water to boiling temperatures. Modern boilers must include a number of safety and control devices, and there are various materials that can be used for construction. Cast iron has been the most common material used for earlier vessels. Types of boilers and their construction will be covered in more detail in Chapter 3, *Boilers*.

Terminal devices, also referred to as *heat emitters*, are located within the conditioned space and emit or transfer heat into the air from the water in order to maintain a specific set point temperature within the space. Types of terminal

Figure 1-3. A—An earlier model of a cast-iron boiler. B—A more advanced model of a condensing boiler.

devices include fin tube baseboard heaters, fan coil units, convectors, unit ventilators, and steel panel radiators—as shown in **Figure 1-5**. Terminal devices and their various uses will be covered later in Chapter 11, *Terminal Devices*.

Piping carries the heated water or fluid away from the boiler to the terminal device and then back to the boiler. Types of hot water piping originated with galvanized, black-iron, and cast-iron piping. Today, there are more modern types of materials used in hydronic heating systems, including copper, chlorinated polyvinyl chloride (CPVC), and cross-linked polyethylene (PEX) piping. Various types of piping and their applications are discussed throughout this textbook, including Chapter 7, *Hydronic Piping Systems*.

These three hydronic heating components have developed significantly throughout time. For instance, early boilers started out by burning wood or coal. At the beginning of each frosty winter's morning, the homeowner or building occupant needed to add fuel or stoke the boiler in order to fire it up. The building occupant may have even needed to light the boiler if the fuel supply for the previous evening ran out before daybreak. These boilers had few or no controls to regulate the temperature and comfort of the home or building. Consequently, this caused the building occupants to experience extreme ***temperature droop***, or the extreme

Figure 1-4. A basic hydronic heating system.

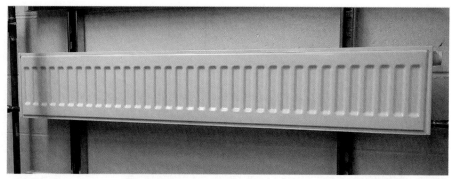

Figure 1-5. This steel panel radiator is an example of a modern terminal unit.

temperature fluctuation between the actual space temperature and the design set point, throughout the day and evening periods. A *set point* is the target value at which a controlling device attempts to maintain the desired temperature.

In addition to the lack of boiler temperature control, early terminal units typically were not equipped with any control devices, such as a thermostat or space temperature sensor. A *thermostat* is a component that senses the temperature of a heating or cooling system and performs actions so that the system's temperature is maintained near a desired set point. Because these early terminal units were sized for a worst-case heating load, they tended to overheat the conditioned space on most days when the outdoor temperature began to moderate. In fact, today in some older buildings with conventional heating systems that have little or no control over the terminal units, the only way to regulate the space temperature is to open a window.

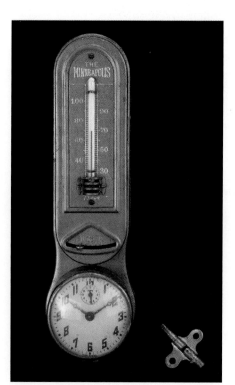

Figure 1-6.

DID YOU KNOW?

History of Thermostats

We often take for granted how instrumental the wall-mounted thermostat is in maintaining a consistent temperature within our home or business. However, where did the thermostat come from and how was it invented? In 1885, Albert Butz, a Swedish immigrant to the United States, filed a patent for an invention he named the *Damper-Flapper*. This device consisted of a damper mounted on the boiler that was connected by a chain to a crank. A damper is a device that opens and closes to restrict the flow of air through a duct or passage. The crank was, in turn, connected to a motor and thermostat.

When a room cooled to below its preset temperature, the thermostat closed its contacts, which energized an armature on the motor. This allowed a crank attached to a main motor shaft to turn a 1/2 revolution. The chain, which was connected to the crank, then opened the damper on the boiler, which allowed more air into the combustion chamber. The addition of air into the furnace made the fire burn hotter, allowing the room temperature to rise above the preset level. When the temperature increased to the room's set point, the thermostat then signaled the motor to turn another 1/2 revolution. This then closed the damper and reduced the fire. Temperature modulation was now automatic.

With his ingenious invention, Albert Butz created the Butz Thermo-Electric Regulator Company of Minneapolis, which eventually became the Minneapolis Heat Regulator Company. Meanwhile, in 1904, a young engineer by the name of Mark Honeywell was perfecting a heat generator as part of his heating and plumbing business. His company, the Honeywell Heating Specialty Company, merged with the Minneapolis Heat Regulator Company in 1927, which eventually became the world's largest producer of residential and commercial thermostats. See **Figure 1-6**.

1.3 Modern Hydronic System Configurations

Hydronic heating systems slowly began to modernize with the onset of controls, such as thermostats, aquastats, and devices that can reset the supply hot water temperature based on the outdoor air temperature. Controls such as these serve to increase comfort within the conditioned space as well as improve the boiler's efficiency. The boiler's fuel supply became modernized as the use of wood as the main source of heating fuel progressed to coal, then to natural gas and propane, and even electricity.

Systems today still depend on a boiler, terminal devices, and piping. There are also improvements that have been made throughout the years that advanced the overall efficiency and comfort of modern hydronic systems. These improvements can be broken down into the following categories:
- Safety
- Controls
- Innovative hydronic systems
- Renewable fuel sources

Today's modern hydronic systems incorporate safety devices, such as pressure relief valves, low water cut-outs, and backflow preventers, to protect boiler operators and building occupants. Boiler fuel systems, such as gas and oil used to heat the boiler water, include their own safety mechanisms to ensure that the boiler ignites cleanly and completes lockout if there is an interruption in its fuel supply, **Figure 1-7**.

Controls continue to advance technologically for improved climate control and overall system efficiency. Devices and technologies that have improved space temperature control and provided great cost savings include web-based thermostats for homes and businesses, direct digital controls for commercial buildings, and wireless communications. Boiler control technologies have resulted in greater system efficiency, lower fuel combustion emissions, and overall cost savings. See **Figure 1-8**.

Goodheart-Willcox Publisher

Figure 1-7. Pressure relief valves are an important safety device on today's boilers.

Goodheart-Willcox Publisher

Figure 1-8. An example of a modern, web-based thermostat.

TROUBLESHOOTING

Sensible Heat Formula

The sensible heat formula for water can be a useful tool to determine whether a boiler is maintaining its rated Btu/hr output. As an example, first calculate what the actual Btu/hr output of the boiler should be based on its efficiency. For example, if the input of a 150,000 Btu/hr boiler were rated at an AFUE efficiency rating of 80%, the actual output would be: 150,000 × 0.80 = 120,000 Btu/hr. **AFUE** is an acronym for **Annual Fuel Utilization Efficiency**, which is a measurement of how efficiently a heating appliance can use the consumed fuel.

Next, determine the flow rate in gallons per minute. This can be determined using a measuring device such as a flow meter or by referencing the circulating pump's specifications. Now measure the temperature difference between the inlet and outlet of the boiler water once it has reached its steady operational state. For instance, a boiler has a flow rate of 10 GPM and a ΔT of 20°F. Using the sensible heat formula, we can now calculate:

$$10 \text{ GPM} \times 20°F \times 500 = 100,000 \text{ Btu/hr}$$

By performing this calculation, we can see that this particular boiler is operating below its rated output. It is now the technician's job to determine why this boiler is not functioning at its peak efficiency.

Figure 1-9. An example of radiant floor heating.

Today, innovations like radiant heating arrangements installed in a building's flooring system are popular for improved comfort and reliability. These systems are used for space temperature control and are helpful for melting snow and ice on sidewalks and driveways in both residential and commercial applications. In the past, galvanized piping was the main source for hydronic heat transfer. Today, copper piping is used, as are innovative products such as PEX piping, which can offer easier installation and greater versatility when used for multiple heating applications (like potable water supply or radiant flooring). See **Figure 1-9**.

For years, natural gas, liquid petroleum (LP), and oil were common boiler fuel heating sources. Today there is a resurgence in the use of renewable sources for boiler fuel operation. The most popular application is solar thermal storage, **Figure 1-10**, in which the sun is used for heating large volumes of water. Solar thermal storage can be used for space temperature control and in such hot water applications as car washes, apartment buildings, and swimming pools. Biofuels or biomass can also be used as alternatives for boiler heating, **Figure 1-11**. Although wood is the most popular source, other biofuels include corn (which can also be used to process ethanol), cherry pits, and other grain crops.

Figure 1-10. A solar collector used to generate hot water.

1.4 Heating and Human Comfort

Space heating is essential to human comfort. The first responsibility of an HVAC (heating, ventilation, and air-conditioning) engineer, technician, or service person is to maintain a comfortable environment within the home or building.

To understand what real comfort means, we must first understand the workings of the human body and how comfort differs between individuals and their surroundings. Core body temperature for the average human being is 98.6°F. Although this temperature may vary slightly depending on the physiology of a person, the goal of human comfort is to maintain its core temperature. When this temperature falls below a person's individual comfort level, additional heat is needed. A body responds to the conditions of its environment (temperature and humidity) by adjusting its core temperature through several different means, **Figure 1-12**.

Heat that is removed from the skin to the surrounding air is considered convective heat transfer. This warmth from the body automatically flows to the cooler surrounding air, thus potentially lowering the core body temperature. If the convective air currents that are created by this heat transfer between the skin and surrounding air continue to remove heat from the body faster than the body can replace it, hypothermia can result. Fortunately, to minimize the risk of hypothermia, humans have adapted to the cold by bundling in layers of warm clothing when the surrounding temperature begins to drop.

Heat transfer through conduction occurs when two objects are in direct contact with each other. When the skin comes in contact with an object that is warmer or colder than the skin temperature, heat transfer takes place. An example of this type of heat transfer is wrapping your cold hands around a hot cup of coffee on a cold day or holding a glass of cold water to your brow on a hot summer's day.

Figure 1-11. An example of an outdoor wood boiler.

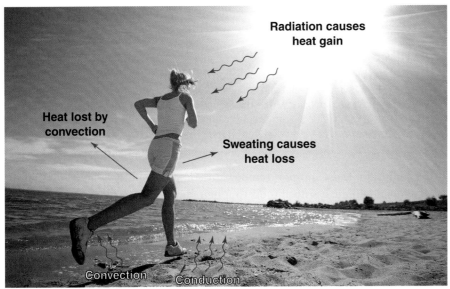

Aleksandr Markin/Shutterstock.com

Figure 1-12. The human body releases heat through conduction, convection, radiation, and perspiration.

Heat transfer by radiation occurs through infrared waves. Radiation warms the surrounding objects, not the air, and can occur between any two objects that differ in temperature. The body can experience warmth through radiation when it is seated next to a room's hot water radiator in the wintertime or when the sun is beating down on a hot summer's day. The human body can also emit heat through radiation by transferring warmth from the skin to surrounding objects. Heat that is lost through radiation from the human body accounts for about 60% of its total heat loss.

If the air temperature surrounding a body rises and causes an increase in core temperature, the body will try to naturally cool itself by perspiring, or sweating. Perspiration is an evaporative cooling process in which heat is transferred from the body to the surrounding air by evaporation, **Figure 1-13**. Because there is a great deal of energy released when water changes form from liquid to vapor, perspiration removes the most amount of heat energy from the skin. This evaporative cooling process is also evident when you exit the shower to towel off. When the water droplets on the skin from the shower are exposed to the cooler, drier room air, evaporation occurs—making the body feel quite a bit cooler.

PorporLing/Shutterstock.com

Figure 1-13. Perspiration is an evaporative cooling process.

1.4.1 Understanding Desired Space Comfort Conditions

Individuals have different indoor comfort preferences. This is due to a number of factors, which can include an individual's metabolism, age, weight, and even gender. In order for HVAC engineers, technicians, and service people to maintain desired comfort levels within residential and commercial buildings, consistent criteria must be established.

According to the standards set forth by the ***American Society of Heating, Refrigerating and Air-Conditioning Engineers (ASHRAE)***, the indoor design conditions for most buildings and homes adhere to the following:
- *For Winter*: A space temperature range of 68.5°F to 75°F and a ***relative humidity*** level between 30% and 40%.
- *For Summer*: A space temperature range of 75°F to 80.5°F and a relative humidity level of 50%.

TECH TIP

ASHRAE
ASHRAE is a global professional association founded in 1894. Its mission is to advance HVAC system design and construction through the development of standards and sustainability within the industry. ASHRAE standards focus on the improvement of building systems, energy efficiency, and indoor air quality. See *Career Connections* at the end of this chapter.

To maintain these proper temperature and humidity conditions, HVAC designers and engineers need tools that help develop these desired indoor conditions regardless of extreme outdoor conditions. In 1904, Dr. Willis Carrier, father of modern air-conditioning, created the first **psychrometric chart**. This tool was developed and is still used as a means of determining proper indoor temperature and humidity conditions based on the properties of air and moisture. Within the psychrometric chart, a **comfort zone** can be established. A comfort zone sets the boundaries in which the average person maintains optimum comfort based on the temperature and humidity of the conditioned space. See **Figure 1-14**.

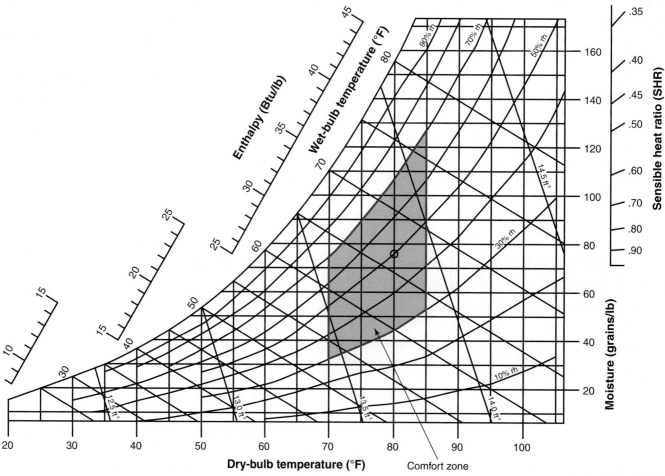

Goodheart-Willcox Publisher

Figure 1-14. The comfort zone as shown on a psychometric chart.

1.5 Benefits of Using Hydronic Heating

There are numerous benefits that a hydronic system has to offer compared to other conventional heating systems, such as forced air heating that uses air to transfer heat from a furnace to the occupied space. These benefits include:
- Comfort factor
- Zoning flexibility
- Health reasons
- Humidity factor
- Higher efficiency
- Installation versatility

Hydronic heating systems provide the most comfort compared to other systems. Because hydronic heating systems do not use forced air as compared to other types of heating systems, there are no cold drafts. In addition, *stratification*, known as a stack effect or layering of warm air from ceiling to floor, is held to a minimum. Depending on the outdoor weather conditions, the temperature difference caused by stratification between multiple floors of a home or building can vary by as much as 20°F, **Figure 1-15**. This uneven layering of warm air between floors results in dissatisfied building owners and occupants. Because hydronic heating systems do not use forced air as a medium for heat transfer, homes and buildings heat much more evenly.

Hydronic heating systems offer the most zoning flexibility because each room can be individually controlled with a thermostat and zone valve, **Figure 1-16**. This feature allows for personalized temperature control throughout the home or building. Thus, there is greater energy savings over conventional heating systems because individual room temperature set points can be set back when not in use.

Hydronic heating offers several health advantages over other systems. Poor indoor air quality can be a serious issue, especially if an individual suffers from dust allergies or asthma. Conventional forced-air heating systems have a greater potential to circulate harmful allergens in the air throughout a home or building. This issue is compounded if the home or building owner fails to change the furnace air filter on a regular basis, **Figure 1-17**. Hydronic heating systems therefore provide a healthier environment for occupants.

Air that has been heated and circulated through ductwork can drastically reduce humidity levels due to increased infiltration of drier outdoor air. Because

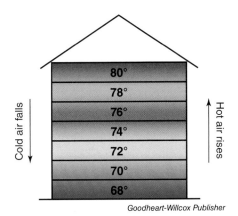

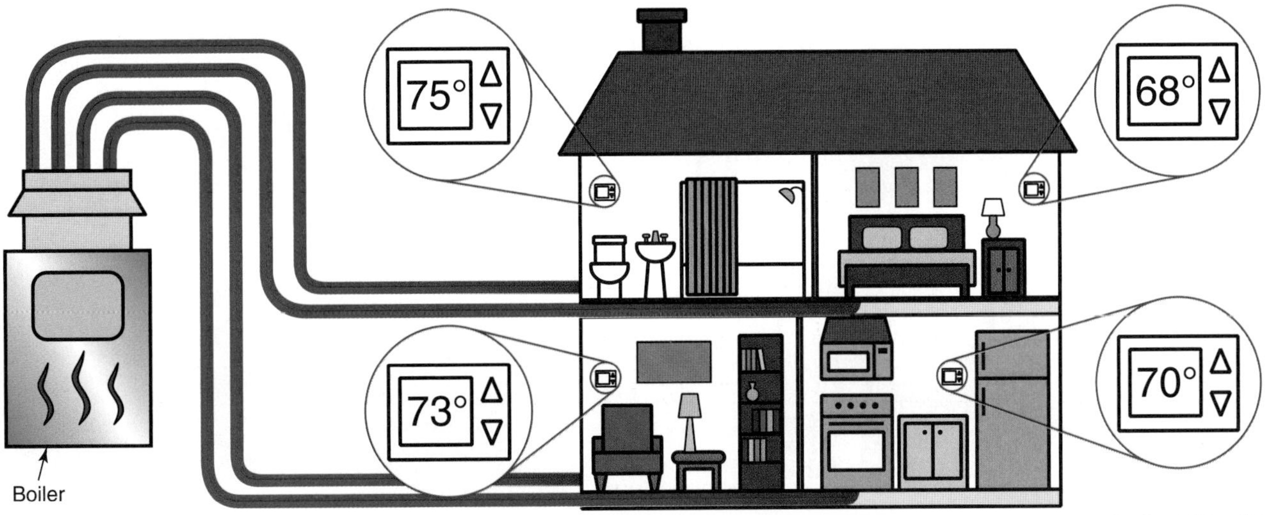

Goodheart-Willcox Publisher

Figure 1-15. Heat stratification creates a large temperature difference in a building, causing two different temperature zones in the same area.

Figure 1-16. Hydronic heating allows for zoning flexibility.

Rob Crandall/Shutterstock.com

Figure 1-17. Forced-air systems have more potential to circulate harmful allergens.

of this, a traditional forced-air heating system tends to dry out a home or building faster than a hydronic heating system. Hydronic systems more readily maintain a balanced humidity level in the home or building because they have reduced airflow circulation, especially in the winter. In addition, there is no need to provide an additional humidifier, which can be costly to purchase, install, and maintain.

Hydronic heating systems have higher efficiency. Water has a much higher density than air, which means it can transfer heat more efficiently. Less energy is needed to operate a hot water circulating pump compared to a forced-air circulating fan and motor. In addition, unlike hydronic systems, a forced-air heating system creates a differential pressure between the indoor and outdoor environments. This means it is more likely for heat to be forced out of the home or building through cracks around doors and windows and other openings, **Figure 1-18**. Buildings with poor insulation and forced-air systems can suffer from greater heat loss than those with hydronic heating systems. In addition, hydronic heating systems have higher cost efficiency. Studies show that on average a hydronic heating system using a standard boiler costs approximately 20% less to operate than a conventional forced-air central heating system.

Hydronic heating systems offer more versatility because less space is needed to install hot water piping, **Figure 1-19**. Thus, hydronic piping can be more easily routed around the building, which can also reduce construction costs. With conventional forced-air heating systems, its extensive ductwork can easily take up valuable square footage within most homes and buildings. Hydronic heating also offers options that are not available with conventional forced-air heating systems, such as radiant floor heat, towel warmers, ice and snow melt features, and more. See **Figure 1-20**.

1.6 Principles of Heat Transfer

Heat transfer occurs when there is a difference in temperature from one substance to another. As mentioned earlier, heat will always transfer from a warmer substance to a colder substance. There are three different methods commonly used in comfort heating in which heat can be transferred:

- Conduction
- Convection
- Radiation

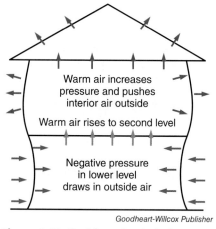

Goodheart-Willcox Publisher

Figure 1-18. Heat forced out of a home through cracks around doors, windows, and other openings.

nikkytok/Shutterstock.com; vchal/Shutterstock.com

Figure 1-19. Hydronic heating pipes (A) take up less room than ductwork (B).

Heat transfer through **conduction** occurs when there is physical contact between two types of materials. An example of heat transfer by conduction is heating a pan on a stove. In order for the pan to warm properly, the pan must be in contact with the heat source. Consider trying to fry an egg in a pan that is held 12″ above the surface of the stove. It would be almost impossible for the egg to ever cook.

Heat transfer through **convection** involves moving heat through a fluid source, such as air or water. Convection heating can occur either through forced or natural means. A boiler is a good example of forced convection heating. Heat is transferred from the combustion source to the water that is circulating through the boiler. This heated water is transferred, or forced, to the terminal units by the circulating pump. The heat is then released by the terminal units through natural convection currents. Heat transfer through natural means occurs through the normal process of heating a room. Because hot air is lighter than cold air, it rises from the terminal unit, where it heats the room before falling back down to the floor, where it recirculates naturally.

Radiation is the result of transferring heat through light waves. A common example of radiation is the sun warming the Earth. Note that radiant heating warms solid objects—not the air. This principle is easy to understand given that there is no air in space, yet we continue to remain warm here on Earth by the heat from the sun.

Radiation relies on heating objects through a direct physical path. This means that radiant heat can travel in any direction and is not dependent on air currents to transfer its heat. An example of this is standing next to a campfire. The side of your body facing the fire is warmer than the side that is turned away because of the light waves from the fire warming a cooler surface. Hydronic heating terminal units also emit radiant heat, warming the objects around them. See **Figure 1-21**.

brizmaker/Shutterstock.com

Figure 1-20. An example of a hydronic heating towel warmer.

TECH TIP

Multiple Heat Transfer Methods

The use of a modern hydronic heating system typically involves more than one method of heat transfer. For example, a gas flame heats the water inside a boiler. This is accomplished by heating the fire tubes, which contain water. The water is physically touching the tubes inside the boiler. This is considered heat transfer by conduction.

The heat transferred by the water circulating through the piping to the terminal units is an example of convective heat transfer. When the warmth from the terminal device heats the objects within the conditioned space, this heat can be re-radiated by the objects in the room in the form of radiant heat. This example of multiple heat transfer demonstrates that a modern heating system used for residential or commercial application never consists of a single method of heat transfer.

1.7 Heat Transfer through Piping

A network of piping is required to circulate heated water throughout a hydronic heating system. It is important for HVAC engineers, technicians, and service people to understand the basic physics behind how heat transfers through this system. Heat transferred through convection requires that it pass through a medium, such as air or water. Water has a much higher density than air, which means it can transfer heat more efficiently. One cubic foot of air weighs about 0.075 lb at 70°F, as opposed to one cubic foot of water, which weighs about 62.4 lb at 70°F. Another example of how water has a much greater heat-carrying capacity than air can be

Heat Transfer Methods

Figure 1-21. Heat can be transferred by conduction, convection, and radiation.

defined by comparing the Btu capacity of a forced-air duct to another material. Btu stands for British thermal unit, which is the US customary unit of heat. It measures the amount of heat required to raise the temperature of 1 lb of water 1°F.

In **Figure 1-22A**, a forced-air duct measuring 12″ × 16″ delivers approximately 20,160 Btu/hr of heat at a velocity rate of 700 feet per minute, and with a *temperature difference*, or the measure of the relative amount of internal energy between two substances, of 20° F. However, this same Btu capacity can be delivered by a 3/4″ copper tube that has water flowing though it at only 4 gallons per minute (GPM), **Figure 1-22B**. In this example, temperature difference is defined as the supply water temperature delivered to a device minus the temperature of the water leaving that device.

Water is an effective medium for heat transfer because it has high specific heat content. *Specific heat* is the amount of heat required to raise the temperature of a material by 1°F. For forced-air heating, the specific heat of air is only 0.24 Btu per pound per degree Fahrenheit compared to the specific heat of water, which is 1 Btu per pound per degree Fahrenheit. By comparison, water has the capability of carrying several thousand times more heating Btu than does air per volume.

These comparisons between the heat carrying capability of air and water were calculated using the *sensible heat formula*. This formula is used throughout the HVACR industry (the "R" stands for refrigeration). A hydronic heating system designer should have a thorough understanding of it for both air and water calculations. The sensible heat formula, and other similar formulas, will be used throughout this book.

The sensible heat formula is different for air and water. The sensible heat formula for air is as follows:

$$Q = CFM \times 1.08 \times \Delta T$$

where

$$\begin{aligned} Q &= \text{Btu/hr} \\ CFM &= \text{Cubic feet per minute of airflow} \\ 1.08 &= \text{Constant} \\ \Delta T &= \text{Temperature difference across a terminal unit} \end{aligned}$$

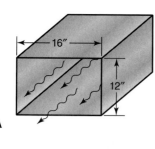
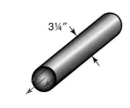

Figure 1-22. An example of how water can deliver a higher level of heating capacity compared to air.

Note that the constant 1.08 is calculated using the following factors:
$$1.08 = 0.075 \times 0.241 \times 60$$
where

- 0.075 = Weight of air in pounds per cubic foot
- 0.241 = Specific heat of air (1 lb = 0.241 Btu/hr)
- 60 = minutes in one hour

Note that 0.075 is the weight of air at a standard temperature of 70°F and 50% relative humidity. The actual weight may vary based on temperature, humidity, and altitude.

The sensible heat formula for water is as follows:
$$Q = GPM \times 500 \times \Delta T$$
where

- Q = Btu/hr
- GPM = gallons per minute of water flow
- 500 = Constant
- ΔT = Temperature difference across terminal unit

Note that the constant of 500 is calculated using the following factors:
$$500 = 8.33 \times 1 \times 60$$
where

- 8.33 = Weight of one gallon of water
- 1 = Specific heat of water
- 60 = minutes in one hour

This constant is based on a water temperature of 60°F and may vary depending on the temperature and viscosity of water. The viscosity of water can change if antifreeze is added, which will be discussed later in this book.

When considering the value of a hydronic heating system, heat is measured in two different ways. The first measurement is known as sensible heat. **Sensible heat** is the amount of heat that can be measured by temperature—usually in degrees Fahrenheit or degrees Celsius—with no change of state of matter (solid, liquid, or vapor). The second way that heat is measured is by latent heat. **Latent heat** is defined as the amount of heat required to change the state of matter—from a liquid to a vapor, or a vapor to a liquid—without a change in temperature. Latent heat cannot be measured by temperature, but rather by Btu. The term *latent* is translated from Latin, meaning hidden.

When studying the physics of heat transfer, the relationship between temperature and enthalpy of water should be understood. Temperature can be described as the degree or intensity of heat present in a substance. **Enthalpy** is defined as the total heat content in the Btu (sensible and latent) of a substance. As the temperature of water is increased, the enthalpy or Btu volume increases. This temperature/enthalpy relationship can be visualized when considering steam as a heating source. As the temperature of water is increased, there is a change in state from liquid to vapor and also an increase in the Btu content. The relationship between specific, sensible, and latent heat in hydronic heating needs to be considered when designing or troubleshooting a system. **Figure 1-23** shows that both sensible and latent heat are changed by the temperature and state of water. It takes only 1 Btu of sensible heat to raise the temperature of water by 1°F. However, it takes 970 Btu of latent heat to change the state of water from a liquid to a vapor (in this case, steam). This illustrates why steam is such a valuable heat source. Steam can hold five or six times as much potential energy as an equivalent mass of water and can be transported by pressure rather than pumping. Therefore, steam can be used extensively in larger buildings where heat must travel long distances.

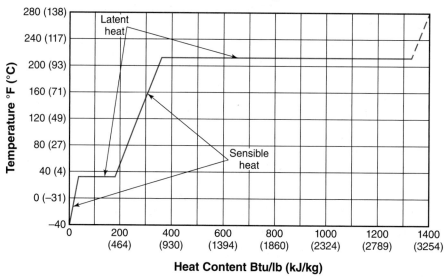

Figure 1-23. This graph shows how both sensible and latent heat can affect the temperature and state of water.

TIMEOUT FOR MATH

Order of Operations

HVAC technicians are required to perform mathematical calculations on the job. In order to compute any answer accurately, calculations must be completed in the proper sequence using the *order of operations*. This mathematical notation refers to which operations should be done first, and in what order. One method of understanding the mathematical order of operations is by referring to the mnemonic PEMDAS, which can be remembered as **P**lease **E**xcuse **M**y **D**ear **A**unt **S**ally. The proper order of operations is as follows:

1. Parentheses
2. Exponents
3. Multiplication and Division (left to right)
4. Addition and Subtraction (left to right)

These rules must be followed in order to obtain the correct answer for a math calculation. Note that calculations are performed left to right for multiplication and division as well as for addition and subtraction in the last steps.

Practice by solving the following equations:

$$4 + 5 \times 8$$

Using PEMDAS, complete the following steps in the correct order:

$$5 \times 8 = 40$$
$$(40 + 4 = 44)$$
$$4 + 5 \times 8 = 4 + 40$$
$$= 44$$

Try another equation:

$$(2 + 3) \times (9 - 3)$$

First, solve the equations within the parentheses:

$$(2 + 3) = 5$$
$$(9 - 3) = 6$$

Now multiply the solutions from within the parentheses:

$$5 \times 6 = 30$$

The sensible heat formula can be modified to solve for other variables, or unknowns. For instance, if we know the Btu output and the water temperature difference of a particular boiler, we can solve for the GPM pump flow through the system. Here is an example: A boiler has an output of 500,000 Btu/hr and a water temperature difference of 25°F. Calculate the flow through the system in gallons per minute:

Q = 500,000 (Btu/hr output)
ΔT = 25°F (temperature difference between supply and return water)
500 = Constant

The sensible heat formula can now be reconfigured as follows:
$$GPM = Q \div (500 \times \Delta T)$$

Following the order of operations, we first solve within the parentheses:
$$(500 \times 25°F) = 12,500$$

Now divide this answer into the B/hr output:
$$500,000 \div 12,500 = 40 \text{ GPM}$$

If the GPM through the system is below the specific flow according to project specifications or the pump manufacturer's performance data, the service technician will need to determine why the pump is not operating properly.

As mentioned earlier, these formulas and others will be used throughout this book and will be of great help in determining proper circulating pump sizing, building load calculations, and boiler sizing.

CAREER CONNECTIONS

HVAC Organizations

For students considering a career in the HVAC industry, there are many professional organizations available that promote continuing education, networking, and career enhancement. Here are several such organizations:

- **ASHRAE**—The American Society of Heating, Refrigerating and Air-Conditioning Engineers is based in Atlanta, GA, and is intended to advance HVAC system design and construction through the development of standards and sustainability within the industry. ASHRAE sets standards in the industry that affect the health, safety, and efficiency of people and equipment. With 125 years of service to the HVAC community, ASHRAE offers technical resources, professional development, and learning opportunities to engineers and tradespeople throughout the industry. Students of four-year institutions that offer HVAC programs have the opportunity to develop and join ASHRAE student chapters as a means of developing their careers in the HVAC industry.
- **ACCA**—The Air Conditioning Contractors of America is a national trade association furthering the interests of the HVAC contracting business and the broader HVAC industry. ACCA members are provided with services that promote business development, employee training, and government legislation. Their annual exposition focuses on helping HVAC contractors optimize business growth and success by learning from the industry's most successful owners, entrepreneurs, and top consultants, while showcasing the top products and services available to the industry.
- **UA**—The United Association of Journeymen and Apprentices of the Plumbing and Pipe Fitting Industry of the United States and Canada, affiliated with the national building trades, represents approximately 355,000 plumbers, pipe fitters, sprinkler fitters, service technicians, and welders in local unions across North America. The UA has been training qualified pipe tradespeople longer than anyone else in the industry. They provide the premier training programs available in the industry today, including five-year apprenticeship programs, extensive journeyman training, a comprehensive, five-year instructor training program, and numerous certification programs. They help signatory contractors grow their market share by identifying new opportunities, providing support, and connecting them to the safest, most skilled, and most highly trained workforce in the industry.
- **USGBC**—The US Green Building Council is committed to a sustainable, prosperous future through LEED (Leadership in Energy and Environmental Design), the leading program for green buildings and communities worldwide. Its vision is that buildings and communities will regenerate and sustain the health and vitality of all life within a generation. Its mission is to transform the way buildings and communities are designed, built, and operated, enabling an environmentally and socially responsible, healthy, and prosperous environment that improves quality of life. Colleges throughout the United States offer student membership in the USGBC. HVAC manufacturers such as Carrier and Johnson Controls are USGBC members.

Chapter Review

Summary

- Heat always travels from a warm location to a colder location. When a substance is heated, its atoms move more rapidly, causing them to collide with slower-moving colder atoms. When faster-moving warmer atoms encounter slower-moving colder atoms, they transfer their kinetic energy.
- A basic hydronic system consists of a boiler, a terminal device, and connecting piping.
- Early systems usually consisted of cast-iron boilers and galvanized, black-iron, or cast-iron piping. They burned wood or coal. Early boilers and terminal units included few to no controls to regulate temperature.
- Improvements have been made throughout the years that have advanced the efficiency and comfort of modern hydronic systems. These include improvements in the areas of safety; controls, such as thermostats and aquastats; innovative hydronic systems, such as radiant heating arrangements and new piping material like copper and PEX; and fuel sources, from natural gas, propane, liquid petroleum, oil, and electricity to renewable sources like solar, geothermal, and biofuels.
- Desired indoor comfort conditions will vary among individuals. Heating is necessary to maintain human comfort and may differ from individuals depending upon their physiological makeup.
- A psychrometric chart is used to determine proper indoor temperature and humidity conditions, based on the properties of air and moisture.
- The comfort zone sets the boundaries in which the average person will maintain optimum comfort based on the temperature and humidity of the conditioned space.
- Hydronic systems offer numerous benefits compared to other conventional heating systems, including comfort factor, zoning flexibility, health reasons, humidity factor, higher efficiency, and installation versatility.
- Heat moves from one point to another by conduction, convection, or radiation. Conduction occurs when there is physical contact between two types of materials. Convection involves moving heat through a fluid source, such as air or water, through forced or natural means. Radiation is the transfer of heat through light waves.
- Water has a greater heat-carrying capability than air because of its higher density.
- Specific heat is the amount of heat needed to raise the temperature of a substance by 1°F.
- Sensible heat is heat that can be measured by temperature.
- Latent heat is the amount of heat measured in Btu to convert water from one state of matter to another.
- The sensible heat formula is used to calculate total Btu based on flow rates and temperature difference.

Know and Understand

1. Heating fuels used in early hydronic heating systems include _____.
   A. geothermal
   B. propane
   C. coal
   D. liquid petroleum
2. The three components of a basic hydronic heating system are _____.
   A. boiler, terminal device, and piping
   B. valves, dampers, and cranks
   C. boiler, thermostat, and radiator
   D. conductor, convector, and radiator
3. *True or False?* Copper piping is among the improvements that have been made to modern hydronic heating systems.
4. *True or False?* Comfort can mean different things to different people based on their metabolism, age, weight, and even gender.
5. A psychrometric chart establishes the _____, the boundaries in which an average person maintains optimum comfort based on the temperature and humidity of the conditioned space.
   A. latent heat
   B. comfort zone
   C. thermal equilibrium
   D. aquastat settings
6. *True or False?* Air is a more efficient means of heat transfer than water.
7. The amount of heat required to change the state of matter without changing the temperature is defined as which of the following?
   A. Sensible heat
   B. Specific heat
   C. Latent heat
   D. Convection
8. When performing mathematical calculations, what is the first step in the order of operations?
   A. Exponents
   B. Division
   C. Multiplication
   D. Parentheses
9. *True or False?* Enthalpy is the latent heat minus the sensible heat content of a substance.

Apply and Analyze

1. How do the early types of heating fuels compare with those used today?
2. Define the term temperature droop. How does it affect human comfort?
3. Describe the difference between the specific heat of air and the specific heat of water.
4. Solve the following math problem using the proper order of operations: $(5 + 4) \times (10 - 3)$.
5. What would be the flow in gallons per minute through a heating system if the heating output capacity of the boiler was 150,000 Btu/hr, the supply water temperature was 160°F, and the return water temperature was 130°F?
6. List and describe the benefits of using a hydronic heating system over other conventional methods of heating.
7. How is heat transferred by conduction, convection, and radiation? Give examples of each.
8. Explain the difference between sensible heat and latent heat.

Critical Thinking

1. What is a psychrometric chart? Explain how it is used in the HVAC industry.
2. Using the specific heat formula for water, calculate the Btu/hr output of a terminal unit that has 4 GPM of water flowing through it and a temperature difference of 20°F.
3. Using the pressure/enthalpy chart, how many Btu are required to convert water (liquid) to steam (vapor)? What term is used for this conversion of water to steam?

2 Safety

Chapter Outline
2.1 Personal Protective Equipment (PPE)
2.2 Fire Safety
2.3 Electrical Safety
2.4 Lockout/tagout
2.5 Safety with Tools and Equipment
2.6 Working around Pressurized Gases
2.7 Chemical Hazards
2.8 Working in a Confined Space
2.9 Safety in the Boiler Room

Igor Sokolov (breeze)/Shutterstock.com

Learning Objectives

After completing this chapter, you will be able to:
- Identify why safety is vital for anyone working on or around boilers.
- Describe the proper protective equipment that should be worn when working around boilers and hydronic equipment.
- List the various types of fire extinguishers used for different types of flammable materials.
- Explain why electricity poses a potential safety hazard when working with boilers.
- Describe how to work safely when using tools.
- Explain how to work safely when using ladders.
- Summarize why pressurized gases can pose a safety hazard.
- Explain how to work safely when using chemicals.
- Discuss how to work safely within a confined space.
- Describe potential boiler room safety hazards.

Technical Terms

American National Standards Institute (ANSI)
arc flash
Canadian Standards Association (CSA)
carbon monoxide (CO)
confined space
decibel (dB)
ground fault circuit interrupter (GFCI)
grounding prong
lockout/tagout
low water cutoff
low-level CO detector
Occupational Safety and Health Administration (OSHA)
personal protective equipment (PPE)
Safety Data Sheet (SDS)
self-contained breathing apparatus (SCBA)
side shield
Toolbox Talk

There is a popular saying that states: "Safety is everyone's business." This practical advice applies to everyone working in boiler rooms as well as to those performing installations, servicing, or repairing boilers and hydronic equipment. Thousands of accidents could be avoided each year if technicians would take the time to work safely. This chapter will cover the proper safety procedures to which all service persons and technicians should adhere as well as review how to safely handle equipment used in the boiler and hydronics industry.

The US Department of Labor estimates that 85% of workplace injuries are avoidable. Most accidents are caused by workers deliberately performing unsafe work practices, not thinking clearly, practicing bad habits, or working around or with hazardous equipment. This is why safety training is of the utmost value for workers and employers alike. Most companies have implemented safety programs that all new employees must participate in before they are allowed to set foot on the job site. Also, weekly safety meetings, known as *Toolbox Talks*, are typically performed during construction projects to keep safe working practices at the forefront of everyone's daily responsibilities. In addition to educating employees on safe construction practices, other Toolbox topics include the correct procedures for handling tools, working safely around chemicals, and what to do should an accident occur. One of the most important aspects of safety on the job is maintaining the proper, proactive attitude. No one wants to be injured or see their coworker injured. A blatant disregard for safety rules shows a lack of respect both for yourself and your fellow workers. Always be aware of potential safety hazards around you, and report any unsafe equipment or job site situations. Take the time to perform tasks correctly rather than rushing to get them done. Often, rushing through tasks leads to taking shortcuts, which can result in accidents. Always ask a supervisor or coworker if you are unsure of the correct or safest way to complete a task on the job site. Above all, do not think, "An accident can never happen to me." If you take too many risks, chances are an accident will eventually find you and injure either you or a coworker.

SAFETY FIRST

Look for Potential Issues before They Become a Problem

The National Board of Boiler and Pressure Vessel Inspectors recommends that when first entering a boiler room, always look for potential problems. Before starting up the boiler, ensure that the area is free of any possible dangerous situations, such as flammable materials not safely stored. Also, look for mechanical or physical damage to the boiler or related equipment. Always make sure that intake and exhaust vents are clear of obstructions, and check for possible vent leaks.

2.1 Personal Protective Equipment (PPE)

Alongside safety awareness, a technician's first line of defense in maintaining a safe work environment is the proper use of *personal protective equipment (PPE)*. Risks due to exposure to various health and physical hazards can be minimized by wearing appropriate safety equipment, covered in this section. PPE includes devices to protect the eyes, ears, head, face, hands, feet, and respiratory system. All service technicians and installers need training on how and when to wear PPE and how to maintain this equipment so that it remains in good working condition.

2.1.1 Eye and Face Protection

The most significant risk for eye injuries is flying particulates, such as dust, metal fragments, and other types of debris. Most of these types of injuries are avoidable with the proper use of protective eye gear. The most common types of eye protection include safety glasses, goggles, and face shields, **Figure 2-1**.

At a minimum, wrap-around safety glasses should be worn on every job site and construction area for eye protection. Confirm they are *Occupational Safety and Health Administration (OSHA)* approved protective eyewear. Prescription glasses *may* be acceptable on some job sites, but in most cases, they must be worn in conjunction with *side shields*, **Figure 2-2**. Full safety goggles are required when working around dusty conditions or with certain power tools such as nail guns. Face shields are required when working with or around chemicals or caustic cleaners. These, along with standard safety glasses, provide the best defense in eye protection. Tinted goggles or face shields are required when soldering, brazing, or welding, **Figure 2-3**. These devices filter out harmful radiation to the eyes. Take care to choose the correctly tinted eye protection based on the job. Different levels of tinting are designed for brazing versus welding, and using the wrong type may cause severe eye injury.

Figure 2-1. Safety glasses, goggles, and face shields are the most common types of eye protection.

Uline

Tom Grundy/Shutterstock.com

Figure 2-2. In some cases, side shields may need to be worn along with prescription glasses.

2.1.2 Hearing Protection

Hearing loss is often insidious, happening in a gradual, subtle way, but with harmful effects. This is why ear protection is so important on the job site. Hearing protection requirements are determined by the loudness of the noise as rated in **decibels (dB)**, the length of time a worker is exposed to the noise, and if the sound is produced by a single source or from multiple sources. The two most common types of hearing protection are:

- Earplugs made from foam or silicone rubber, which are inserted directly into the ears.
- Headphones fitted around the ears, covering them completely, **Figure 2-4**.

The OSHA guidelines for allowable exposure to noise are outlined in **Figure 2-5**. These requirements state that hearing protection is mandatory whenever noise levels meet or exceed 90 dB for 8 hours per day. Keep in mind that noise from a combination of sources can have an additive effect on hearing loss.

Ukki Studio/Shutterstock.com

Figure 2-3. Tinted goggles are required when soldering, brazing, or welding.

2.1.3 Head Protection

Injuries to the head can be devastating and can cause permanent disability or even death. Therefore, head protection is essential to job site safety. Hard hats provide the necessary head protection against falling materials or bumping into low hanging objects. Most hard hats feature a shock-absorbing lining and adjustable straps to ensure a proper, secure fit, **Figure 2-6**.

2.1.4 Body Protection

Hand protection on the job consists of gloves for various tasks to protect against cuts, abrasions, extreme heat and cold, and certain chemicals, as well as to provide an improved grip. Always be sure to choose the correct type of gloves for the given task, **Figure 2-7**.

Footwear must provide the proper support and allow for secure footing on hazardous surfaces. Safety shoes, such as high-top boots with steel toes, are sometimes a mandatory requirement on construction sites where heavy objects could be dropped, causing injury to the toes. Footwear should be made from

Loko/Shutterstock.com

Figure 2-4. A variety of ear protection options are available, including earmuffs and earplugs.

Maximum Permissible Noise Exposure		
Duration of Exposure in Hours per Day	Decibel Level	Loudness Comparison
8	90	Lawnmower, electric drill, shouted conversation
6	92	
4	95	
3	97	Train, motorcycle, factory machinery
2	100	
1.5	102	Dance party, sporting event, snowmobile
1	105	
0.5	110	Leaf blower, car horn, rock concert, emergency vehicle siren
0.25 or less	115	

Goodheart-Willcox Publisher

Figure 2-5. Table of permissible noise exposure levels.

24 Hydronic Heating: Systems and Applications

Tap10/Shutterstock.com

Figure 2-6. Hard hats provide the best protection against falling objects.

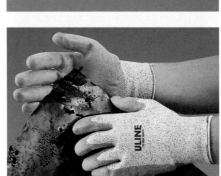 

Uline

Figure 2-7. A variety of work gloves for different uses.

Harry Cabance/Shutterstock.com

Figure 2-8. Steel-toed boots may be required on some job sites.

nonconductive heavy leather, have heat- and slip-resistant soles, and fit comfortably, **Figure 2-8**.

Back injuries account for nearly 20% of all incidents reported on the job site. The proper back support in the form of an approved back brace will minimize injuries whenever lifting or moving heavy objects, **Figure 2-9**. Do not attempt to lift heavy or large objects alone. Always ask for assistance when trying to lift objects that may be too heavy for one person. In addition, always lift with your legs—not your back—and keep your back straight when lifting. Devices such as hand trucks, pry bars, and portable dollies can assist in moving heavy objects. Know how they operate, and learn the proper steps for lifting heavy objects.

PROCEDURE

Lifting Heavy Objects

1. Position feet shoulder-width apart.
2. Bend at your knees.
3. Lift the load smoothly, without any sudden movements.
4. Keep the load close to your body while lifting. Do not extend your arms away from your body.
5. Keep your body straight while lifting and holding the load. Do not twist your back or turn to the side.
6. Lift heavy objects with a partner. Do not try to lift a heavy load by yourself.

Some workplaces require the use of a safety harness when working above certain heights, **Figure 2-10**. Safety harnesses limit a potential fall within a safe distance above the ground, protecting the worker from internal injuries. These harnesses should be securely fastened or clipped to an immovable object such as scaffolding or a lift bucket. They are not, however, designed to suspend the

Uline

Figure 2-9. A back brace can provide the proper back support when much lifting is required.

Uline

Figure 2-10. An example of a safety harness.

worker for an indefinite period. Being suspended by a safety harness for too long can result in a restriction of blood flow to the legs, a worker passing out, or an even worse situation. If you fall when wearing a safety harness, be sure to keep your legs moving periodically to keep your blood flowing to your extremities until help arrives.

2.1.5 Respiratory Protection

When working around non-ventilated areas, or in areas where there is a high concentration of dust or airborne chemicals, workers must use respiratory protection. The two most common types of respirators are air-purifying and atmosphere-supplying. Air purifying breathing protection devices can be a basic, lightweight type of mask, **Figure 2-11**, or they can be respirators with replaceable cartridge filters, **Figure 2-12**. Respiratory protection devices are necessary in a variety of conditions, and workers should refer to the product's instructions for guidance on the correct application.

When working in a *confined space*—where there is the risk for lack of oxygen—or in areas where there may be a high concentration of harmful airborne particulates, an atmosphere-supplying respirator may be necessary. These types of devices provide supplemental oxygen to the worker as well as filter out toxic chemicals, dust, or other particulates.

Regardless of the type of respiratory protection in use, always ensure that it fits snugly to the face to prevent leakage from the outside air and always keep this equipment clean and in good working order. Filters must be replaced whenever they are damaged, soiled, or cause noticeably increased breathing resistance.

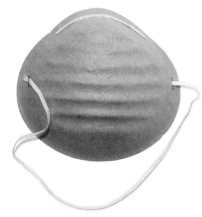

Stock Up/Shutterstock.com

Figure 2-11. A basic, lightweight dust mask.

Anton Starikov/Shutterstock.com

Figure 2-12. A respirator with replaceable cartridge filters.

SAFETY FIRST

Proper Ventilation

Set up proper ventilation before working in non-ventilated areas or in a confined space. Fans can be used to push and pull fresh air through a space, which will help remove any toxic fumes or harmful airborne particulates.

2.2 Fire Safety

A boiler room can contain many fire hazards, including the boiler's fuel source, chemicals, solvents, fuels for torches, and electricity. All boiler technicians, service personnel, and installers should receive periodic training in fire prevention and fire safety. Training should include the proper steps to take if a fire breaks out and the proper procedure for safely putting out a fire. Every boiler room should be equipped with a portable fire extinguisher suited to the flammable material present. All boiler room personnel should familiarize themselves with the different types of fire extinguishers and how to use them. Whenever a fire breaks out, time is of the essence, as the average fire extinguisher only holds 10 seconds of extinguishing material.

Be sure that the fire extinguisher in your boiler room is large enough to handle the appropriately sized fire. Too often, a fire becomes too large for the average-sized extinguisher to handle. Periodic inspections can help eliminate the stockpiling of combustible material in your boiler room. Also, staging multiple extinguishers in the proper location will ensure easy identification, and if more than one is required, they will be available. Your local fire department or fire inspector can help with your selection.

Fires have different classifications based on the type of material that is being consumed, and fire extinguishers are labeled with symbols that identify the type of fire they will extinguish. Always be aware of these different classifications and use the correct extinguisher for the fire at hand. Fire extinguishers classified as Type A extinguish wood, paper, or trash fires. Type B is for flammable liquids such as grease, gasoline, paints, and oils. Type C is for electrical equipment fires. Some fire extinguishers are rated for all three types of fires and are usually the dry-chemical type. Know which type to use on specific types of fires, **Figure 2-13**. Failure to do so can lead to the fire spreading rather than being extinguished and can also result in serious personal injury, death, or property damage.

> **SAFETY FIRST**
>
> **Choosing the Correct Fire Extinguisher**
>
> Never use water on oil fires or electrical fires. Remember that oil and water do not mix. Dousing an oil fire with water can cause the oil droplets to spread rapidly, making the fire even worse. In addition, the vaporized water will also carry oil droplets with it, which can further intensify the fire.
>
> Water is very conductive to electricity. Using water on an electrical fire can create a high risk of electric shock. Only use Class B extinguishers on flammable liquids such as oil, and only use Class C fire extinguishers on electrical fires.

> **GREEN TIP**
>
> **Environmentally Friendly Extinguishing Solutions**
>
> To protect against fires, boiler rooms in commercial and industrial settings often require automatic fire suppression systems. These systems automatically detect fire and release a chemical fire extinguishing agent. Traditional fire extinguishers contain halon, a chemical that can cause ozone depletion when released into the atmosphere. New potassium-based extinguishing agents are more environmentally friendly. They do not cause ozone depletion, are nontoxic, and are safer and easier for workers to clean up.

2.3 Electrical Safety

Injury from electricity can result in shocks, burns, and even death. For these reasons, it is of the utmost importance that proper electrical safety is always practiced when working in the boiler room. Human skin is very conductive, and when a body comes in contact with an electrical circuit, it can become the path of least resistance through which electricity flows. Current, which is the flow of amperes through an electrical circuit, can damage the heart—stopping it from beating and possibly resulting in death if not remedied quickly. Because of this, it is wise for technicians and service persons to know first aid techniques, including CPR (cardiopulmonary resuscitation).

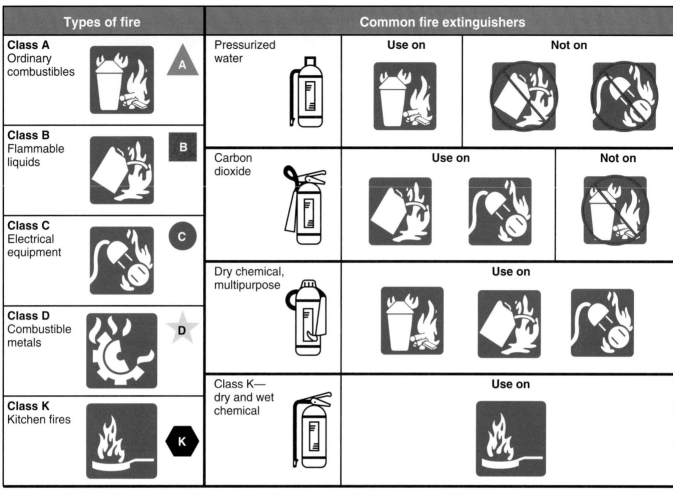

Figure 2-13. Know which fire extinguishers are compatible with which types of fires.

The first step in electrical safety when working on equipment is to ensure that the power is off. Know the location of the breaker or service disconnect for the circuit being serviced, and use the appropriate lockout/tagout procedure before beginning. At times it may be necessary to work on live circuits for troubleshooting purposes. When this is the case, use caution so that your arms or other parts of your body do not come in contact with live, exposed circuits. When using an electrical test meter, be careful that the meter probes do not slip to avoid short-circuiting equipment. Always know the voltage rating of the circuit you are servicing. Never use a screwdriver or other tools in and around the electrical enclosure that has power applied. One slip from a line voltage terminal to a grounded terminal can create a short circuit between the two terminals, or a short to ground, resulting in a possible arc flash, shock, or electrical burn, **Figure 2-14**. Utilizing insulated hand tools will reduce exposure to this type of hazard while working on live circuits, **Figure 2-15**.

If a breaker panel or main disconnect panel requires servicing, use the proper PPE and work safely to prevent *arc flash*. An arc flash (also known as a flashover) is the light and heat produced as part of an arc fault, which is a type of electrical explosion or discharge that results from a low-impedance connection through air to ground in an electrical system. An arc flash happens when electrical current flows through an air gap between conductors within the breaker panel or main disconnect. This can be a real safety concern. The most common cause of

SAFETY FIRST

Watch Where You Are Standing

Never stand in water or on wet surfaces when checking live electrical circuits. Water is very conductive to electricity, and you could become part of the circuit. When working on live electrical circuits, it is good practice to wear insulated-sole shoes. Many workplaces display signs warning that proper footwear and eye protection be worn in certain areas.

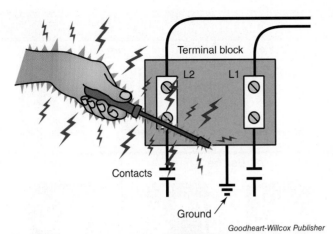

Figure 2-14. If a screwdriver slips when working on a live circuit, it can come in contact with a grounded terminal, resulting in a short to ground—a potentially dangerous occurrence.

Figure 2-15. Always use tools with insulated handles when working on live circuits.

arc flash is the touching of test probes to the wrong surface within a panel, or when a tool such as a screwdriver slips and makes contact with the ground bar. Other causes of an arc flash include:

- Sparks from a break or gap within the wire's insulation.
- Equipment failure from using substandard parts.
- Dust, corrosion, or other impurities found on the conductor's surface.

Most companies have rules for working on breaker panels or main service disconnects to prevent arc flash. Safety training companies can provide additional arc flash training. It is best to allow only trained, qualified technicians to work on service panels and main disconnects, **Figure 2-16**.

When working with electrical power tools, ensure they are properly grounded. This means that the electrical cord connected to the tool has a three-prong connector and the circuit in use is properly grounded as well. Failure to do so may result in the technician becoming part of the electrical circuit, which can cause shocks, burns, and potentially more significant injuries, **Figure 2-17**.

When using an extension cord with any electrical power tool, it must be connected to a ***ground fault circuit interrupter (GFCI)*** receptacle. A GFCI receptacle or breaker will detect the faintest short to ground and immediately break the circuit, interrupting the flow of power to the tool. Most of today's worksites require GFCIs with the use of extension cords, **Figure 2-18**. Tools equipped with electrical cords require proper care. Inspect cords before each use for signs of cuts, frays, nicked insulation, or loose connections. Never carry a tool by its power cord or yank the cord to disconnect it from a power source. Always unplug power tools when not in use, before servicing, or when changing or installing accessories. Be sure the power tool's switch is off before plugging it into a power outlet.

Figure 2-16. A worker wearing arc flash protection.

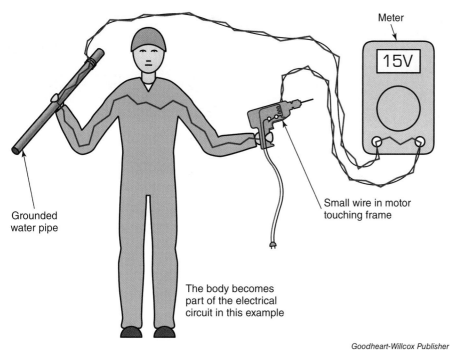

Figure 2-17. A worker becoming part of the electrical circuit because of an improperly grounded power tool.

Figure 2-18. An example of a GFCI receptacle.

> **CODE NOTE**
>
> **Shut It Down!**
>
> Codes now require that on commercial boiler rooms, a "manually operated remote shutdown switch shall be located just outside the boiler room door and marked for easy identification. If the boiler room door is on the building's exterior, the switch should be located just inside the door. If there is more than one door to the boiler room, there should be a switch located at each door. Activation of the emergency shutdown switch shall immediately shut off the fuel or energy supply," **Figure 2-19**.
>
> This is in accordance with ASME CSD-1, 2009. ASME is the *American Society of Mechanical Engineers*, and CSD stands for *Controls and Safety Devices*.

Figure 2-19. An example of an emergency stop button outside of a boiler room.

2.4 Lockout/tagout

To ensure that electrical work can be performed safely on boilers and hydronic equipment, most job sites implement a *lockout/tagout* procedure. This safety practice requires that the electrical power to any piece of machinery must first be shut off and locked out at the disconnect switch before work begins on the equipment, **Figure 2-20**. Secondly, the lock should be tagged to identify the company or technician working on the equipment. If multiple companies or technicians are working on the same piece of equipment, then multiple tags should be used. The last person to finish their work is responsible for removing their tag—and only then may the equipment be re-energized.

Various forms of electrical lockout/tagout equipment are available. If the equipment being serviced does not have the capability of being locked out locally, there are other items for locking out the equipment at the breaker panel, **Figure 2-21**.

30 Hydronic Heating: Systems and Applications

A

B

C

Ideal Industries, Inc.; Master Lock

Figure 2-20. These are examples of lockout/tagout components for various devices. They include lockouts for ball valves (A), gate valves (B), and a multi-lock hasp that allows several different trades workers to install their own lockout/tagout locks (C).

Master Lock

Figure 2-21. This image shows a lockout/tagout kit for different types of electrical switches. This kit includes the following lockout/tagout devices: lockout hasp, adjustable cable lockout, gate valve covers, electrical plug cover, ball valve lockout, devices for circuit breaker, and danger tags. Include the company name and pertinent information on tagout devices.

CODE NOTE

Lockout/tagout

OSHA is the code authority for lockout/tagout. They set specific standards for lockout/tagout procedures, so it's important you fully understand and follow your company's safety policy regarding this practice.

PROCEDURE

Steps for Lockout/tagout

Here are the proper steps for a lockout/tagout procedure:

1. Identify all sources of energy when servicing equipment.
2. Disable any backup energy sources, such as batteries and generators.
3. Identify shut-offs for each energy source.
4. Notify all other workers of the lockout.
5. Shut off energy sources and lock switches in the *Off* position. If multiple workers are servicing the equipment, each should place a lock on each energy source.
6. Place a tag on the lockout device.
7. Deplete any stored energy sources, such as capacitors.
8. Test the equipment and circuit to make sure the power has been disconnected.

Copyright Goodheart-Willcox Co., Inc.

2.5 Safety with Tools and Equipment

Another important aspect of safety on the job is in working safely when using tools and equipment. Today's installation and service technicians have a wide variety of tools and equipment to choose from to accomplish the tasks they are required to perform, and it is important to understand how to use these items safely and efficiently.

2.5.1 Hand Tools

Technicians should first understand how to use a tool and acquire the proper training before attempting to use it for the first time. This will reduce the chance of personal injury and damage to the tool. Do not use a tool for something other than its intended purpose—always use the right tool for the job. When working with hand tools, it is crucial to keep the following in mind:

- Always wear the appropriate PPE, including eye protection, steel-toed boots, and possibly gloves.
- When transporting tools, carry them in a safe and secure manner. Take care to store sharp instruments in their designated holders, and never carry sharp or pointed tools in your pocket.
- Always keep track of tools when working at heights—a falling tool can injure or kill a coworker.
- When working from a ladder, transfer tools to the work area by rope or bucket.
- When using hand tools, make sure your grip and footing are secure.
- Always keep tools in good working condition, and do not use damaged or defective tools. Make sure they are clean and that cutting tools are kept sharp. Dull cutting tools or screwdrivers with worn tips can cause personal injury.

2.5.2 Power Tools

Before using power tools, first read and understand the owner's manual to ensure appropriate use. These guidelines will also include safety and warning labels. Other important things to keep in mind include:

- Always choose the correct tool for the job. Do not use a tool that is not intended for the task at hand.
- Wear the appropriate PPE, including appropriate hearing and eye protection.
- Avoid wearing loose-fitting clothing or jewelry and tie back long hair, as these could catch in moving parts.
- Always be aware of your surroundings and identify any hazards in your work area, such as power lines, electrical circuits, water pipes, and other mechanical hazards that may be hidden from view or below the work surface.
- Use clamps or vises to secure your work. Loose workpieces can become airborne projectiles and cause injury.
- Inspect power tools for defects and signs of wear before using. Keep tools clean, sharp, and lubricated. Make sure handles are clean, dry, and free from oil and grease.
- Change power tool accessories according to the instructions and hire qualified service personnel to make necessary repairs.

In addition, check electric power tools to ensure they have the proper **_grounding prong_** and are in good condition. Failure to use properly grounded tools can result in electrical shock and injury to the technician. When using

Alexander Remy Levine/Shutterstock.com

Figure 2-22. An example of a three-to-two-prong electrical adapter.

three-prong power tools on two-prong receptacles, use a three-to-two prong adapter to ensure proper grounding. The grounding clip requires fastening under the receptacle wall plate screw for safe operation, **Figure 2-22**.

2.5.3 Ladder Safety

Ladders are essential on any construction site or when working on elevated levels within a boiler room. Typically, ladders are constructed from aluminum, wood, or fiberglass, each with their own characteristics.

- Aluminum ladders are lightweight and easy to maneuver. However, they are not recommended for HVAC and boiler room applications because they are highly conductive to electricity.
- Wooden ladders are nonconductive but can be bulky and hard to maneuver. Care must be taken to ensure that the wooden rungs are not damaged before using these ladders.
- Fiberglass ladders are the ladder of choice in the HVAC industry. They are relatively lightweight, nonconductive, and more durable than aluminum and wooden ladders.

SAFETY FIRST

Grounding Prongs

Never operate an electrical power tool if the grounding prong is missing. Failure to do so can result in the worker becoming the grounding conductor, which leads to electrical shock or possibly another injury.

SAFETY FIRST

Ladder Safety

A worker may inadvertently raise a ladder into a power line or place it against a live electrical hazard without realizing it. Care must be taken to observe any overhead obstruction, including live electrical lines, to ensure the ladder does not come in contact with these hazards, **Figure 2-23**. This is one reason why aluminum ladders are not the preferred choice on the job site.

Always adhere to the following safety guidelines when working with or around ladders:

1. Use only **Canadian Standards Association (CSA)** or **American National Standards Institute (ANSI)** approved ladders.
2. Keep ladders in good working order, free from damage and deterioration. Never use a damaged or broken ladder. This is especially true of wooden ladders, which are susceptible to damaged or broken rungs.
3. Ladders must be inspected periodically to ensure they are in sound working order. Many companies have safety officers who are in charge of inspecting any ladders that you may be carrying in your service van.
2. Make sure all ladders have nonslip feet and are placed on a stable surface, free from oil and grease.
3. Note the maximum carrying capacity of the ladder. Take into account the weight of both the worker and any materials transported up the ladder.
4. Place the ladder away from the wall at a distance equal to one-quarter of the height of the building. Ensure the top of the ladder extends at least three feet above the top of the roof, **Figure 2-24**.
5. Always tie off or otherwise secure the top of the ladder where it meets the building to ensure that it does not slip sideways.
6. Always face the ladder and stay centered when climbing or descending.
7. When using stepladders, make sure they remain in the fully open position.

Brent Wong/Shutterstock.com

Figure 2-23. Watch for overhead power lines when working on a ladder.

2.6 Working around Pressurized Gases

Whether installing, servicing, or repairing boiler piping or other hydronic equipment, technicians may be required to connect hydronic supply and return lines by either brazing or soldering. These tasks require the use of pressurized gases

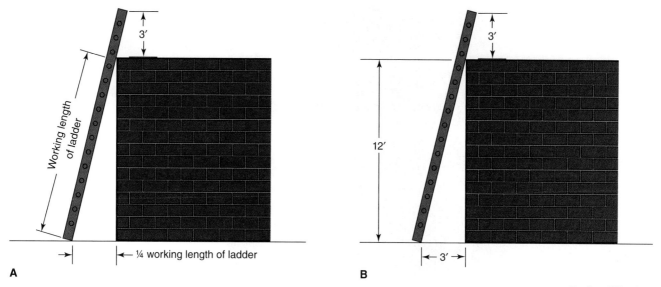

Figure 2-24. This illustration shows the proper placement of a ladder against a building.

such as propane, oxygen, and acetylene. There are essential steps to follow when working with or around pressurized gases to ensure safety. Propane and acetylene are extremely flammable, so first and foremost, they should never be used around open flames. In addition, the following safety precautions must be followed when working with pressurized gases:

- When not in use, seal pressurized tanks with protective caps, **Figure 2-25**.
- Transport larger tanks using approved carts, making sure tanks are properly secured to the carts with chain.
- Always store pressurized tanks in an upright position.
- Do not allow tanks to drop or accidentally fall over. They could become projectiles if a valve is damaged or breaks off.
- Oil in the presence of pressurized oxygen is highly explosive. Keep oxygen lines free of any oil. Oil residue in an oxygen regulator could cause an explosion.
- Check pressure regulators for damage and keep them in good working order.
- Know the proper pressure settings for oxygen and acetylene when using an oxyacetylene torch, **Figure 2-26**.

2.7 Chemical Hazards

Various chemicals are used in and around the boiler room for cleaning and water treatment. Although most chemicals used for cleaning are rather mild, water treatment chemicals are often harsh and may irritate the skin and eyes, **Figure 2-27**. Certain chemicals are either acidic or alkaline and can react differently if not handled properly. Always follow the manufacturer's directions for safe handling of any chemical, and remember to use the proper protective gear, which includes rubber gloves and protective face shields. Know where the nearest eyewash station is located and post *Safety Data Sheets (SDS)* in a conspicuous area wherever chemicals are used. A Safety Data Sheet lists information on various chemicals, including their properties and potential hazard to human health and the environment. Furthermore, an SDS defines the protective measures and safety precautions required for proper handling, storage, and transportation of such chemicals, **Figure 2-28**.

Figure 2-25. Be sure that unused tanks always have protective caps.

SAFETY FIRST

Remove Regulators for Tank Transport

Many states require the removal of regulators on oxygen and acetylene tanks before transport in any motorized vehicle. The reason for this is that if an accident occurs involving your truck or service van, these regulators—if left on the tanks—could be broken off, causing a fire or explosion. It is always a sound safety practice to remove regulators on all tanks before transporting them.

Mut Hardman/Shutterstock.com

Figure 2-26. Know the proper pressure settings for oxygen and acetylene when using an oxyacetylene torch.

Millionstock/Shutterstock.com

Figure 2-27. This sign indicates that a particular chemical can cause burns to the skin. Warning signs like this need to be posted where caustic chemical exposure is possible.

Safety Data Sheet (SDS) Sections	
Section	**Description**
Section 1. Identification	Includes a product identifier, contact information for the manufacturer or distributor, recommended use, and restrictions on use.
Section 2. Hazard(s) identification	Lists all hazards regarding the chemical and required label elements.
Section 3. Composition/information on ingredients	Lists information on chemical ingredients and trade secret claims.
Section 4. First-aid measures	Describes potential symptoms or effects and required treatment.
Section 5. Fire-fighting measures	Describes potential fire hazards and lists suitable extinguishing techniques and equipment.
Section 6. Accidental release measures	Lists emergency procedures, required protective equipment, and proper methods of containment and cleanup.
Section 7. Handling and storage	Lists precautions for safe handling and storage, including incompatibilities.
Section 8. Exposure controls/personal protection	Specifies exposure limits, including OSHA's Permissible Exposure Limits (PELs), ACGIH Threshold Limit Values (TLVs), and any other exposure limits recommended by the manufacturer, importer, or employer. Also includes recommended engineering controls and personal protective equipment (PPE).
Section 9. Physical and chemical properties	Specifies the chemical's characteristics.
Section 10. Stability and reactivity	Lists the stability of the chemical and the possibility for hazardous reactions.
Section 11. Toxicological information	Describes the toxicity of the chemical, including routes of exposure, symptoms of toxicity, acute and chronic effects of toxicity, and numerical measures of toxicity.
Section 12. Ecological information	Lists potential ecological impacts of the chemical.
Section 13. Disposal considerations	Explains how to dispose of the chemical safely.
Section 14. Transport information	Describes transportation requirements and safety considerations.
Section 15. Regulatory information	Lists regulations pertaining to the chemical.
Section 16. Other information	Lists the date of preparation or last revision of the SDS.

Adapted from OSHA

Figure 2-28. Safety Data Sheets (SDS) should be posted in a conspicuous area wherever chemicals are in use.

2.8 Working in a Confined Space

OSHA defines a **confined space** as "an area that has limited or restricted means for entry or exit and is not designed for continuous occupancy." Examples of confined spaces include tanks, storage bins, pits, tunnels, and pipelines, **Figure 2-29**. Confined spaces are typically closed off from an outside source of ventilation, and, therefore, may require the worker to wear a supplemental oxygen device in the form of a ***self-contained breathing apparatus (SCBA)***, **Figure 2-30**. Before entering a confined space, the worker should verify that the atmosphere is void of any hazardous vapors. According to OSHA, the only way to safely do this is with a calibrated direct reading instrument that will detect hazardous substances in the air. When technicians are trained on the operation of this instrument and the procedures for confined space monitoring, the risks involved with this type of work are greatly reduced. Even if the confined space checks out, remember that certain gases such as nitrogen or refrigerants will displace oxygen and can cause suffocation should there be a leak from their tanks. Some situations may require that a "spotter" stand outside the confined space to ensure the safety of the worker inside.

2.9 Safety in the Boiler Room

The first thing to remember when working in any boiler room is that the boiler is a pressurized vessel that could—if large enough—level the entire building. Of course, boilers are required to have built-in safety devices to prevent them from exploding. However, the next person to work in a boiler room may not be aware of what tasks the last person performed. Keep in mind, the National Board of Boiler and Pressure Vessel Inspectors reports that 127 people have lost their lives and 720 people have been injured over the last 10 years as a result of boiler accidents, **Figure 2-31**. Over 80% of these accidents and fatalities were the direct result of human error or poor maintenance on the part of the operator or technician. A defective ***low water cutoff*** is the leading mechanical cause of accidents in the boiler room. A low water cutoff could fail due to improper

> **SAFETY FIRST**
>
> **Safety Data Sheets**
>
> When purchasing chemical products for use in the boiler room, always request a Safety Data Sheet (SDS) from the vendor. Most job sites require an SDS on each chemical product used or stored in the boiler room. One way to maintain these critical pieces of information is to store them in a large, clearly marked envelope kept handy in your service vehicle.

zulkamalober/Shutterstock.com

Figure 2-29. An example of a worker descending into a confined space.

Technicsorn Stocker/Shutterstock.com

Figure 2-30. When working in a confined space, a self-contained breathing apparatus may be needed.

High Simple/Shutterstock.com

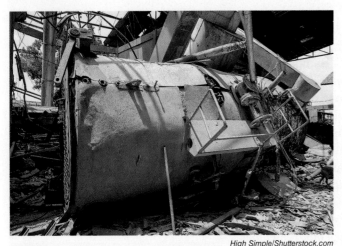

Figure 2-31. Boiler explosions have the capability of severely damaging or leveling an entire building.

maintenance, or a maintenance or service technician could bypass it, not understanding the importance of this safety device, **Figure 2-32**. The bottom line is: never assume that the last person to work on a boiler knew exactly what they were doing.

Another safety consideration is the proper operation of the boiler's combustion equipment. If not maintained properly, the burner assembly and controls on the boiler could potentially produce and release deadly carbon monoxide gas into the boiler room atmosphere. ***Carbon monoxide (CO)***, is an odorless, colorless gas resulting from incomplete combustion in the boiler's burner equipment. It can kill in minutes by inhibiting the blood's ability to carry oxygen throughout the body. According to the Centers for Disease Control and Prevention (CDC), CO poisoning is responsible for over 1500 deaths and over 40,000 trips to the hospital emergency room annually. In fact, carbon monoxide poisoning is the leading cause of accidental poisoning deaths in the United States. **Figure 2-33** shows the effects of carbon monoxide on the human body when exposed to various levels in parts per million (ppm). The best approach to

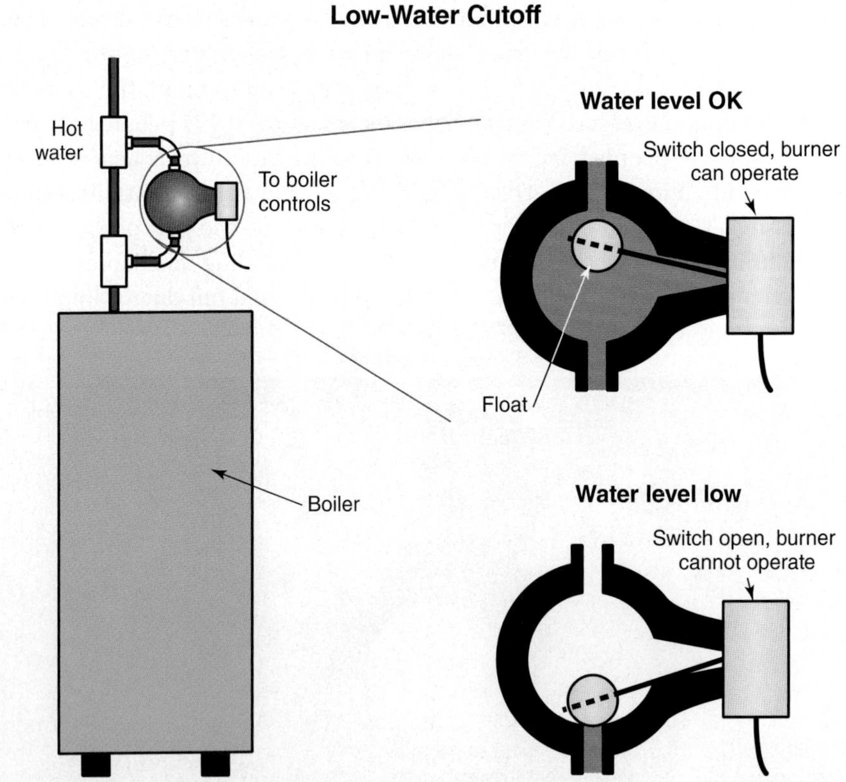

Goodheart-Willcox Publisher

Figure 2-32. A low-water cutoff shuts off electrical power to the oil burner or gas burner if the water level or pressure in the heating system falls below a safe level.

Effects of Carbon Monoxide Exposure

Parts Per Million (ppm)	Health Effects
9 ppm	ASHRAE maximum recommended safe level for 8-hour exposure
50 ppm	OSHA maximum recommended safe level for 8-hour exposure
200 ppm	Slight headache, fatigue, dizziness, and nausea within 2–3 hours
400 ppm	• Headache within 1–2 hours • Life threatening within 3 hours • EPA and AGA maximum ppm in flue gas
800 ppm	• Headache, dizziness, nausea, and convulsions within 45 minutes • Loss of consciousness within 2 hours • Fatal within 2–3 hours
1,600 ppm	• Headache, dizziness, and nausea within 20 minutes • Fatal within 1 hour
3,200 ppm	• Headache, dizziness, and nausea within 5–10 minutes • Fatal within 30 minutes
6,400 ppm	• Headache, dizziness, and nausea within 1–2 minutes • Fatal within 10–15 minutes
12,800 ppm	Fatal within 1–3 minutes

Goodheart-Willcox Publisher

Figure 2-33. Effects of carbon monoxide exposure.

preventing accidental carbon monoxide poisoning is through proper maintenance practices. Each boiler should undergo a thorough combustion analysis on an annual basis, ensuring that CO levels are within tolerance and also determining whether the boiler is operating at peak efficiency. Another step is to inspect all breaching, flues, and chimneys for any cracks, rust, or abnormalities. Repair any defects to the boiler's venting system immediately upon detection.

SAFETY FIRST

Preventing Carbon Monoxide Poisoning

The most logical preventive step in guarding against carbon monoxide poisoning is to install a *low-level CO detector* in every boiler room. Ordinary carbon monoxide detectors available at most hardware stores are not intended to alarm at low CO levels. Even though these types of carbon monoxide detectors are UL approved (UL is an OSHA-approved safety certification company), they are not required to trigger an alarm until the CO levels reach 150 ppm for up to 50 minutes or 400 ppm for up to 15 minutes. This means that a UL-listed alarm would allow you to breathe air with up to 350 ppm of CO for 45 minutes! A quality, low-level CO detector will alarm when levels rise above 6 ppm. Early signs of carbon monoxide poisoning include a dull headache, dizziness, nausea, and drowsiness.

Chapter Review

Summary

- It is important for all boiler workers to know that safety is everyone's business. Most accidents are caused by workers deliberately performing unsafe work practices, not thinking clearly, practicing bad habits, or working around or with hazardous equipment. Eighty-five percent of workplace injuries are avoidable with the proper safety measures in place.
- Personal protective equipment (PPE) includes devices to protect the eyes, ears, head, face, hands, feet, and respiratory system.
- Fire safety includes proper training on what to do when a fire occurs and knowing the correct fire extinguisher to use on different types of fires. A Type A extinguisher is used to put out wood, paper, or trash fires. Type B is for flammable liquids such as grease, gasoline, paints, and oils. Type C is for electrical equipment fires. Some fire extinguishers are rated for all three types of fires and are usually the dry-chemical type.
- Injury from electricity can result in shocks, burns, and even death.
- Workers must know how to work safely around electrical equipment, especially when troubleshooting live circuits.
- An arc flash happens when electrical current flows through an air gap between conductors within the breaker panel or main disconnect, causing an electrical explosion or discharge. The most common causes of arc flash are the touching of test probes to the wrong surface within a panel, or when a tool such as a screwdriver slips and makes contact with the ground bar. Sparks from a break or gap within the wire's insulation; equipment failure from using substandard parts; and dust, corrosion, or other impurities found on the conductor's surface can also cause arc flash.
- When working with electrical power tools, be sure that they are grounded.
- Extension cords should be used with ground fault circuit interrupters.
- Practice the proper lockout/tagout procedure when working on equipment.
- Always wear the proper protective equipment when working with hand tools and power tools.
- Proper ladder safety includes using CSA- or ANSI-approved ladders with nonslip feet on a stable surface, noting maximum carrying capacity, placing the ladder appropriately in relation to the building and tying it off, facing the ladder and staying centered while climbing, and fully opening stepladders.
- Pressurized gases must be handled carefully to avoid an explosion. Some gases are highly flammable, and tanks can become projectiles if broken.
- Safely using chemicals involves wearing the proper protective gear—rubber gloves and protective face shields.
- Working in a confined space may require the use of a self-contained breathing apparatus (SCBA) or a "spotter" standing outside the confined space to ensure the safety of the worker inside.
- Boiler safety includes servicing and maintaining low-water cutoff devices.
- Carbon monoxide hazards can occur when the boiler's burner equipment is not maintained properly.

Know and Understand

1. What topic would generally *not* be discussed during a weekly Toolbox Talk?
   A. The correct procedures for handling tools
   B. Working safely around chemicals
   C. Scheduling the weekly work schedule
   D. None of the above.
2. *True or False?* Electrical shock can result in burns to the skin and even death.
3. Name the phenomenon that occurs when electrical current flows through an air gap between conductors within a breaker panel or main disconnect.
   A. Arc strike
   B. Arc flash
   C. Electrical flash
   D. Short circuit
4. What can happen if loose-fitting clothing or jewelry is worn when using power tools?
   A. It can be caught in moving parts.
   B. It can cause personal injury.
   C. It can become ripped or damaged.
   D. All of the above.
5. Why are aluminum ladders *not* recommended for use on jobsites?
   A. They are bulky and hard to maneuver.
   B. Their rungs are easily damaged.
   C. They are conductive to electricity.
   D. They are too expensive.
6. What is the worst that could happen if a pressurized gas cylinder were to fall over without its protective cap in place?
   A. It could become a projectile if the valve is damaged.
   B. It could leak its contents and cost extra money.
   C. It could injure someone's foot.
   D. None of the above.
7. What information regarding a chemical substance would a Safety Data Sheet (SDS) typically *not* cover?
   A. The chemical's potential hazard to human health
   B. The weight and volume of the chemical
   C. The protective safety precautions when handling the chemical
   D. The proper storage of the chemical
8. *True or False?* Before entering a confined space, a worker should verify that the atmosphere has been checked for hazardous vapors.
9. With regard to mechanical failure, what is the leading cause of accidents in the boiler room?
   A. A gas or chemical leak
   B. A defective gas shutoff valve
   C. A defective pressure relief valve
   D. A defective low water cutoff
10. What is the best maintenance practice for preventing accidental carbon monoxide poisoning in the boiler room?
    A. Ensure that the gas pressure to each boiler is correct.
    B. Make sure the boiler room's exhaust fans are working properly.
    C. Have each boiler undergo a thorough combustion analysis on an annual basis.
    D. Make sure there is a low-level carbon monoxide detector installed above each boiler.

Apply and Analyze

1. Define PPE, and list the different types for each part of the body.
2. List the proper steps for lifting heavy objects.
3. Define what a confined space is, and explain why respiratory protection may be needed when working in one.
4. List the different classifications for fire extinguishers and which types of materials they are designed to extinguish.
5. Explain how a human body can become a path of current flow in an electrical circuit.
6. Explain how and why electrical power tools must be properly grounded.
7. Define a ground fault circuit interrupter and where it is used.
8. List the proper safety practices that should always be followed when working with or around ladders.
9. Explain why extra care needs to be taken when working with pressurized gases.
10. What is the by-product of incomplete combustion in the boiler's burner equipment, and how can it affect humans?

Critical Thinking

1. Explain the term *lockout/tagout* and why it is vital to job safety.
2. Give some examples of what can happen when hand tools are not used for their intended purpose.

3 Boilers

Chapter Outline
3.1 Operating Pressures
3.2 Types of Fuel
3.3 Construction Materials and Boiler Efficiency
3.4 Condensing Boilers
3.5 Modulating Boilers
3.6 Electric Boilers
3.7 Low NO_x Boilers

EvijaF/Shutterstock.com

Learning Objectives

After completing this chapter, you will be able to:
- Identify the different types of boiler classifications.
- Differentiate the three boiler operating pressures.
- Describe the different types of fuels used for heating boiler water.
- List the different types of materials used in boiler construction and the advantages and disadvantages of each.
- Describe the difference between a wet-base and dry-base boiler design.
- Explain the difference between a fire-tube and water-tube boiler.
- Summarize what constitutes a condensing boiler design.
- Explain how latent heat is used in a condensing boiler.
- Describe how a "mod/con" boiler works.
- Define turndown ratio.
- Describe the advantages of an electric boiler.
- Discuss the application of low NO_x boilers.

Technical Terms

American Society of Mechanical Engineers (ASME)
Annual Fuel Utilization Efficiency (AFUE)
biomass
condensing boiler
cord
dew point
dry-base boiler
fire-tube boiler
flue
heat exchanger
heating load
high-mass boiler
high-pressure boiler
hydrocarbon
kilowatt (kW)
liquefied petroleum (LP)
low NO_x boiler
low-mass boiler
low-pressure boiler
medium-pressure boiler
mod/con boiler
modulating boiler
Scotch Marine boiler
turndown ratio
water-tube boiler
wet-base boiler

By definition, a boiler is an enclosed pressurized vessel in which water or other fluid is heated. The name *boiler* is deceiving. Some boilers are designed to simply heat water while others are designed to generate steam—only in the latter is there actual "boiling" action. The boilers discussed in this book do not actually "boil" but are used simply to heat water to a design temperature. The normal operating temperature of a hot water boiler can range from 90°F to 200°F. Boilers vary in size, shape, and complexity, **Figure 3-1**. Hot water boilers can be classified in several ways:

- Operating pressure
- Type of fuel
- Construction material
- Wet- or dry-base
- Fire-tube or water-tube
- Efficiency (conventional and condensing)

Any particular boiler may fit a combination of these classifications, but for clarity's sake, we discuss each quality separately as well as several other special types of boilers.

3.1 Operating Pressures

Hot water boilers can be classified according to their normal operating pressure. The three common operating pressures are low pressure, medium pressure, and high pressure. It is important to remember that the water in a hot water boiler is not designed to actually boil, even though the operating temperature according to its classification may exceed 212°F. The boiling point of water at sea level is 212°F. As the pressure in some boilers (per classification) is increased, so is the water's boiling point.

3.1.1 Low-Pressure Boilers

By definition, a ***low-pressure boiler*** has an operating pressure of up to 160 psi and an operating water temperature of up to 250°F. However, most residential boilers are designed with a normal operating pressure of 12 psi and a normal operating

A
Lubos Chlubny/Shutterstock.com

B
Goodheart-Willcox Publisher

C
Goodheart-Willcox Publisher

Figure 3-1. Hot water boilers are available in many different configurations. A—A medium-efficiency industrial boiler. B—A wall-hung, high-efficiency boiler. C—A floor-standing, high-efficiency boiler.

Goodheart-Willcox Publisher

Figure 3-2. An example of a boiler's rating plate, or nameplate, with an input rating of 45,000 Btu/hr.

water temperature of up to 180°F. Most low-pressure boilers are used for residential and commercial comfort heating applications, with the majority of residential boilers having an input rating of below 300,000 Btu/hr. The *input rating* is the measurement of the amount of fuel the boiler will burn, stated in either Btu/hr or Mbh (1000 Btu/hr). Low-pressure boiler ratings are dictated and tested according to the *American Society of Mechanical Engineers (ASME)* code requirements, and every boiler is required to list these ratings on its nameplate, **Figure 3-2**.

3.1.2 Medium-Pressure Boilers

Medium-pressure boilers have an operating pressure range of 160 to 300 psi and an operating water temperature of between 250°F and 350°F. These boilers are typically used for manufacturing process applications such as for food, paper, and chemicals. They may also be used for the dry cleaning and laundromat industries.

3.1.3 High-Pressure Boilers

High-pressure boilers have an operating pressure above 300 psi and typically have an operating water temperature above 350°F. Applications for high-pressure boilers include metal plating and anodizing tanks, laminating processes, food processing, plastic manufacturing, and parts washing systems. One advantage of high-pressure boilers is the lower initial equipment cost as compared to high-pressure steam applications. In addition, a high-pressure hot water boiler can save on maintenance costs as compared to steam due to the elimination of steam trap discharge losses and boiler water treatment. They have virtually no make-up water costs. Furthermore, piping distribution systems are easier to install and have fewer complications than do steam heating systems.

SAFETY FIRST

Boiler Operator License

Local, state, and federal agencies require employees and technicians to obtain a boiler operator's license in order to operate larger boilers. This license typically applies to commercial boilers and boilers located in public and municipal buildings. Only those employees and technicians who hold the appropriate license may operate or repair any such boiler. Smaller boilers designed for residential or light commercial use are typically exempt.

3.2 Types of Fuel

The earliest manufactured boilers burned either coal or wood, **Figure 3-3**. With the exception of outdoor wood boilers (discussed in Chapter 20, *Outdoor Wood Boilers*), most of today's boilers utilize fuel oil, natural gas, propane (LP), or electricity as a source of fuel.

3.2.1 Fuel Oil

Fuel oil is considered a fossil fuel, as it is derived from crude oil through distillation. There are six grades of fuel oil, with the lightest grade being #1 and the heaviest being #6. The fuel oil grade most often used in residential and light commercial boilers is #2. This fuel oil is the same fuel used in automotive vehicles that run on diesel. Fuel oil has a heating value of 140,000 Btu/gallon. Although the demand for fuel oil for boilers has reduced somewhat in the upper Midwest region of the United States due to the increased popularity of propane fuel, it is still quite popular along the Northern Atlantic states and up into New England. Fuel oil is delivered by tanker truck to homes and businesses, where it is stored in tanks either above ground or below, **Figure 3-4**. This is covered in greater detail in Chapter 5, *Oil Systems*.

nata-lunata/Shutterstock.com

Figure 3-3. Many original boilers were coal fired.

3.2.2 Natural Gas

Natural gas results from the decay of organic material from plants and animals over millions of years, forming deep pockets within the Earth's surface. Natural gas is primarily comprised of methane (CH_4), but it can also contain small quantities of ethane, propane, and butane. All of these gases are classified as **hydrocarbons**, which are combustible, efficient, clean-burning gases. In its native state, natural gas is odorless and colorless. Gas utility companies add sulfur compounds, called *odorants*, to natural gas to make gas leaks more easily detectable. The energy content of natural gas is approximately 1050 Btu/cubic foot. Natural gas is distributed across the United States by a network of pipelines, **Figure 3-5**, and delivered to homes and businesses through underground piping.

Marcel Derweduwen/Shutterstock.com

Figure 3-4. Fuel oil is stored in tanks either above or below ground, where it is piped to the boiler.

3.2.3 Liquefied Petroleum (LP)

Propane is the most common type of **liquefied petroleum (LP)** gas and can include a mixture of butane and propane. Propane has a heating content of

Figure 3-5. Natural gas is distributed by a network of piping, from which it is delivered to homes and businesses.

Figure 3-6. LP gas is shipped to homes and businesses by tanker truck and stored in above-ground tanks such as this.

2500 Btu/cubic foot, or approximately 92,000 Btu/gallon. As the name implies, LP gas will condense into a liquid state when pressurized. This makes it more easily transportable into rural areas, where natural gas may not be available. Propane is shipped by truck and stored in above-ground tanks, where pressure regulators step down the pressure, converting it from a liquid to a gaseous state as it enters a building, **Figure 3-6**. Liquefied petroleum is a derivative of crude oil, produced as a by-product of the oil refining process. LP gas is heavier than air, with a specific gravity of 1.52. (The specific gravity of air is 1. Gases with a higher specific gravity are heavier than air, and their vapor will drop when released into an open space.) For this reason, take care when working with propane in basements, crawl spaces, or low-lying spaces to guard against the buildup of dangerous pockets of explosive gases.

TECH TIP

Converting from Natural Gas to Propane

Most residential and commercial boilers are shipped from the factory designed for use with natural gas. It is possible to convert these boilers to utilize LP gas with the proper changes. Note the requirements for converting a boiler from natural gas to LP:

1. Change out the orifices on the burner manifold.
2. Change the regulator pressure spring on the gas valve.
3. Set the proper operating pressure on the gas valve upon startup.

3.2.4 Electricity

Although not as predominant as natural gas and LP, electricity can be used to fuel boilers. Electricity is almost 100% efficient because there are no heat exchangers or flue pipes used with electric boilers, as compared to boilers burning fossil fuels. However, because electricity has a much higher cost per Btu when compared to fossil fuels, it can have a prohibitively high operational cost. A *heat exchanger* is a device that transfers heat from one fluid to another without allowing the two fluids to make physical contact. The heat exchanger in a fossil fuel boiler works by converting the energy (heat in this case) from fuel over to water. The water that passes through the heat exchanger in the boiler is heated up in the process. The water is then fed into the central heating system via piping. A *flue* is a duct, pipe, or opening in the boiler's chimney for conveying exhaust gases from the fire side of the heat exchanger to the outdoors.

The heat content of electricity is about 3.41 Btu/watt. A watt is a unit of power and is calculated by multiplying the input voltage of the appliance by the amperage used by the appliance. Most commercial electric boilers are rated in *kilowatts (kW)* of energy. One kilowatt of electricity is equal to 1000 watts of power, or approximately 3410 Btu of energy.

3.2.5 Biomass

Biomass is organic matter—such as wood, corn, and other biological materials—that is burned as fuel. The burning of biomass releases carbon emissions, but it is considered a renewable energy resource since trees and other organic materials can be replaced with new growth. The Btu content of wood varies according to the particular species of tree, with hardwood varieties containing the highest Btu content per *cord*. A cord of wood is the unit of measurement in the United States and Canada, equal to 128 cubic feet of volume, or approximately $4' \times 4' \times 8'$. In comparison, a cord of oak contains a heating content of about 36 million Btu while cottonwood possesses only about 16.8 million Btu/cord. Biomass boilers may be placed in basements, but most are located outdoors, **Figure 3-7**. Outdoor wood boilers are discussed in Chapter 20, *Outdoor Wood Boilers*.

Goodheart-Willcox Publisher

Figure 3-7. An example of an outdoor wood boiler, or biomass boiler.

> **GREEN TIP**
>
> **The Natural Carbon Cycle**
> Even though the burning of different types of biomass does generate greenhouse gases, burning wood releases about the same amount of carbon dioxide (CO_2) into the atmosphere as does the decaying of that same amount of wood as part of the natural carbon cycle—only at a faster pace. The EPA has determined that wood burning is carbon neutral, yet environmentalists feel that this accelerated release of greenhouse gases contributes to climate change.
>
> One method of reducing wood-burning emissions is gasification technology, which is used in many of today's wood-burning boilers. In the gasification process, exhaust gases from burning wood are forced through a secondary heat exchanger in the wood boiler and reburned at temperatures up to 2300°F. This results in a cleaner wood-burning boiler that is virtually smoke free. These types of boilers have been independently verified to meet EPA phase 2 emissions output for reduced greenhouse gases. More information on EPA wood-burning standards can be found in the *Burn Wise* section of the EPA's website.

3.3 Construction Materials and Boiler Efficiency

Boiler construction materials must hold up to high operating pressures and temperatures, and the evolution of boiler materials used over time reflects the technological advances seen throughout manufacturing, such as in the automotive industry. Today's construction materials must adhere to local and national codes, and boilers must meet ASME standards. Several types of materials are used in the construction of modern-day hot water boilers, based on boiler efficiency.

Boiler efficiencies and their corresponding types of construction materials can be grouped into the following categories:

Lower-efficiency Boilers:
- Natural draft that creates a flow of combustion gases
- Continuous pilot light
- Heavy-duty, cast-iron heat exchanger
- Metal flue pipe material
- Higher return water temperatures
- 56 to 70% AFUE

Mid-efficiency Boilers:
- Combustion exhaust fan that controls the flow of combustion air and gases more precisely
- Electronic ignition (no pilot light)
- Compact size and lighter weight to reduce cycling losses

- Steel and copper heat exchangers
- Metal flue pipe material
- Small-diameter flue pipe
- Higher return water temperatures
- 80 to 83% AFUE

High-efficiency Boilers:
- Combustion exhaust fan that controls the flow of combustion air and gases more precisely
- Condensing flue gases pass through a second heat exchanger for extra efficiency
- Stainless steel heat exchangers
- Sealed combustion
- PVC flue pipe material
- Lower return water temperatures
- 90 to 98.5% AFUE

3.3.1 Cast Iron

One of the earliest boiler construction materials was cast iron, and it is still used today. Cast-iron boilers are often found in residential and light commercial applications. *Heat exchangers* in cast-iron boilers are built in vertical sections, similar to the slices in a loaf of bread. This design allows them to be expanded or enlarged for future use. Cast-iron boilers are very heavy and typically hold between 15 and 30 gallons of water. These attributes classify cast-iron boilers as *high-mass boilers*, which means they take longer to heat up than other types, such as low-mass boilers. However, they tend to hold their heat longer. Low-mass boilers typically have copper heat exchangers that allow them to heat up quickly, but they do not hold their heat as well over longer periods of time. Because of this, cast-iron boilers typically have longer run times and longer off cycles than other types of boilers. One drawback to cast-iron boilers is their susceptibility to corrosion if excessive air is trapped in the system over long periods. See **Figure 3-8**.

Cast-iron boilers come in wet-base and dry-base designs. In a *dry-base boiler*, the area under the combustion chamber is dry—in other words, the boiler water is contained in a space *above* the combustion chamber. In a *wet-base boiler*, the water that is heated is located throughout the combustion chamber. See **Figure 3-9**.

3.3.2 Steel and Stainless Steel

Steel and stainless steel boilers, **Figure 3-10**, include bundles of welded heat exchanger tubes submerged in water. These tubes are often electroplated, or clad with nickel or other corrosion-resistant materials, to protect against corrosion and increase the life of the heat exchangers. Another method of reducing corrosion is the inclusion of an anode rod made of aluminum or magnesium. This "sacrificial" rod is consumed by oxidation much faster than the treated steel heat exchanger, thus prolonging the life of the heat exchanger. Steel boiler heat exchangers are typically classified as either the fire-tube or water-tube type.

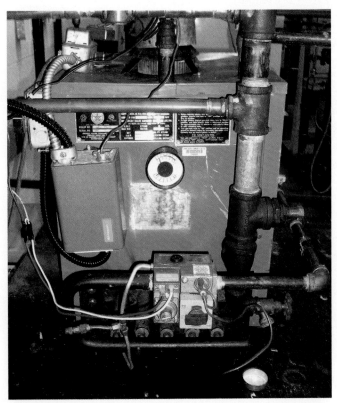

Goodheart-Willcox Publisher

Figure 3-8. Cast-iron boilers are considered high-mass boilers.

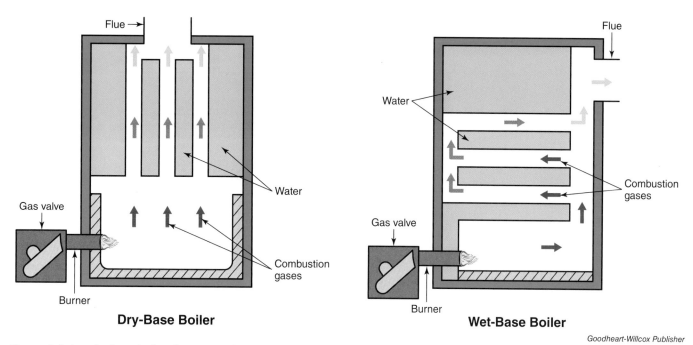

Figure 3-9. In a dry-base boiler, the area under the combustion chamber is dry. In a wet-base boiler, the water that is heated is located throughout the combustion chamber.

Figure 3-10. Steel boilers such as this (A) have bundles of welded heat exchanger tubes (B) that are submerged in water.

Fire-tube boilers contain tubes surrounded by water. The more commonly found type of boiler heat exchanger, fire-tube boilers are used in both residential and commercial applications, **Figure 3-11**. Combustion gases from the burner pass through the inside of the tubes, which heats the water. These tubes contain baffles, designed to slow down the speed of the flue gases as they pass through the tubes. Without these baffles, the hot flue gases would pass through the boiler too quickly, failing to transfer much of the heat from the gases to the water. By reducing the speed of the combustion gases through the tubes, boiler efficiency is increased. Each set of tubes makes several *passes* through the boiler before exiting through the flue outlet, located at the rear of the boiler. For example, a three-pass boiler will have three sets of tubes that pass back and forth within the boiler three times, **Figure 3-12**. The most common type of fire-tube boiler used for heating applications is known as the ***Scotch Marine boiler***. This type of boiler

Figure 3-11. An example of a fire-tube boiler.

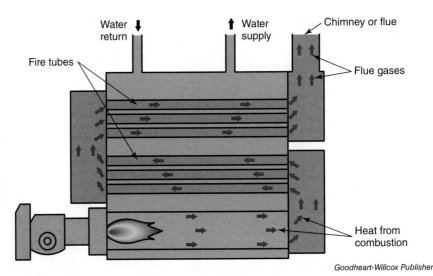

Figure 3-12. Fire-tube boilers have tubes that make several passes through the boiler before they exit through the rear of the boiler.

DID YOU KNOW?

The Most Famous Scotch Marine Boiler

The *Titanic* had 29 boilers: 24 double-ended and 5 smaller single-ended. The larger boilers were 15′-9″ in diameter and 20′ long. The smaller were 11′-9″ in length. All had three corrugated furnaces of 3′-9″ diameter, 159 furnaces in total, and a working pressure of 215 pounds per square inch (psi).

was notably used for marine service on ships because of its compact size as well as its efficiency in operation and ability to work in any type of water.

Advantages of fire-tube boilers include their simple design and the fact that they require less stringent water treatments. In addition, they are easy to clean, typically compact in size, and well suited for space heating as well as for industrial process applications. The disadvantage of these types of boilers is that they require extensive time to raise the water temperature due to the large ratio of water volume to heating surface. Because of this, fire-tube boilers do not respond quickly to changes in the building's heating load. The **heating load** is defined as the heating requirement necessary to maintain a desired temperature in a building. This topic is covered in detail in Chapter 13, *Building Heating Loads and Print Reading*.

In a **water-tube boiler**, water is contained inside tubes within the heat exchanger, and the combustion gases from the burner pass around the outside of these tubes. Heat from the flue gases is transferred to the water through the walls of the tubes, **Figure 3-13**. The more tubes added to a boiler, the greater volume of heating surface, thus increasing the boiler's volume and output. This design is the exact opposite of the fire-tube design. Water-tube boilers are used for both residential and commercial applications and can provide a faster response to heating load than fire-tube boilers. In addition, water-tube boilers are available in a greater range of sizes than fire-tube boilers and can reach very high temperatures. The disadvantages of water-tube boilers are higher initial capital costs, difficulty of cleaning, and their sometimes-daunting physical size.

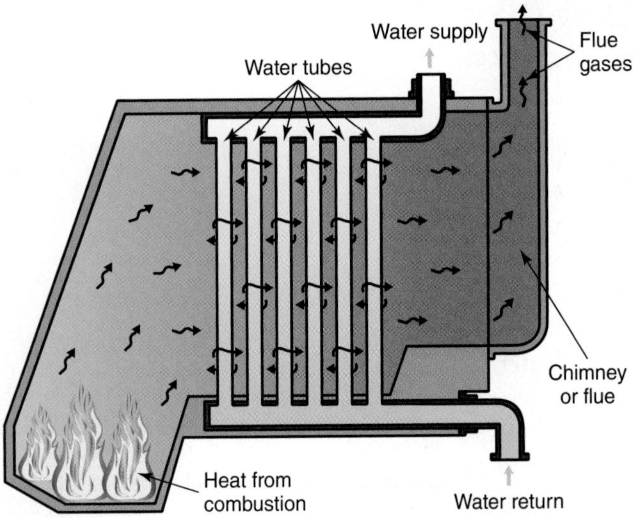

3.3.3 Copper

Copper-tube boilers are considered **low-mass boilers** because of their lightweight design and minimal volume of water they contain, **Figure 3-14**. The heat exchanger on a copper-tube boiler is constructed as a series of fins surrounding the tubing, **Figure 3-15**. These fins are designed to increase the surface area of the heat exchanger, thus

Figure 3-13. This illustration shows the configuration of a water-tube boiler.

increasing heat transfer. Water inside the tubing is heated by combustion gases from the burner located inside the surrounding chamber. Because copper-tube boilers have less surface area than cast-iron boilers, they do not hold their heat as long. But because copper is a greater conductor of heat than cast iron, they are able to heat water more quickly and efficiently. When replacing an old cast-iron boiler in a residence or commercial building, a copper-tube boiler can operate more efficiently, saving the owner hundreds of dollars per year. Some copper-tube boilers are rated at and above 80% AFUE. AFUE stands for *Annual Fuel Utilization Efficiency*, which can result in a significant reduction in heating bills.

Copper-tube boilers can be a good fit for snowmelt applications. They generate a large volume of heat in a small package. Furthermore, copper-tube boilers can run at a reduced rate with a constant load, conducive to the lower water temperature required in snowmelt applications.

When choosing a copper-tube boiler, the water quality feeding the boiler is critical. Hard water tends to create a buildup of scale quickly on the tube-type heat exchanger, diminishing the boiler's performance and reducing its lifespan. Always check with the boiler manufacturer regarding their water treatment specifications. Because copper-tube boilers need adequate water flow, ensure the circulating pump is sized correctly. Maintain the proper minimum water temperature through a copper-tube boiler; it is critical that the return water temperature is above the dew point (130°F) to avoid condensation on the heat exchanger, as discussed in the next section.

Goodheart-Willcox Publisher

Figure 3-14. Copper-tube boilers are considered low mass because of their lightweight design.

Mykhailo Motov/Shutterstock.com

Figure 3-15. An example of a heat exchanger commonly used in a copper-tube boiler.

3.4 Condensing Boilers

A non-condensing, medium-efficiency boiler needs to maintain a return water temperature above its condensing *dew point*. By definition, dew point is the temperature below which water vapor begins to condense. **Figure 3-16** shows the relationship between boiler return water temperature, dew point, and boiler efficiency. Conventional lower-efficiency and medium-efficiency boilers tend to condense the products of combustion when the return water temperature approaches 130°F. When this occurs, it can create major operational, installation, and safety issues. Flue gas condensation is very acidic, with a pH between 4 and 5. This acidic condensate can reduce efficient heat transfer, damage the heat exchanger, and corrode the venting system, causing premature failure. This last condition can create a potentially dangerous spillage of flue gases.

To remediate this situation, the minimum return water temperatures of lower- and medium-efficiency hot water boilers need to be maintained above 130–140°F. This can be accomplished by changing the supply water temperature set point on the boiler's temperature controls. A 20°F temperature difference is typically desired on lower- and medium-efficiency boilers. This would require that the boiler supply

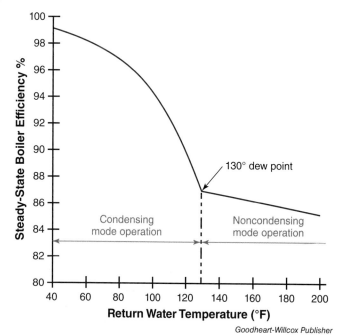

Goodheart-Willcox Publisher

Figure 3-16. This illustration shows the relationship between boiler return water temperature, dew point, and boiler efficiency.

water temperature be set between 160°F and 180°F. This topic is discussed further in Chapter 7, *Hydronic Piping Systems* and Chapter 11, *Terminal Devices*. However, raising the return water temperature tends to reduce the boiler's efficiency because not as much heat will have been extracted from the water. For these reasons, the industry has developed **condensing boilers**, which are considered high-efficiency boilers, **Figure 3-17**. High-efficiency boilers contain a secondary heat exchanger that "squeezes" additional heat from the flue gases before they exit the boiler. To understand how a condensing boiler operates, first review the by-products of perfect combustion: heat, carbon dioxide (CO_2), and water.

The boiler's combustion process can be expressed as follows:

$$[C + H \text{ (fuel)}] + [O_2 \text{ (Air)}] \rightarrow \text{(Combustion Process)} \rightarrow [CO_2 + H_2O + \text{(Heat)}]$$

where

C = Carbon
H = Hydrogen
O = Oxygen

The heat of combustion is either transferred into the boiler water or is passed through the boiler and up the chimney. The goal in increasing boiler efficiency is to reduce the amount of heat that passes through the chimney and increase the amount of heat transferred into the boiler water. However, in order to decrease the amount of heat passing through the chimney, we must "squeeze" more heat out of the fuel, which will create condensation, or water, in the flue gases. As

A

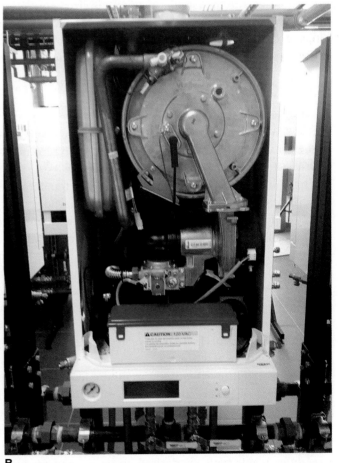
B

Goodheart-Willcox Publisher

Figure 3-17. The outside and inside view of a condensing boiler.

mentioned earlier, this situation is not good for lower- and medium-efficiency boilers using conventional metal flue vents.

Condensing boilers extract additional heat by adding a second heat exchanger to the system, **Figure 3-18**. The flue gases pass through the secondary heat exchanger after they leave the primary heat exchanger, extracting more heat from the burned fuel and increasing the efficiency of the boiler above 90%. When this additional heat is removed by the secondary heat exchanger, part of the combustion gases condenses into liquid, releasing latent heat into the boiler's water. (Recall from Chapter 1, *Human Comfort and Heat Transfer* that latent heat is the amount of heat required to change the state of matter without a change in temperature.) This secondary heat exchanger is made of noncorrosive stainless steel, which resists the effects of flue gas condensate. Condensing boilers do not use metal for their flue vents; rather they use plastic piping such as PVC (polyvinyl chloride) or PP (polypropylene), which resist corrosion caused by acidic condensate from the products of combustion, **Figure 3-19**. Because of this, condensing boilers may be vented through the wall, roof, or existing chimney. Drain connections also attach to their heat exchangers for condensate removal.

CODE NOTE

PP versus PVC Pipe

When installing a condensing boiler, note that more states are requiring the use of polypropylene (PP) instead of polyvinyl chloride (PVC) as vent piping material. PVC will not withstand higher flue gas temperatures compared to polypropylene. PP piping may be more expensive than PVC, but it does have more flexibility when it comes to flue pipe installation, as it does not need to be glued.

3.5 Modulating Boilers

In order to match the heat output of a boiler to the corresponding load on a building, the industry has developed what is known as ***modulating boilers***. This type of boiler is capable of automatically changing its heat output by varying or "modulating" the input of fuel and air used for combustion, **Figure 3-20**. The

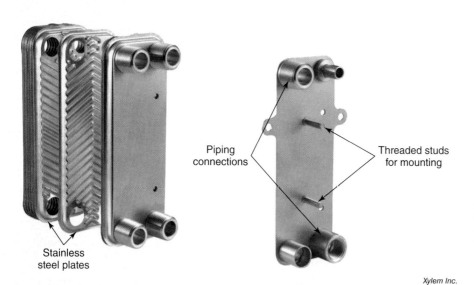

Xylem Inc.

Figure 3-18. A cutaway view of a secondary heat exchanger used in a condensing boiler.

Goodheart-Willcox Publisher

Figure 3-19. Because condensing boilers have acidic flue gases, vents must be polyvinyl chloride (PVC) or polypropylene (PP), which is shown here.

Goodheart-Willcox Publisher

Figure 3-20. A—A floor-mounted boiler. B—A wall-hung modulating boiler.

action of a modulating boiler is similar to that of a gas pedal on an automobile. As the outdoor temperature drops, the heat load on a building increases. The modulating boiler responds to this change by increasing its heat output through increasing the fuel and air input used for combustion. Conversely, as the outdoor air temperature increases, the modulating boiler responds by decreasing its heating output, thus saving money on the building owner's heating bill. Modulating boiler controls are typically only found on condensing boilers, **Figure 3-21**. This is due to the fact that as the boiler's output decreases, the hot water return temperature decreases, therefore increasing the chance of flue gases reaching their dew point. These combination modulating and condensing boilers are known in the industry as ***mod/con boilers***.

The rate at which a modulating boiler can throttle its output is known as the ***turndown ratio***. To understand this concept, think of the minimum and maximum firing rate, or Btu output, of a modulating boiler to correspond to 0–100%. Because today's combustion technology only allows for a minimum firing rate of 20%, the best turndown ratio in the industry is currently 100:20 or 5:1. A mod/con boiler that has the capability of modulating between 100% and 30% would have a turndown ratio of 3:1, and so on.

Some boilers operate using a two-stage temperature control, also known as high and low fire control. This control is a fixed Btu burner output, which usually equates to 100% of the maximum Btu output capacity at high stage and approximately 66% of the maximum Btu output capacity at low stage. Two-stage controls are governed by the boiler's electronic circuit board and are controlled based on the supply water temperature in relation to the actual supply water temperature set point. Although this type of control does not have the full efficiency capability of a fully modulating boiler, it provides adequate water temperature control and is much more efficient than conventional lower-efficiency and medium-efficiency single-stage boilers. Both two-stage and fully modulating controls are governed by the outdoor air temperature, which is also known as *outdoor reset control*. This concept is covered more thoroughly in Chapter 8, *Boiler Control and Safety Devices*.

The application of mod/con boilers as dual-purpose appliances has generated much interest over the past several years. By dual purpose, we mean using the boiler for both comfort heating and heating water used in domestic applications. A fair share of these boilers is of the "tankless" type. A complete explanation and application of dual-purpose boilers will be covered in Chapter 18, *Domestic Hot Water Production*.

3.6 Electric Boilers

In areas where other fuels such as natural gas are not readily available and electric utility prices are competitive, the use of an electric boiler may be attractive. Just as gas and oil are used to generate heat for heating the water, electricity is used to heat an element submersed inside the boiler. Electric boilers are available in many sizes and can be used for residential, commercial, and even industrial applications, **Figure 3-22**. One of the biggest selling points for electric boilers is their efficiency. Because they do not use a conventional heat exchanger or flue venting, there is no heat lost through wasted gases. They are virtually 100% energy efficient. Electric boilers are typically more compact in size than gas- or oil-fired boilers, which results in greater flexibility when locating and positioning the boiler for use with comfort heating applications. Because electric boilers do not require a separate flue for combustion gases, initial installation costs can be cheaper than for a gas- or oil-fired boiler. Furthermore, there are no costs involved in running gas piping throughout the building or any requirements for setting an oil or propane tank on the premises. Because they do not contain as many mechanical parts, electric boilers are durable and require very little maintenance, with only the electric heating elements needing replacement.

Electric boilers come with certain disadvantages, however. There is a limit to the amount of water they can heat at any one time. This means that larger buildings may need multiple boilers in order to meet heating demands. Although

Goodheart-Willcox Publisher

Figure 3-21. An inside view of a modulating boiler.

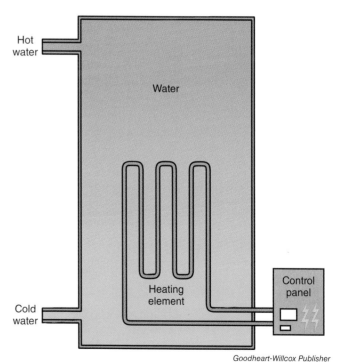

Goodheart-Willcox Publisher

Figure 3-22. This illustration shows an electric boiler used for an industrial application.

electric boilers have close to perfect energy efficiency, electricity costs in most areas outweigh the cost of other fossil fuels such as natural gas, oil, or LP. Remember to compare the costs of various heating fuels in your area before making a purchasing decision.

> **GREEN TIP**
>
> **Are Electric Boilers Truly Environmentally Friendly?**
> While electric boilers are generally considered environmentally friendly since they do not produce combustion gases, there are concerns that the initial process of generating the source electricity at its fossil fuel-generated power plant can create as much pollution as a gas-fired boiler system, therefore negating the environmentally friendly aspects. This does not, of course, take into account that wind, solar, or nuclear power may be used to generate the electricity. However, when contemplating sources of renewable energy for electrical power generation, you should consider that the wind does not always blow, the sun does not always shine, and the last nuclear power plant in the US was built 30 years ago. Possibly the best approach to environmental stewardship relies on individual choices that best fit individual applications.

3.7 Low NO_x Boilers

Nitrogen Oxides (NO_x) belong to a family of poisonous, highly reactive gases. These types of gases form when fuels are burned at high temperatures. NO_x is emitted from various sources, including automobiles, power plants, and industrial boilers. It often appears as a brownish gas, with a strong oxidizing agent that can play a major role in atmospheric reaction with organic compounds, causing smog and acid rain. For this reason, certain parts of the United States and Canada require the specifications of new boilers to include low-NO_x burner emission levels. The key to producing a *low-NO_x boiler* is effective burner design. Low NO_x burners control the mixing of air and fuel at each burner, creating larger flame patterns. This reduces the peak flame temperature, which in turn significantly reduces the level of NO_x produced by each burner, **Figure 3-23**.

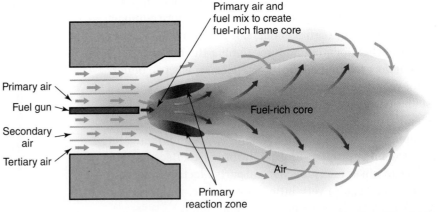

Goodheart-Willcox Publisher

Figure 3-23. This illustration shows how controlling the mixing of air and fuel at each burner results in low NO_x emissions.

CAREER CONNECTIONS

Sales Representative

A boiler sales representative sells boilers and other equipment to businesses, institutions such as schools and churches, and to government facilities on behalf of a manufacturer or wholesaler. The sales representative works directly for the manufacturer or wholesaler, or he or she may work for an independent sales agency whose clients are manufacturers or wholesalers.

The sales representative demonstrates products to customers—showing them the benefits of improving building comfort and performance while reducing energy costs. There are generally two types of sales reps: an inside sales rep who typically works in an office, and an outside sales rep who travels directly to the customer's facility.

Both types of sales representatives work independently or with teams to develop sales programs that exceed revenue goals. Sales reps should have experience developing and closing potential sales leads and a background in the mechanical construction industry—including HVAC, plumbing, and electrical, preferably with residential and commercial boilers. They should possess experience in inventory control and be capable of generating sales proposals based on mechanical and architectural plans and specifications. Effective written and verbal communication skills are a must.

The annual salary for an experienced boiler sales representative can range from $50,000 to over $100,000 per year and include a compensation package with insurance benefits and a 401(k) retirement plan. There may also be opportunities for inside sales training on new products.

Chapter Review

Summary

- There are several different types of boilers classified by pressures, construction materials, and efficiencies.
- Boilers can be classified as low-pressure, medium-pressure, and high-pressure applications.
- Various types of fuels may be used as heating sources for modern-day hot water boilers, including fuel oil, natural gas, propane, electricity, and biomass.
- Boilers are constructed from cast iron, steel and stainless steel, and copper. The types of materials used in boiler construction can influence their efficiency. Cast-iron boilers are very heavy and take longer to heat up, but they tend to hold their heat longer. Steel fire-tube boilers have a simple, compact design, require less stringent water treatments, clean easily, and are well suited for space heating as well as for industrial process applications. However, they require extensive time to raise the water temperature and do not respond quickly to changes in load. Steel water-tube boilers provide a faster response to heating load, are available in a greater range of sizes, and can reach very high temperatures. The disadvantages of water-tube boilers are higher initial capital costs, difficulty of cleaning, and their often-challenging physical size. Copper-tube boilers do not hold their heat as long, but they are able to heat water more quickly and efficiently.
- In a dry-base boiler, the boiler water is contained in a space *above* the combustion chamber. In a wet-base boiler, the water that is heated is located throughout—both above and below—surrounding the combustion chamber. The boiler's condensing dew point can influence the boiler's efficiency.
- Condensing boilers are classified as high-efficiency boilers and extract additional heat by adding a second heat exchanger to the system.
- A condensing boiler uses latent heat as a means to achieve higher efficiencies. When additional heat is removed by the secondary heat exchanger, part of the combustion gases condense into liquid, releasing latent heat into the boiler's water.
- Modulating/condensing (mod/con) boilers can modulate their firing rate to achieve higher efficiencies and comfort.
- Turndown ratio is the rate at which a modulating boiler can throttle its output.
- Electric boilers are used where other potential fossil fuels are scarce. They are virtually 100% energy efficient, durable, and relatively small in size. However, there is a limit to how much water they can heat at one time, and most areas' electricity costs outweigh the cost of other fossil fuels.
- Low NO_x boilers are required in various areas of the United States and Canada in an effort to reduce toxic emissions that contribute to smog and acid rain.

Know and Understand

1. What is the purpose of an anode rod in a steel boiler?
   A. It improves boiler efficiency by adding chemicals to the water.
   B. It prolongs the life of the heat exchanger by consuming oxidation.
   C. It is a safety feature that reduces boiler water pressure.
   D. All of the above.
2. A Scotch Marine boiler is a _____ type boiler.
   A. water-tube
   B. fire-tube
   C. tube-in-tube
   D. None of the above.
3. At what approximate temperature does a non-condensing boiler reach its dew point?
   A. 110°F
   B. 120°F
   C. 130°F
   D. 140°F
4. What is the approximate pH of flue gas condensate?
   A. Between 4 and 5
   B. Between 5 and 6
   C. Between 8 and 10
   D. Approximately 12
5. What are the three by-products of perfect combustion?
   A. Heat, water, and carbon monoxide
   B. Heat, carbon monoxide, and carbon dioxide
   C. Nitrogen, hydrogen, and carbon dioxide
   D. Heat, water, and carbon dioxide
6. What type of vent piping is normally used with a condensing boiler?
   A. Polypropylene
   B. Polyvinyl chloride
   C. PEX
   D. Both A and B.
7. *True or False?* Condensing boilers must be equipped with a condensate drain.
8. What are the typical low- and high-fire Btu output rates on a two-stage boiler?
   A. 100% and approximately 66%
   B. 100% and approximately 50%
   C. 75% and approximately 25%
   D. 100% and approximately 25%
9. What is the approximate efficiency of an electric boiler?
   A. 80%
   B. 90%
   C. 100%
   D. 110%
10. What effect does NO_x have on the environment?
    A. It can react with nitrogen in the air, causing increased CO_2 levels.
    B. It can react with organic compounds in the air, causing smog and acid rain.
    C. It can react with humidity in the air, causing elevated levels of greenhouse gases.
    D. All of the above.

Apply and Analyze

1. Describe the different types of boiler classifications.
2. Explain where low-, medium-, and high-pressure boilers are used.
3. List the types of fuels and their heating content used for heating a boiler.
4. Explain why a cast-iron boiler is considered a high-mass boiler.
5. Describe the difference between a wet-base and dry-base boiler design.
6. Describe the difference between a fire-tube and water-tube type boiler.
7. Explain why a copper-tube boiler is considered a low-mass boiler.
8. What determines the AFUE of a boiler?
9. What is meant by turndown ratio?
10. Explain the operation of a mod/con-type boiler.

Critical Thinking

1. Describe how a modulating boiler is similar to the gas pedal on an automobile.
2. Explain the role of a secondary heat exchanger in a condensing boiler.

4 Gas Burners and Ignition Systems

Chapter Outline
4.1 Principles of Combustion
4.2 The Gas Train
4.3 Gas Control Valves
4.4 Types of Gas-Fired Burners
4.5 Ignition Systems
4.6 Safety Devices for Burner Ignition Controls

yevgeniy11/Shutterstock.com

Learning Objectives

After completing this chapter, you will be able to:
- List the elements needed for combustion and the products of stoichiometric combustion.
- List the components of a boiler gas train and identify the purpose of each.
- Identify the purpose of a drip leg.
- Explain the difference between atmospheric burners and power burners.
- Explain the difference between FM and IRI gas manifolds.
- Describe the different types of gas valves used on a boiler.
- Describe the different types of ignition systems.
- Describe the purpose of a flame rod.
- Compare the different types of ignition control modules and their various levels of lockout.
- Describe the sequence of operation of an ignition control module.
- Identify ignition safety components.
- Explain how flame rectification works.

Technical Terms

air pressure switch
atmospheric burner
automatic gas shutoff valve
combination valve
combustion
combustion air
diaphragm valve
direct spark ignition system
drip leg
flame-proving device
flame rectification system
flame rod
flame rollout switch
FM gas manifold
gas ignition module
gas manifold
gas train
glow coil
high temperature limit
hot surface ignition system
integrated ignition control module
intermittent pilot ignition system
IRI gas manifold
modulating gas valve
nonintegrated ignition module
normally closed valve
orifice
pilot gas line
pilot generator
pilot valve
power burner
pressure regulator
primary air
redundant gas valve
secondary air
solenoid valve
spark igniter
spud
standing pilot
stoichiometric combustion
thermocouple
thermopile
turndown ratio
two-stage gas valve
venturi

At the heart of any gas-fired boiler is the burner and ignition system. Together, the burner and ignition controls make up the combustion system that heats the boiler water. The gas-fired burners and ignition systems found on modern hot-water boilers look quite different than those used many years ago. Today's hot-water boiler combustion systems offer a much cleaner, efficient arrangement, resulting in very economical heating. In order for you to effectively troubleshoot and repair today's boiler combustion systems, a thorough understanding of how gas-fired combustion works is a must. This chapter covers the different types of gas-fired burners and ignition systems and their applications in residential and commercial hot-water boiler systems.

4.1 Principles of Combustion

In order to thoroughly understand the workings of a gas-fired burner and ignition system used on a hot-water boiler, we must first discuss the principles of combustion. By definition, *combustion* is the rapid expansion of gases resulting in oxidation of fuel and oxygen, producing heat and light. Combustion requires three components:
- Fuel source
- Oxygen
- Heat source

Notice the third component is a heat source, not specifically a flame, **Figure 4-1**. A spark or hot surface igniter at the correct temperature can ignite the mixture of fuel and oxygen. Combustion also requires a proper mixture of fuel and oxygen. If this mixture is out of balance, combustion cannot occur. Too much fuel and not enough oxygen results in a mixture that is considered "too rich." Too much oxygen and not enough fuel produces a mixture classified as "too lean." For example, a perfect balance of oxygen to natural gas would be a ratio of 2:1. This condition would create perfect combustion, or ***stoichiometric combustion***. By definition, stoichiometric combustion, or theoretical combustion, is the ideal combustion process in which fuel is burned completely. However, this 2:1 ratio

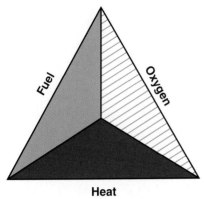

Figure 4-1. The three components necessary for combustion are fuel, oxygen, and heat.

will not work using the air that we breathe, which contains only 21% oxygen and 78% nitrogen. Nitrogen is nonflammable, so it actually prohibits combustion. Because of this, there needs to be an excess amount of air for proper combustion to take place. For combustion, a more realistic ratio of air to natural gas would be closer to 10:1.

Perfect, or stoichiometric, combustion creates only three by-products:
- Heat
- Carbon dioxide
- Water

This process can be defined in the following formula:

$$CH_4 \text{ (Natural Gas)} + O_2 \text{ (Oxygen)} + \text{Heat} = CO_2 \text{ (Carbon Dioxide)} + H_2O \text{ (Water)} + \text{Energy}$$

In reality, today's hot-water boilers cannot achieve perfect, or stoichiometric, combustion, and the real products of combustion also include carbon monoxide, nitrogen dioxide, and soot. However, it is important for you to understand these basic combustion principles in order to properly maintain the boiler's ignition and burner system as well as effectively perform a combustion analysis so that the boiler will work at peak efficiency.

4.2 The Gas Train

To understand the combustion process in a boiler, first it is important to be familiar with the components of a *gas train*. The gas train consists of a series of components that safely feeds fuel into the burner. This section will cover these components and how they have been modernized through the years, from early burner gas trains to today's modern commercial and industrial boiler fuel gas trains.

4.2.1 Residential and Light Commercial Gas Train Components

Early-model gas trains consisted of simple components that delivered fuel from its source to the burners, as shown in **Figure 4-2**. These components include the following:
- Drip leg
- Manual shutoff valve
- Gas pressure regulator
- Pilot line with manual shutoff valve and solenoid gas valve
- Automatic gas shutoff valve

Modern gas trains have similar components but are streamlined to include the following:
- Drip leg
- Manual shutoff valve
- Combination gas valve that includes a built-in solenoid valve, pressure regulator, and, in some cases, a pilot valve

Following is a description that compares the components of early gas trains with those used today.

The first component in early-model gas trains we will discuss is a *drip leg*, a short gas pipe extending from a tee at the beginning of the gas train, **Figure 4-3**. A drip leg is still required today by code on all gas-fired appliances. It serves as a trap to prevent moisture and debris from entering the gas valve. The second component is the manual shutoff valve. Early gas shutoff valves would sometimes

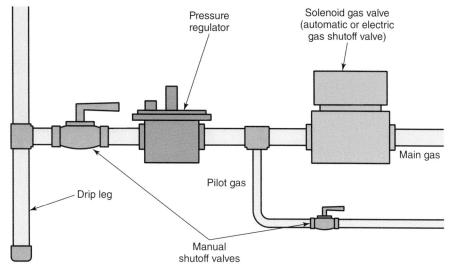

Figure 4-2. An older-style gas train in which the gas passes through a separate pressure regulator to the gas valve.

Figure 4-3. A drip leg installed in the gas supply piping is used to trap moisture and debris.

include a 1/4″ threaded plug, allowing for testing of the incoming gas pressure to the gas train. Whether used on earlier gas trains or on modern types, the handle of this shutoff valve should be oriented so that it is parallel to the gas flow when open and perpendicular to the gas flow when closed, allowing the user to quickly visualize whether the valve is opened or closed.

The next component in the early-model gas train is a ***pressure regulator***, **Figure 4-4**. This device is used to reduce the inlet gas pressure to a level specified by the boiler manufacturer. The gas pressure at the inlet of the gas valve—whether it be natural gas or LP gas—is typically higher than what is required at the burner's combustion zone. The purpose of the gas pressure regulator is to reduce the gas pressure delivered to the burner and maintain this pressure automatically. This gas pressure regulator consists of a valve, diaphragm, pressure spring, and regulator adjusting screw. There are two pressures at work within the device: the inlet gas pressure, which presses against the diaphragm and tries to close the valve, and the spring pressure, which presses down on the diaphragm opening the valve. The proper gas pressure is delivered to the burner and automatically maintains this pressure by opening the valve against the spring. If the outlet or downstream pressure is reduced, the pressure that is exerted against the diaphragm reduces, causing the valve to open. This balance of pressures between inlet and outlet causes the regulator to modulate and maintain the proper outlet pressure at a constant rate at all times. The proper gas pressure is either set by turning the adjusting screw clockwise, which increases the outlet pressure to the desired rate, or counterclockwise, which decreases the outlet pressure, **Figure 4-5**. Pressure regulators also have a small orifice known as a vent located on the top of the regulator body. This vent should be piped either to the outside of the building or to the pilot light. In the event that the diaphragm is damaged, the vent will prevent unburned gas from seeping into the boiler room, averting a hazardous and explosive situation.

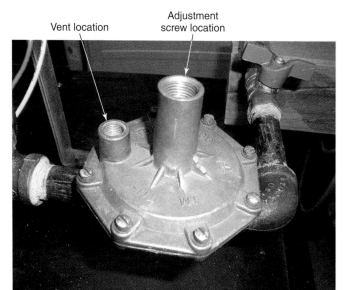

Figure 4-4. The gas pressure regulator is used to reduce gas pressure to the proper working level.

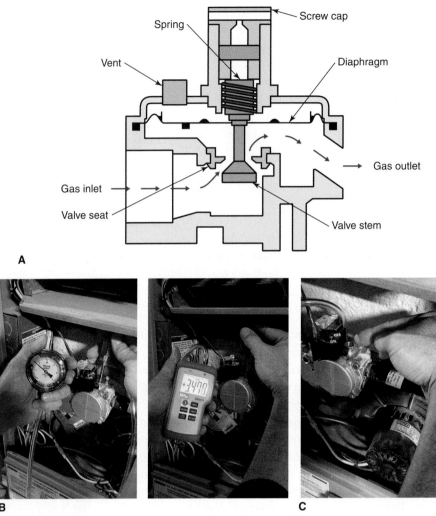

Figure 4-5. A—This diagram shows how a standard gas pressure regulator functions. B—The gas pressure on a regulator can be adjusted using a standard pressure gauge and an electronic pressure gauge. C—A close-up view of the adjustment screw. *Note: Although this adjustment is being performed on a combination gas valve, the procedure is the same using a standard pressure regulator.*

TECH TIP

Gas Pressure

In most residential and light commercial boilers, the gas pressure exiting the automatic gas valve should be:

- 3.5 in. WC for natural gas
- 11 in. WC for propane or LP

The term *in. WC* is an abbreviation for "inches water column." This is a unit of measurement used for small increments of pressure. A gas pressure reading of 3.5 in. WC is the amount of pressure exerted on a column of liquid that would raise the level of that liquid 3.5 inches.

This pressure is easily checked using a U-tube manometer or digital pressure gauge. Adjustments can be made either at the pilot pressure regulator or at the adjustment screw found on a combination gas valve. Be sure to follow the manufacturer's installation instructions for setting the gas pressure. Also check the inlet gas pressure to the automatic valve to ensure it is higher than these settings. If not, check with the gas supplier to see if there is a problem at the outdoor gas pressure regulator.

The next component found on earlier gas trains is a *pilot gas line*, which tees off the main supply line. The pilot line has its own manual shutoff valve and sometimes includes a separate pilot pressure regulator. In earlier boilers, the gas pressure for the pilot flame was adjusted according to the desired flame's size and shape, **Figure 4-6**. A pilot gas line may be found on some modern boilers. However, this line will run from the boiler's combination gas valve directly to the pilot light assembly. Combination gas valves are discussed later.

Next in the gas train is the *automatic gas shutoff valve*. This is typically a 24-volt powered solenoid-type gas valve that opens when there is a call for heat from the space thermostat or the boiler's aquastat, **Figure 4-7**. The automatic valve is the last component of the gas train before the gas enters the manifold, which is connected to the burners.

4.2.2 Commercial and Industrial Gas Train Components

Depending on the size and Btu output of the boiler, some larger commercial and industrial boilers may require additional components on the gas train to ensure they comply with local, state, and national code requirements as well as the owner's insurance requirements. Insurance companies that cover losses resulting from boiler fires and explosions have formed organizations called underwriters. These underwriters assess the risk of coverage offered by insurance policies and develop safety standards and codes for boilers rated for 400,000 Btu/hr and greater. The two types of gas trains

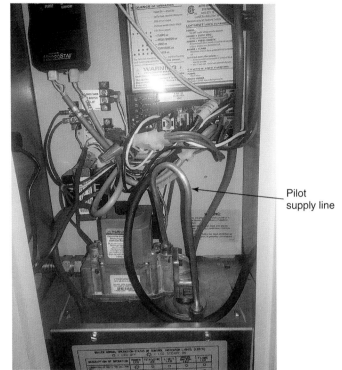

Goodheart-Willcox Publisher

Figure 4-6. The pilot supply line may be run separately on both older-model gas trains and larger boilers.

SAFETY FIRST

Gas Pressure Regulator Positioning

Be sure that the inlet gas pressure regulator is installed upstream of any automatic shutoff valves. Positioning the regulator downstream of the electric shutoff valve could cause the flame to rollout when the burner fires or create delayed ignition, both hazardous situations. In this scenario, when the automatic shutoff valve is powered closed, the gas pressure regulator will open wide due to the differential pressure between inlet and outlet. After the gas valve is energized open, the pressure regulator will be wide open, causing an overfeed of gas to the burner. The pressure will regulate itself to its proper setting, but not before the possibility of flames shooting out of the boiler's combustion zone.

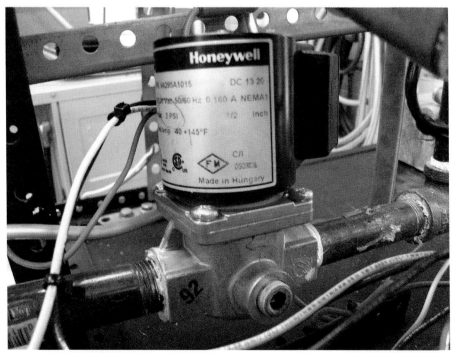

Goodheart-Willcox Publisher

Figure 4-7. An automatic gas shutoff valve opens when there is a call for heat from the space thermostat or from the boiler's aquastat.

that are typically found on these larger commercial and industrial boilers are compliant with FM Approvals or Industrial Risk Insurers. FM Approvals, known historically as Factory Mutual, is the independent testing arm of the international insurance carrier FM Global. Industrial Risk Insurers is a company whose business includes the underwriting of fire, marine, and life insurance.

Factory Mutual or **FM gas manifolds** require additional components as opposed to those found on standard boilers and those rated below 400,000 Btu/hr. Depending on the boiler's size, these components will include:
- Additional manual shutoff valves for leak testing purposes
- Low gas pressure switches
- Two electric safety shutoff valves

Industrial Risk Insurers or **IRI gas manifolds** are the strictest standards for boiler gas trains. If a gas train requires IRI specifications, it will easily meet FM requirements, **Figure 4-8**. Required components found on IRI gas manifolds include:
- Additional manual shutoff valves for leak testing purposes
- High and low gas pressure switches
- Two electric safety shutoff valves
- A normally open bleed valve between the safety shutoff valves, which purges any excess gas to outside the building when the safety shutoff valves are closed

Both FM and IRI gas trains will also include a standard drip leg, manual shutoff valve, and an automatic control valve—typically called a butterfly valve. This automatic control valve is a modulating gas valve that regulates the flow of gas into the burner combustion zone to adjust the Btu output of the boiler, based on the building's heating demand.

4.3 Gas Control Valves

The boiler's main gas control valve, found on the gas train, is used to regulate the flow of fuel into the burner. Several different types of gas valves can be found on

Goodheart-Willcox Publisher

Figure 4-8. IRI gas manifolds are the strictest standards for boiler gas trains.

residential and commercial boilers. All are used to stop or start the flow of gas when there is a call for heating by the boiler controls. Among these types of gas valves are:

- Solenoid valves
- Diaphragm valves
- Combination valves
- Two-stage valves
- Modulating gas valves

4.3.1 Solenoid Gas Valves

A *solenoid valve* can be found on earlier gas trains, both for the pilot line and for the main burner. The solenoid valve consists of a valve stem or plunger, valve disc or seat, spring, and solenoid coil. This type of valve is considered a *normally closed valve*, which means it is always closed unless power is applied to the solenoid coil. A solenoid coil is a spooled coil of copper wire that acts like an electromagnet when it is powered up. This magnetic force lifts the valve off its seat, allowing gas to pass through. A spring is used to automatically close the valve whenever it is deenergized, **Figure 4-9**. A solenoid valve is a fast-acting valve, in that it opens quickly whenever power is applied to the coil.

4.3.2 Diaphragm Gas Valves

Diaphragm gas valves are found on earlier gas trains and may be on some later-model gas trains. A *diaphragm valve* uses differential pressure to hold shut the gas valve when it is deenergized and to push open the valve when there is a call for heat. This is accomplished through the use of a *pilot valve*. When there is no call for heat, the pilot valve opens a port to the upper chamber of the valve. The inlet gas pressure flows through this port and floods the upper chamber of the valve, pushing down on the diaphragm and keeping the main valve closed. This occurs because the pressure from the inlet gas supply is greater than the atmospheric condition downstream of the main valve, **Figure 4-10A**. Upon a call for heat, the pilot valve closes the port to the upper chamber and the inlet gas pressure is allowed to push against the diaphragm, opening it from the lower chamber of the valve and allowing it to flow through the main valve into the burner. When the valve is opened during a call for heat, a bleeder port allows the residual gas from the upper chamber to be vented or bleed off into the pilot. This creates a reduced pressure on the upper side of the diaphragm, **Figure 4-10B**. The pilot operator may be a small solenoid, bimetal, or heat motor-type valve.

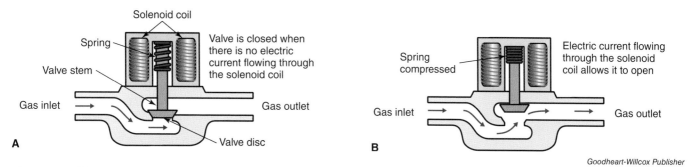

Goodheart-Willcox Publisher

Figure 4-9. This illustration shows how a solenoid gas valve operates. A—The valve closes when the solenoid valve is deenergized, with the spring holding the valve closed. B—A magnetic force is created when the solenoid valve is energized, causing the valve stem to be drawn upward, opening the valve and allowing the gas to flow.

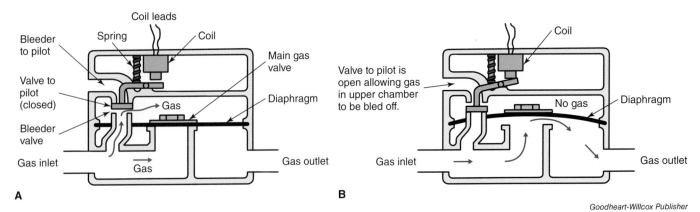

Figure 4-10. A—An illustration of a diaphragm gas valve without power applied. B—When power is applied to the diaphragm valve, the gas in the upper chamber is bled off, which reduces its pressure, allowing the valve to open.

4.3.3 Combination Gas Valves

Today's residential and light commercial boilers utilize a *combination valve*, which incorporates all the features usually found separately on older-model gas trains. These components include:

- A solenoid valve
- A built-in regulator for adjusting the gas pressure to the burner
- A pilot supply line and controls to adjust the pilot flame
- Ignition controls
- A safety shutoff feature
- A manual on-off control

Because these combination gas valves typically have dual shutoff seats for extra safety protection, they are also referred to as *redundant gas valves*, **Figure 4-11**.

A variety of gas ignition systems use combination gas valves, including:

- Standing pilot ignition systems
- Intermittent pilot ignition systems
- Direct burner ignition systems

These ignition systems are discussed in detail later in the chapter.

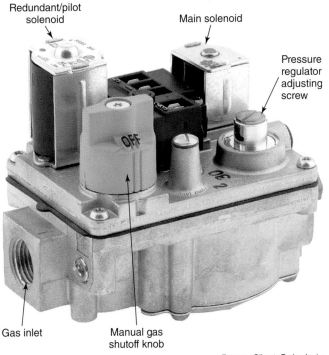

Figure 4-11. A combination gas valve combines the function of a solenoid valve with a gas pressure regulator.

SAFETY FIRST

Don't Repair—Replace

Do not attempt to repair a defective gas valve or pressure regulator. Unless given specific authorization from the valve manufacturer to perform repairs, it is safer to simply replace a defective valve. Liability is too great for a technician to disassemble and reassemble a gas valve or pressure regulator and expect it to work properly.

4.3.4 Two-Stage Gas Valves

As a means of increasing fuel efficiency, today's boiler manufacturers sometimes include a *two-stage gas valve* in the burner ignition system. These valves provide a dual firing rate: a first stage, or low-fire stage, which maintains a lower gas manifold pressure, and a second stage, or high-fire stage, that provides a higher gas pressure to the burner. During milder weather when there is less demand for heating, these automatic two-stage gas valves operate on their low-fire stage for greater fuel economy. They provide maximum output capacity when a greater heating demand is required. In addition, a two-stage gas valve initially fires on its first stage before it ramps up to its high-fire or second stage. This allows for longer boiler run times, which reduce startup and shutdown heat losses, **Figure 4-12**.

Two-stage gas valves include two pressure regulators with separate adjustment screws, two independently powered solenoid valves, and typically three wires, in contrast to single-stage valves, which usually require two wires for operation. Two-stage valves do not have a 100/50% proportionally rated output. Instead, the first stage, or low-fire stage, provides approximately two-thirds of the rated high-fire output. For instance, a boiler with a rating of 60,000 Btu/hr will have a first stage rating of approximately 39,600 Btu/hr. Two-stage valves are designed to work with various gas burner ignition systems, including standing pilot, intermittent pilot, direct ignition, and hot-surface ignition systems. The space thermostat, the boiler's aquastat, or an integrated combustion control board can control high- and low-fire sequencing.

Figure 4-12. A two-stage gas valve allows for a dual firing adjustment for better fuel efficiency.

TECH TIP

Setting the Inlet Gas Pressure

Setting the inlet gas pressure on a two-stage gas valve is a two-step process. First, set the second stage gas pressure by forcing the boiler to high fire. This may require a higher than normal thermostat setting. Next, set the first stage gas pressure by allowing the boiler to trim back to a lower heating demand.

With natural gas, the high-fire setting is normally 3.5 in. WC pressure and 1.75 in. WC for low fire. Propane or LP gas will typically have a high-fire gas pressure setting of 11 in. WC and for low fire a setting of 5.5 in. WC.

These are typical industry settings. As always, consult the manufacturer's installation and startup guide for any deviations.

4.3.5 Modulating Gas Valves

Today's higher efficiency, modulating boilers typically incorporate a *modulating gas valve* as a means of improving fuel efficiency and providing advanced comfort from the overall heating system, **Figure 4-13**. On residential and light commercial boilers, the modulating gas valve receives a signal from the solid-state control module, which monitors system temperatures from various sources such as room temperature, boiler hot water temperature, and even a domestic hot water tank. This signal causes the valve to modulate from fully open down to 20% of its maximum operating capacity. This equates to a 5:1 *turndown ratio*. For instance, if the maximum output capacity of the boiler is 80,000 Btu/hr, the

Figure 4-13. A modulating gas valve is used on high-efficiency modulating boilers to increase fuel efficiency.

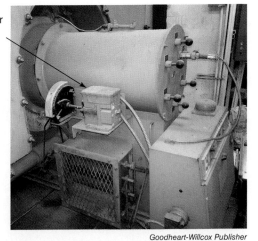

Figure 4-14. This image shows a modulating motor linking the gas valve with the combustion blower on a large commercial boiler.

modulating valve will have the capability of reducing the air and fuel mixture downward to generate an output as low as 16,000 Btu/hr. This reduction in capacity in response to the changing heating loads can result in a boiler fuel efficiency rating of up to 95%.

On larger commercial and industrial boilers, the modulating valve is typically mechanically linked in parallel with the input combustion blower to modulate both the input gas supply and the air needed for proper combustion. As the boiler heating demand reduces, a modulating motor drives the gas valve closed, along with the dampers or vanes, on the combustion blower outlet. This results in a lower fuel-air mixture going into the burner compartment, which reduces the heating output. The modulating motor receives its signal to proportionally open and close from the boiler's control module, which is monitoring space temperature, outdoor air temperature, and the boiler's supply hot water temperature, **Figure 4-14**.

4.4 Types of Gas-Fired Burners

Gas burners are devices that mix *combustion air* with fuel to create heat in the boiler's heat exchanger. The type of fuel gas used for combustion flows from its source through the *gas manifold*, a pipe that distributes the fuel into multiple burners through small sockets called *spuds*. These threaded spuds contain pre-drilled holes called *orifices*, which meter the flow of gas at a predetermined rate into the burner. Here, the primary combustion air mixes with the gas to create combustion.

There are two types of gas burners:
- Atmospheric burners
- Power burners

Atmospheric burners fall into three categories: ribbon burners, slotted or stamped burners, and single port burners, **Figure 4-15**. Ribbon burners have slots that run laterally down the length of the burner, creating a solid flame along its top. Slotted or stamped burners consist of a series of small, narrow slots running horizontally along the burner, which vary in size according to the burner capacity. Single port burners can be classified *upshot* or *inshot*, depending on their orientation. Both ribbon and slotted burners are found on boilers with no combustion blower. Sometimes referred to as "natural draft" combustion

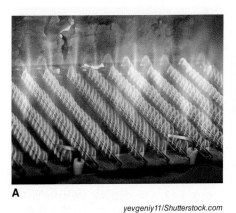

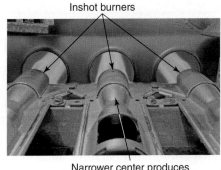

Figure 4-15. Atmospheric burners can be ribbon burners (A), slotted burners (B), or single-port configurations (C).

systems, atmospheric burners rely on gravity or the natural buoyancy of heated air to vent the products of combustion through the flue or chimney.

Single port burners, on the other hand, are the most common type of burners found on today's modern boilers, typically with an inshot burner configuration. These burners receive their combustion air from a combustion blower, which is also used to vent the products of combustion out of the boiler mechanically rather than naturally. Combustion blowers are classified as either *forced draft*—in which they are located on the inlet side of the heat exchanger, or *induced draft*—in which they are found on the discharge, or downstream, side of the heat exchanger, **Figure 4-16**. Regardless of the type used, combustion blowers are an integral part of improving the boiler's efficiency rating.

Atmospheric burners introduce ***primary air*** at their inlet, which is mixed with the gas that passes through the manifold spud. Regardless of their unique burner designs, all atmospheric burners incorporate an hourglass shape near their opening known as a ***venturi***. The venturi creates a narrow constriction, causing the air and gas to accelerate as they pass along through the burner. The venturi also mixes or swirls the air and gas mixture, which promotes good combustion. Once the fuel is ignited when it leaves the burner, it is mixed with ***secondary air***, which improves combustion efficiency by burning the fuel more completely and entraining the combustion gases through the boiler's vent or flue, **Figure 4-17**.

Power burners found on commercial boilers incorporate a blower to force both the primary and secondary air and the fuel into the burner tube. Inside this burner tube, angular deflector plates mix the air and gas by swirling or spinning them, creating a more efficient combustion process. Most power burners use a damper to control the amount of combustion air as it enters the inlet side of the blower assembly. Boilers utilizing a power burner typically also incorporate a modulating gas or butterfly valve to control the flow of gas into the combustion chamber, **Figure 4-18**.

A

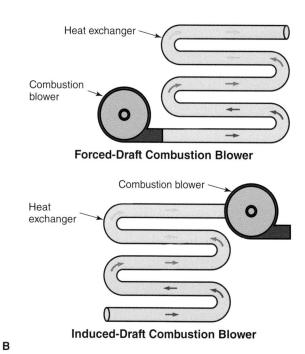
B

Goodheart-Willcox Publisher

Figure 4-16. A—An example of a combustion blower. B—A combustion blower can be either induced draft or forced draft, depending on its position relative to the heat exchanger.

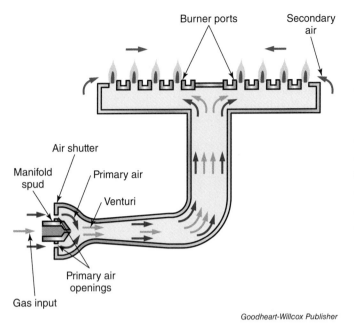

Figure 4-17. An illustration of an atmospheric burner showing primary air, secondary air, and venturi locations.

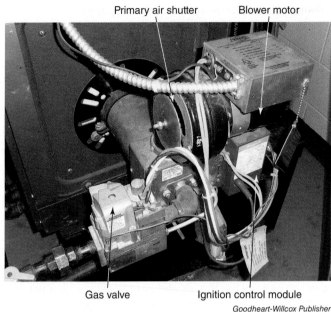

Figure 4-18. An example of a power burner that mixes primary and secondary air with fuel gas.

4.5 Ignition Systems

The combustion of fuel within the boiler would not be possible without an ignition system, which safely lights the fuel and monitors the flame for safe, continuous operation. These systems are sometimes referred to as *flame-proving devices*. The ignition system is controlled and monitored by the boiler's control system or module. This section discusses four different types of flame-proving devices for boiler ignition systems:

- Standing pilot ignition systems
- Intermittent pilot ignition systems
- Direct spark ignition systems
- Hot surface ignition systems

4.5.1 Standing Pilot Ignition Systems

The *standing pilot* is one of the earliest types of ignition systems used on boilers that utilize natural gas or LP as a fuel source. In these systems, the pilot flame is used as a heat source to ignite the burner whenever there is a call for heat. It burns continuously whether the burner is firing or the boiler is on standby. The fuel source for the pilot flame travels from the main gas line through a dedicated pilot supply line or from the combination gas valve. The gas then passes through a small orifice located inside the pilot hood, which reduces the gas pressure before igniting. On combination gas valves that utilize a standing pilot, a small adjustment screw is provided on the valve to adjust the size of the pilot flame. This is typically a manual/visual adjustment. The operator sets the size of the pilot flame so that it is large enough to sustain the ignition of the main burner but not so large that it wastes fuel, **Figure 4-19A** and **Figure 4-19B**.

The standing pilot flame typically is supervised by either a *thermocouple* or a *thermopile*. A thermocouple is made up of two dissimilar metals that "bend" when heated to make contact. One end of the thermocouple is immersed in the

---

TECH TIP

Clear the Cobwebs

Both pilot lines and main burner orifices can become plugged by spider webs. These little critters like to crawl through boiler vent pipes and look for spots near a gas odor. This usually happens over the summer when a boiler is shut down for the season. Telltale signs include a pilot that will not light, an unusual pilot flame shape, or individual burners that will not fire. Fortunately, this situation is easily remedied by disconnecting the gas lines and blowing them out with compressed air or nitrogen. Afterward, check to see that the pilot flame shape looks correct and that the gas pressure is at its proper setting.

Figure 4-19. A—An example of a standing pilot assembly showing pilot hood and thermocouple. B—An image of the pilot flame when lit.

pilot flame. The other end is connected to the body of the gas valve. When the thermocouple is heated by the pilot flame, the heated end bends to make contact (or complete the circuit) and generate a small voltage—15 to 30 millivolts direct current (dc mV)—which sends a signal back to the gas valve to keep the pilot valve held open, **Figure 4-20**. As long as the pilot flame remains lit, the main gas valve is allowed to open whenever there is a call for heat. This is usually accomplished through a 24-volt circuit, which is wired in series with the space thermostat or the boiler's aquastat. Upon a call for heat, the thermostat contacts close and 24 volts are delivered to the gas valve—opening the main valve and igniting the burners. Inside the combination gas valve is a low-resistance, spring-loaded solenoid coil, which acts as the pilot valve. When the pilot is lit, the thermocouple generates enough dc voltage to hold the solenoid coil open, allowing the valve to continuously deliver gas to the pilot flame. However, if the pilot flame is extinguished for any reason, the thermocouple contacts open, interrupting the voltage to the pilot valve and causing the pilot valve inside the gas valve to close. This system provides flame supervision safety for the burner assembly.

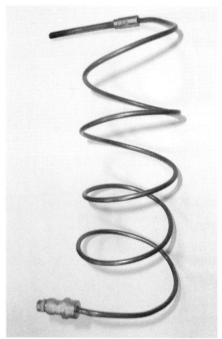

Figure 4-20. A thermocouple generates an electrical signal when heated by the pilot flame.

GREEN TIP

Standing Pilot versus On-Demand Ignition
Today, standing pilots are found only on older boilers. Because the pilot flame burns continuously—even when there is no call for heat—standing pilots have become obsolete, as they waste energy. Today's modern boilers incorporate more efficient ignition systems that light only on a call for heat, which saves fuel—thus saving money for the customer and reducing their impact on the environment.

A thermopile is simply a series of thermocouples joined together. Sometimes known as a ***pilot generator***, the thermopile is capable of generating a higher dc

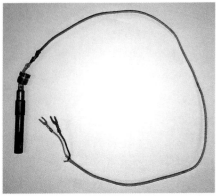

Goodheart-Willcox Publisher

Figure 4-21. The thermopile or pilot generator is a series of thermocouples joined together.

voltage output—usually 500–750 dc mV, **Figure 4-21**. Thermopiles were originally designed to control gas appliances—such as fireplaces—without the need for an external electrical circuit. When the thermopile has generated enough millivolts to hold open the pilot valve, the main valve can be opened by closing a set of contacts on the gas valve—thus eliminating the need for an additional electrical circuit such as those found on gas-fired boilers. Typically, the contacts that open the main gas valve are connected to the space thermostat. Whenever the pilot flame is lit and there is a call for heat, the thermostat contacts close, which in turn closes the main valve contacts and allows the main valve to open.

TROUBLESHOOTING

Checking the Thermopile

When checking a thermopile, there may be some instances where the millivolt reading is high enough to keep the pilot flame lit, but not high enough to open the main gas valve.

Before condemning the valve as defective, always take a millivolt reading across the thermopile terminals on the gas valve. A reading of 450–500 millivolts may be enough to keep the pilot lit, but that is all. When this happens, try relocating the thermopile so it is better immersed in the pilot flame. This may mean situating it above the flame, with the flame making better contact with the end of the thermopile. Doing so will allow the thermopile to generate a higher voltage signal—high enough to open the main valve.

There is a distinct difference between gas valves that are used with thermocouple applications and those used with thermopiles. Gas valves utilizing a thermocouple have a single port into which the end of the thermocouple is attached. Thermopiles have a two-wire end, which attaches to two screw-down terminals on their respective gas valves. Both types of valves employ the same method for lighting the standing pilot.

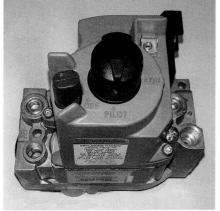

Goodheart-Willcox Publisher

Figure 4-22. This gas valve uses a thermocouple and requires that the red button be depressed to light the pilot.

PROCEDURE

Lighting the Pilot

1. On most standing pilot gas valves, turn the on-off knob to the "pilot" position and depress it.
2. On other models, depress and hold the red pilot light button when lighting the pilot, **Figure 4-22**.
3. By pushing down and holding this button, the pilot valve inside the combination gas valve opens and the pilot can be lit manually.
4. Once lit, continue to hold down this button for approximately 30 seconds, allowing the thermocouple or thermopile enough time to generate the proper amount of voltage to hold open the internal pilot valve solenoid.
5. When the pilot flame remains lit after letting up on the pilot light button, the main valve is now operable.

TROUBLESHOOTING

Diagnosing a Defective Thermocouple

There are two methods to diagnose a defective thermocouple. The first method involves the following steps:

1. Remove the thermocouple threaded end from the gas valve.
2. Using a voltmeter set on direct current voltage (DCV), connect one lead to the end of the thermocouple where it screwed into the gas valve and one lead anywhere on the side of the thermocouple (this will read a completed circuit).
3. Expose the sensing end of the thermocouple to an open flame and allow it to warm up.
4. The output reading should reach 30 dc mV. If it does not, the thermocouple is probably defective.

The second diagnostic method is to install a thermocouple adapter into the gas valve. To do this, remove the thermocouple from the valve and install the adapter. Once it is in place, screw the thermocouple threaded end into the adapter, **Figure 4-23**. When the pilot is then lit, simply place the leads of the voltmeter across each side of the adapter using the same procedure as above. This will save having to disconnect the thermocouple from the gas valve anytime it needs to be tested.

Goodheart-Willcox Publisher

Figure 4-23. A thermocouple adapter can be added to a gas valve for checking the dc voltage.

4.5.2 Intermittent Pilot Ignition Systems

Advancements in gas-fired ignition system technology led to the development of the ***intermittent pilot ignition system***. This ignition system offers two distinct advantages over the original standing pilot system:

1. It is more economical because the pilot is only lit when there is a call for heat.
2. It is safer than a standing pilot because the response time to close the main gas valve after there is a pilot failure is much faster.

Another distinct feature of the intermittent pilot system is an additional component called the ignition control module. This module performs several different tasks and is covered in more detail later in the chapter. In order for the boiler's ignition system to operate, there must be a separate source of ignition for the pilot to light. On most boilers with intermittent pilot ignition systems, this source is a ***spark igniter***. In addition to lighting the pilot, there also must be a method for flame recognition or "proving" in order for the main burner to be energized, **Figure 4-24**. Both of these features are found in the ignition control module. This ignition module contains a step-up transformer that generates approximately 10,000 volts at very low amperage. This voltage is used to produce the spark as part of the pilot ignition process. The ignition module also includes a ***flame rod***, which is immersed in the pilot flame when ignited and sends a signal back to the ignition module, proving the flame is lit and allowing the main gas valve to open, **Figure 4-25**.

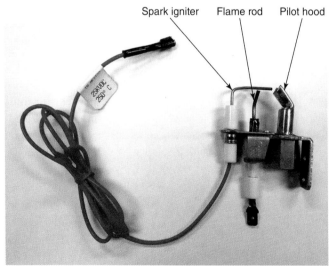

Goodheart-Willcox Publisher

Figure 4-24. This image shows a pilot assembly showing the pilot hood, igniter, and flame sensing rod.

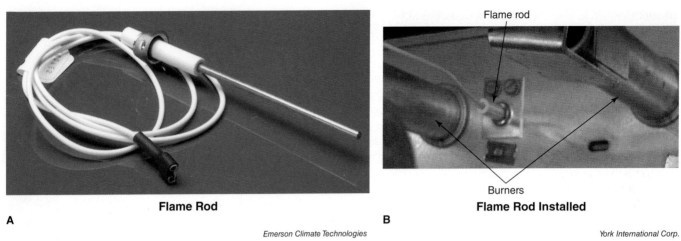

Figure 4-25. A—An example of a flame rod. B—This illustrates the location of where the flame rod is inserted into the burner chamber. The flame rod senses heat from the burner and sends a microamp signal back to the ignition module to confirm that the burner is lit.

Following is the sequence of operation for burner ignition using the intermittent pilot feature:

1. When the boiler is shut off, or in standby mode, the pilot is not lit.
2. Upon a call for heat, two things occur: the pilot valve opens and the ignition system generates a spark.
3. The pilot flame is ignited by this spark, and the flame rod senses the flame.
4. When the flame rod proves that the pilot is lit, it sends a signal back to the control module and the main valve is allowed to open.
5. The main burner is ignited by the pilot flame and continues to burn until the thermostat is satisfied.

On some systems, the pilot is extinguished once the main burner is lit. On other ignition systems, the pilot remains lit throughout the burner cycle.

4.5.3 Direct Spark Ignition Systems

A *direct spark ignition system* is similar to the intermittent pilot ignition system except it does not utilize a pilot light. Instead, a direct spark ignition system ignites the main burner directly from a spark generated from a transformer located within the ignition module. Just like the intermittent pilot system, the spark is created between the electrode and ground. This grounding is connected directly to the burner itself or to a grounding strap as part of the spark assembly. In addition, similar to the intermittent pilot ignition system, the direct spark ignition system utilizes a flame rod to prove ignition. However, in some direct spark ignition systems the igniter doubles as the flame rod. This is sometimes referred to as "local sensing."

One characteristic of the combination gas valve on a direct spark ignition system is that it is classified as a "slow-opening" valve. Because gas is being introduced directly to the burner rather than to the pilot during ignition, the valve is prevented from opening too quickly, which could cause delayed ignition, flame rollout into the vestibule, and possibly a loud explosion from excessive gas igniting too quickly. Normally, this slow-opening period is between 4 and 20 seconds. Another safety feature of the direct spark ignition system is that if the flame is not proven within a certain period of time (generally 4–10 seconds), the ignition module will lockout and prevent a trial for ignition to reoccur. In order to reset the ignition module, the power to the module must be shut down and

SAFETY FIRST

Safety When Working with Spark Ignition Circuits

Although the spark ignition circuit has very low amperage, it does generate approximately 10,000 volts. This may not be considered a major safety hazard, but it can cause a great deal of discomfort if you should come into contact with it while working in the area!

re-energized. This manual reset function is used to prevent excessive unburned fuel from flooding the combustion chamber due to ignition failure.

> **TECH TIP**
>
> **Grounding the Electronic Ignition Module**
> The correct grounding of an electronic ignition module is absolutely essential to proper functionality. The grounding wire should be secured to the burner or to a grounding strap as part of the spark assembly. It also must originate from the boiler's main ground that comes from the electrical panel in conjunction with the inlet power supply. This ensures that the 24 volts delivered to the module is not a "floating" 24 volts. This means that the 24 volts cannot float from one side of the transformer to the other. When grounded properly, the 24-volt power supply to the ignition module will be read from only one side of the transformer to ground. Many times, a technician will condemn the ignition module as defective when all it needs is a good grounding connection.

4.5.4 Hot Surface Ignition Systems

An alternative to the direct spark ignition system is the *hot surface ignition system*. This system is very similar to the direct spark system except rather than generating a high voltage spark to ignite the burners, the hot surface system utilizes an ignition device made of silicon carbide. This igniter is sometimes referred to as a *glow coil* because of the way it illuminates when energized. Similar to the direct spark ignition system, hot surface igniters do not use a pilot light but rather ignite the burner directly. They use the same operating sequence when igniting and provide the same system of safety lockout if ignition does not occur, **Figure 4-26**.

The hot surface igniter produces a great deal of heat when energized due to the high level of current passing through it, but not enough current to cause damage to the coil. Although this igniter is sturdy and quite durable, it is not completely impact resistant and can break if not handled appropriately. Most igniters operate off 120 volts, but some can use 24 volts.

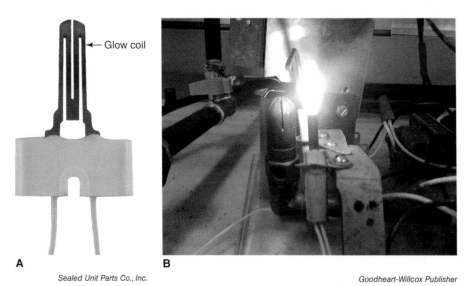

Sealed Unit Parts Co., Inc. Goodheart-Willcox Publisher

Figure 4-26. A—Hot surface igniters are made up of silicon carbide. B—When current is applied, they glow hot to ignite the burner.

> **PROCEDURE**
>
> **Diagnosing the Hot Surface Igniter**
>
> Diagnosing the hot surface igniter (or HSI) for defects if there is a flame failure is fairly simple:
>
> 1. Examine the HSI for any cracks or other damage.
> 2. Disconnect the power cord leading to the HSI.
> 3. Using a voltmeter, connect the leads into the power cord leading to the HSI.
> 4. Energize the burner circuit. If there are 120 volts present at the power cord, the HSI is probably defective.
> 5. Also check the ohm resistance through the HSI with it disconnected from the power cord. There should be approximately 40–90 ohms if the HSI is not defective.

> **TECH TIP**
>
> **Cleaning the Flame Rod or Hot Surface Igniter**
>
> A typical service call at the beginning of the heating season might sound like this: "The boiler fires, but immediately shuts down." This scenario could simply be a dirty flame rod or hot surface igniter. To clean these components, use a nonabrasive cloth or even a dollar bill—but not sandpaper—it is too abrasive. Using an abrasive such as sandpaper can damage the exterior of the sensor and create even more problems. Also, be careful when handling the flame rod or hot surface igniter. Oils from your skin can be transferred to these components, which can shorten their lives.

4.5.5 Gas Ignition Modules

Gas-fired boilers utilize various controls to ensure safe and reliable burner ignition at all times. At the heart of modern boiler combustion systems is the ***gas ignition module***. The types of ignition modules found on gas-fired boilers vary depending on their sequence of operation and lockout requirements. Most electronic ignition modules indicate this series of events and their intended lockout or shutoff specifications on their faceplates.

Older boilers may possess ***nonintegrated ignition modules***. These less complicated ignition modules control the spark or hot surface igniter, operate the gas valve opened and closed, and monitor the burner using a flame rod and lockout feature. Newer gas boiler systems feature more sophisticated ***integrated ignition control modules***, which offer more advanced electronics for greater control and functionality. One feature of the integrated ignition module is its ability to perform self-diagnostics. The control board includes a built-in LED display, which can generate a trouble code for greater troubleshooting efficiency.

It is important for you to understand the type of ignition control module the boiler is utilizing in order to identify its control sequence and level of safety lockout control. Some of the most common sequences used for gas ignition control modules follow.

4.5.5.1 100% Shutoff

A system that employs an ignition module with 100% shutoff locks out both the pilot valve and main burner in the event of a flame failure or failure to prove pilot upon the ignition sequence. This type of safety feature is employed with boilers using LP or propane as a fuel source. Because liquid petroleum fuel or propane is heavier than air, it will not vent naturally through the boiler flue. If the boiler incorporated a gas ignition module that utilized a continuous retry for ignition, the boiler area could flood with dangerous unburned gas, causing an explosive situation.

4.5.5.2 Non-100% Shutoff

An ignition module classified as non-100% shutoff will close the main gas valve but allow the pilot valve to remain open if the flame rod fails to detect a flame. By doing this, the ignition system is allowed to try to relight the burner, but only after a delayed period—usually about 5 minutes. This type of ignition system prevents permanent lockouts and provides a delay in reigniting the flame in order to purge any remaining unburned gas from the combustion chamber when there is a flame failure, **Figure 4-27**.

4.5.5.3 Continuous Retry with 100% Shutoff

This ignition module configuration will shut off both the main gas valve and pilot valve in the event of a flame failure with a 90-second trial for ignition. However, it will continue to try to relight the pilot and main burner once there is a delayed shutoff period without completely locking out, **Figure 4-28**.

Goodheart-Willcox Publisher

Figure 4-27. An example of a non-100% shutoff ignition module.

Goodheart-Willcox Publisher

Figure 4-28. An example of a continuous retry 100% shutoff ignition module.

> **TECH TIP**
>
> **Ignition Module Wiring Terminals**
>
> Most gas ignition modules have specified wiring terminals, which make it easy for the technician to install and rewire a new replacement module. These wiring terminals may vary between different manufacturers, so always refer to the manufacturer's installation instructions for clarifications. Most of the common abbreviations used for wiring an ignition module are as follows:
>
> - MV: Main gas valve
> - PV: Pilot gas valve
> - MV/PV: Connection terminal for the common wire between main valve and pilot valve
> - GND: Ground wire connection to the burner
> - 24V: Primary connection from the 24-volt transformer, providing main power to the module
> - 24V GND: Secondary connection or grounded side of the 24-volt transformer
> - TH or TH-W: Thermostat lead to energize the ignition sequence
> - SENSE: Flame rod connection
> - SPARK: Spark ignition source to light the flame

4.5.5.4 Gas Ignition Circuit Boards

Larger gas-fired hot water boilers used for commercial and industrial applications also include circuit board-type ignition modules for burner control. Found on natural gas and LP-fired boilers, these microprocessor-based integrated controllers manage automatic burner sequencing, flame supervision, system status indication, and self-diagnostics. Wiring terminals on these boiler/burner ignition controls are similar to those used on residential and light commercial boiler applications, but they also include terminals for safety and control devices found on commercial and industrial boilers with FM and IRI gas manifolds, as discussed earlier. In addition to LED readouts, some of these modules provide a text display for ease of troubleshooting as well as a personal computer interface and remote text display features, **Figure 4-29**.

Goodheart-Willcox Publisher

Figure 4-29. A microprocessor-based integrated controller used for automatic burner sequencing, flame supervision, system status indication, and self-diagnostics.

4.5.6 Ignition System Sequence of Operation

Whether used for residential, commercial, or industrial applications, today's gas ignition modules follow a sequence for each startup and shutdown procedure to ensure safety and reliability. Depending on the type of ignition module and boiler application in use, some or all of the following operations may be incorporated in the sequence of operations as dictated by the ignition module:

- Pre-purge
- Trial for ignition
- Pilot proves
- Main burner ignites
- Post-purge
- Inter-purge

Upon a call for heat, the ignition module energizes a combustion blower to pre-purge or vent any products of combustion or unburned gas from the combustion chamber.

Next, there is a trial for ignition in which the ignition device (spark or glow coil) is energized along with the pilot valve opening. Upon lighting the pilot, the flame rod senses the flame and that the pilot is proven.

Once the pilot proves, the main gas valve is opened and the main burner ignites. The burner remains ignited until the call for heat is satisfied and the thermostat opens.

After the main gas valve closes, the gas ignition module keeps the combustion blower energized for a short period of time until any remaining products of combustion have been fully vented from the combustion chamber. This is known as post-purge.

In the event of a flame failure either midway through the burner cycle or during the trial for ignition, the module may energize the combustion blower for a period of time to vent any remaining gases from the combustion chamber. This cycle is known as the inter-purge stage.

4.6 Safety Devices for Burner Ignition Controls

The gas burner ignition system would not be complete without the necessary safety devices used to prevent hazardous situations. Safety devices used for burner ignition controls are wired in series with the burner circuit to ensure that if any single safety switch trips, the entire circuit becomes deenergized. The following safety devices are most commonly used with gas boiler ignition circuits:

- High temperature limit
- Flame rollout switch
- Air pressure switch

The **high temperature limit** safety switch can be found on all types of ignition systems, including those utilizing a standing pilot. This high limit switch is located inside the boiler's combustion vestibule, with the sensing element protruding into the heat exchanger. A variety of high temperature situations can cause this limit switch to trip: excessive fuel gas pressure, lack of water flow through the boiler's heat exchanger, or a blocked exhaust vent or flue. The high temperature limit switch will usually reset automatically when it cools down below its set point, **Figure 4-30**.

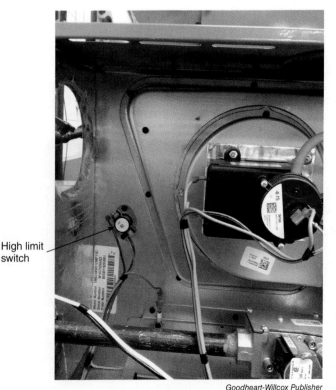

Goodheart-Willcox Publisher

Figure 4-30. An example of a high-limit switch.

Another issue that can result from a blocked exhaust vent is a lack of combustion air. When this situation occurs, the burner flame will seek combustion air wherever it can be found—including the boiler's vestibule. As a result, the burner flame can rollout into the vestibule, causing damage to the control wiring and other devices. To mitigate the damage to these devices, a *flame rollout switch* can be found inside the boiler vestibule near the gas burners. If a flame rollout occurs, the rollout switch will trip due to an excessive increase in temperature. These switches usually require manual resetting to prevent excessive damage, so if they lockout the technician is required to look for reasons for the failure. The flame rollout switch is typically found on newer boilers that possess electronic ignition controls, **Figure 4-31**.

Older, lower-efficiency boilers with atmospheric burners typically utilized gravity and the natural buoyancy of warm air as a means of venting the products of combustion. However, newer, higher-efficiency boilers require a combustion blower to vent the products of combustion. To ensure the flue vents are unobstructed, an *air pressure switch* must prove positive air pressure throughout the venting system before the ignition system is allowed to be energized, **Figure 4-32**.

Rollout Switch
White-Rodgers Division, Emerson Climate Technologies

Rollout Switch Installed
York International Corp.

Figure 4-31. An example of a flame rollout switch. Notice the manual reset button and where the switch is located within the burner vestibule, where it will detect heat from the burner and trip out, shutting down the burner circuit.

4.6.1 Flame Rectification

Technically, the flame rod could also be considered a safety device as part of the burner ignition controls. Whether a separate flame rod is used for flame sensing or the igniter assembly is used, both systems incorporate a *flame rectification system* for flame detection. Pilot and main burner flames actually conduct electricity. This is because the flames contain ionized combustion gases made up of positively and negatively charged particles. In a flame rectification system, the gas burner ignition module generates an ac signal that passes through the flame—from the flame rod to ground. Because the flame rod is smaller than the grounding assembly, the current flow becomes greater in one direction. This ac signal is rectified to create a dc pulsating current that flows directly to ground. The dc amperage is the only signal the ignition module will recognize, and it keeps the main gas valve open as long as a flame is present. If there is an interruption in the flame signal, the ignition module recognizes this and immediately shuts down the main burner circuit. The ignition circuit then either tries to reignite or locks out and must be reset by cycling power to the controller, **Figure 4-33**.

Goodheart-Willcox Publisher

Figure 4-32. An example of an airflow proving switch used to ensure that there is no blockage through the venting system.

PROCEDURE

Measuring the Signal Strength of the Flame Rod
To measuring the signal strength of the flame rod, follow these steps:
1. Disconnect the flame rod lead from the ignition module.
2. Using a voltmeter, change the setting to microamps (µA).
3. Connect one lead from the voltmeter to the end of the flame rod.
4. Connect the other lead to the flame rod terminal on the ignition module.
5. Fire the boiler, and read the microamps on the voltmeter. It should read between 1 and 25 µA.
6. If there is no microamp reading, the flame rod is usually defective.

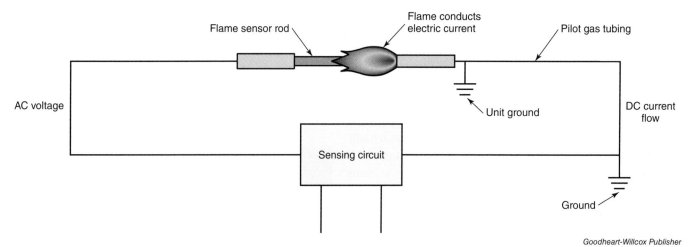

Figure 4-33. This illustration shows how flame rectification is achieved through the burner circuit.

CAREER CONNECTIONS

Employee Efficiency

An efficient employee is a critical asset to any successful business. Employee efficiency will lead to increased productivity and maximize the business's profitability. There are several traits of an efficient employee that will keep each task flowing smoothly and improve the overall workplace environment.

1. **Daily Work Planning:** Set goals at the beginning of each day. Plan your work, schedule according to the number of tasks that are required, and prioritize those tasks to ensure that they are achieved. Look at the list of daily projects and anticipate what tools will be needed for each job. Organize your tools based on what is used most often and keep them within easy reach.
2. **Job Performance:** Take pride in a job well done. An efficient employee will learn to work with minimal supervision and is willing to take on additional responsibilities. However, employees will learn to improve efficiency by knowing when to ask for help in order to complete the job in a timely manner. Efficient employees take ownership in their work and derive satisfaction from productivity.
3. **Customer Relations:** An efficient employee develops effective communication skills. By listening to the customer's needs and accurately diagnosing the problem, an employee can finish tasks faster and more efficiently.
4. **Going the Extra Mile:** When a project is complete, an efficient employee will check the work environment before leaving to ensure the workplace is clear, clean, and all tools are collected and properly put away. Furthermore, an efficient employee explains to the customer the work that was performed and offers ideas on how the equipment can be properly maintained and kept in good working order.

The business's management can play an active role in employee efficiency by extending a positive work environment and working to develop the employees' professional development. Feedback is essential between employees and supervisors to improve efficiency by ensuring the workload is balanced and well maintained. Everyone benefits from employee efficiency through improved performances and customer satisfaction.

Chapter Review

Summary

- The burner and ignition system are the integral components that make up the combustion system, which heats boiler water.
- *Combustion* is the rapid expansion of gases resulting in the production of heat and light. For combustion to take place, there must be fuel, heat, and oxygen.
- The products of complete, or *stoichiometric*, combustion are energy, carbon dioxide, and water.
- For combustion to occur, there must be the correct ratio of fuel and oxygen.
- The gas train is a series of components that safely feeds fuel into the burner.
- The components of a simple gas train include a drip leg, manual shutoff valve, pressure regulator, pilot line, and automatic gas valve.
- Large commercial and industrial gas trains may include additional safety and control components and may be FM or IRI compliant. FM safety standards include additional manual shutoff valves for leak-testing purposes, high- and low-gas pressure switches, and two electric safety shutoff valves. Stricter IRI standards include these and a normally open bleed valve between the safety shutoff valves.
- There are several different types of gas valves, including solenoid, diaphragm, combination, two-stage, and modulating. Solenoid gas valves are normally closed valves. Diaphragm gas valves use a pressure differential in which to operate. A combination gas valve includes a pressure regulator, a manual shutoff valve, a pilot supply line, and multiple safety shutoff valves. A two-stage gas valve has high- and low-fire capabilities. Modulating gas valves incorporate a turndown ratio for maximum fuel efficiency.
- Gas-fired burners utilize primary and secondary air to provide proper fuel combustion. The gas manifold delivers fuel to the burners through small sockets called spuds. The spud contains predrilled holes called orifices that meter the gas into the burner.
- Atmospheric burners rely on gravity or the natural buoyancy of heated air to vent the products of combustion through the flue or chimney. They can be ribbon type, slotted type, or single port. They utilize a venturi, a device that creates a narrow constriction, causing air and gas to be accelerated as they pass through the burner.
- Power burners mix air and fuel before they enter the combustion zone, and are mostly found on larger commercial and industrial application boilers.
- The gas ignition system is used to safely light the fuel being burned and to monitor the flame. A standing pilot ignition system has a continuously burning pilot flame that is monitored by a thermocouple or thermopile. Intermittent pilot ignition systems ignite the pilot by either a spark or hot surface device only when there is a call for heat. Direct spark ignition systems are used to light the burner without the need for a pilot flame.
- A flame rod is used to prove ignition and to monitor the pilot light or burner.

(continued)

- Gas ignition modules can have various levels of shutoff and can offer LED displays and diagnostic capability. These include 100% shutoff, non-100% shutoff, continuous retry with 100% shutoff, and circuit board ignition modules.
- Safety devices for burner ignition controls include high-limit switches, rollout switches, and air pressure switches for proving positive flow through the venting system.
- Flame rectification is used to verify the presence of a flame.

Know and Understand

1. What are the three items required for combustion to take place?
   A. Fuel, oxygen, flame
   B. Fuel, oxygen, heat source
   C. Fuel, nitrogen, heat source
   D. Gas, carbon monoxide, oxygen
2. What is meant by *stoichiometric combustion*?
   A. It is incomplete combustion.
   B. It is combustion that takes place in every boiler.
   C. It is combustion that requires very little oxygen.
   D. It is the ideal combustion process in which fuel is burned completely.
3. *True or False?* In older boilers, the pilot flame gas pressure is non-adjustable.
4. Which device is *not* required on a commercial or industrial FM-approved gas train?
   A. A normally open bleed valve between the safety shutoff valves
   B. Additional manual shutoff valves
   C. Two electric safety shutoff valves
   D. High- and low-gas pressure switches
5. What is the function of a drip leg used in a gas train?
   A. It is used to catch oil from the system.
   B. It is used to support the gas train.
   C. It is used as a trap to prevent moisture and debris from entering the gas valve.
   D. It is used to purge excess air from the gas train.
6. Explain the difference between an atmospheric and power burner.
   A. A power burner is used to force primary air, secondary air, and fuel into the burner tube.
   B. Atmospheric burners introduce primary air through natural means at the inlet, which is mixed with the gas that passes through the manifold spud.
   C. Both A and B are correct.
   D. None of the above.
7. What device creates a narrow constriction at the burner inlet causing the air and gas to be accelerated as they pass through?
   A. A spud
   B. A venturi
   C. An oblique
   D. A thermopile
8. What device supervises a standing pilot flame?
   A. A thermocouple
   B. A thermopile
   C. An oblique
   D. Both A and B are correct.
9. What should the ohm reading be through a hot surface ignition coil?
   A. Approximately 40 to 90 ohms
   B. Approximately 50 to 100 ohms
   C. Approximately 100 to 200 ohms
   D. Approximately 150 to 250 ohms
10. *True or False?* Automatic gas valves are typically considered normally open valves.

Apply and Analyze

1. Explain the function of a gas pressure regulator.
2. Describe how a diaphragm gas valve uses differential pressure to open and close the valve.
3. Explain what is meant by a redundant gas valve.
4. List the features that are found on a combination gas valve compared to other valves.
5. Explain how turndown ratio is used to describe a modulating gas valve.
6. What is the difference between primary air and secondary air, and how are they utilized within the gas-fired burner?
7. Explain the purpose of a flame rod when used with an intermediate pilot ignition system.
8. Why is a slow opening gas valve used on a direct spark ignition system?
9. Describe the different levels of shutoff used on some electronic ignition modules.
10. What is the purpose of pre-purge and post-purge used on the sequence of operation for some ignition control modules?
11. Explain why a flame rollout switch is used as a safety device on hot water boilers.
12. Describe how the flame rectification system works with a boiler flame rod.

Critical Thinking

1. Why would the size of the burner orifices be larger on some boilers as opposed to others?
2. What are the benefits of using a two-stage gas valve versus a single-stage valve?

5 Oil Systems

Chapter Outline
5.1 Fuel Oil Characteristics
5.2 Fuel Oil Combustion
5.3 Fuel Oil Piping Systems
5.4 Oil Burner Components
5.5 Fuel Oil Boiler Venting
5.6 Combustion Analysis

Thaiview/Shutterstock.com

Learning Objectives

After completing this chapter, you will be able to:
- Describe the various grades of fuel oil.
- Identify the characteristics and refining process of fuel oil.
- Explain how the combustion process takes place in an oil-fired boiler.
- Describe how primary air and excess air are used in the combustion process.
- Explain how fuel oil is atomized in an oil burner.
- Compare how one-pipe and two-pipe delivery systems are used with fuel oil boilers.
- Explain how an oil deaerator is used in a fuel oil system.
- Describe the function of each component that makes up an oil burner.
- List the differences between a stack relay and a cad cell relay when used with a primary control unit.
- Discuss why draft is important to an oil-fired boiler.
- Explain how a draft controller is used in a fuel oil system.
- Describe the various components measured in a combustion analysis.

Technical Terms

ash content
American Society for Testing and Materials (ASTM)
atomization
cad cell relay
carbon dioxide reading
carbon monoxide reading
carbon residue
combustion analysis
cracking
dilution air
distillation quality
draft
draft test
electrodes
excess air
flash point
fossil fuel
fuel line filter
fuel oil
fuel oil pump
gun-type burner
hydrocarbon
ignition point
ignition transformer
net stack temperature
nozzle
oil burner
oil burner motor
oil deaerator
one-pipe delivery system
oxygen reading
pour point
primary control unit
Saybolt Seconds Universal (SSU)
stack relay
two-pipe delivery system
viscosity

In addition to natural gas and LP (propane) gas, fuel oil is a viable option as a heating fuel source for boiler operations. Although not utilized as much in the midwestern United States as it once was, fuel oil is still commonly used in boilers in the northeastern and mid-Atlantic regions. This chapter focuses on the characteristics of fuel oil as well as the combustion and safety equipment used for the effective adaptation of fuel oil for boiler heating applications.

5.1 Fuel Oil Characteristics

Fuel oil, or *heating oil*, is a petroleum product refined from crude oil. Petroleum formed millions of years ago from plant and animal remains buried under dirt and rock. Heat and pressure from inside the earth converted these plant and animal remains into oil, natural gas, and coal, which are characterized as *fossil fuels*, **Figure 5-1**.

angkrit/Shutterstock.com

Figure 5-1. Fossil fuels were created from plant and animal remains buried deep inside the earth, where heat and pressure converted these plant and animal remains into crude oil.

Fuel oil is composed of liquid **hydrocarbons**, which are made up of hydrogen and carbon in chemical composition. The process of refining fuel oil from crude oil is a distillation process known as *cracking*, **Figure 5-2**. There are six grades of fuel oil, categorized as 1 through 6 to identify each grade. The US Department of Commerce establishes these grades, which comply with specifications set forth by the ***American Society for Testing and Materials (ASTM)***.

Lower-numbered fuel oils are identified as *light oils* because of their lower weight-per-gallon. The lightest of the fuel oils is Number 1, which is referred to as *kerosene*. Number 6 fuel oil is a heavy, viscous fluid sometimes used in marine engine applications and in some diesel locomotives. Higher-numbered oils have a higher Btu per gallon content since they have a higher carbon content. With hydrocarbon fuels, a higher carbon content equals a higher Btu content and therefore a higher heat content. The most common type used for boiler applications is Number 2 fuel oil, which is also referred to as *pentane*.

Avigator Fortuner/Shutterstock.com

Figure 5-2. Fuel oil is refined from crude oil into six different grades. Number 2 fuel oil is used most predominantly for oil-fired boilers.

DID YOU KNOW?

Red Fuel

Number 2 fuel oil is also used as diesel fuel for the automotive industry. To differentiate the use of fuel oil between heating and automotive applications, heating oil is dyed red. The Internal Revenue Service (IRS) requires that heating oil—which is not used for automotive purposes—be colored with a red dye to identify it as exempt from the federal, state, and local taxes applied to fuels sold for use on public roads. Fuel oil designed for use in boiler applications is illegal for use in vehicles operating on public roads. If you get caught running your diesel car or truck on Number 2 heating fuel, you risk serious penalties for breaking the law!

5.1.1 Heating Values

Number 2 fuel oil has a Btu content of approximately 140,000 Btu/gallon. The actual Btu per gallon value can range from between 135,000 to 142,000, depending on the carbon content within different batches of oil. When classified as a hydrocarbon, fuel oil or heating oil is typically made up of about 85% carbon and 12% hydrogen. Various other elements make up the remaining 3% of heating fuels.

5.1.2 Flash Point

The ***flash point*** is the lowest temperature at which fuel oil vapors will ignite above a pool of liquid for a short period of time when exposed to a flame. Lighter-weight oils have a lower flash point. The flash point for Number 2 fuel oil is about 100°F.

5.1.3 Ignition Point

A fuel oil's ***ignition point*** is slightly higher than its flash point. At its ignition point, fuel oil will ignite and continue to burn as a vapor as it rises from a pool of liquid.

5.1.4 Distillation Quality

The ***distillation quality*** of fuel oil describes the ability of the oil to become vaporized. Lighter fuel oils tend to vaporize more easily than heavier oils. Since fuel oil must become a vapor or gas in order for it to combust, the distillation quality is important to engineers and manufacturers because a higher distillation quality can result in a more efficient boiler.

5.1.5 Viscosity

Fuel oil ***viscosity*** can be defined as the measurement of the oil's resistance to flow. It is also a measurement of the thickness of the oil under normal temperatures. Compare the thickness of lighter fuel oils to heavier fuel oils in **Figure 5-3**. Heavier oils tend to be thicker than lighter oils when at the same temperature.

The measurement of an oil's viscosity is expressed in ***Saybolt Seconds Universal (SSU)***. This measurement describes the time elapsed for an oil to flow through a calibrated orifice at a defined temperature, usually 100°F. The total

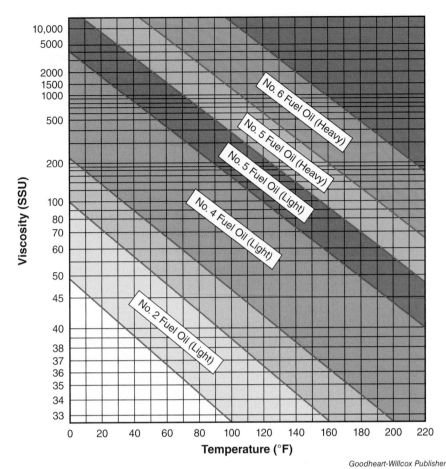

Goodheart-Willcox Publisher

Figure 5-3. This graph shows how different grades of fuel oil have different viscosities based on SSU units. The different grades of oil are circled from Number 2 through Number 6. The viscosity of the various grades of oil increases as their temperature decreases.

number of seconds determines the viscosity rating. As shown in **Figure 5-3**, Number 2 fuel oil is required to have an SSU rating between 37.9 and 32.6 at 100°F, according to the ASTM Standard D396. A higher SSU rating translates to a thicker, higher viscosity oil. This means fuel oil that exceeds the upper limit of 37.9 SSU can result in poor atomization, delayed ignition, and incomplete combustion within the boiler's combustion chamber.

Figure 5-3 also illustrates that higher grades of fuel oil have a higher viscosity. Fuel oil viscosity determines the appropriate nozzle size to use with the oil burner. This is usually defined by the boiler manufacturer. If the technician determines that there may be a problem with the nozzle size as a result of the oil's viscosity, he or she should contact the fuel oil supplier to find out the actual oil viscosity, then consult the manufacturer to determine if there should be changes to the nozzle size.

5.1.6 Pour Point

The lowest temperature at which fuel oil can be stored and handled is called the *pour point*. This temperature is usually about 5°F above the point at which the oil thickens into a solid mass. Number 2 fuel oil has a pour point of approximately 20°F. This means that colder regions of the United States are more likely to have problems with oil flow below this threshold temperature. In colder temperatures, fuel oil companies mix Number 1 distillate fuel (kerosene) with Number 2 fuel oil to effectively lower the pour point. It is important to understand that below these temperatures the oil does not freeze, but certain waxes within the oil will become so thick that the oil will not readily flow. This can result in a loss of combustion within the boiler and the burner control locking out. Pour point can also determine the location of the fuel oil tank. In warmer climates, the tank can safely be located outside, but in colder climates it is a good idea to locate the tank either underground or in the basement of the building. In instances where tanks must be outside in the Northeast, kerosene is used in place of Number 2 heating oil.

5.1.7 Carbon Residue

Carbon residue is simply the amount of carbon left in a sample of oil after it is converted from a liquid to a vapor by boiling the oil in an oxygen-free atmosphere. Ideally, the amount of carbon residue left after combustion is zero. However, if enough oxygen is not available for proper combustion, carbon residue may show up as soot. A properly tuned oil-fired boiler will not have any appreciable carbon residue.

5.1.8 Ash Content

In fuel oil, the ***ash content*** indicates the amount of noncombustible particulate found within the oil. These noncombustible inorganic materials are considered contaminants and can cause wear and abrasion on oil-fired boiler components. The fuel oil refinery is responsible for maintaining minimum tolerances of ash content.

5.1.9 Water and Sediment Content

Oil refineries do their best to ensure that their product is pure and free of any water or sediment content. However, even under ideal conditions, issues with these contaminants can crop up once the fuel oil is delivered to the customer and stored for extended periods of time. When the fuel oil tank is not filled to capacity, moisture can condense into the oil and form sludge. Sediment can form from rust on internal piping and tank surfaces. Both sludge and sediment can result in clogged oil lines, oil filters, and burner nozzles. These impurities can also cause improper flame characteristics and incomplete fuel combustion. To minimize these conditions, change the fuel filter on a regular basis and strive to keep the tank as full as possible.

5.2 Fuel Oil Combustion

The combustion process for fuel oil is similar to that of natural gas and propane fuel. As described in Chapter 4, *Gas Burners and Ignition Systems*, combustion is the rapid expansion of gases resulting in the oxidation of fuel. Burning fuel by combustion generates heat, regardless of the fuel being burned. In order for combustion to occur, there must be three components:
- Fuel source
- Oxygen
- Heat source

The combustion process also requires that these three components be supplied in the proper proportion. If there is too much fuel in proportion to the amount of oxygen, the mix will be too rich and combustion will not occur. Conversely, too much oxygen or air and not enough fuel results in a mixture that is too lean for combustion to take place in the fuel oil burner. As mentioned previously, fuel oil is comprised of carbon and hydrogen. For proper combustion to take place in a fuel oil burner, 1 pound of fuel oil must be combined with approximately

Copyright Goodheart-Willcox Co., Inc.

3 pounds of oxygen. Because air contains only 21% oxygen and 79% hydrogen, we would need 14.4 pounds of air in order to obtain the 3 pounds of oxygen required for complete combustion. Here is how this requirement is determined:

3 lb of oxygen ÷ 0.21 lb of oxygen per lb of air = approximately 14.4 lb of air

In order to calculate how many cubic feet of air would be required to burn 1 pound of fuel oil, we will first determine that the volume of 1 pound of air is equivalent to 13.33 cubic feet. Using the 14.4 pounds of air needed for combustion, we can now use the following formula:

13.33 cubic feet of air × 14.4 lb of air = 192 cubic feet of air

This shows that 192 cubic feet of air per pound of fuel oil is needed to achieve proper combustion. Note, however, that this formula represents the amount of primary air for *combustion only*. As noted in Chapter 4, *Gas Burners and Ignition Systems*, in order for the products of combustion to be vented through the appliance properly, there must be **excess air** in the form of secondary and **dilution air**. Without the correct amount of both combustion air and excess air, incomplete combustion will occur. Most oil-fired boilers are designed to operate with approximately 50% excess air for proper combustion. To understand the total amount of air needed (both primary and excess air) for burning 1 pound of fuel oil, observe the following formulas:

14.4 lb primary air + 50% excess air = 21.6 lb total air
21.6 lb total air × 13.33 cu ft / lb = 287.9 cu ft of air / 1 lb fuel oil

If 1 gallon of Number 2 fuel oil weighs approximately 7 pounds, then:

287.9 cu ft of air × 7 lb per gal = 2015.3 cu ft / 1 gal fuel oil

These figures emphasize the importance of proper air supply to ensure complete fuel oil combustion.

When perfect combustion takes place, the three by-products are:
- Heat
- Carbon dioxide (CO_2)
- Water

Realistically, perfect combustion only exists in theory, **Figure 5-4**. In the real world, the burning of fuel oil in oil-fired boilers results in products of combustion that include soot, nitrogen dioxide, and carbon monoxide, **Figure 5-5**. To ensure that an oil-fired boiler is operating at peak efficiency, combustion analysis is essential. This topic is covered later in this chapter.

To ignite fuel oil, it must be delivered to the burner as a vapor in order for it to reach its flash point and ignition point. To achieve this, the oil burner atomizes the fuel oil before it is ignited. First, the oil burner pressurizes the fuel oil through a high-pressure pump. The fuel oil is then forced under pressure through a nozzle. This nozzle breaks the oil into tiny droplets, vaporizing the oil, which

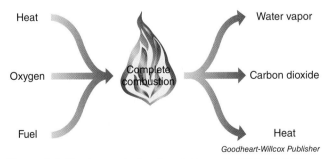

Figure 5-4. The by-products of perfect combustion are heat, carbon dioxide, and water vapor.

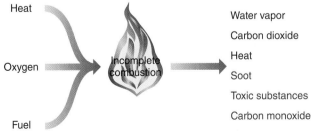

Figure 5-5. The by-products of incomplete combustion include soot, carbon monoxide, and other toxic substances.

then mixes with air, **Figure 5-6**. The lighter carbon molecules form pockets of gas around the droplets, which are ignited by a spark created by electrodes. When combustion initially takes place, the heat that is generated vaporizes the droplets of fuel oil for continuous ignition.

5.3 Fuel Oil Piping Systems

Fuel oil is delivered to the customer and stored in a large tank, which can be located indoors or outdoors, above ground or below, **Figure 5-7**. The delivery of fuel oil from the tank to the oil burner requires a one-pipe or two-pipe delivery system.

5.3.1 One-Pipe System

As the name implies, a *one-pipe delivery system* uses a single supply line between the oil tank and fuel oil burner. This type of delivery system is typically used when the storage tank is located above the fuel oil burner. A one-pipe system utilizes gravity to allow fuel to flow to the fuel oil burner, **Figure 5-8**. Because the fuel oil pump is capable of pumping more oil than is necessary for proper combustion, the pump used in a one-pipe system includes an internal bypass valve that allows any excess oil to be recirculated through the pump. Advantages of using a one-pipe system include less material needed during installation, reduced installation costs, and a reduced risk of the formation of sludge as a result of the fuel oil reacting chemically to the copper tubing. The main disadvantage of a one-pipe system is the risk of air entering the fuel line should the fuel tank run dry. This issue will prohibit the flow of fuel oil to the burner. When this situation occurs, the fuel line will need to be bled before the system will operate properly.

5.3.2 Two-Pipe System

In a *two-pipe delivery system*, a second line is routed from the fuel oil burner back to the storage tank. Instead of recirculating oil through the pump, any excess oil that is not used for combustion is returned back to the storage tank. The oil pump used in a two-pipe system must have the capability of operating in a vacuum since it must pump oil upward when the storage tank is located below the oil burner, **Figure 5-9**. The main advantage of a two-pipe system is the virtual elimination of any air in the delivery lines. The two-pipe delivery system is self-priming and needs no bleeding if the system is opened to the atmosphere. Because the oil is constantly recirculated between the supply and return lines, air is eliminated automatically. However, because the return line on a two-pipe system is pressurized, any leak in this line could go unnoticed and cause damage to the boiler and oil tank facilities.

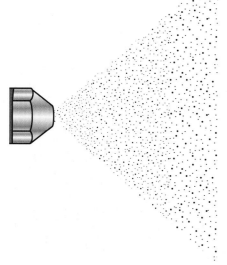

Goodheart-Willcox Publisher

Figure 5-6. Fuel oil is atomized by pressurizing it through an oil burner nozzle.

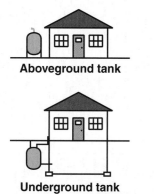

Goodheart-Willcox Publisher

Figure 5-7. The local oil supplier delivers fuel oil from the refinery to residential and commercial buildings, where it is stored in tanks either above or below ground.

GREEN TIP

Underground Storage Tanks and the EPA

According to the Environmental Protection Agency (EPA), approximately 550,000 underground storage tanks store petroleum and other hazardous substances nationwide. The greatest potential threat from these tanks is the contamination of groundwater from leaks. This threatens the source of drinking water for nearly half of all US citizens. The EPA, along with states and territories, is working with industry to protect the environment and human health from potential underground tank leaks by revising underground storage tank regulations and by revising existing requirements for secondary containment and operator training.

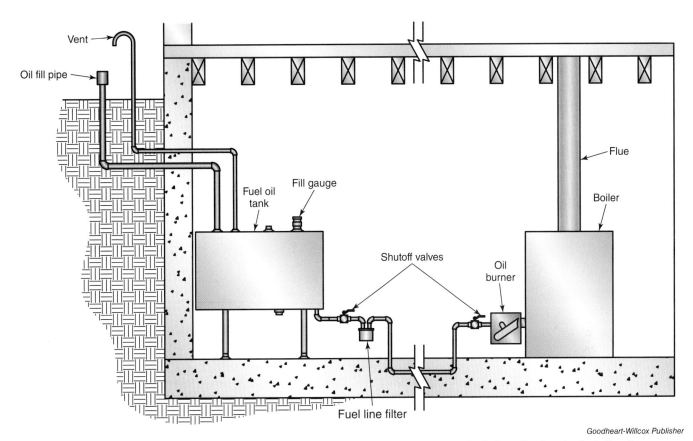

Figure 5-8. A one-pipe fuel delivery system is used when the fuel oil tank is located above the boiler, allowing the fuel oil to flow toward the burner by gravity.

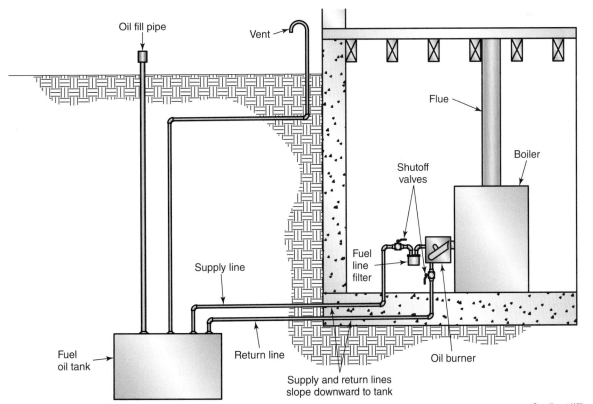

Figure 5-9. A two-pipe fuel delivery system is used when the fuel oil tank is located below the boiler.

Note the following installation tips when installing and servicing fuel oil piping systems:
- Supply and return lines should be sized at a minimum of 3/8″ OD copper.
- Material used for supply and return lines can be copper, wrought iron, steel, or brass. Never use PVC, galvanized, or cast-iron piping—PVC can become brittle and leak; galvanized and cast-iron piping will oxidize, causing the fuel oil to become contaminated.
- Use only flared connections on supply and return lines. Do not use compression connectors.
- Use only approved pipe-joint compound for joining connections.
- Keep line lengths as short as possible and avoid kinking any lines.
- Provide shutoff valves at the outlet of the tank and at the inlet of the oil pump.
- Insulate any lines that are run out of doors.
- Provide protective sleeves or conduits on lines that are run underground.
- Provide proper sloping of lines toward the oil burner to prevent any oil trapping.

CODE NOTE

Fuel Oil and Piping Rules

Chapter 13 of the International Mechanical Code (IMC) and Chapter 8 of NFPA (National Fire Protection Association) 31 contain information on fuel oil piping and storage practices. They also outline specific information on acceptable piping materials, pipe installation practices, piping support, gauges, valves, and testing. Installers should be familiar with the content of these manuals or make sure they are available for referencing.

5.3.3 Oil Deaerators

In oil-fired systems that suffer from chronic problems with air in the fuel delivery lines, it may be necessary to install an ***oil deaerator*** in the fuel oil supply line, **Figure 5-10**. An oil deaerator automatically removes air from the fuel oil system in both one-pipe and two-pipe delivery systems. When used in a two-pipe system, the deaerator can take the place of the oil return line. In a one-pipe system, the deaerator allows the system to function as a two-pipe system. The oil deaerator has three fuel line connections:
1. A connection from the fuel tank to the deaerator.
2. A connection from the deaerator to the fuel pump.
3. A connection from the fuel pump back to the deaerator.

Figure 5-11 shows how to pipe a deaerator into the fuel oil system. A deaerator works similarly to an air vent on a boiler. Any air that is in the system will be trapped in the deaerator and automatically removed into the atmosphere. On a two-pipe system, the deaerator can eliminate any potential leaks that may develop in the return line and exclude the need for long oil return lines. In addition, any excess oil not used for combustion will automatically be recirculated through the deaerator.

When using a deaerator with a one-pipe system, the bypass plug on the oil pump must be installed in order for the pump to operate properly, just as with a two-pipe system (see the following Procedure). The bypass plug will direct the oil from the pump return port into the deaerator return inlet. On a two-pipe system, the existing return line from the oil pump is rerouted back to the deaerator rather than back to the fuel tank. Always install a fuel oil filter between the oil tank and the deaerator inlet to eliminate the chance of damage to the deaerator from contaminates in the oil.

Westwood Products, Inc.

Figure 5-10. When incorporating an oil deaerator into a system, be sure to check each port for proper installation.

PROCEDURE

Converting a One-Pipe System to a Two-Pipe System

When converting a one-pipe system to a two-pipe system, a bypass plug must be installed in the pump. There may be a separate port for the bypass plug or it may need to be installed in the return port of the pump. Reference **Figure 5-12** for this conversion. Here are the proper steps:

1. Remove the cap from the return port on the pump.
2. Insert the bypass plug inside the return port. (This is usually an Allen screw-type plug.)
3. Install the tubing fitting onto the return port.
4. Connect the return line to the return port fitting.

Remember: the bypass plug is removed for a one-pipe system and must be installed for a two-pipe system.

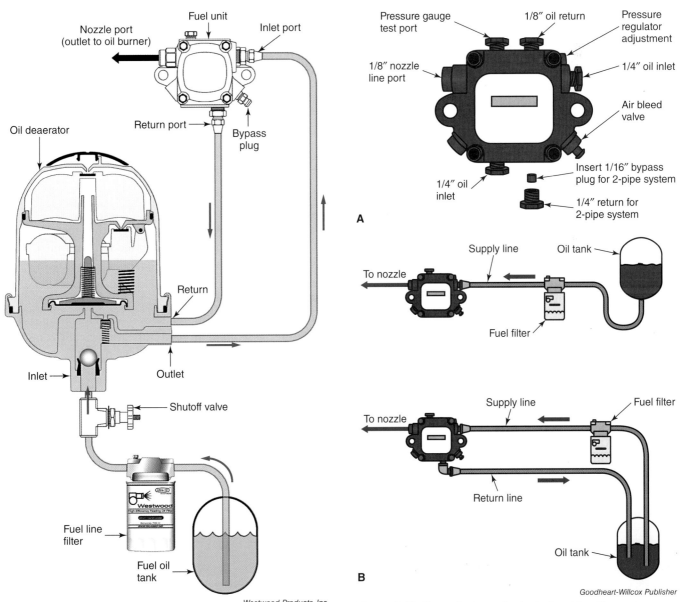

Figure 5-11. This illustration shows the proper connections for an oil deaerator.

Westwood Products, Inc.

Figure 5-12. Figure A shows the typical port locations on a fuel oil pump. Figure B shows how the fuel oil pump would be connected to a one-pipe and two-pipe system.

Goodheart-Willcox Publisher

5.3.4 Fuel Line Filters

A *fuel line filter* is an essential component of the fuel oil system. Just as oil filters are used in automobiles, a fuel line filter traps any particulates and impurities that could damage the system before the oil reaches the pump. Most filters are of a canister-type with replaceable cartridges that can consist of different types of filter media, **Figure 5-13**. Some fuel oil systems include dual in-line filtration systems, with a sludge/water isolator located before the main filter. This device is simply an empty canister with a drain plug on the bottom. Because water and sediment are heavier than oil, they fall to the bottom of the canister, where they can be periodically drained from the system, **Figure 5-14**. When installing fuel line filters, make certain the arrows on the oil filters are pointing in the direction of the oil flow. Also, always replace any gaskets on the filter canister when replacing filter elements (cartridges).

> **SAFETY FIRST**
>
> **Clearing Clogged Lines**
>
> If fuel lines become clogged and need to be blown out to remove obstructions, use only dry nitrogen. Never use compressed air or oxygen to blow out fuel lines because extreme explosions could result. Remember always to use approved nitrogen tanks equipped with pressure regulators and pressure relief valves.

5.4 Oil Burner Components

The *oil burner* is the device that controls the combustion of the fuel oil within the boiler's heat exchanger. It is, in fact, a collection of devices that work together to ensure safe and efficient burning of the oil within the boiler, **Figure 5-15**. The devices that make up the oil burner assembly include:

- Oil burner motor
- Oil burner blower
- Fuel oil pump
- Nozzle
- Electrodes
- Ignition transformer
- Primary control unit
- Cad cell relay or stack relay

There are two types of oil burners: pot burners and gun burners. Pot-type burners are rarely used today but may be found on older model oil burners. Most of today's oil-fired burners are classified as *gun-type burners*, **Figure 5-16**.

5.4.1 The Oil Burner Motor

The *oil burner motor* provides power to both the blower and fuel pump simultaneously within the oil burner assembly. These motors are typically classified

Figure 5-13. This illustration shows a typical dual in-line fuel oil filter system with a sludge/water isolator.

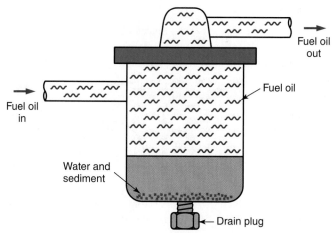

Figure 5-14. Diagram of a sludge/water isolator.

Carlin Combustion Technology, Inc.

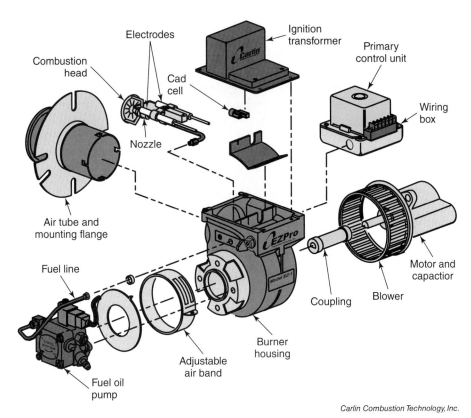

Figure 5-15. This illustration shows the components that make up a typical gun-type burner.

as either split-phase motors or permanent split-capacitor motors (PSC). Both motors are characterized as having both start and run windings. A split-phase motor uses its start windings to generate enough torque to allow the motor to operate. Torque is a measurement of force that causes the motor shaft to rotate. The PSC motor operates using a single-run capacitor, which improves its overall efficiency and performance, **Figure 5-17**. The oil burner motor is powered by a 120 V electrical circuit and typically operates at 3450 RPM. Be aware that older model oil burner motors operate at 1725 RPM. If an older motor needs replacement, be sure to check for proper operating speeds.

Carlin Combustion Technology, Inc.

Figure 5-16. A typical oil-fired gun burner used in residential and light commercial oil-fired boilers.

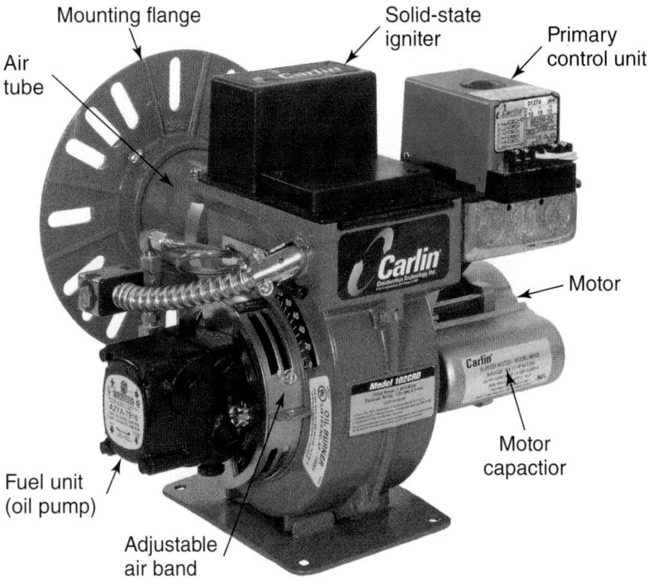

5.4.2 The Oil Burner Blower

The blower or fan that is used with the oil burner assembly is a forward-curved centrifugal type blower, also known as a squirrel-cage fan. It is directly mounted to the shaft of the oil burner motor. The blower housing surrounding the fan has adjustable air inlet openings called air bands, located in a collar attached to the housing, **Figure 5-18**. These openings are designed to adjust the amount of primary air and secondary air needed for proper combustion and should be adjusted as part of the combustion analysis (discussed at the end of this chapter). Air passes through the blower and into an air tube, where it enters the combustion chamber and mixes with atomized fuel oil to provide the proper mix of fuel and air for complete combustion.

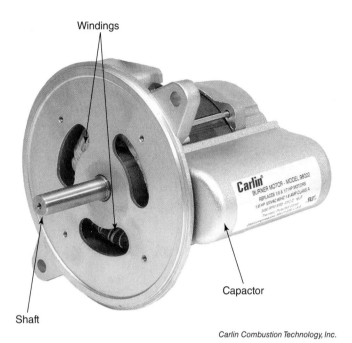

Figure 5-17. A typical oil burner motor. *Carlin Combustion Technology, Inc.*

Figure 5-18. This image of an adjustable air band shows that the airflow setting is expressed as a percentage of the total potential airflow. *Carlin Combustion Technology, Inc.*

PROCEDURE

Adjusting the Air Bands

There are times when a service technician may need to make adjustments on the oil-fired boiler without having immediate access to combustion analysis equipment. When this is the case, adjustment can be made to the air bands on the oil burner in order to make it operational.

Here are the proper steps:

1. Loosen the locking screw and open the air bands far enough to inspect the burner housing and blower wheel. There may be a buildup of soot clogging the blower wheel, which would reduce the flow of combustion air into the burner housing. If necessary, disassemble the burner housing and clean the blower wheel.
2. Close the air opening to its original position and energize the burner. Open the air bands as necessary in order to get the burner to fire.
3. After the burner has run for several minutes, observe the flame characteristics through the burner observation port: with proper combustion the flame should appear luminously bright yellow. A flame that has incomplete combustion will have a dull orange or even red appearance. If this is the case, the air bands should be opened.
4. Adjust the air band openings until the desired flame characteristics are achieved. Be patient! Small adjustments can have a large impact on burner performance.
5. Another test for air adjustment is to observe the amount of smoke. If the burner exhaust can be observed, adjust the air band openings until almost all smoke has disappeared.

Remember: These adjustments are not intended to be used as a substitute for performing a proper combustion analysis.

5.4.3 The Fuel Oil Pump

Fuel oil pumps are also directly connected to the oil burner motor by a flexible coupling, **Figure 5-19**. They rotate at the same RPM as the oil burner blower. These are positive displacement-type pumps. A positive displacement pump has an expanding cavity on the suction side and a decreasing cavity on the discharge side. Fluid flows into the pump as the cavity on the suction side expands and flows out of the discharge side as the cavity collapses, similar to a piston in an engine. The fuel oil pump serves two purposes for the oil system. First, it delivers oil from the storage tank to the combustion chamber. Second, it regulates the pressure at which the oil is delivered into the burner for ignition. Most of today's oil pumps are designed to deliver oil at a pressure of 100 to 140 psig. This is the same pressure at which the nozzle is rated. Check with the pump manufacturer to determine the correct oil pump pressure. To check and adjust the oil pump pressure, first connect a standard pressure gauge to the pressure gauge port as shown in **Figure 5-12A**. Adjust the pump pressure as necessary by rotating the pressure adjustment screw clockwise to increase, or counterclockwise to decrease. The location of this adjustment screw may vary between manufacturers. It is important to check the oil burner manufacturer's specifications before making any pressure adjustments.

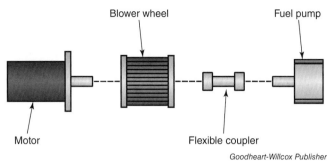

Figure 5-19. An illustration of how the oil burner motor, pump, connector, and blower assembly fit together.

Fuel oil pumps may be single stage or two stage. Movement of the oil through the pump is facilitated by a set of gears located within the pump housing. Single-stage pumps have one set of gears and may be used on either a one-pipe or two-pipe system so long as the level of vacuum required to operate the system does not exceed 12 in. WC. Two-stage pumps contain two sets of gears. One set is used to draw oil from the storage tank, and the second set is used to deliver the oil into the combustion chamber. This second set of gears acts as the "suction pump" gears, which allow for greater vacuum to lift the oil in systems where the storage tank is below the level of the oil burner.

TROUBLESHOOTING

Do Not Be a Parts Changer
Fuel pumps typically are equipped with inlet screens designed to prevent foreign particulates from entering the pump. Before replacing a pump that is thought to be defective, check this screen to see if it is clogged.

5.4.4 The Nozzle

The oil burner *nozzle* is a device made of stainless steel or brass that atomizes the fuel oil. *Atomization* is the process of breaking the fuel oil into tiny droplets in order to allow it to ignite within the combustion chamber. As mentioned earlier, fuel oil will not burn as a liquid; it must be atomized into a vapor for ignition to take place. The nozzle is designed with a fine mesh filter at the inlet and a small orifice at the discharge end, **Figure 5-20**. As the oil moves through the nozzle it passes through tiny holes, which are drilled at an angle to the nozzle. These holes create a swirling motion, which causes turbulence. When the oil passes through the orifice at a high pressure, the turbulent droplets mix with air from the blower and are ignited in the combustion chamber.

Each fuel oil nozzle has a distinct angle of spray designed to be used with specific types of combustion chambers. Short, rounded combustion chambers typically use a nozzle with a 70°–90° spray angle, whereas long, narrow combustion chambers require a nozzle with a 30°–70° spray angle, **Figure 5-21**.

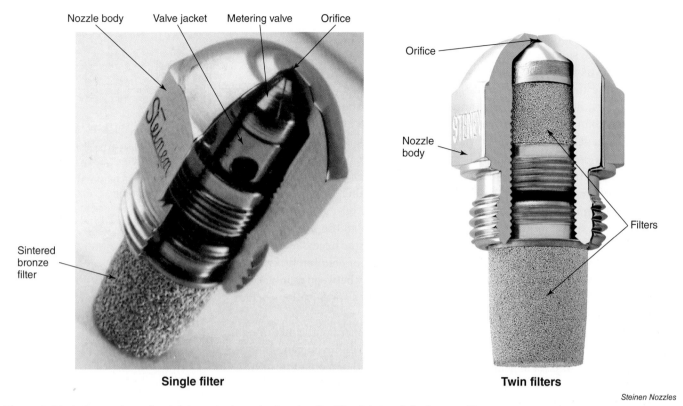

Figure 5-20. Cutaway view of a stainless steel nozzle showing the filter inlet and discharge orifice.

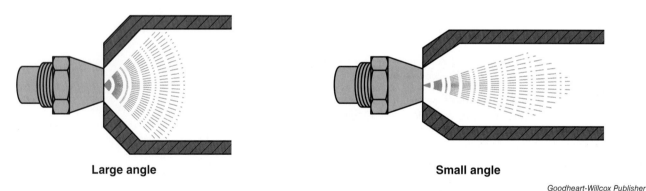

Figure 5-21. Large spray angles are used for shallow combustion chambers. Smaller spray angles are used for deeper combustion chambers.

Nozzles are also identified by their spray patterns, **Figure 5-22**. The three most widely used spray patterns are:
- Solid cone
- Hollow cone
- Semisolid cone

Solid cone nozzles distribute a spray pattern with uniform fuel oil droplets across the cone's shape. These types of nozzles are typically used with larger oil-fired boilers that require a longer firing rate. Hollow cone nozzles spray no droplets within the center of the cone. These are used with smaller boilers that have lower firing rates. They are quieter than solid cone nozzles. Semisolid cones can be used with different types of boilers and are good replacements for solid cone nozzles. Always follow the manufacturer's recommendations when replacing nozzles, and never install a nozzle with a higher firing rate than is recommended.

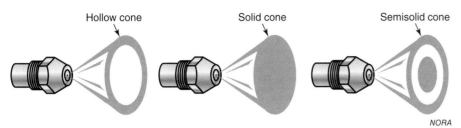

Figure 5-22. This illustration shows the three primary types of spray patterns used with oil-fired boilers.

Oil burner nozzles are identified with three distinct markings:
- Spray pattern
- Spray angle
- Output capacity

The nozzle's spray pattern is typically identified with a specific letter depending on the manufacturer of the nozzle. The spray angle is identified as the degree of spray, and the output capacity is expressed in gallons per hour. For instance, a nozzle that has a rating of 0.75 gallons per hour will simply be marked as ".75," **Figure 5-23.**

> **TECH TIP**
>
> **Clean or Replace Nozzles?**
>
> Do not attempt to clean nozzles with such devices as wire brushes or sandpaper. These items can damage the precisely machined surfaces of the nozzle and distort their performance, which can lead to the incorrect flow of oil into the burner and poor combustion. It is usually cheaper and more effective simply to replace the nozzle rather than risk damaging the oil burner.

> **TIMEOUT FOR MATH**
>
> **Finding the Right Size Nozzle**
> To determine the correct-sized nozzle for a fuel oil system, follow these simple steps:
> 1. Determine the Btu capacity of the boiler.
> 2. Divide this number by 140,000. (This is the Btu equivalent of a gallon of Number 2 fuel oil.)
> 3. The result is the proper gallons per hour (GPH) rating of the nozzle to be used with this boiler.
>
> For instance, consider a boiler with a Btu capacity of 105,000 Btu/hr. When this figure is divided by 140,000, the result is 0.75. This is the correct-sized nozzle for use with this boiler.

Figure 5-23. Oil burner nozzles are identified according to their spray angle, spray pattern, and GPH flow.

5.4.5 The Electrodes

Two *electrodes*, located near the end of the nozzle orifice, ignite the fuel oil in the combustion chamber. These electrodes are made of stainless steel and insulated with a ceramic material to prevent them from grounding to the burner, **Figure 5-24.** An electric current generated by the ignition transformer creates a high-voltage spark, which jumps between the two electrodes and ignites the vaporized fuel oil, **Figure 5-25.** Just as with a spark plug on an automobile, determining the proper gap between the electrodes is very important. There are three specific gaps to check closely on oil burner electrodes to ensure that they work properly:

1. The *electrode gap* is the distance between the two electrodes. This gap should be spaced between 1/8″ and 3/16″.
2. The *above gap* is the vertical distance between the center of the nozzle orifice and the tips of the electrodes. This gap should be approximately 1/2″ to 5/8″ above the nozzle.

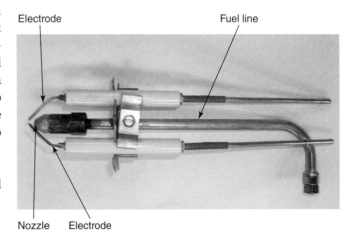

Figure 5-24. This image shows how the electrodes are aligned with the oil burner nozzle.

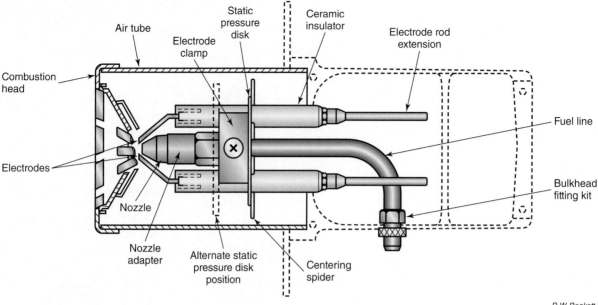

Figure 5-25. This cutaway of a gun burner air tube shows the burner nozzle and electrode assembly.

3. The *front gap* is the distance between the front of the nozzle and the tips of the electrodes. This gap should be between 5/16″ and 1/2″.

Figure 5-26 shows where these gaps are located between the electrodes and the nozzle.

TECH TIP

Setting the Electrode Gap

An electrode setting gauge tool can help set electrode gaps, **Figure 5-27**. To use this device, loosen the electrode assembly from the burner. Place the gauge over the end of the nozzle, with the corresponding spray degree rating of the nozzle in the upright position against the tip of the nozzle. Adjust the electrodes so they are seated against the gauge tool. Tighten the electrodes in place and remove the gauge tool. Simple as that!

5.4.6 The Ignition Transformer

The oil burner uses a step-up *ignition transformer* to generate the spark used by the electrodes to ignite the fuel oil in the combustion chamber. A step-up

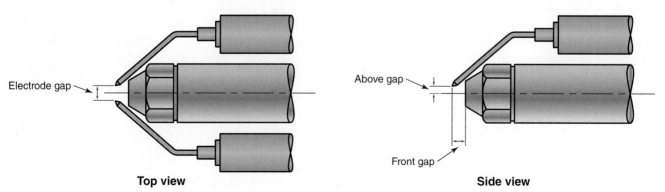

Figure 5-26. This diagram shows top and side views of electrode adjustment gaps.

transformer increases its voltage from the primary source to its secondary source. In older model oil burners, the ignition transformer located on top of the oil burner is constructed with an iron core and uses 120 V as its primary source of electricity, **Figure 5-28A**. This step-up transformer configuration is needed to generate the 10,000–14,000 V required to create the spark across the electrodes. The ignition transformer usually has springs that come in contact with the ends of the electrodes. Older model transformers are hinged so that the springs and the back ends of the electrodes can be accessed, **Figure 5-28B**.

Newer model oil-fired burners incorporate a solid-state igniter rather than the older, heavier iron core transformer. The solid-state igniter works in the same way as the older model ignition transformer, but it is much lighter and uses solid-state circuitry instead of multiple coils of wire to generate the same results. These newer, solid-state igniters usually generate a higher output voltage than the earlier models—typically between 14,000 and 20,000 V, **Figure 5-29**.

A

B

Goodheart-Willcox Publisher

Figure 5-27. A–An electrode setting gauge tool. B–The gauge tool fits over the nozzle.

SAFETY FIRST

Using the Right Tool for the Job

Be careful when working on high-voltage ignition generating equipment. The electrical arc produced by these devices is similar to that of a spark plug on a car's engine. It normally will not injure you, but the shock could cause you to react in a way that may result in an injury. When testing ignition transformers for proper voltage, use only a special high-voltage ignition transformer tester designed for this type of field examination. Do not use a conventional volt-ohm meter, or it could be irreparably damaged. Special igniter testers are also available for field testing solid-state igniters. Be sure to use the proper instrumentation when testing both conventional ignition transformers and solid-state igniters, and follow the manufacturer's instructions.

5.4.7 The Primary Control Unit

Every oil-fired burner has a ***primary control unit***, which is used both to control the boiler combustion process and to ensure a proper safety shutdown, **Figure 5-30**. The first duty of the primary control unit is to energize the burner circuitry when there is a call for heat. When the contacts close on the thermostat or aquastat, the primary control unit responds by energizing the oil burner's

A

B

Goodheart-Willcox Publisher

Figure 5-28. Figure A shows an older style ignition transformer on an oil burner. Figure B shows how this ignition transformer is hinged so that the springs and the back ends of the electrodes can be accessed.

Carlin Combustion Technology, Inc.

Figure 5-29. A solid-state igniter typically used with an oil-fired boiler.

Figure 5-30. Illustration of a primary control unit showing the red reset button.

blower and fuel pump, while the ignition transformer generates a spark across the electrodes. Upon ignition of the fuel oil, the task of the primary control unit is to prove ignition and maintain the combustion process. If for some reason the burner should fail to light, the primary control unit must sense this failure and lock out the burner circuitry to prevent large volumes of unburned fuel oil from accumulating within the combustion chamber. The primary control unit utilizes two different types of devices for proving ignition and maintaining constant combustion: the stack switch relay and the cad cell relay.

5.4.8 The Stack Relay

A safety device found more often on older boilers, the *stack relay*, or stack switch, is located in the flue (chimney) between the boiler and the barometric flue damper. It is a heat-sensing device used to de-energize the burner if it fails to detect heat within the flue piping or stack. The stack relay utilizes a bimetal strip that extends into the flue pipe and senses the flue gas temperatures to determine whether the burner has ignited, **Figure 5-31**. Upon sensing an increase in flue gas temperature, the bimetal strip warps, causing one set of electrical contacts to close and another set of contacts to open. These sets of contacts are respectively referred to as the *hot contacts* and *cold contacts* because they respond to whether there is heat or no heat in the flue. If the stack relay is energized by a call for heat from the burner and does not sense heat within 15 seconds, the hot contacts fail to close and the burner is locked out. **Figure 5-32** shows the respective wiring diagrams for a stack relay when there is an initial call for heat and when the ignition operation has been confirmed.

5.4.9 The Cad Cell Relay

Modern fuel oil boilers utilize a *cad cell relay* as part of the primary ignition controls rather than a stack relay. The cad cell is a light-sensitive semiconductor that is made up of a chemical compound known as cadmium sulfide. This compound responds to light by changing its ohms resistance. When exposed to darkness, the cad cell exhibits a high ohms resistance reading (100,000 Ω). When exposed to light, this resistance reduces substantially (under 1000 Ω). The cad cell is connected to the primary ignition control by two wires, **Figure 5-33**. When the oil burner ignites, the light from the flame causes the resistance through the cad cell to drop, allowing current to flow through the primary control circuit. This current is diverted around a triac and a safety switch heater within the primary controller, which allows the burner to maintain combustion. If the burner fails to ignite, the resistance through the cad cell remains elevated, which allows the safety switch to open after a period of time, shutting down the burner circuitry, **Figure 5-34**. A red button must be reset on the primary control unit before the burner is allowed to fire again.

Cad cells are mounted within the air tube of the burner control housing and should be checked periodically to ensure the "eye" remains clean. If the eye is obstructed—perhaps by soot from the burner—it can be cleaned with a soft cloth.

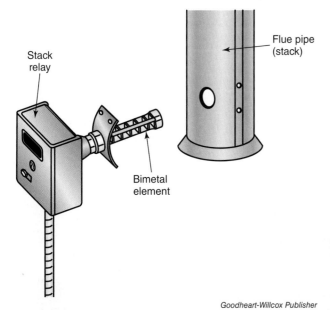

Figure 5-31. A stack relay is located in the flue pipe, where it senses heat from combustion.

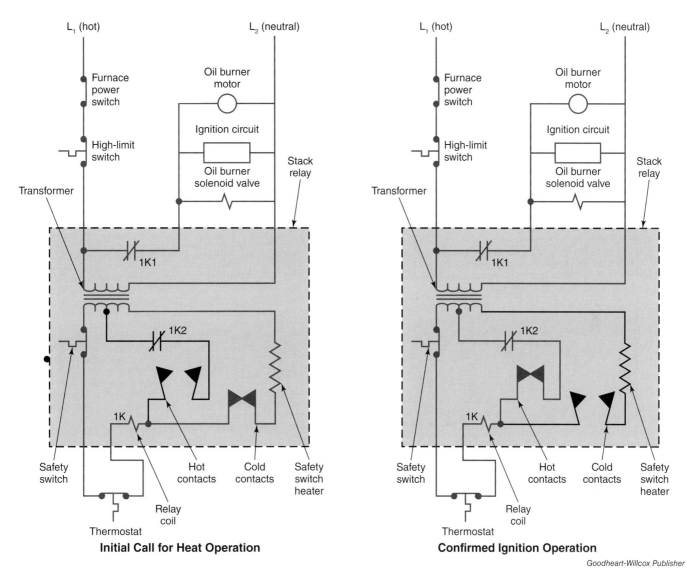

Figure 5-32. Diagrams that show the sequence of operation for a stack relay when there is (l) an initial call for heat, and (r) when there is confirmed ignition from the oil burner.

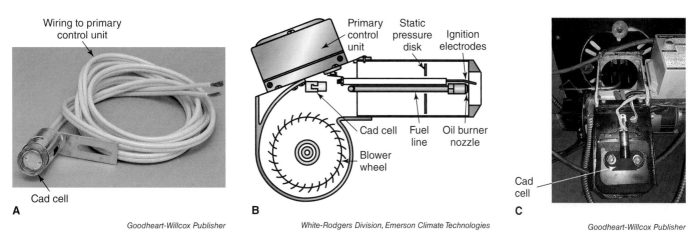

Figure 5-33. Figure A shows what a cad cell looks like before installation. Figure B shows the proximity of the cad cell placement in relation to the primary control unit and blower wheel. Figure C is a photo showing the position of the cad cell mounted on the inside of the ignition transformer.

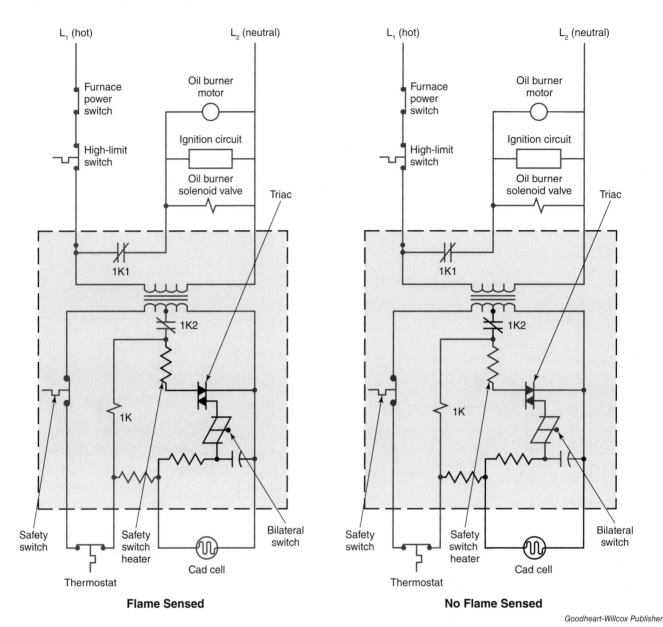

Figure 5-34. Diagrams that show the sequence of operation for a cad cell relay when the flame is sensed and when there is no flame sensed.

DID YOU KNOW?

Triac

A triac, or triode for alternating current, is a three-terminal electronic component that conducts current in either direction when triggered by a positive or negative signal at the gate electrode. Its formal name is *bidirectional triode thyristor*.

SAFETY FIRST

Beware of Excessive Fuel Accumulation

Do not attempt to reset the primary control unit if there have been several consecutive trials for ignition that have resulted in lockouts without ignition. When this happens, there is more than likely a large accumulation of oil within the combustion chamber or fire pot. If the oil burner happens to ignite when this condition exists, a large explosion can occur from an excess of unburned fuel. Remove the oil burner gun from the boiler's combustion chamber and allow the excess fuel to evaporate while checking to see why the primary control unit is locking out.

5.5 Fuel Oil Boiler Venting

Because of the large volume of air required for proper fuel oil combustion and venting (approximately 2000 cu ft per gallon), it is important that the flue piping from the boiler to the chimney and beyond be sized properly. In addition, oil-fired boilers must maintain a constant *draft* to allow for efficient operation. The term draft refers to the amount of vacuum or suction that exists inside the heating system. Draft is measured in inches of water column pressure (in. WC). This is the same measurement used to determine such things as natural gas or propane pressure measured at the boiler burner or the amount of pressure exerted within the ductwork of a forced-air furnace. Draft is affected by several factors:

- Outside air temperature
- Temperature of the flue gases
- Height of the flue or chimney

Because the first two factors constantly change, the amount of draft through the exhaust system will not remain constant. For instance, when the boiler is first started up, the flue is filled with cool gases and the amount of draft is minimal. After the boiler has operated for a period of time, the gases in the flue or chimney warm up, which increases the amount of draft. When the outside air temperature falls, the draft will again increase. Therefore, the level of draft within the boiler needs to be monitored and controlled. Without draft control, several issues arise:

- Too little draft reduces the delivery of combustion air to the burner, which can cause excessive smoke.
- Excessive draft can increase the volume of air delivery from the oil burner blower, which can raise the stack temperature and reduce operating efficiency.
- Increased levels of draft during the burner off-cycle increases standby heat losses through the flue and chimney, also reducing efficiency.

To fix these problems, the oil-fired boiler relies on a draft regulator. The most common type of draft regulator used with oil-fired systems is a barometric damper, which is installed in the flue stack between the boiler and the chimney or outside wall, **Figure 5-35**. This type of device is referred to as a bypass draft regulator. The barometric damper works by opening and closing its damper automatically using a counterweight. When the amount of draft becomes excessive, the damper swings open, allowing more ambient air to enter the flue. This indoor ambient air mixes with the flue gases and cools them, which reduces the temperature difference between the outdoor air and the flue gases. This, in turn, reduces the amount of draft through the system. When the level of draft decreases, the damper closes as needed. The boiler manufacturer will determine the guidelines for setting the proper amount of draft.

Field Controls, LLC

Figure 5-35. A barometric damper acting as a bypass draft regulator. The adjustable weight is used to set the appropriate amount of draft.

5.6 Combustion Analysis

As mentioned earlier, the by-products of perfect combustion are heat, water, and carbon dioxide. Because perfect combustion cannot be achieved in the field, a technician must obtain the suitable testing equipment to perform a proper

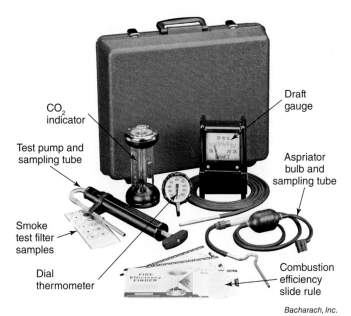

Figure 5-36. An older testing kit used to test products of combustion.

Figure 5-37. A technician using a modern combustion analyzer to measure products of combustion.

combustion analysis. The purpose of testing the products of combustion is to ensure that the boiler is operating safely and at peak efficiency.

In the past, technicians relied on several different devices to accurately perform a combustion test, **Figure 5-36**. These included a draft gauge, a CO_2 tester, a smoke tester, and a high temperature thermometer. Today, however, modern combustion analyzers are available that will provide accurate combustion testing results using a single piece of equipment, **Figure 5-37**. To perform a combustion analysis, the technician drills a small hole in the flue pipe at least 12 inches above the boiler, but below the draft regulator, and inserts the analyzer probe. When the boiler temperature reaches its steady state, the following tests and readings are performed and analyzed to ensure proper combustion is being achieved:

- Draft test
- Net stack temperature
- Oxygen reading
- Carbon dioxide reading
- Carbon monoxide reading

The ***draft test*** is performed to ensure that the products of combustion pass through the boiler at the proper rate and that enough air is supplied to support proper combustion. Draft is the movement of flue gases through the combustion chamber and venting. An excessive amount of draft can reduce efficiency by causing too much heat to travel through the venting too quickly. Too little draft can cause delayed ignition, resulting in excessive smoke that can fill the boiler room when the burner fires. Most residential and commercial oil-fired boilers require a negative flue draft between –0.04 in. WC and –0.06 in. WC. vacuum. The barometric flue damper can be adjusted to ensure the proper level of draft is achieved by adjusting the counterweight. If the draft reading is too low, the counterweight should be adjusted inward to increase the draft. If the reading is too high, move the weight outward to reduce the draft.

Net stack temperature is an indicator of proper combustion efficiency. The proper stack temperature on a medium efficiency oil-fired boiler should read between 330°F–450°F. Manufacturers will provide proper stack temperatures for individual boilers. The net stack temperature is determined by subtracting the ambient air temperature around the boiler from the measured flue gas temperature. To accurately measure the stack temperature, be sure that the boiler has operated for at least 5 minutes or until the stack thermometer has reached a steady state. A higher than normal stack temperature can indicate that the boiler is not operating at peak efficiency. This may be caused by excessive soot on the heat exchanger (which will reduce the amount of heat being transferred from the flame) or from a nozzle that is too large for the burner's heating capacity. Lower than normal stack temperatures—which may be due to an undersized burner nozzle—will result in a reduced firing rate.

The *oxygen reading* should be between 3.0% and 6.0%. To adjust this reading, open or close the air bands on the oil burner to deliver the least amount of oxygen for the best efficiency. If the burner has too much air or an inadequate amount of fuel, the mixture will be too lean, and efficiency will be lowered. If the burner has inadequate air and too much fuel, the mixture will be too rich, which can result in excessive soot and high levels of carbon monoxide.

Carbon dioxide readings are displayed as a percentage of flue gas volume and indicate the efficiency of the burner. Most oil-fired boiler manufacturers desire a CO_2 reading of between 10.0% to 12.5%. Lower than normal CO_2 levels indicate incomplete combustion. When this occurs, the technician should check for the proper nozzle size, the proper amount of combustion air, and the proper oil pressure. Causes of higher than normal CO_2 readings include insufficient draft or the burner overfiring due to an incorrect nozzle size.

The *carbon monoxide reading* is measured in parts per million (ppm) and should be less than 100 ppm—air free. "Air-free" CO readings are calculated by the combustion analyzer to determine what the CO concentration in the flue gas would be if all of the excess air were removed. Higher than normal CO readings indicate incomplete combustion. This can occur due to an oversized nozzle, inadequate combustion air, or excessive oil pressure.

Most boiler manufacturers publish the desired combustion analysis data in their installation and startup guide. In addition, many manufacturers of combustion analyzers can provide data for proper combustion results based on the size and efficiency standards for most boilers.

TECH TIP

Filling the Hole after the Combustion Test

Once the combustion analysis procedure is complete, the technician will be left with a hole in the boiler's exhaust flue that must be properly sealed.

One method to safely seal this hole is by using a stainless-steel plug. First, coat the threads of the plug with high-temperature silicone sealant. Then insert the plug until it is firmly seated.

Another suggestion is to use a hex-head countersunk plug, **Figure 5-38**. Again, coat the threads of the plug before inserting it into the hole. Then, as an added safety measure, coat the plug hole with the same silicone sealant. This will not make it tamperproof, but it will be tamper resistant.

Sabelskaya/Shutterstock.com

Figure 5-38. This is an example of a hex-head countersunk plug used to close the hole in the vent pipe after combustion analysis has been completed.

Chapter Review

Summary

- Fuel oil is considered a fossil fuel and is refined from crude oil.
- Number 2 fuel oil is the most common grade used in residential and light commercial boilers. It has a heating value of approximately 140,000 Btu/gallon.
- Important fuel oil characteristics include: flash point, ignition point, distillation quality, viscosity, pour point, carbon residue, ash content, and water and sediment content.
- In order for correct combustion to take place, there must be the correct proportion of fuel, oxygen, and heat.
- The correct amounts of primary and excess air are important to proper combustion and venting in a fuel oil system.
- Fuel oil must be atomized through a nozzle before it can mix with air and ignite for proper combustion.
- A one-pipe fuel delivery system is used when the storage tank is above the burner. One-pipe systems are more prone to air-related problems but require less tubing and labor for installation.
- A two-pipe fuel delivery system has separate supply and return lines and is used when the storage tank is below the burner. Two-pipe systems are self-bleeding and less likely to experience air-related problems.
- An oil deaerator is used to remove air from the fuel lines and can be used to convert a two-pipe system to a one-pipe system.
- Fuel line filters use various types of filter media and are used to remove impurities from the fuel oil.
- Oil burners consist of a motor, blower, fuel pump, nozzle, electrodes, ignition transformer, and primary control unit.
- The fuel oil pump may be a single-stage or two-stage pump. Two-stage pumps are used to assist in lifting the oil to the burner when the fuel oil storage tank is located below the burner.
- The nozzle is used to atomize the oil for proper combustion and is rated by the type of spray pattern, spray angle, and gallons per hour flow.
- Two electrodes create a spark used to ignite the fuel oil in the combustion chamber.
- The ignition transformer is a step-up type transformer that produces approximately 10,000 volts used to generate a spark between the electrodes.
- The primary control unit is used to control and supervise the combustion process in the fuel oil heat exchanger.
- Two types of relays are used to supervise fuel oil flames: a stack relay and a cad cell relay. A stack relay uses a bimetal element mounted in the flue to sense heat produced by combustion. The cad cell relay uses a light-sensitive device to determine when combustion occurs by sensing the flame.
- The appropriate amount of draft through the boiler is essential to ensure that proper combustion takes place and that the products of combustion are safely vented from the building.
- A barometric damper—located in the flue pipe just above the boiler outlet—is used as a bypass draft regulating device in most modern fuel oil systems.
- Proper combustion analysis is important to ensure that the boiler is operating safely and at peak operating efficiency. The elements analyzed during a combustion test are: draft levels, oxygen levels, carbon dioxide levels, carbon monoxide levels, and net stack temperature.

Know and Understand

1. The process of refining fuel oil from crude oil is a distillation process known as _____.
   A. fracking
   B. cracking
   C. filtering
   D. None of the above.
2. *True or False?* Lower numbered fuel oils are identified as *heavy oils*.
3. The most common type of fuel oil used for residential and light commercial boilers is _____.
   A. Number 1
   B. Number 2
   C. Number 3
   D. Number 6
4. The heating value of Number 2 fuel oil is approximately _____.
   A. 120,000 Btu/gallon
   B. 120,000 Btu/cubic foot
   C. 140,000 Btu/gallon
   D. 140,000 Btu/cubic foot
5. The ability of fuel oil to become vaporized is referred to as the fuel oil's _____.
   A. ash content
   B. distillation quality
   C. flash point
   D. viscosity
6. *True or False?* In order for combustion to occur, there must be the correct amounts of oxygen, carbon dioxide, and a fuel source.
7. Excess air for combustion consists of _____ and _____ air.
   A. secondary and dilution
   B. primary and secondary
   C. primary and dilution
   D. carbon monoxide and carbon dioxide
8. *True or False?* In order for fuel oil to be ignited, it must be delivered to the burner as a vapor.
9. A(n) _____ can be used to convert a two-pipe delivery system to a one-pipe delivery system.
   A. fuel line filter
   B. atomizer
   C. cad cell
   D. oil deaerator
10. Oil burner motors are typically _____ or _____ motors.
    A. single-phase or three-phase
    B. capacitor-start or capacitor-run
    C. split-phase or PSC
    D. semi-hermetic or shaded-pole
11. Most of today's fuel oil pumps are designed to deliver _____ psig.
    A. 50 to 100
    B. 100 to 140
    C. 150 to 200
    D. 200 to 250
12. *True or False?* Solid cone nozzles distribute a spray pattern with fuel oil droplets that are uniform across the cone's shape.
13. Solid-state ignition transformers are designed to deliver _____ from the secondary side of the transformer.
    A. 1000 to 10,000 volts
    B. 10,000 to 15,000 volts
    C. 14,000 to 20,000 volts
    D. 20,000 to 25,000 volts
14. *True or False?* The stack relay circuitry changes resistance with a change in temperature.
15. What type of draft regulator is most often used with today's fuel oil boilers?
    A. Barometric damper
    B. Pressure switch
    C. Face and bypass damper
    D. Modulating motorized damper
16. *True or False?* When performing a combustion analysis, CO_2 levels that are higher than normal indicate incomplete combustion.

Apply and Analyze

1. What two chemical elements make up fuel oil?
2. What is the difference between fuel oil's flash point and ignition point?
3. What is the measurement by which fuel oil's viscosity is expressed?
4. How many pounds of oxygen are required to combust one pound of Number 2 fuel oil?
5. Explain what role excess air plays in the combustion and venting process.
6. Describe how a nozzle atomizes fuel oil.
7. Which type of piping delivery system should be used if there is a problem with constant air in the lines?
8. What device is used to remove water and sediment from the fuel oil supply tank?
9. List and describe the three different markings that are found on a fuel oil nozzle.
10. Explain the three specific gaps that need to be checked when setting up the oil burner electrodes.
11. How do solid-state igniters compare to older type ignition transformers?
12. What is the approximate ohms resistance of a cad cell when it is exposed to a light source?
13. What items should be checked when a combustion analysis reveals that there is incomplete combustion through the boiler?

Critical Thinking

1. Would a one-pipe or two-pipe delivery system be more advantageous if the fuel oil tank were a great distance away from the boiler? Explain why.
2. What would be the appropriate nozzle rating in GPH for a fuel oil boiler that is rated for 112,000 Btu/hr? Explain your answer.
3. Which device do you think is more reliable, a stack relay or a cad cell relay? Explain.

6 Boiler Fittings and Air Removal Devices

Chapter Outline
6.1 Pipe Fittings
6.2 Pressure and Temperature Gauges
6.3 Expansion Tanks
6.4 Air Removal Devices

Mironmax Studio/Shutterstock.com

Learning Objectives

After completing this chapter, you will be able to:
- List the different types of fittings used in hydronic heating systems.
- Identify different types of copper tubing and ways in which copper tubing is joined together.
- Explain how PEX tubing is used in a hydronic system.
- Define the specialty fittings used in a hydronic system and how they are installed.
- Explain why PVC piping may not always be the proper material used for boiler venting.
- Describe how pressure/temperature gauges are used for system monitoring.
- Compare the different types of expansion devices used with hydronic heating systems.
- Calculate expansion tank sizing and proper expansion tank pressure.
- Describe why air removal is essential to the hydronic heating system.
- Compare the difference between an air separator and an air scoop.
- Differentiate between types of air vents.

Technical Terms

adapter
air lock
air scoop
air separator
air vent
ASHRAE Handbook of Fundamentals
cavitation
cross-linked polyethylene (PEX) tubing
dielectric union
drain-waste-vent (DWV) piping
expansion tank
galvanic corrosion
national pipe thread taper (NPT)
polyvinyl chloride (PVC) piping
pressure/temperature gauge
Schedule 40 piping
water hammer
waterlogged

Hot water boilers incorporate different types of piping material. A single type of material may be used exclusively for a particular boiler application, or different types of piping material may be used together to complete a particular boiler project. Whether a boiler application calls for copper, black pipe, or PEX piping, each type uses similar pipe fittings to complete the job.

Once the boiler is completely installed and filled, one of the key factors to a successful startup is the removal of air from the system. Several different devices can isolate and remove air from a hydronic system.

This chapter discusses the various types of fittings used in boiler installation and compares the use of compression tanks and expansion tanks for water expansion. Furthermore, it analyzes how trapped air is separated and removed from the hydronic system.

6.1 Pipe Fittings

Pipe fittings used in hot-water residential and commercial boiler applications are similar to those used for domestic water supply systems and gas piping and venting systems. However, different types of piping and tubing require different joining methods, and the installation technician should be familiar with each material as well as the correct procedures for joining and fitting copper, steel, and PEX tubing. Proper joining and fitting practices are covered in Chapter 15, *Boiler Installation*. In this chapter, we cover various types of fittings used in boiler installations.

6.1.1 Iron and Steel Pipe Fittings

The most common type of steel piping used for boiler installations is Schedule 40 black steel pipe. ***Schedule 40 piping*** can be used for water or gas piping. This type of piping uses ***national pipe thread taper (NPT)*** as a standard for the tapered threads used in the joining of pipe fittings. These tapered threads pull together tightly when the fitting and piping are joined together, thus making a fluid-tight seal. The fittings used for this type of piping include tees, elbows, couplings, caps, plugs, unions, bushings, and reducers, **Figure 6-1**.

- **Tees** come in standard and reducing forms. On a standard tee, all three ports are the same size. A reducing tee includes one or two ports of different sizes and is named for its *side*-by-*side*-by-*center* (or top) dimensions. For instance, a 3/4″ by 3/4″ by 1/2″ tee has left and right diameters that measure 3/4″ and a top diameter that measures 1/2″.
- **Elbows** have either a 45° or 90° radius and are categorized as either standard elbows or street elbows. Both ports on a standard elbow have female threads, or threads on the inside of the fitting. A street elbow has one port with male threads (threads on the outside of the fitting) and one port with female threads.
- **Couplings** are used to join two pipes of the same size. Couplings have female threads on both ends.
- **Caps** are used to seal the male thread end of a single pipe, such as on the end of a drip leg.
- **Plugs** are used to seal the end of a female fitting, such as a tee or coupling.
- **Unions** are very similar to a coupling, except that they can be unscrewed into two separate pieces and then screwed back together when both separate ends are attached to the two pipe ends that need to be joined.
- **Bushings** connect two dissimilar dimensioned pipes. They have both internal and external threads (male and female threads).

Mueller Industries, Inc.

Figure 6-1. These images show various types of steel pipe fittings used with hydronic lines and gas piping.

- **Reducers**, or *bell reducers*, are also used for joining two dissimilar dimensioned pipes but are bell-shaped couplings and have internal or female threads on both ends. They are also known as *reducing couplings*.

> **CODE NOTE**
>
> **Pipe Reducers versus Bushings**
>
> Although bushings may be more convenient to use than bell reducers in some circumstances, be aware that the International Fuel Gas Code (IFGC) prohibits the use of bushings with gas piping installations. The threaded bushing increases both friction loss and turbulence through the pipe, and there is a greater chance for the threads in a bushing to break down, causing a gas leak.

6.1.2 Copper Tubing Fittings

Copper tubing is available in three different wall thicknesses: K (heavy wall), L (medium wall), and M (light wall). Type M copper tubing is acceptable for most boiler water piping applications. Copper tubing is available in soft and hard tempers (hardness). Soft copper is usually purchased in 50′ and 100′ rolls, while hard copper is available in 10′ and 20′ lengths, **Figure 6-2**. Most supply-and-return water lines used for boiler applications incorporate hard-drawn copper. Joining copper tubing is accomplished by flaring, soldering, or brazing. Soldering, or sweating, is the most widely practiced method of joining copper tubing for hydronic heating systems. Installation techniques will be covered in Chapter 15, *Boiler Installation*.

Copper fittings are available with solder-, or sweat-type socket ends, threaded ends, and a combination of both. Fittings that have both socket and threaded ends are referred to as or ***adapters*** or *connectors*, **Figure 6-3**. These can join together two dissimilar piping materials. Just as with Schedule 40 black steel pipe, copper fittings are available as tees, elbows, caps, plugs, unions, bushings, and reducers, **Figure 6-4**.

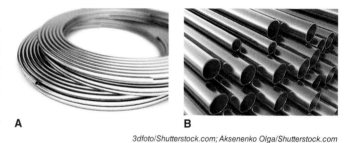

3dfoto/Shutterstock.com; Aksenenko Olga/Shutterstock.com

Figure 6-2. A—Soft rolls. B—Hard-drawn lengths.

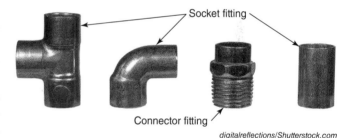

digitalreflections/Shutterstock.com

Figure 6-3. Copper fittings that have both socket and threaded ends are referred to as adapters, or connectors, and are used where two dissimilar piping materials are to be joined together.

6.1.3 PEX Tubing Fittings

PEX is the reference name for ***cross-linked polyethylene tubing***, **Figure 6-5**. Very popular for hydronic applications such as in-floor radiant heating, PEX is very durable and can withstand repeated heating and cooling cycles. It is also resistant to chemicals, which can cause scaling and corrosion in iron piping and copper tubing. PEX is also very flexible and relatively easy to join and install.

PEX tubing offers a wide variety of flexible fittings that can join PEX-to-PEX tubing as well as PEX to dissimilar materials. PEX fittings are available as tees, elbows, couplings, and reducers. PEX can be connected to respective fittings by:

- Crimp connections: these fittings require a crimping tool and crimp rings.
- Push-to-connect fittings: these fittings are the simplest to install and need no special tools.

Figure 6-4. Copper fittings are similar to those used for hydronic piping and gas lines.

Figure 6-5. PEX tubing has become very popular because of its durability and flexibility.

- Expansion connections: these fittings require an expander tool and PEX or brass sleeves.
- Press connections: these fittings require a press tool and steel sleeves.
- PEX fittings are available in brass, bronze, and poly (plastic) materials, **Figure 6-6**.

6.1.4 Specialty Fittings

In some hydronic systems, it may be desirable to join copper tubing directly with Schedule 40 black piping. Although this practice is generally acceptable in a sealed hydronic system—in which the same water is used continuously—in some instances, it could cause problems. If the boiler water is slightly conductive, a process called *galvanic corrosion* (also known as *bimetallic corrosion*) could occur. Boiler water can become conductive if it contains sodium, chloride, calcium, or magnesium ions. These ions increase the water's ability to conduct electricity. Galvanic corrosion is an electrochemical process in which one metal is more likely to corrode when it is in contact with a dissimilar metal in the presence of this conductive water. This phenomenon will cause iron or steel components to prematurely corrode and eventually fail.

To avoid this situation, the use of a *dielectric union* is recommended. This special fitting has a threaded steel female fitting on one end and a female copper or brass sweat fitting on the other end. The fittings are joined with steel to steel

Figure 6-6. PEX fittings are available in several materials. A—Brass. B—Bronze. C—Poly (plastic) type materials.

on one side and copper to copper (or copper to brass) on the other side. The two sides of the union are separated by a synthetic O-ring and a plastic sleeve. The sleeve and O-ring create a barrier so the two dissimilar metals cannot come into direct contact with each other, **Figure 6-7**. Besides allowing the joining of copper to steel, the dielectric union provides easy disassembly of the lines for any future work on the system.

Photo courtesy of Watts

Figure 6-7. Dielectric unions are used to join dissimilar metals together.

DID YOU KNOW?

Joining Dissimilar Metals

Joining copper to black pipe within a hydronic system is more common than not, and it generally does not create a problem in a closed system. This is due to a lack of oxygen in the water, meaning an electrolytic reaction cannot occur. Because the same water typically is used for many years, any dissolved oxygen in the system is quickly depleted and expelled. In fact, there are homes with black steel pipes from around 1910 that were mixed with copper in the 1960s that still have no problems today!

6.1.5 PVC Pipe Fittings

Polyvinyl chloride (PVC) pipe is also available as tees, elbows, caps, plugs, unions, bushings, and reducers, **Figure 6-8**. PVC piping is typically used for condensate drains on high-efficiency boilers, and in some cases, it is still used for venting materials on high-efficiency boilers. Specifically, the type of PVC piping that is used for condensate drains and for venting is classified as *drain-waste-vent (DWV) piping*, **Figure 6-9**. PVC is bonded with a glue that consists of either cyanoacrylates or UV-curable adhesives. Just as with copper fittings, PVC fittings are manufactured with glue-type socket ends, threaded ends, or a combination of both. Boiler manufacturers may offer threaded or glued connectors for condensate drains, which typically protrude from the boiler for easy installation.

Nor Gal/Shutterstock.com

Figure 6-8. This image shows the different types of PVC fittings available.

A

B

Goodheart-Willcox Publisher

Figure 6-9. PVC piping is used for both of these components. A—Condensate drains. B—Venting on high-efficiency boilers.

> ## CODE NOTE
>
> ### PVC versus Polypropylene
> Although the use of PVC as a venting material for boilers is still approved in most states in the United States, some states as well as Canada have outlawed this practice. One reason for this is because the maximum recommended operating temperature for PVC piping is 140°F, which actual boiler flue gas temperatures may approach in some situations. To remedy this issue, some manufacturers are requiring the use of *polypropylene vent piping* (PPV) in lieu of traditional PVC piping. As stated in the International Fuel Gas Code: "*Plastic pipe and fittings used to vent appliances shall be installed in accordance with the appliance manufacturer's installation instructions.*" Be sure to follow these instructions when installing or replacing any boiler.

6.2 Pressure and Temperature Gauges

The ability to quickly reference the supply temperature and pressure within the boiler at any given time is essential. To accomplish this, most boiler manufacturers provide pressure and temperature gauges—either as a combination gauge or separately—with the boiler package. The ***pressure/temperature gauge*** provides readings that indicate whether the boiler is operating normally, **Figure 6-10**. These gauges may come preinstalled on the boiler or shipped loose to be installed separately. When installing the pressure/temperature gauge, mount it near the outlet or on the supply line leading out of the boiler. These gauges must be easily within sight of the operator for monitoring of the proper temperature and pressure.

Pressure/temperature gauges can be round or square in shape, and they measure water pressure in pounds per square inch (psi) and kilopascal (kPa) and temperature in degrees Fahrenheit (°F) and degrees Celsius (°C). Each gauge should be encased in brass or bronze material that is durable and corrosion resistant. Boiler pressure/temperature gauges have a minimum accuracy of +/–3%. These gauges assist in diagnosing such problems as pressure relief valve leaks, boiler leaks, and heating system piping leaks.

Goodheart-Willcox Publisher
Figure 6-10. Pressure/temperature gauges provide readings that determine whether the boiler is operating normally.

> ## GREEN TIP
>
> ### Trusting Your Gauges
> In order for technicians to make sound judgments regarding troubleshooting diagnoses, they must rely on the accuracy of their gauges and instrumentation. Instruments in poor working condition or out of calibration will not accurately measure system conditions and can lead to false determinations. Technicians should periodically calibrate their test instruments and gauges to ensure they are in sound working order. This will ensure that the system is functioning at maximum efficiency. Similarly, tools and equipment in poor condition can result in improper repairs and increased waste.
>
> Trust your gauges, but first make sure they are accurate!

6.3 Expansion Tanks

A typical hydronic system is considered a closed-loop system, meaning it is not open to the atmosphere and ideally is free of any excess air. However, even in an "air-free" system, when the boiler water is heated, it naturally expands. If the system does not allow for this expansion, a pressure relief valve will pop open whenever the boiler operates—or worse, piping could burst or rupture. To prevent these situations, an *expansion tank* must be installed with every hydronic system to allow for the expansion of water as it is heated.

6.3.1 Compression-Type Expansion Tanks

Older hydronic systems incorporated a compression, or standard, expansion tank. This vessel is nothing more than a steel tank with a cushion of air at the top, **Figure 6-11**. These types of tanks are designed to absorb the expansion forces of heated water and to control the pressure within the hydronic system. They are usually mounted above the boiler, close to the ceiling of the mechanical room. When the boiler is initially filled, air is allowed to take up about 1/3 to 1/2 of the area at the top of the standard expansion tank. A sight glass located on the side of the tank will indicate the water level. As the boiler water is heated and expands, it moves into the tank and squeezes against this cushion of air, **Figure 6-12**.

One issue with the compression tank is that the water and air within the tank are in direct contact with each other. Because of this, each time the boiler cools down after cycling off, some of the air in the compression tank is reabsorbed into the boiler water's system. Eventually the water takes the place of the air cushion and the tank completely fills with water. This condition is known as a *waterlogged* tank. When this situation occurs, there is no longer room for the heated boiler water to expand. One of the main symptoms of a waterlogged compression tank is a pressure relief valve that intermittently leaks water after the boiler has run for a period of time. This loss of water will be replaced by the boiler's makeup water system; however, if this situation is allowed to continue over a long period of time, this oxygen-enriched makeup water can eventually lead to excessive corrosion within the boiler

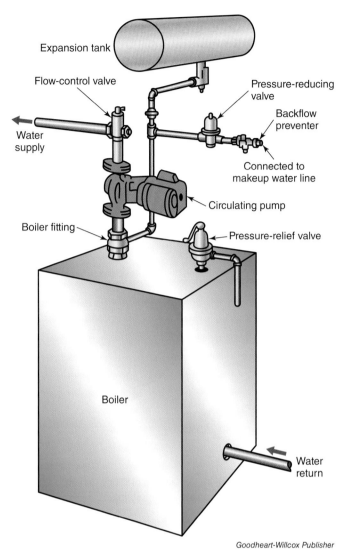

Goodheart-Willcox Publisher

Figure 6-11. Older compression-type expansion tanks (as seen here) are basically a steel tank with a cushion of air at the top.

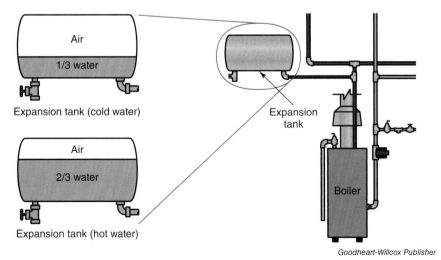

Goodheart-Willcox Publisher

Figure 6-12. Conventional compression tanks are usually located near the top of the ceiling in the boiler room. As the boiler water is heated, it expands against the cushion of air at the top of the tank.

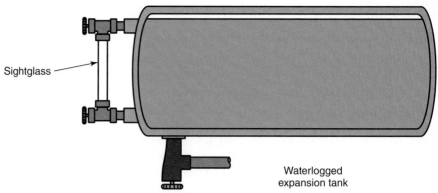

Figure 6-13. Air in the compression tank is reabsorbed into the boiler water's system. The water takes the place of the air cushion and the tank can become waterlogged.

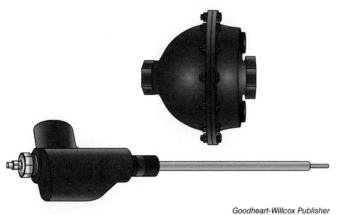

Figure 6-14. This image shows an Airtrol fitting, which is used on a compression tank to alleviate waterlogging.

piping, **Figure 6-13**. Fresh makeup water has a much higher level of oxygen because it has not been removed over time by the system's air removal devices, which are discussed later in this chapter.

To remedy this waterlogging situation, a special fitting is used with compression expansion tanks. This fitting—sometimes called an *Airtrol* fitting—provides a dip tube that extends from the body of the fitting up into the top of the compression tank, **Figure 6-14**. Any air that is trapped within the boiler water is allowed to separate and is drawn up the dip tube to maintain the proper air level within the tank. Furthermore, an air vent at the bottom of the fitting allows the technician to replenish the tank with additional air as necessary, **Figure 6-15**.

DID YOU KNOW?

How an Airtrol Fitting Works

The Airtrol fitting is used in older compression-type expansion tanks to prevent waterlogging. As shown in **Figure 6-15**, one end of the fitting is connected to the supply water piping, and the end with the dip tube fits into the tank. Any air migrating from the supply water into the fitting will be forced up the dip tube and into the top of the expansion tank. This not only purges unwanted air from the boiler water but also helps maintain the proper amount of air in the top of the tank. If too much water enters the tank, a second tube inside the dip tube is connected to an air vent at the bottom of the fitting to allow for more air to be introduced to the tank once the excess water has been drained off.

6.3.2 Diaphragm-Type Expansion Tanks

Modern hydronic systems incorporate a diaphragm-type expansion tank to allow for hot water expansion, **Figure 6-16**. A diaphragm expansion tank separates air from the boiler water with a flexible rubber membrane within its shell. On one side of the membrane, the tank contains compressed air. The other side of the membrane is exposed to the boiler water. As the boiler water is heated, it pushes against the membrane and flexes against the compressed air, **Figure 6-17**.

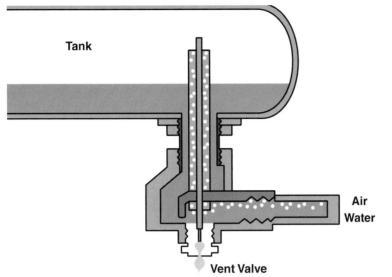

Figure 6-15. On the compression tank Airtrol fitting, any trapped air in the boiler is drawn up through the dip tube to maintain the proper air level within the tank.

Figure 6-16. Illustration of a modern-type expansion tank.

Diaphragm expansion tanks have several advantages over conventional compression tanks:

- Because the air and water are not in physical contact with each other, there is no risk of the diaphragm tank becoming waterlogged.
- Because waterlogging is eliminated, there is a reduced risk of pipe corrosion due to excessive makeup water entering the system.
- Diaphragm tanks have an air pressure valve that allows the technician to check the expansion tank and fill it with air if necessary.
- No additional special vent valve fittings are necessary on a diaphragm tank.
- The diaphragm tank can be mounted in any position.

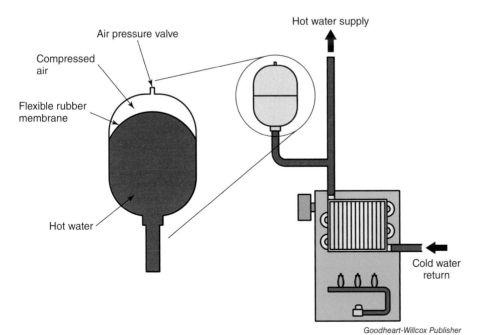

Figure 6-17. A diaphragm expansion tank separates air from the boiler water by using a flexible rubber membrane within its shell.

> **PROCEDURE**
>
> **Recharging the Diaphragm-Type Expansion Tank with Air**
> If the expansion tank should need to be recharged with air, the proper steps to complete this task include the following:
> 1. Shut off the boiler and close the isolation valve between the tank and the boiler water supply.
> 2. Allow the tank to cool down before proceeding.
> 3. Open the water drain valve on the tank and drain the tank until it is empty.
> 4. Close the drain valve.
> 5. Attach a tire pressure gauge to the air fitting connected to the expansion tank and check the pressure reading against the manufacturer's requirements.
> 6. If the air pressure reading is low, add air using an air compressor or hand pump and fill the tank to the required pressure level.
> 7. Open the isolation valve and refill the tank with water.
> 8. Re-energize the boiler and allow it to run for several hours.
> 9. Check the air pressure on the expansion tank again to ensure that it is within the proper range.
>
> Note: Should the diaphragm within the tank fail, the air cushion will be eliminated, which can lead to the pressure relief valve popping open each time the boiler fires.

6.3.3 Sizing Expansion Tanks

Hydronic equipment suppliers will usually select the correct expansion tank to match the system being installed. However, there may be instances when the existing expansion tank needs to be checked to ensure the right one has been selected. To effectively size an expansion tank, you need to know the system operating pressure and the boiler capacity in gallons.

While manufacturer's selection tables and rule-of-thumb sizing approximations may help you arrive at a general size range, these sizing methods require additional safety margins, often leading to the selection of a tank larger than what is needed. To accurately size a diaphragm expansion tank, use the following formula, which is based on the ***ASHRAE Handbook of Fundamentals***:

$$Vt = \frac{(0.00041t - 0.0466) \times Vs}{1 - (Pf + 14.7 / Po + 14.7)}$$

where

Vt = minimum expansion tank volume in gallons
t = maximum average operating temperature
Vs = system water volume in gallons
Pf = minimum operating pressure at the tank expressed in psia (absolute)
Po = maximum operating pressure at the tank expressed in psia (absolute)
14.7 = added to Pf and Po to obtain absolute pressure

Example:

A given system holds a total of 500 gallons of water at a maximum average operating temperature of 180°F. The minimum operating pressure at the tank is 12 psig, and the maximum operating pressure is 30 psig.

> **TECH TIP**
>
> **Understanding PSI, PSIG, and PSIA**
> Pounds per square inch (psi) is the unit that is used most of the time in the United States for measuring air, water, gas pressure, and so on, and is the unit displayed on pressure gauges.
>   Pounds per square inch in gauge, or psig, is the pressure measured using a gauge or other pressure measuring device. It is the difference between the pressure inside a pipe or tank and the atmospheric pressure.
>   The term used to describe absolute pressure in psi is psia, which includes atmospheric pressure (14.7 psi at sea level). It is sometimes referred to as total pressure.

$$Vt = \frac{[(0.00041 \times 180) - 0.0466] \times 500}{1 - (12 + 14.7/ 30 + 14.7)}$$

$$Vt = \frac{13.6 \text{ (gallons of expanded water)}}{0.403 \text{ (acceptance factor)}}$$

$$Vt = 33.7 \text{ gallons}$$

Notes on estimating water volumes:
- The volume of water in the boiler can be referenced from the manufacturer.
- Following are some estimations for calculating water volumes in various terminal units:
    - Cast-iron radiators require about 1.08 gallons of water per 1000 Btu
    - Fan-coil units require about 0.6 gallons of water per 1000 Btu
    - Baseboard heaters require about 0.46 gallons of water per 1000 Btu
    - Radiant-floor heating requires about 1.8 gallons of water per 1000 Btu
- Following are guidelines for estimating the volume of water in piping:
    - 1/2″ water lines = 1.02 gallons per 100′ of pipe
    - 3/4″ water lines = 2.06 gallons per 100′ of pipe
    - 1″ water lines = 3.46 gallons per 100′ of pipe
    - 1-1/4″ water lines = 5.16 gallons per 100′ of pipe
    - 1-1/2″ water lines = 7.00 gallons per 100′ of pipe
    - 2″ water lines = 12.34 gallons per 100′ of pipe

6.3.4 Calculating Expansion Tank Pressure

Following are the proper steps for determining the required pressure for a diaphragm expansion tank:
1. Measure the vertical distance in feet between the inlet of the expansion tank and the highest piping in the system.
2. Use the following formula to calculate the required tank pressure:

$$P_{tank} = (H \div 2.31) + 5$$

where

P_{tank} = required expansion tank pressure in psi
 H = the vertical distance in feet between the inlet of the expansion tank and the highest pipe in the system
 2.31 = the amount of vertical lift in head for each pound of pressure (review Chapter 7, *Hydronic Piping Systems* to understand the relationship between system pressure and feet of head)
   5 = the amount of additional pressure added to the system to ensure there is enough positive pressure at the highest point in the system

Example:
A residential boiler is located in the basement of a house. The opening of the expansion tank is 5′ from the basement ceiling. The residence has two floors, and each floor has 8′ ceilings. The highest floor uses radiant wall panels with the piping extending 3′ from the floor.

$H = 5' + 8' + 3' = \mathbf{16'}$
$P_{tank} = (H \div 2.31) + 5$
$P_{tank} = (16 \div 2.31) + 5$
$P_{tank} = 6.93 + 5$
$P_{tank} = 11.93 \text{ psi}$

This example shows why most expansion tank manufacturers size their tanks based on 12 psig for residential and light commercial usage.

6.4 Air Removal Devices

A perfect hydronic system is free of all air. However, even with the most sophisticated air eliminating devices installed in a hot water system, trapped air can still become an issue. Air might enter the hydronic system for several reasons:

1. When the hydronic system is initially filled, it must be properly purged of any excess air. If not, this air can become trapped within the system and cause problems.
2. The water used to fill the system is rich with oxygen. This oxygen separates itself from the water and becomes trapped air within the system.
3. Whenever water escapes from the system—either from piping leaks, the relief valve popping open, or simply from attrition—makeup water replaces the lost water and more oxygen enters the system.

If trapped air is not removed from the hydronic system, several issues can result. Trapped air can:

- Block water from being circulated
- Cause undesirable noise
- Cause corrosion in pipes
- Result in pump cavitation

Sometimes referred to as an *air lock*, trapped air within the hydronic system can be very frustrating to resolve. This issue can lead to poor water circulation within the system or even a complete loss of flow within a zone, which results in poor heat transfer.

Trapped air causing undesirable noise within the system is a condition known as *water hammer*. It is caused by a pressure surge when the water within the system is forced to stop or suddenly change direction. Air pockets within the hydronic piping system can act as acoustic chambers, elevating these noise levels as the water flows through the piping.

Air contains oxygen. When dissolved oxygen comes in contact with a ferrous metal such as iron, oxides (rust) form. This can lead to pitting on the inside wall of the hydronic piping, resulting in corrosion. If left unchecked, these oxides can lead to sludge within the system, which can cause even greater problems.

Cavitation is a result of bubbles or cavities forming within the boiler water. This form of cavitation in the water develops at the inlet of the circulation pump, where pressure is low. Water pressure increases as it travels through the circulator pump. As the air bubbles travel through the pump, the increased water pressure causes them to collapse. This leads to powerful shock waves inside the pump, causing damage to the pump's impeller and pump housing. Pump cavitation can be identified by an increased level of noise, which can sound like marbles rattling around through the pump.

In order to prevent these problems, an effective hydronic system includes a two-step approach to removing air: separate the air and then remove the air by venting it from the system.

6.4.1 Air Separators and Air Scoops

There are two types of devices commonly used to separate air from water within the hydronic heating system:
- Air separators
- Air scoops

As the name implies, an ***air separator*** is used to separate air from the water as it flows through the hydronic system. The latest generation of air separators incorporates a metal mesh screen inside. This screen causes turbulence within the circulated water. This agitation causes oxygen within the water to form tiny bubbles. These microbubbles then adhere to the mesh screen. As the microbubbles build up on the screen, they join together, making larger bubbles, and eventually break loose. Because air is lighter than water, the bubbles travel up the separator and through the built-in air vent, where they are vented from the system, **Figure 6-18**.

Air scoops are air separation devices with a series of deflectors or baffles, which also create turbulence as the circulated water passes through them. When the water comes in contact with these baffles, turbulence is created, which results in a drop in pressure. This pressure drop causes dissolved air to be released from the water. Just as with an air separator, the air forms tiny bubbles that adhere to the baffles. As smaller bubbles merge, they become larger bubbles, which rise to the top of the scoop, where they are removed by an air vent, **Figure 6-19**. Air scoops are installed in horizontal piping and include air vents mounted on their top ports, **Figure 6-20**.

For the optimum effectiveness of both air separators and air scoops, the inlet water velocities should be kept at 4′ per second or less. If the water velocity exceeds this rate, small air bubbles will remain trapped within the fluid flow and be carried through the hydronic system without being separated.

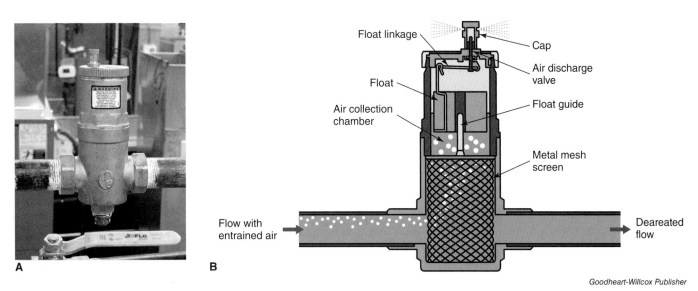

Goodheart-Willcox Publisher

Figure 6-18. A—An air separator in a hydronic system. B—An air separator is used to separate air from the water as it flows through the hydronic system.

124 Hydronic Heating: Systems and Applications

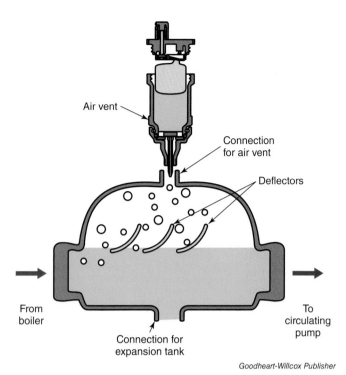

Figure 6-19. Air scoops contain baffles that separate the air from the boiler water.

Figure 6-20. Air scoops typically are mounted on horizontal piping on top of the expansion tank.

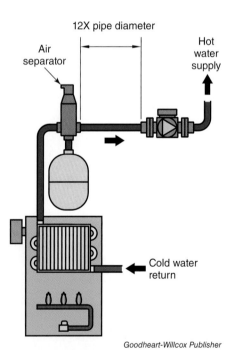

Figure 6-21. This illustration shows the best placement for the air separator, which is approximately 12 pipe diameters from the inlet of the circulator pump.

TECH TIP

Air Separator Placement

The best place to locate an air separator is near the outlet of the boiler, approximately 12 pipe diameters from the inlet of the circulator pump, **Figure 6-21**. At this location, the water is at a relatively low pressure and is still at an elevated temperature away from the boiler—resulting in a low solubility of dissolved gases within the system.

6.4.2 Air Vents

Once separated from the water, air is eliminated from the system through an ***air vent***. Several types of air removal devices or air vents are used in residential and light commercial hydronic systems. As mentioned earlier, because air is lighter than water, air vents are ideally located at the highest point of the system, where air is most likely to accumulate. However, air vents can also be located at the top of air scoops and are automatically built into some models of air separators.

Three types of air vents are commonly used in residential and light commercial hydronic heating systems:

- Manual air vents
- Hygroscopic air vents
- Automatic air vents

Manual air vents are usually installed at the highest point in the hydronic system and can be found on top of terminal devices such as fan coil units or baseboard heaters. These small valves thread into 1/8″ or 1/4″ FPT (female pipe thread) openings. They may be installed directly onto the terminal device or mounted onto a special fitting called a baseboard tee. This tee resembles a brass elbow but with a threaded port on top where the manual air vent is fastened, **Figure 6-22**. To bleed

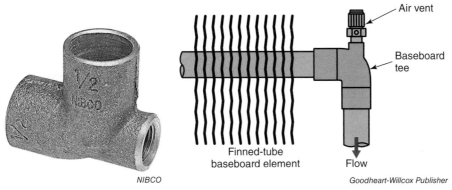

Figure 6-22. Baseboard tees are used in conjunction with manual air vents.

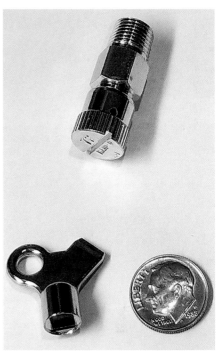

Figure 6-23. Manual air vents are used to bleed air from terminal devices using a screwdriver, a coin, or a slotted tool as shown here.

air from the system, manual air vents are opened manually using a screwdriver or the edge of a coin, such as a dime, **Figure 6-23**. They are often referred to as *coin vents* or *air bleeders*. When opened, the operator can bleed any excess air from the system until water appears. Manual air vents are often utilized at the beginning of each heating season to bleed any excess system air that has accumulated during the off-season and when the hydronic system is first filled at startup.

Hygroscopic air vents contain an internal cellulose fiber disc that swells when in contact with water. This swelling seals off the air vent port. However, as air accumulates around the disc it becomes dry and shrinks, allowing the air to pass through the vent port. As the vented air is replaced by water, the disc once again swells, closing off the port. Hygroscopic air vents can be used in either automatic or manual mode. When the knob is opened one turn from its fully closed position, it responds in the same manner as a manual air vent. Any pressurized air at the base of the vent exits through a small hole at the side of the vent body. When the knob is closed completely, an internal O-ring seals off the side port. However, if any air is present within the vent, the fiber disc will shrink and allow air to pass through it, **Figure 6-24**. This air is discharged under the vent's knob. Hygroscopic air vents are typically mounted on heating terminal units, **Figure 6-25**.

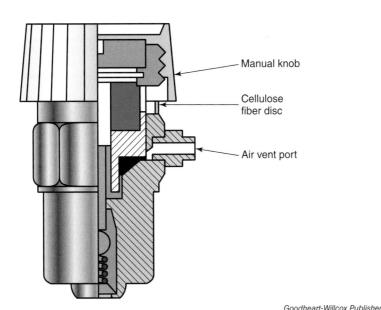

Figure 6-24. This illustration shows how a hygroscopic air vent works.

Figure 6-25. Hygroscopic air vents are usually found on terminal units such as radiators.

Automatic, or float-type, air vents provide fully automatic air release from the hydronic system, responding quickly to any air found within the system. These types of air vents consist of an air chamber, float assembly, and an air valve. The float responds to the amount of air within the vent chamber. When no air is present, the float rises to the top of the chamber and the air valve is closed. When air accumulates within the chamber, the float falls, opening the air valve and removing the air from the system. As the air is released, water flows into the air chamber, lifting the float once again and closing off the valve, **Figure 6-26**.

TECH TIP

Checking the Cap on an Air Vent

Most float-type air vents are equipped with a cap that protects the valve from any debris. It is important that this cap remains loosened when the vent is used during normal operation. If the cap is fully closed, the vent cannot remove any air.

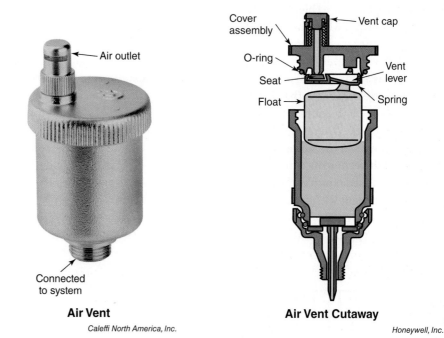

Air Vent
Caleffi North America, Inc.

Air Vent Cutaway
Honeywell, Inc.

Figure 6-26. This image and illustration show how a float-type air vent works.

CAREER CONNECTIONS

HVAC Wholesaler: Account Manager, Purchasing Agent, Customer Service Representative, Inside Sales Representative

HVAC technicians, service people, and installers all need to purchase parts and supplies. These purchases are typically done through HVAC wholesalers. As part of these organizations, HVAC wholesalers employ various positions such as account managers, purchasing agents, customer service representatives, and inside sales representatives.

- *Account Managers* are the primary contact for their assigned contractor accounts. They are responsible for seeking new accounts and promoting new business while managing and growing existing assigned accounts. They assist in developing long-term customer relationships and maximize customer retention with contractors and consulting engineer/architect accounts.
- *Purchasing Agents* are responsible for ordering stock inventory for wholesale branch locations. They maintain adequate stock levels and manage those levels to ensure an efficient flow of inventory while working within budget and department guidelines.
- *Customer Service Representatives* initiate, obtain, maintain, and promote sales of both stock and non-stock materials and services that are offered by the HVAC wholesaler. Customer service representatives work closely with the sales manager, account managers, and inside salespeople to generate sales by utilizing the bid department, private bids, negotiated bids, or promotion of commonly stocked products.
- *Inside Sales Representatives* are responsible for selling and promoting HVAC products in person at sales counters and over the phone. They also request quotations and lead times from customers. In addition, inside sales representatives quote product prices on standard items, enter orders into the computer system, and relay pertinent order information to the customer.

Successful candidates for these positions should possess the following:
- Experience in HVAC or the construction industry. Related education may be considered as a substitute for experience.
- Familiarity with light commercial and residential HVAC product sales
- Wholesale and retail sales experience
- General HVAC product knowledge
- Analytical skills necessary to make decisions and solve problems
- Ability to multitask while working in a fast-paced environment
- Excellent organization and time-management skills
- Strong written and verbal communication skills
- Excellent interpersonal skills for maintaining professional working relationships with vendors as well as other company personnel

Educational requirements for these positions include:
- High school diploma or GED at a minimum
- Additional college or trade school experience is a plus

Benefits offered by most wholesalers include:
- Medical and dental insurance
- Flexible Spending Account (FSA) for health care and dependent care
- Paid holidays
- Paid time off based on years of service
- Retirement
- Short- and long-term disability insurance
- Life insurance
- Employee discounts on purchases

Starting wages for work at an HVAC wholesaler are typically between $12.00 and $16.00 per hour for inside sales representatives and customer service representatives. Purchasing agents can receive salaries of up to $50,000 per year. Advanced positions such as account managers can earn as much as $60,000 per year with experience.

Chapter Review

Summary

- There are various types of piping materials and fittings used in hydronic heating installations. Types of fittings used with various piping materials include tees, elbows, caps, couplings, plugs, unions, bushings, and reducers.
- The most common type of steel piping used in hydronic systems is Schedule 40 black pipe.
- Copper tubing is available in three different wall thicknesses: K (heavy wall), L (medium wall), and M (light wall). Type M copper tubing is acceptable for most boiler applications. Copper tubing can be purchased as either soft- or hard-drawn types. Joining copper tubing is accomplished by flaring, soldering, or brazing.
- PEX is the reference name for cross-linked polyethylene tubing. It is popular for hydronic applications such as in-floor radiant heating because PEX is very durable and can withstand repeated heating and cooling cycles. It is also resistant to chemicals, which can cause scaling and corrosion in iron piping and copper tubing. PEX is also very flexible and relatively easy to join and install.
- Dielectric unions are considered specialty fittings and are used to join dissimilar metals.
- PVC piping may be used for either condensate draining or for flue venting in high-efficiency boilers. The maximum recommended operating temperature for PVC piping is 140°F, which flue gas temperatures may approach in some situations. PVC should not be used for venting in these instances or in Canada and individual US states where its use for venting is outlawed.
- Pressure/temperature gauges provide readings to determine whether the boiler is operating normally.
- Expansion tanks are used in hydronic systems as a cushion for the expansion of heated water. Older hydronic systems incorporated a compression, or standard, expansion tank, which is a steel tank vessel with a cushion of air at the top. Modern systems use a diaphragm expansion tank, which separates air from the boiler water with a flexible rubber membrane within its shell.
- It is important to correctly calculate the volume and operating pressure of the hydronic system when sizing expansion tanks.
- Air removal devices are critical to the proper functioning of a hydronic heating system. Air trapped in the system can block water from being circulated, cause undesirable noise, cause corrosion in pipes, and result in pump cavitation.
- Air removal is categorized as the separation and venting of unwanted air from the system. An air separator uses a mesh screen to separate air from the water as it flows through the hydronic system. Air scoops separate air from the water with a series of deflectors or baffles.
- There are several types of air vents used on hydronic heating systems, including manual, hygroscopic, and automatic air vents.

Know and Understand

1. The most common type of steel piping used for boiler installations is _____.
   A. Schedule 40 black pipe
   B. Schedule 40 galvanized pipe
   C. Cast iron
   D. None of the above.
2. *True or False?* A tee that measures 1″ by 1″ by 3/4″ would have the left and right diameters measure 3/4″ and the top diameter would be 1″.
3. What is the difference between a bushing and a reducer?
   A. Bushings have both male and female threads.
   B. Reducers are bell-shaped couplings.
   C. Both A and B are correct.
   D. None of the above.
4. Copper tubing with the greatest wall thickness is _____.
   A. type K
   B. type L
   C. type M
   D. type P
5. Copper fittings that have both socket and threaded ends are referred to as _____.
   A. couplings
   B. connectors
   C. adapters
   D. Both B and C are correct.
6. *True or False?* PEX tubing can be joined to dissimilar materials.
7. The specialty fitting that is used to connect dissimilar metals is known as a _____.
   A. galvanic union
   B. dielectric union
   C. dissimilar union
   D. None of the above.
8. What is the difference between a compression tank and a diaphragm expansion tank?
   A. A compression tank is prone to waterlogging.
   B. A diaphragm expansion tank uses a rubber membrane.
   C. In the compression tank, the water and air are in direct contact.
   D. All of the above.
9. *True or False?* Air scoops are air separation devices that have a series of deflectors or baffles, which create turbulence as the circulated water passes through them.
10. Which type of air vent is also known as a *coin vent*?
    A. Manual air vent
    B. Automatic air vent
    C. Hygroscopic air vent
    D. Separator air vent

Apply and Analyze

1. What is the purpose of having tapered threads in steel pipe fittings?
2. List the types of fittings commonly used for joining various piping materials.
3. Name three techniques for joining copper tubing.
4. What application are adapter-type pipe fittings used for?
5. Explain what PEX stands for and what applications it is used for.
6. Describe what a dielectric union is and where it is used.
7. Name the two applications where PVC is used for boilers.
8. Describe the type of fitting used to prevent waterlogging in compression tanks.
9. Explain what can happen if the diaphragm in an expansion tank ruptures.
10. Describe the formula used for sizing expansion tanks.
11. Explain the difference between an air separator and an air scoop.
12. What types of devices would use a coin-type vent?

Critical Thinking

1. Explain what steps should be taken to properly retrofit the piping on a new boiler system that has existing galvanized piping.
2. How and why would the calculated pressure for an expansion tank differ in a 4-story building versus a 2-story building?

7 Hydronic Piping Systems

Chapter Outline
7.1 The Physics of Water
7.2 Steps for Proper Pipe Sizing and System Layout
7.3 Piping Arrangements

Dmitry Kalinovsky/Shutterstock.com

Learning Objectives

After completing this chapter, you will be able to:
- Summarize the fundamentals of how fluid flows through pipe.
- Calculate fluid flow using the sensible heat formula.
- Explain how the system's fluid velocity can affect the water flow.
- Explain the difference between laminar and turbulent fluid flow.
- Differentiate between static pressure and system pressure.
- Describe how system head is affected by static pressure.
- Outline the steps involved in designing a hydronic heating system.
- Explain the differences between various hydronic piping configurations.
- Describe how a series-loop piping arrangement operates.
- Design a one-pipe hot water system.
- Explain how a diverter tee using a venturi works.
- Explain the difference between a two-pipe direct piping arrangement and a two-pipe reverse return piping arrangement.
- Explain the advantages of a primary-secondary piping system.
- Define the purpose of closely spaced tees and low-loss headers used with primary-secondary piping configurations.

Technical Terms

balancing valve
closely spaced tees
deadheading
differential pressure bypass valve
diverter tee
equivalent length
feet of head
feet per second (FPS)
friction loss
gallons per minute (GPM)
hydraulic separation
laminar flow
load calculation
low-loss header
one-pipe system
pipe friction chart
primary-secondary piping
pump affinity law
pump performance curve
series loop arrangement
standby losses
static pressure
system pressure
thermal mass
turbulent flow
two-pipe direct return
two-pipe reverse return

In order for the hydronic heating system to work effectively, it must be sized properly. To accomplish this, the hydronic system designer must have a clear understanding of proper heat transfer, fluid flow rates, various piping arrangements, correct pipe sizing, and hydronic resistance through fittings and other devices. This chapter focuses on the various types of piping arrangements and system configurations used for residential and light commercial hot water heating systems. It also discusses the proper procedure for sizing a functional hydronic heating system based on pipe sizing and resistance through hydronic fittings and other devices. Emphasis is placed on formulas used for sizing hydronic systems and fluid flow rates based on industry standards.

7.1 The Physics of Water

The first step in configuring an effective piping system is in understanding the fundamentals of how fluid flows through a pipe. When laying out a complete piping system, the designer must consider a number of factors in order to create a system that is efficient, economical, and most of all comfortable for the end user. Hydronic system design is not very complicated; however, a working knowledge of terms and definitions used in system development needs to be acquired in order for the layout and design to be operational.

Note: The generic term water will be used for clarity throughout this section when discussing fluid flow. The physics of fluid mechanics can also relate to glycol or a glycol/water mix and will be discussed in further chapters.

7.1.1 Fluid Flow

The rate of water flow through a pipe is expressed in *gallons per minute (GPM)*. This flow rate indicates the *volume* of water flowing through the pipe at a given time. It is important that the system's flow rate be accurately calculated when designing a system. Water that is flowing too quickly through a system will reduce the heating output at the terminal unit and can create excessive noise issues. Water that is flowing too slowly through the system can affect the return water temperature back at the boiler. As outlined in Chapter 1, *Human Comfort and Heat Transfer*, the systems flow can be calculated using the sensible heat formula:

$$GPM = Q \div (500 \times \Delta T)$$

where

- GPM = gallons per minute of water flow
- Q = Btu/hr
- 500 = constant
- ΔT = temperature difference across terminal unit

Using this formula, the system flow can be calculated by knowing the total heating load of the building and the desired temperature difference. Flow will also need to be calculated through individual terminal units if the designer chooses to use separate circulating pumps for each zone.

TIMEOUT FOR MATH

Calculating GPM

To calculate the correct water flow in GPM through a terminal unit, determine the heating capacity of the terminal unit (reference the manufacturer's specifications) and choose the desired temperature drop across the unit. For example, a fan coil unit has a heating capacity of 20,000 Btu/hr at a 20°F temperature drop. Using the sensible heat formula, the correct flow through the fan coil unit would be as follows:

$$GPM = 20{,}000 \div (500 \times 20°F\ \Delta T)$$

Correct flow = 2 GPM

7.1.2 Fluid Velocity

The speed of the flow of water or other fluids through a pipe is known as the *velocity* of that fluid, and is expressed in ***feet per second (FPS)*** when sizing hydronic heating systems. The system's fluid velocity can affect the water flow by making it either laminar or turbulent. Visualize ***laminar flow*** as the water flowing in more of a straight line, whereas water in a ***turbulent flow*** becomes chaotic or moves in more of a swirling motion as it travels through the pipe, **Figure 7-1**.

In a typical hydronic system, laminar flow is more desirable since it reduces the amount of internal resistance the circulating pump must overcome. Conversely, turbulent flow is more desirable through a heat exchanger because it enhances heat transfer. The velocity of hot water through the system plays an important role in choosing the correct piping size and circulating pump.

Generally, the desired velocity through the hydronic system should be maintained between 2 FPS and 4 FPS. Too much velocity can result in laminar flow issues and cause undesirable noise in the piping. Not enough velocity can result in a reduced heating load to the terminal units as well as a reduction in the amount of air purged from the system. (Recall from Chapter 6, *Boiler Fittings and Air Removal Devices*, that air separators rely on turbulence to function.) Determine the correct fluid velocity using a ***pipe friction chart*** or flow table, covered later in this chapter.

Laminar flow pattern

Turbulent flow pattern

magnetix/Shutterstock.com

Figure 7-1. This illustration shows the difference between laminar and turbulent flow.

7.1.3 System Pressures

A hydronic heating system utilizes two types of pressure. The first type is the *system pressure*, which is the amount of force generated by the hot water circulating pump, covered in Chapter 10, *Circulating Pumps*. The second type is static pressure. **Static pressure** in a hydronic system is the pressure generated by the weight of the water within the system. The term *static* means lacking movement, or a body at rest. Remember that water has weight—approximately 8.33 lb per gallon at 60°F, and this weight is exerted downward by gravity. The static pressure within a hydronic system is measured in pounds per square inch, or psi. As an example, a column of water that is 2.31' high (approximately 28") will exert a downward pressure of 1 psi. It does not matter whether the cross-sectional diameter of the column is wide or narrow—if the column is 2.31' in height, the pressure at the bottom of the column will always be 1 pound per square inch.

Now consider what happens to the static pressure if we increase this column of water to 11.55'. At this height, the column of water now exerts 5 psi at the bottom of the column (11.55 ÷ 2.31 = 5). If we increase the column height to 23.1', our pressure is now at 10 psi and so on, **Figure 7-2**. The occurrence of increased pressure based on column height is evident when scuba divers feel a greater increase in pressure as they dive deeper into the ocean. It is also the basis for sizing the boiler's pressure reducing valve, as discussed in Chapter 9, *Valves*.

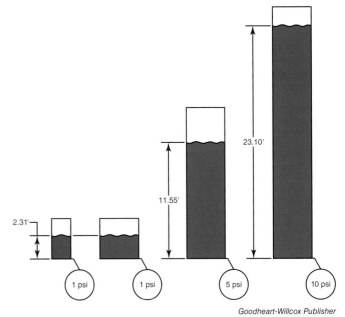

Figure 7-2. This figure shows the relationship between a column of water and its static pressure.

TIMEOUT FOR MATH

Calculating Pressure Requirements

If 1 psi of pressure is needed to lift water to a height of 2.31', how much pressure is needed to lift the same amount of water to 16'?

Answer: 16 ÷ 2.31 = 6.93 psi

7.1.4 System Head

By definition, *head* is a concept that relates the amount of energy in a fluid (water) to the height of an equivalent static column of that same fluid or water. System head takes into account the internal energy of a fluid due to the pressure, velocity, and elevation exerted onto the pipe by that fluid.

Head is expressed in ft-lb (foot-pound) units and is normally stated as *feet of head* when used for calculating hydronic system piping and circulating pumps.

Useful formulas to remember are:

Feet of head ÷ 2.31 = pressure in psi
Pressure in psi × 2.31 = feet of head

It is easy to see now where the constant 2.31 is derived from, **Figure 7-3**. As stated earlier, a column of water that is 2.31' high will exert a downward pressure of 1 psi. The concept of pump head is used in Chapter 10, *Circulating Pumps*.

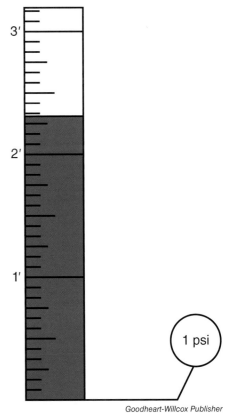

Figure 7-3. This illustration shows the relationship between system pressure and feet of head.

> ### TECH TIP
>
> **Static Pressure versus Pump Pressure**
> Keep in mind that static pressure and pump pressure are completely independent of each other. Static pressure deals only with gravity and the weight of a column of water. Pump pressure (or head) is dependent on the number of fittings and valves in the system, the height at which the fluid is elevated, and also the size of the building's piping layout.

7.2 Steps for Proper Pipe Sizing and System Layout

Following are the concepts used when designing a hydronic heating system:

- *Load calculation.* A mathematical design tool used to determine the heat loss of a building based on the design outdoor temperature for a geographic location. Load calculations are used to properly size the Btu output of the heating system.
- *Equivalent length* of pipe. Each hydronic pipe fitting (elbow, tee, valve, and so on) creates a resistance to flow in the piping circuit. The resistance of each fitting must be calculated in terms of an amount of straight pipe of the same dimension with equivalent resistance.
- *Friction loss* through piping. Losses from friction result from the roughness of the piping material and from the turbulence of the fluid as it flows through the pipe.
- *Feet of head.* Sometimes referred to as *static pressure head*, feet of head is a unit of measurement that takes into account the internal energy of a fluid due to the pressure, velocity, and elevation exerted onto the pipe by the fluid and is used for sizing the circulating pump.
- *Pump performance curve.* A pump curve is a graphical representation of a pump's performance based on testing conducted by the manufacturer. Each pump will have its own pump performance curve, which varies from pump to pump. This performance is based on the pump's motor horsepower and the size and shape of the impeller.
- *Pump affinity laws.* These are used in hydronic design to indicate the influence of the pump performance based on volume capacity, feet of head, and power consumption.

> ### PROCEDURE
>
> **Hydronic System Piping Design**
> In order to achieve an effective hydronic system piping design, the following procedure should be followed to develop a workable system. Keep in mind that this is a general procedure for a basic system design. It is subject to modification based on the type of piping arrangement chosen:
>
> 1. Each system design starts with an accurate heat load calculation on the building. This can be done by following the ACCA Manual J, as detailed in Chapter 13, *Building Heating Loads and Print Reading*. Once the building's heat load has been calculated, a zone-by-zone calculation can be completed. This allows the designer to select proper terminal units for the respective zones according to the calculated Btu/hr heat loss.
> 2. Select a preliminary supply water temperature based on the anticipated type of terminal units to be used. This supply water temperature is usually based on the "worst case scenario" for the outdoor winter design temperature.

(continued)

3. Determine the appropriate temperature drop through each area being heated based on the type of terminal unit to be used for each respective zone.
4. Calculate the required hot water flow through each zone using the sensible heat formula for water: GPM = Btu/500 × ΔT
5. Size and select the proper terminal unit for each zone based on Btu/hr requirements and GPM flow. Consult the manufacturer's data when selecting the proper units.
6. Decide which piping arrangement to utilize for the given building.
7. Lay out the piping arrangement by sketching a scaled drawing of the building's floor plan, showing the location of each terminal unit.
8. Select the proper type and sizes of piping for the system based on minimum and maximum flow rates and fluid velocities, **Figure 7-4**.
9. Calculate the equivalent length of pipe for each fitting located in the system using the table in **Figure 7-5**.
10. Calculate the friction loss through each of the piping circuits. This procedure is used to calculate head loss per 100 feet of pipe and is based on the size of the pipe and the water flow in GPM, **Figure 7-6**.
11. Determine the pressure losses of the heating equipment (boiler) and terminal units from the manufacturer's data, **Figure 7-7**.
12. Calculate the total feet of head loss in the total piping circuitry.
13. Plot a pump system curve on the pump manufacturer's performance curve using the pump affinity laws as outlined in *Chapter 10, Circulating Pumps*, **Figure 7-8**.
14. Select the proper circulating pump based on the manufacturer's performance data, **Figure 7-9**.
15. Select the proper motor horsepower for the circulating pump from the manufacturer's performance data.
16. Review the calculated Btu/hr output of each terminal unit and make adjustments if necessary, based on the choices made for types of piping, terminal units, and pump selection.

Tubing size/type	Minimum flow rate (based on 2 ft/sec) (GPM)	Maximum flow rate (based on 4 ft/sec) (GPM)
3/8" copper	1.0	2.0
1/2" copper	1.6	3.2
3/4" copper	3.2	6.5
1" copper	5.5	10.9
1.25" copper	8.2	16.3
1.5" copper	11.4	22.9
2" copper	19.8	39.6
2.5" copper	30.5	61.1
3" copper	43.6	87.1
3/8" PEX	0.6	1.3
1/2" PEX	1.2	2.3
5/8" PEX	1.7	3.3
3/4" PEX	2.3	4.6
1" PEX	3.8	7.5
1.25" PEX	5.6	11.2
1.5" PEX	7.8	15.6
2" PEX	13.4	26.8

Goodheart-Willcox Publisher

Figure 7-4. This table shows flow rates for different sizes and types of tubing based on velocity flow.

Copper Tube Sizes									
Fitting or valve	3/8"	1/2"	3/4"	1"	1.25"	1.5"	2"	2 1/2"	3"
90° elbow	0.5	1.0	2.0	2.5	3.0	4.0	5.5	7.0	9
45° elbow	0.35	0.5	0.75	1.0	1.2	1.5	2.0	2.5	3.5
Tee (straight run)	0.2	0.3	0.4	0.45	0.6	0.8	1.0	0.5	1.0
Tee (side port)	2.5	2.0	3.0	4.5	5.5	7.0	9.0	12.0	15
B&G Monoflo® tee	n/a	n/a	70	23.5	25	23	23	n/a	n/a
Reducer coupling	0.2	0.4	0.5	0.6	0.8	1.0	1.3	1.0	1.5
Gate valve	0.35	0.2	0.25	0.3	0.4	0.5	0.7	1.0	1.5
Globe valve	8.5	15.0	20	25	36	46	56	104	130
Angle valve	1.8	3.1	4.7	5.3	7.8	9.4	12.5	23	29
Ball valve	1.8	1.9	2.2	4.3	7.0	6.6	14	0.5	1.0
Swing-check valve	0.95	2.0	3.0	4.5	5.5	6.5	9.0	11	13.0
Flow-check valve	n/a	n/a	83	54	74	57	177	85	96
Butterfly valve	n/a	1.1	2.0	2.7	2.0	2.7	4.5	10	15.5

Goodheart-Willcox Publisher

Figure 7-5. This table shows the equivalent lengths of common copper fittings and valves used in determining circulating pump selection.

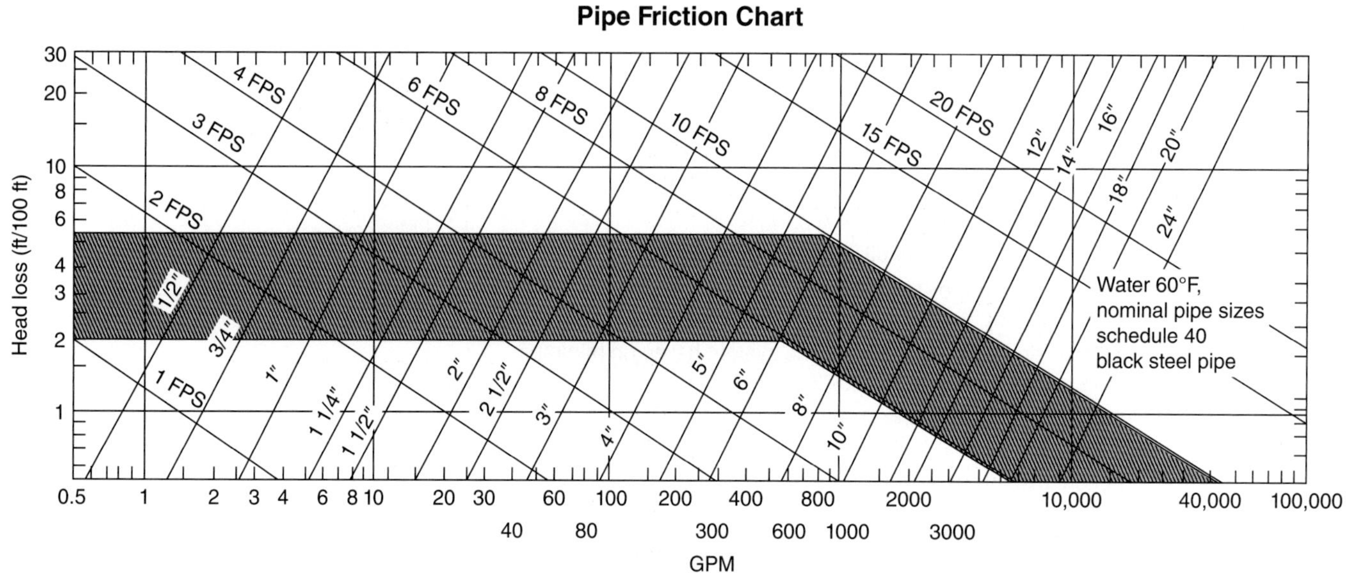

Pipe Friction Chart

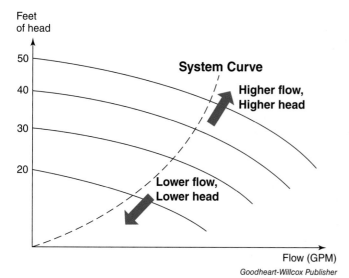

Figure 7-6. This pipe friction chart is used to calculate head loss based on the size of the pipe and the amount of flow in GPM.

Water Flow in GPM	Pressure Loss in Feet of Water		
	4″ Models	6″ Models	8″ Models
0.25	0.044	—	—
0.50	0.160	0.070	0.046
1	0.597	0.270	0.167
2	2.220	1.047	0.616
3	—	2.260	1.367
4	—	3.793	2.380
5	—	—	3.673

Goodheart-Willcox Publisher

Figure 7-7. An example of how to determine the pressure losses through heating equipment based on manufacturer data.

Figure 7-8. An example of plotting a pump system curve using the pump manufacturer's performance curve.

7.3 Piping Arrangements

Several factors will determine the type of piping arrangement chosen by the system designer. In most cases, the building's configuration and available space will drive the decisions on routing the piping and thus the type of piping arrangement to be used. Other factors include the building's heating load, how the system will be controlled, any necessary installation requirements, and the overall installation and operational costs. In order to meet these requirements, it is important to select the proper terminal units and the best piping arrangement. As defined in earlier chapters, a terminal unit is any type of heat emitter located in the controlled zone used to transfer heat into the conditioned space. Common types of terminal units include convectors, radiant panels, fan coil units, and finned tube baseboards. In this section, we focus on the following piping arrangements:

- Series loop arrangement
- One-pipe system with multiple zones
- Two-pipe direct-return system
- Two-pipe reverse-return system
- Primary-secondary piping system

7.3.1 Series Loop Arrangement

The *series loop arrangement* is the simplest piping system. Usually found on smaller-sized buildings, it typically consists of one boiler, one pump, and one zone, **Figure 7-10**. The main characteristic of a series loop is that it consists of a single, continuous pipe that circulates water through each terminal unit consecutively from the supply side of the boiler to the return side. One of the main disadvantages of this type of piping arrangement is that there is typically no method for individual zone control—the entire loop is usually controlled

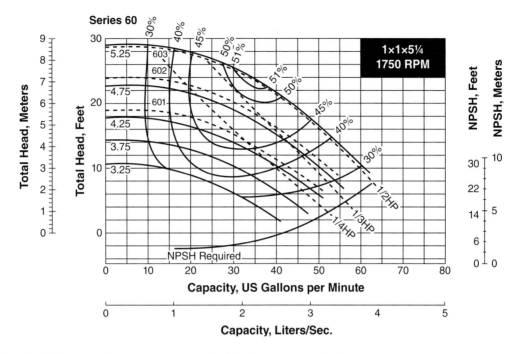

Xylem, Inc.

Figure 7-9. This is a typical pump performance curve provided by the pump manufacturer.

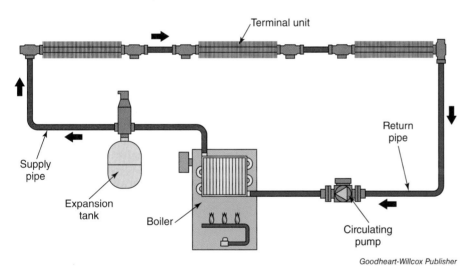

Goodheart-Willcox Publisher

Figure 7-10. An example of a series-loop system.

from one thermostat. Another drawback of the series loop arrangement is the sometimes extreme temperature difference of the water from the supply side of the boiler to the return side. The temperature output of the last terminal unit can sometimes be much colder than the output temperature of the other terminal units. This condition can lead to a return water temperature that is too low for the boiler to handle.

The temperature differential used for calculation purposes between the system supply side and return side water is typically 20°F. For instance, on a cold day the supply side boiler water temperature set point may be as high as 180°F. With a 20°F temperature difference, this example would allow for a return water temperature of 160°F. Some boiler manufacturers require that the return water temperature be maintained above 140°F to avoid potential condensation of corrosive flue gases within the boiler. Return water temperatures lower than that could cause premature failure of the boiler's heat exchanger as well as ruin the chimney and flue piping. Conversely, today's modern-type condensing boilers can tolerate much lower return water temperatures. Always follow the boiler manufacturer's recommendations for supply and return water temperatures when designing a system. These topics are covered at length in other chapters.

Look at an example in which the boiler's return water temperature can affect system performance in **Figure 7-11**. In this illustration, three terminal units are piped into a typical series loop configuration. Each terminal unit has an output rating of 30,000 Btu/hr, and each requires a system flow of 5 GPM. Using the sensible heat formula for water, we can calculate the return water temperature at the outlet of each terminal unit and the return water temperature back to the boiler inlet:

$$\text{Terminal \#1:}$$
$$\Delta T = \text{Btu/GPM} \times 500$$
$$\Delta T = 30{,}000/5 \times 500$$
$$\Delta T = 12°F$$

With a temperature drop of 12°F, the outlet water temperature at Terminal #1 is now 168°F. This will be the inlet temperature at Terminal #2. With the same 12°F temperature drop across Terminal #2, the inlet temperature at Terminal #3 is now 156°F. If Terminal #3 maintains the same temperature drop as the other two terminal units, the return water temperature at the boiler inlet is now 144°F; which is well below the desired 20°ΔT of a typical boiler. In addition, if the boiler has a conventional cast-iron heat exchanger with a metal exhaust flue, the return water temperature of 144°F at the boiler inlet is dangerously close to reaching the condensing temperature of the flue gases.

The series loop arrangement typically consists of several terminal units. These terminal units may even be of different types such as convectors, fan coil units, and finned-tube baseboards. If several different types of terminal units are being incorporated into the series loop system, then the designer should consider choosing terminal units with similar **thermal mass**. Terminal units that are made up of steel or cast iron typically have a higher thermal mass—such as cast-iron radiators. These types of terminal units possess a greater ability to absorb and store more heat energy. Terminal units that are made up of copper—such as finned tube baseboard heat emitters—are classified as low-mass terminal units.

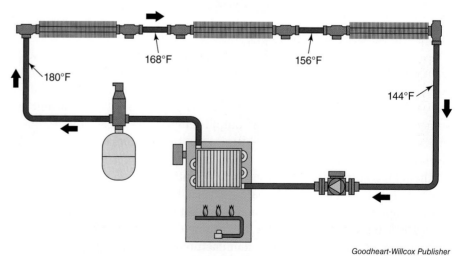

Goodheart-Willcox Publisher

Figure 7-11. This illustration shows the potential temperature drop across a series-loop system.

Although these types of units heat up quickly, they tend to lose heat faster than units of a higher mass construction.

Another variation of the series loop piping arrangement found on larger systems is to divide the entire system into separate loops, or split loops, as shown in **Figure 7-12**. This type of configuration reduces the overall length of each loop and allows the designer to designate each loop to different zones for better temperature control. In addition, separate circulator pumps can be utilized for each section of the split loop, or *balancing valves* can be incorporated for better flow control.

7.3.1.1 Considerations When Designing a Series Loop Piping Arrangement

- Always conduct a proper heat load calculation for each heating zone.
- Select the proper piping sizes based on minimum and maximum flow rates and fluid velocities, **Figure 7-4**.
- Because the terminal units are piped in series, the supply temperatures and temperature drop across the terminal units may vary depending on the types of units selected. Keep this in mind when selecting the respective terminal units.
- With a series loop arrangement, it may be necessary to vary the sizes of the terminal units based on whether they are located at the beginning or end of the heating loop, due to the fact that there is an appreciable drop in water temperature across the respective units.

7.3.2 One-Pipe System with Multiple Zones

A one-pipe piping arrangement with multiple zones is similar to the series loop arrangement in that all of the main water supply passes through only one pipe between the boiler's supply and return flow inlets. However, in the *one-pipe system*, individual terminal units are piped in parallel to the main primary loop as branch circuits. This means that a portion of the water flowing through the main circuit is diverted through each terminal unit by supply and return tees, **Figure 7-13**. In order to maintain the proper water flow through each terminal

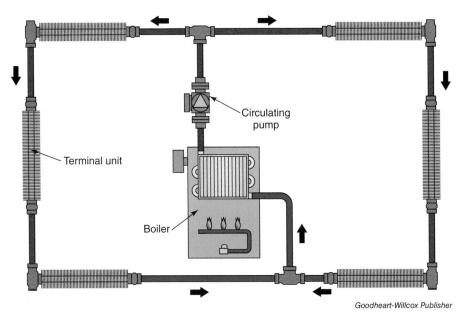

Goodheart-Willcox Publisher

Figure 7-12. An example of a split-series loop, with a circulating pump controlling both loops.

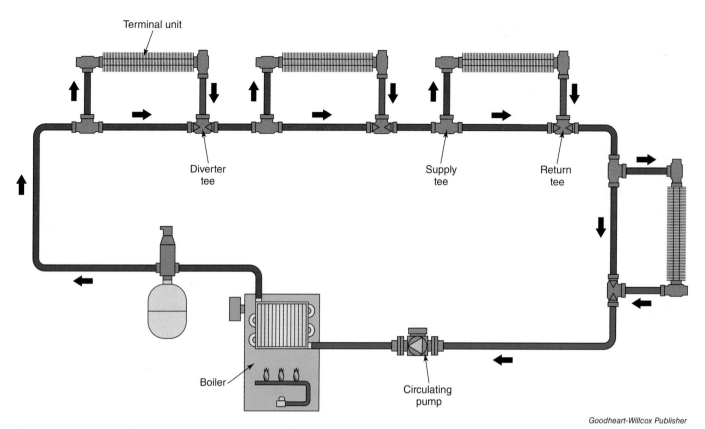

Figure 7-13. An example of a one-pipe loop system with terminal units piped in parallel to the main loop as branch circuits. This system would incorporate the use of diverter tees.

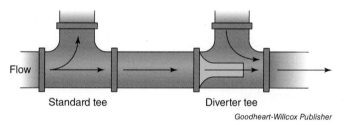

Figure 7-14. This illustration shows the difference between a standard tee and diverter tee.

unit, special ***diverter tees*** are installed, usually on the return side of the terminal unit, **Figure 7-14**. This type of tee is sometimes called a *Monoflo* or *venturi* tee, **Figure 7-15A** and **Figure 7-15B**. They contain a special cone inside that acts as a venturi to restrict the flow of water through the tee outlet, **Figure 7-16**. The purpose of this tee is to create a pressure drop through the main circuit, which allows a certain amount of fixed water flow through the branch terminal unit.

Although the one-pipe system with multiple zones does create a temperature drop between the first and last zones on the main circuit, the use of diverter tees significantly decreases this temperature drop, as compared to a series loop arrangement.

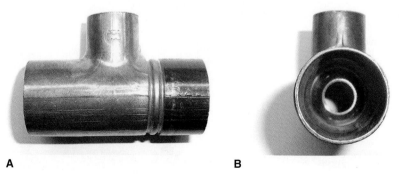

Figure 7-15. An example of a B&G Monoflo tee.

7.3.2.1 Piping Recommendations Using Diverter Tees

When providing diverter tees for a one-pipe system using multiple zones, follow these proper piping practices:

- For most installations in which the terminal unit is located above the main circuit (such as on the first floor when the boiler is in the basement) and there is minimal resistance through the terminal device, only one diverter tee per terminal unit is needed and should be installed on the return side of the terminal unit, **Figure 7-17**.
- For terminal units located above the main circuit but with a high resistance factor, as in installations with very long supply and return lines (when the terminal unit is on the second floor and the boiler is in the basement, for instance), two diverter tees should be installed. Install one diverter tee on the supply side and another on the return side of the terminal unit. In addition, the supply piping leading to the terminal unit should be one size larger than the return piping. Space the diverter tees at the same width as the terminal unit, **Figure 7-18**.
- For terminal units installed below the main circuit (if the terminal unit is in the basement and the boiler on the first floor, for example), two diverter tees should be installed—one on the supply side and one on the return side of the terminal unit. Ensure the two diverter tees are facing in opposite directions and spaced at the same width as the terminal unit, **Figure 7-19**.
- If terminal units are to be piped both above and below the main supply water circuit, it is recommended that the diverter tees be staggered or alternated between the respective units, **Figure 7-20**.

Always follow the manufacturer's installation instructions for diverter tees to ensure they are sized correctly and installed in the correct direction of flow.

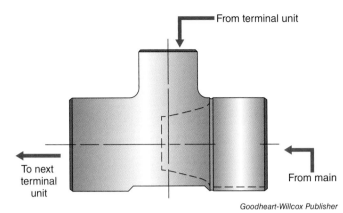

Figure 7-16. This illustration shows how a diverter tee restricts the flow of water through the tee outlet.

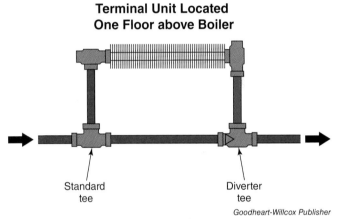

Figure 7-17. This example shows how diverter tees would be installed for terminal units piped above the main hot water supply line.

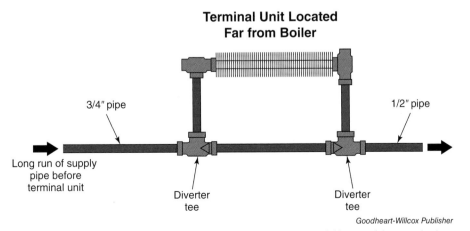

Figure 7-18. This illustration shows how two diverter tees would be used for terminal units piped above the main supply line with high resistance.

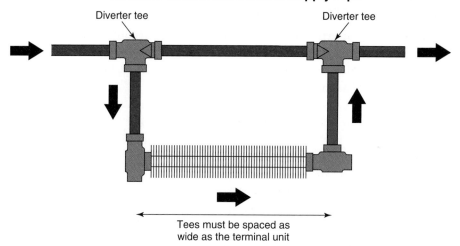

Figure 7-19. This illustration shows how a terminal unit should be piped if it is located below the main hot water supply line. Note how the diverter tees are piped in opposite directions.

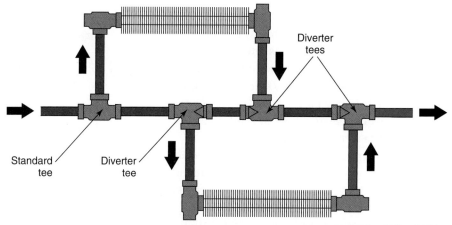

Figure 7-20. This is how terminal units should be piped if they are located both above and below the main supply water circuit.

TECH TIP

How Do I Size a Diverter Tee?

Manufacturers of diverter tees have sizing charts that help you determine which size tee you need for a given branch terminal unit. To select the proper tee, you need to know the correct GPM flow through the branch circuit. This can be calculated using the sensible heat formula. For instance, say you need 24,000 Btu/hr of heat flowing through the branch circuit at a 20°F temperature difference:

$$GPM = Btu / 500 \times \Delta T$$
$$GPM = 24,000 / 500 \times 20$$
$$GPM = 2.4$$

If the main heating circuit requires 10 GPM of water flow based on the total system heating capacity, we can use the manufacturer's sizing charts to select a tee according to the pipe size for the main circuit and the pipe size of the branch circuit.

7.3.2.2 Considerations When Designing a One-Pipe System with Multiple Zones

- Perform the proper heat load calculations for each heating zone.
- Determine the proper temperature drop across the system loop.
- Calculate the total heat load (Q) of the loop by adding the heating load of the individual heating zones.
- Once the total load calculation is determined for the entire heating circuit, determine the loop's flow rate using the sensible heat formula: GPM (loop) = $Q \div (500 \times \Delta T)$
- Select the proper terminal units and diverter tees based on individual zone heating loads and the loop's flow rate.
- Select the proper piping sizes for each zone based on minimum and maximum flow rates and fluid velocities, **Figure 7-4**.
- Pay special attention to the resistance through each fitting and diverter tee when calculating equivalent length of piping and total feet of head for pumping purposes.
- A good control strategy is to incorporate thermostatic radiator valves on the inlet of each terminal unit. This allows for better individual zone temperature control, **Figure 7-21**.

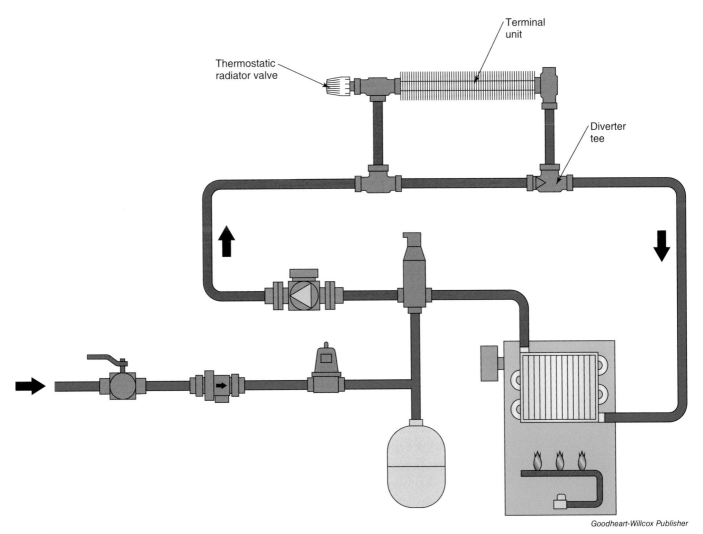

Goodheart-Willcox Publisher

Figure 7-21. Using a thermostatic radiator valve on the inlet of a terminal unit.

> ### DID YOU KNOW?
>
> **The Diverter Tee Made Hot-Water Heating Competitive**
> Diverter tees became popular at a time when steam was the standard heating system for both residential and light commercial buildings. Because these steam heating systems only required a one-pipe system to operate, it was difficult for businesses trying to sell a hot-water heating system to stay competitive because their system required two pipes. The idea of creating a one-pipe zoned hot water system actually came from a heating engineer from Cincinnati, Ohio, by the name of Oliver Schlemmer. He designed and patented his "O-S" fitting in the early twentieth century and used it on gravity hot water systems. Later on, the "Monoflo Tee" was developed by Bell & Gossett, and the "Venturi Tee" was introduced by Taco Comfort Solutions after hot water circulation pumps came on the market. These products made hot water heating systems less expensive to install than steam and eliminated the need for larger diameter piping. Furthermore, hot water heating was marketed as "less dangerous" compared to steam systems because there was less chance of a boiler explosion.

7.3.3 Two-Pipe Direct Return System

The *two-pipe direct return* piping system is similar to the one-pipe multizone system in that there is one pipe supplying water from the boiler to the terminal units and one pipe returning water to the boiler. However, in this system the terminal units are piped in a parallel arrangement instead of a series arrangement, **Figure 7-22**. By using this "first in/first out" piping arrangement, each terminal unit is essentially receiving the same inlet hot water temperature as delivered by the boiler. This may appear to require less piping across the main supply and return lines; however, the main disadvantage of using the two-pipe direct return system is the amount of pressure drop created across each loop, or terminal unit. Because water follows the path of least resistance, the last terminal unit on the system will experience a decreased flow in GPM that can affect its heating performance. This is especially true if each terminal unit has an identical Btu capacity.

To alleviate this pressure drop across the individual terminal units, the two-pipe direct return system typically incorporates balancing valves, or flow control valves, on each terminal unit, **Figure 7-23**. By closing off each individual balancing valve by a predetermined amount (based on the calculated pressure drop across each terminal unit), the terminal unit at the end of the loop can achieve the proper flow according to its Btu capacity. This usually means that the balancing valve on the first terminal unit in the piping system is closed the most compared to the consecutive terminal units on the system loop. As the balancing valve is closed off, the resistance across the terminal unit becomes higher. By properly designing the individual terminal units and correctly adjusting the balancing valves, the velocity throughout the piping system can be maintained below a 4'/second level to avoid any potential noise problems. A description of how balancing valves are installed and operated is discussed in Chapter 9, *Valves*.

7.3.3.1 Considerations When Designing a Two-Pipe Direct Return System

- Perform the proper heat load calculations for each heating zone.
- Select the terminal units for each zone based on heat loss calculations.
- Calculate the required flow rate through each terminal unit based on the desired temperature difference using the sensible heat formula: GPM (zone) = $Q \div (500 \times \Delta T)$.

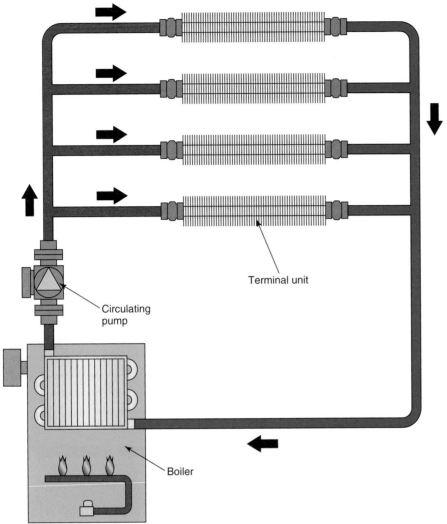

Goodheart-Willcox Publisher

Figure 7-22. An example of a two-pipe direct return system.

- Select the proper piping sizes for each zone based on minimum and maximum flow rates and fluid velocities, **Figure 7-4**.
- Size the circulator pump based on flow rate and overall system feet of head.
- Balance the individual zones according to the manufacturer's directions for the flow setting devices.
- The use of a variable-speed circulating pump controlled by a pressure transducer is a good control strategy for optimum flow control.
- If the individual terminal units are being controlled by either a thermostatic radiator valve or motorized zone valve, a pressure-actuated bypass valve should be piped between the supply and return mains to prevent the pump from *deadheading*. (Pump deadheading occurs when the pump is forced to run when there is no supply water to the pump—the pump impeller continues to rotate the same volume of water in the pump casing without allowing this water to pass through. This situation can result in the pump overheating and possible damage.)

7.3.4 Two-Pipe Reverse Return System

The *two-pipe reverse return* piping system is similar to the two-pipe direct return system except that in this configuration, the first terminal unit on the

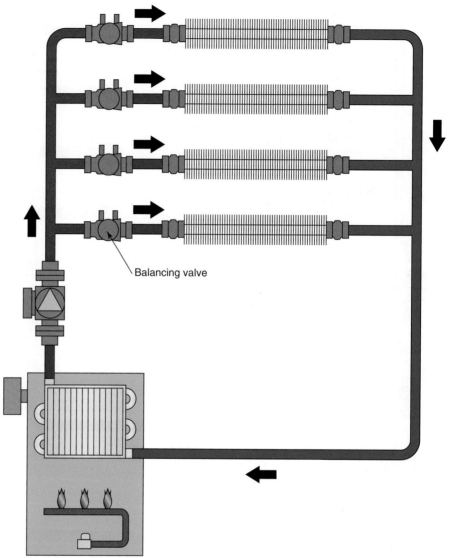

Goodheart-Willcox Publisher

Figure 7-23. An example of a two-pipe direct return system showing balancing valves.

system closest to the boiler is the first to receive inlet water but the last unit to return water back to the boiler, **Figure 7-24**. This "first in/last out" piping configuration lends itself to several advantages over other piping arrangements. When the piping layout for the two-pipe reverse return piping system is sized correctly—meaning that the supply piping is sized incrementally smaller from the first to last terminal units—this system can be considered *self-balancing*. In other words, there is minimal need to provide additional balancing valves for individual terminal units—especially if each zone has the same Btu/hr requirement.

7.3.4.1 Considerations When Designing a Two-Pipe Reverse Return System

- Use the sensible heat formula to identify the zone with the highest expected amount of flow. This allows the designer to determine the proper amount of head loss through this circuit.
- Size the circulating pump based on the zone with the highest amount of head loss. This ensures that all zones receive an adequate amount of water flow.

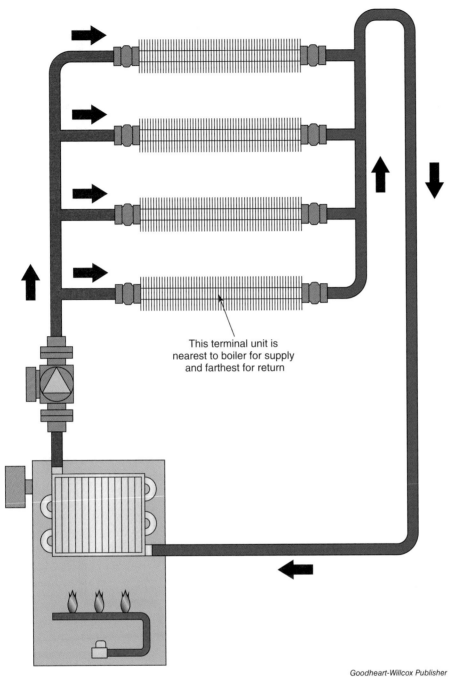

Goodheart-Willcox Publisher

Figure 7-24. An example of a two-pipe reverse return system.

- If subsequent zones are experiencing excessive flow, consider incorporating flow control or balancing valves on each zone.
- When utilizing individual zone valves for temperature control of each terminal unit in the two-pipe reverse return system, consider incorporating a variable speed circulation pump or a ***differential pressure bypass valve*** to lessen the chance of pump deadheading if all of the zone valves are closed, **Figure 7-25**.

7.3.5 Primary-Secondary Piping

The ***primary-secondary piping*** system involves the installation of two different piping circuits. The first, or primary, circuit simply circulates water throughout

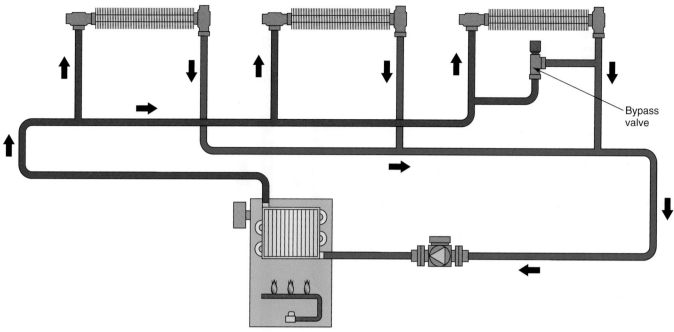

Figure 7-25. A two-pipe reverse return system utilizing a bypass valve to maintain flow through the system.

the boiler and provides a main supply of water for the individual terminal circuits. The secondary circuits are used to feed water through their individual terminal units, and each secondary circuit acts independently of the others, **Figure 7-26**. This piping arrangement is quite popular for several reasons:

- The primary piping circuit to and from the boiler does not need to be a fixed length, and in some cases it may need to consist of only 10′ of piping.

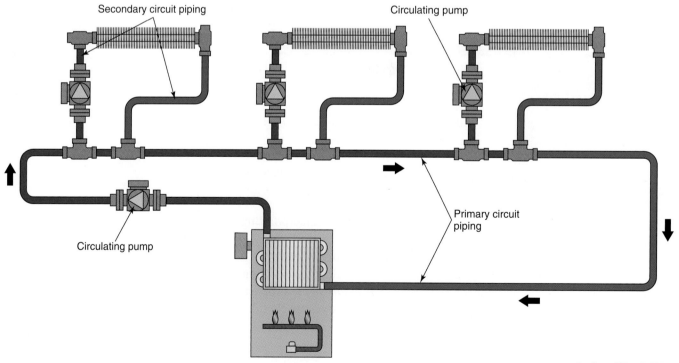

Figure 7-26. An example of a primary-secondary piping circuit.

- This type of piping arrangement offers a simple way to protect the boiler against a lower-than-desired return water temperature. As mentioned earlier, in some boilers a return water temperature below 140°F can result in flue gas condensation that can damage the boiler's heat exchanger.
- The designer can accurately size the system to provide for a 20°ΔT across the piping system.
- The secondary branch circuits can be piped with standard tees rather than diverter tees.
- The individual secondary circuits can be utilized for different types of heating loads such as domestic hot water, radiant floor heating, or even swimming pool heating.

Another advantage of utilizing the primary-secondary piping arrangement is that each secondary branch circuit can be individually controlled with a zone thermostat. This thermostat can energize the branch circuit pump upon a call for heating.

7.3.5.1 Hydraulic Separation

In the primary-secondary heating system, it is essential that there be **hydraulic separation** between the primary and secondary heating loops. Hydraulic separation occurs when the primary and secondary loop pumps operate independently of each other and act as if the other pump does not exist. Without it, the result would be incorrect flow rates between the primary and secondary loops. This could lead to improper return heating water temperatures back to the boiler and an overall inefficient heating system. There are several methods to achieve the proper hydraulic separation within a system, two of which are the use of **closely spaced tees** or the installation of a **low-loss header**. Closely spaced tees are the branch tees located on the boiler supply header. These tees deliver the supply and return heating water to the individual terminal units, with the spacing between the two branch tees known as common piping, **Figure 7-27**. To understand how closely spaced tees work in a primary-secondary heating loop, check out *Did You Know: The Path of Least Resistance*.

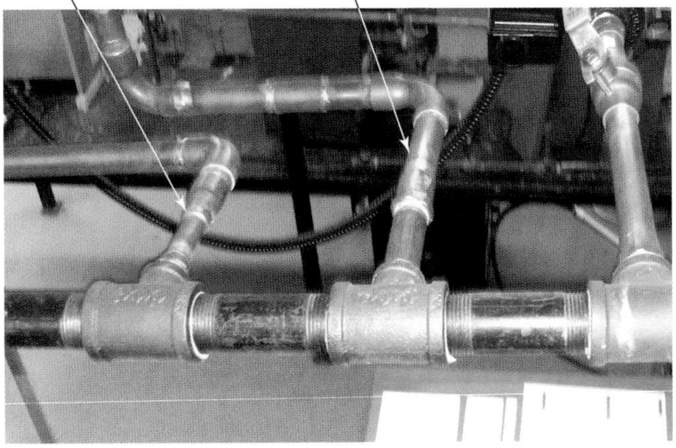

Another method for generating hydraulic separation is to install a low-loss header. A low-loss header works by separating the primary and secondary piping circuits. This allows the boiler to operate at a constant flow rate in the primary circuit, even if the flow rate and temperature varies in the secondary circuit. This avoids any interaction between the two piping circuits and enables the primary circuit return temperature to represent the load on the system.

Low-loss headers offer several advantages over closely spaced tees and are quite popular with today's modern condensing boilers, **Figure 7-28**. Because of the reduction in flow velocity within the low-loss header, air bubbles are more easily separated and automatically removed from the system by means of an air vent at the top of the low-loss header. In addition, because of the reduction in flow velocity, debris and foreign particulate are more inclined to fall to the bottom of the low-loss header, where they can periodically be flushed from the system. Proper boiler flow rates are more easily achieved from the use of a low-loss header, especially with systems that incorporate branch circuits that have variable flow rates caused by system control valves or thermostatic radiator valves.

Figure 7-27. An example of piping incorporating closely spaced tees.

HydroCal™ (2-6″ANSI flange, ASME/CRN), © Caleffi North America, Inc.

Figure 7-28. This photo shows a low-loss header.

7.3.5.2 Considerations When Designing a Primary-Secondary Piping System

- Size the primary loop pump to meet the heating demand of all secondary branches and the proper water flow through the respective terminal units.
- Each branch or secondary circuit must have its own circulating pump. This will allow the hot water to be injected into the individual branches or secondary circuits.
- The spacing between the closely spaced tees must be limited to four pipe diameters apart, with a maximum of 12″ between the two tees. This limited spacing will reduce the flow resistance between the two tees. For example, if the main loop is 2″ in diameter, the maximum spacing between the closely spaced tees would be 8″, **Figure 7-29**.
- Low-loss headers should be sized to minimize the amount of pressure losses when the system is at full flow.
- Low-loss headers should be piped so that the supply water from the boiler and to the secondary circuits is located at the top and the return water piped at the bottom. This ensures that the proper supply water temperature to the secondary circuits is maintained and avoids temperature dilution within the low-loss header, **Figure 7-30**.

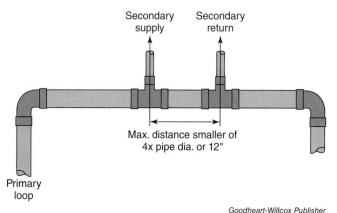

Goodheart-Willcox Publisher

Figure 7-29. This illustration shows the proper spacing for closely spaced tees.

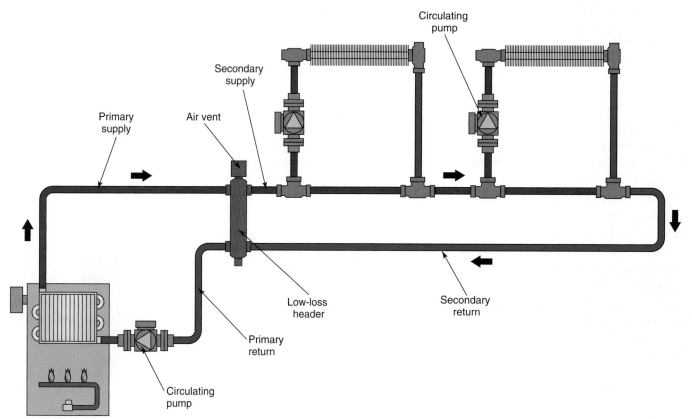

Goodheart-Willcox Publisher

Figure 7-30. This illustration shows the proper piping method for a low-loss header.

DID YOU KNOW?

The Path of Least Resistance
When piped properly, the primary-secondary piping arrangement reduces *standby losses* to the boiler. This means that no water will migrate through the secondary circuit and its terminal unit when the zone is not calling for heat. The arrangement works because of what happens when water flows through a tee. Whatever water enters the tee must come out of the tee. As the primary water enters the tee, it has two possible paths: either go straight through the tee or divert around to the terminal unit—whichever way is easier. With the secondary branch circulating pump deenergized, the path of least resistance is straight through the tee. When the pump is energized, the path of least resistance is diverted through the terminal unit. See **Figure 7-31**. This is why a maximum of four pipe diameters between the branch tees is so important. If the tees are spaced too far apart, the branch circuit will be the primary hot water flow's path of least resistance and two things can happen:

- Less water may travel through the terminal unit when the branch pump is energized.
- Water may begin to travel through the terminal unit when the branch pump is deenergized, rather than passing straight through the tee.

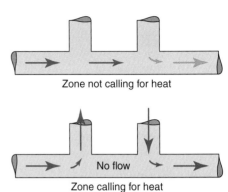

Goodheart-Willcox Publisher

Figure 7-31. This illustration shows why the proper spacing between closely spaced tees is important for reducing standby losses to the boiler.

CAREER CONNECTIONS

Hydronic System Designer
A hydronic system designer plays an important role in the building industry to ensure that the proper climate of homes, businesses, and other facilities stays comfortable despite extremes in outdoor weather.

The hydronic system designer's duties include overseeing the creation, installation, maintenance, and repair of various heating and cooling systems. Hydronic system designers can create and design systems for residential, commercial, and industrial buildings. Institutional buildings such as schools and health-care facilities are also included. The designer typically works with other engineering colleagues, as well as with installers and clients, to design new systems or find solutions to existing system issues. They may work for consulting or design firms, government agencies, facilities management offices, the military, or with HVAC mechanical contractors.

Entry-level jobs for hydronic system designers usually require a bachelor's degree in either mechanical engineering or HVAC engineering technology. Some states and municipalities require licensing for system designers. Professional certification may be obtained through such organizations as the American Society of Heating, Refrigerating, and Air-Conditioning Engineers (ASHRAE) for specific applications in health-care facilities, high-performance building design, and commissioning process management.

Education requirements include classes in HVAC theory and application, fluid flow dynamics, advanced math, HVAC system analysis and design, and computer-aided design. The skills required to become a hydronic system designer include solid problem-solving skills in HVAC system technology, computer skills, good verbal and written communications skills, CAD software training, and a robust ability to pay attention to details.

The US Bureau of Labor and Statistics reported that the median annual salary in 2017 for jobs involving hydronic design was $55,300 per year. Employment of system designers is projected to grow about 5% over the next several years, and the opportunities are great for candidates who can master new software and technology in addition to traditional skills.

Chapter Review

Summary

- A hydronic heating system must be sized correctly in order for it to work properly.
- The sensible heat formula can be used to determine the flow of fluid through a system.
- The system's fluid velocity can affect the water flow.
- Laminar fluid flow through a piping system is more desirable than turbulent flow.
- The system's static pressure takes into account the weight of water within the system.
- A number of factors need to be considered when designing a piping system. These include the heat load on the building, the size of the piping, the type of piping arrangement, the friction and head losses through the piping, the pump performance, and pump horsepower.
- The series-loop piping arrangement is the simplest type of arrangement to install, but it does not allow for individual zone control.
- The one-pipe system has multiple zones, which require special diverter tees to allow for proper water flow through each terminal unit.
- In the two-pipe direct return piping arrangement, individual terminal units are piped in parallel to the supply and return piping from the boiler. Balancing valves should be used at each terminal unit on a two-pipe direct return system to compensate for pressure drop across the system.
- A two-pipe reverse return piping system is sometimes considered a self-balancing system because of its unique piping arrangement. Variable-speed circulating pumps or a pressure differential bypass valve should be incorporated to avoid deadheading the pump.
- A primary-secondary piping system requires the incorporation of two separate piping circuits—one for the boiler water circulation and one to serve as the feed water circuit to the branch terminal units.
- Primary-secondary piping circuits have several advantages over other types of piping systems, including the ability to utilize different types of heating applications through the individual branch circuits. Closely spaced tees are required between the supply and return lines of terminal units on a primary-secondary piping system.

Know and Understand

1. In what increment is the rate of fluid flow through a pipe expressed?
   A. Gallons per hour
   B. Gallons per minute
   C. Feet per second
   D. Feet per minute
2. What would be the fluid flow of water through a terminal unit that has a heating capacity of 15,000 Btu/hr and a temperature drop of 10°F?
   A. 3 gallons per minute
   B. 5 gallons per minute
   C. 10 gallons per minute
   D. 15 gallons per minute
3. When designing a hydronic heating system, where can one find the determined pressure losses through heating equipment and terminal units?
   A. On the equipment rating plate
   B. From the manufacturer's sales brochure
   C. From the manufacturer's equipment data
   D. All of the above.
4. Cast-iron radiators are considered terminal units with _____.
   A. a high thermal mass
   B. a medium thermal mass
   C. a low thermal mass
   D. no thermal mass

5. What is the cone-shaped device located within a diverter tee?
   A. A bypass
   B. A funnel
   C. A biflow
   D. A venturi
6. How many diverter tees should be used on a terminal unit that is installed below the main piping circuit on a one-pipe system?
   A. One diverter tee on the supply side
   B. One diverter tee on the return side
   C. Two diverter tees—one on the supply side and one on the return side
   D. Two diverter tees—both on the return side
7. What type of piping arrangement is the two-pipe direct-return system?
   A. Series
   B. Parallel
   C. Split loop
   D. Equivalent length
8. Why should a balancing valve be located on the outlet of each terminal unit on a two-pipe direct return piping system?
   A. To reduce the fluid temperature through the terminal unit
   B. To increase the fluid temperature through the terminal unit
   C. To adjust the pressure drop across the terminal unit
   D. To convert the piping system to a reverse return
9. Which type of piping arrangement is said to be *self-balancing*?
   A. Series
   B. Parallel
   C. Two-pipe direct return
   D. Two-pipe reverse return
10. What is an advantage of using a low-loss header instead of closely spaced tees on a primary-secondary piping system?
    A. Air is easily removed from the system.
    B. Debris can be flushed from the system.
    C. Proper flow rates are more easily achieved.
    D. All of the above.

Apply and Analyze

1. Which type of flow is more desirable through a hydronic heating system, laminar or turbulent? Why?
2. Calculate the feet of head in a hydronic loop if the system pressure equals 12 psi.
3. Explain what is meant by determining the "equivalent length of pipe" for each fitting when designing a hydronic heating system.
4. Explain why a series-loop piping arrangement does not lend itself to individual zoning control.
5. Why would the differential water temperature across the boiler on a series-loop piping system be greater than on a primary-secondary type of piping system?
6. Explain why diverter tees are important on terminal units of a one-pipe system with multiple zones.
7. Explain why the distance of closely spaced tees on the secondary branch of a primary-secondary piping circuit should be limited to a maximum of four pipe diameters apart.
8. Explain why a variable-speed circulating pump might be advantageous to use on a two-pipe reverse return system.
9. Describe the advantage of a two-pipe reverse-return system over other types of piping arrangements.
10. Explain the difference between static pressure and system pressure.

Critical Thinking

1. Why should the hydronic system designer consider choosing terminal units that have a similar thermal mass on a series-loop piping system?
2. How would a high-rise building be designed to overcome the high static pressure from piping that could extend upward through multiple stories?

8 Boiler Control and Safety Devices

Chapter Outline
8.1 Pressure Relief Valves
8.2 Low-Water Cutoff
8.3 Flow Switch
8.4 Backflow Preventer
8.5 Aquastat Relay
8.6 Outdoor Reset Controller
8.7 Electronic Controls

Bespaliy/Shutterstock.com

Learning Objectives

After completing this chapter, you will be able to:
- Describe the difference between a boiler control device and boiler safety device.
- Explain how a boiler pressure relief valve functions and how to test it.
- Describe how a low-water cutoff switch helps protect a boiler from damage.
- List the different types of hydronic flow switches and their functions.
- Explain how a backflow preventer helps protect the domestic water source.
- Compare the different functions of an aquastat relay.
- Describe the sequence of operation for an aquastat relay.
- Explain how an outdoor reset controller operates and list its benefits.
- Identify how modern electronic boiler controls differ from conventional controls.
- Describe the difference between two-stage boiler controls and fully modulating controls.

Technical Terms

aquastat relay
backflow preventer
control device
current relay
differential pressure switch
flow switch
low-water cutoff (LWCO)
outdoor reset controller
paddle switch
pressure relief valve
safety device
solid-state circuitry

Residential and light commercial boiler packages are equipped with a number of different devices. Most hydronic heating boilers include two distinct categories of devices:
- Safety devices
- Control devices

Safety devices for hydronic systems protect the boiler, hot water piping, mechanical room, and most of all building occupants from dangers that can occur when controlling a pressurized vessel for comfort heating and hot water generation. Safety devices include:
- Pressure relief valves
- Low-water cutoff switches
- Flow switches
- Backflow preventers

Control devices used with hydronic systems can increase system efficiency, reduce energy consumption, and make the controlled space more comfortable for the building occupants. Control devices include:
- Aquastat relays
- Outdoor reset controllers
- Electronic controls

This chapter describes the design and functionality of control and safety devices used with boilers and hydronic heating systems.

Note: There are a number of different types of valves used in conjunction with the control devices found on many boilers and hydronic systems. These control valves are discussed in detail in Chapter 9, *Valves*. Please refer to Chapter 9 and these types of valves when reviewing the control devices mentioned in this chapter.

8.1 Pressure Relief Valves

One of the most important safety devices found on any boiler is the ***pressure relief valve***, **Figure 8-1**. In fact, pressure relief valves are a code requirement for any closed-loop hydronic heating system. Consider the consequences of a defective control device failing to de-energize the boiler's burner circuit once

Goodheart-Willcox Publisher
Figure 8-1. An example of a pressure relief valve.

the hot water control set point is reached: the system's water temperature would continue to increase, and the system pressure would rise as well. Without the safety of the pressure relief valve in place, the weakest device within the system would eventually rupture and the consequences could be disastrous.

Most hydronic devices found on a residential or light commercial hot water boiler system are rated for at least 60 pounds per square inch (psi) and typically can withstand pressures up to double this rating before rupturing. The ASME Boiler and Pressure Vessel Code requires that every boiler have at least one properly rated pressure relief valve that is set to open at or below the maximum allowable working pressure for each low-pressure boiler. Pressure relief valves used for residential and light commercial boiler systems are usually rated to open at 30 psi. However, valves with higher pressure ratings as well as valves with adjustable pressure ratings are available for larger-sized boilers.

The internal workings of a pressure relief valve are illustrated in **Figure 8-2**. They consist of a valve body, a valve seat and spring, a stem with a fixed or adjustable pressure setting, an inlet nozzle, and a seal. They are similar to the makeup of a pressure-regulating valve. When the pressure exerted upon the valve seat through the inlet nozzle exceeds that of the rated pressure spring, the valve seat lifts off the seat holder and the water passes through the valve. Pressure relief valves contain a tag or rating plate that lists the rated opening pressure and the maximum heating capacity of the device for which the valve is rated, **Figure 8-3**.

> **CODE NOTE**
>
> **Pressure Relief Valves and Discharge Tubes**
> Every pressure relief valve shall have a discharge tube connected to the threaded outlet of the valve that shall be the same diameter as the valve outlet—never smaller. Copper is the preferred construction material for the discharge tube. The discharge tube should be terminated not more than 6 inches above the floor or floor drain. Do not connect the discharge tube directly into a floor drain. Also, all pressure relief valves should be installed in a vertical position. This will reduce the chance of sediment accumulating inside the valve, which could prevent the valve from seating properly, resulting in leaks.

8.1.1 Testing a Pressure Relief Valve

Most manufacturers of pressure relief valves recommend inspecting and testing them annually. Every pressure relief valve contains a lever at the top of the valve that allows the valve to be opened manually. This lever serves two purposes:
It allows for the purging of air from the system when the boiler is initially filled.
It can be used to perform a service test to ensure the valve is working properly.

To test the valve, first lift the lever manually. This will result in hot water being flushed through the valve and down through the discharge tube. Be careful: the discharged hot water could cause scalding of exposed skin. After manually opening the valve, release the lever and allow it to close. It should automatically shut off and seal tightly. If it seals, the test is complete. If the valve closes but continues to drip, there may be sediment obstructing the seat. This problem may be alleviated by lifting the lever several times in an effort to flush the valve seat. If this does not work, gently tapping on the center of the valve body may reseat the valve washer. If the valve does not shut off and continues to drip, it may need to be replaced.

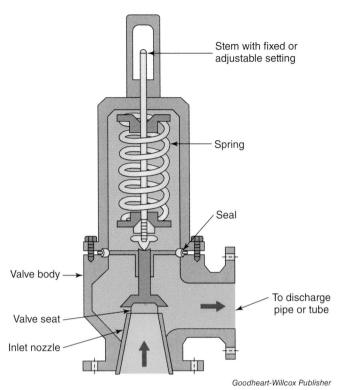

Figure 8-2. This illustration shows the internal workings of a pressure relief valve.

Goodheart-Willcox Publisher

Figure 8-3. A pressure relief valve rating plate lists the rated opening pressure and maximum heating capacity of the device for which the valve is rated.

DID YOU KNOW?

Pressure Relief Valve versus T&P Valve

Do you know the difference between a boiler pressure relief valve and a water heater T&P valve? Boiler pressure relief valves are typically rated for 30 psi and respond to excessive pressure only. Temperature and pressure (T&P) valves are designed to protect both gas and electric water heaters and are rated for 150 psi, at a temperature of 210°F. These two safety devices are designed for separate purposes and should not be interchanged.

SAFETY FIRST

Never Cap off a Pressure Relief Valve

Under no circumstances should the valve or discharge tube be capped off to prevent a leak! Doing so could result in a boiler explosion that could prove deadly!

8.2 Low-Water Cutoff

The *low-water cutoff (LWCO)* switch is another important safety device used for hot water boiler systems, **Figure 8-4**. As the name implies, a low-water cutoff is designed to de-energize the boiler's burner circuit if the water level within the boiler falls below a predetermined point. If, for instance, there is an unseen leak somewhere in the heating system, the makeup water controls would normally replenish the boiler water to its correct level. However, if that leak is so extreme that the makeup water system cannot keep up, the boiler could experience a condition known as *dry firing*—operating the boiler below its proper water level. This condition can turn a boiler into a ticking time bomb, and the consequences could be disastrous.

Original low-water cutoff designs used a mechanical float switch, similar to the float arm on a toilet tank, **Figure 8-5**. As the water level within the boiler dropped, the arm fell, opening the electrical contact switch connected to the heating system circuit. Newer LWCO controls use a conductivity sensor. This

Goodheart-Willcox Publisher

Figure 8-4. A low-water cutoff switch protects the boiler if the water level becomes too low.

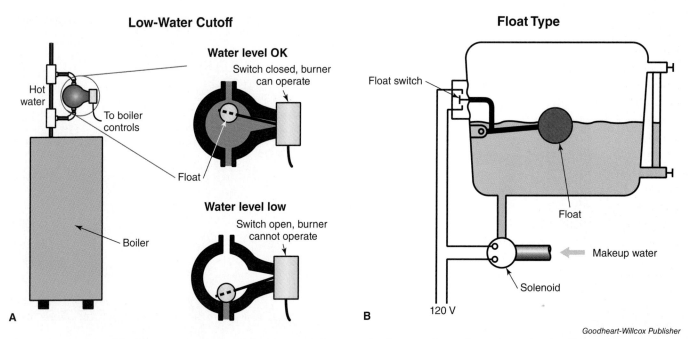

Figure 8-5. A—An original low-water cutoff switch that uses a float. B—The float arm in a toilet tank. Notice that if the water level drops, the float valve shuts off the switch to the boiler controls.

sensor uses a moisture bridge that closes the contacts of the LWCO when it is submerged in water. When the sensor becomes dry, the contacts open. This type of sensor reduces the chance of a cutoff malfunction, as seen in the older float-type controls. Low-water cutoff controls can be either automatic reset or manual reset. Manual reset controls are considered safer because they prompt the operator to investigate the system to determine the cause of the low-water situation. When installing low-water cutoff controls, always consult the manufacturer's installation instructions for proper placement and correct wiring. Controls are normally installed directly on the boiler; however, some manufacturers allow for installation on the piping near the outlet of the boiler supply, **Figure 8-6**.

CODE NOTE

Low-Water Cutoff Requirements
Certain states require by law that a low-water cutoff be installed on every boiler. In other states, it is governed by the size of the boiler. Whether or not it is mandatory by state, city, or local code requirement, it is always a good safety practice to equip all boiler systems with a low-water cutoff.

TECH TIP

Low-Water Cutoffs with Radiant Flooring
Low-water cutoff devices are a good idea for use with radiant floor heating systems. Because the hot water tubing is buried in the concrete floor, the water level is below the level inside the boiler. If a leak should occur in the floor's tubing, it could cause the system to drain itself.

8.3 Flow Switch

Hydronic heating systems require that the correct water flow be maintained through the boiler in order for the system to operate safely and properly. To accomplish this, boilers incorporate *flow switches* to ensure proper water flow through the boiler before the burner is activated. The flow switch is wired in series with the heating circuitry. When the circulating pump is first activated, water flow through the piping is sensed by the flow switch. When the correct amount of flow is generated, the switch contacts close and the burner controls energize. If the boiler were allowed to fire when the system had no flow, it could overheat the water, which would result in the high temperature limit switch tripping or even possible damage to the boiler's heat exchanger.

Several types of flow switches are available. Only one is required per system. The most common ones used with hydronic heating systems are:

- Paddle switches
- Differential pressure switches
- Current relays

Goodheart-Willcox Publisher

Figure 8-6. The LWCO switch should be mounted on the boiler, or as close to the outlet piping as possible.

Paddle switches are water flow switches that have a blade suspended within the water piping, **Figure 8-7**. When the water flows through the pipe, it pushes on the paddle, causing a set of contacts to close. Paddle flow switches are very good at proving water flow, but they are susceptible to residue buildup from foreign objects in the water, which can cause them to stick open or shut. They do need periodic maintenance to make sure they operate properly, and they should be checked for wear due to their constantly moving parts.

When water flows through a circulating pump, a difference in pressure is generated across the pump inlet and outlet. *Differential pressure switches* measure this difference in pressure to prove flow through the piping circuitry, **Figure 8-8**. Similar to the paddle switch, the differential pressure switch contains a set of normally open contacts that close when the switch senses the proper difference in pressure. Unlike the paddle switch, the differential pressure switch may be adjusted to account for differing pump sizes.

A *current relay*, or current sensing switch, proves flow through the hydronic circuit using a different method, **Figure 8-9**. The current relay attaches to the electrical leads of the circulating pump in much the same way an ammeter clamps onto a wire to measure current. When the circulating pump is energized, the current that passes through the electrical wire is detected by the current relay. Upon detecting this current, the relay contacts close, allowing the boiler's burner circuit to energize. What makes the current sensing relay so reliable is that there are no moving parts to fail, and the current must be strong enough for the relay contacts to close.

The flow switch should be installed near the boiler inlet for accuracy and tested on a regular basis. Test it by disconnecting the wires leading to the burner circuit while the boiler is operating. A properly operating flow switch will immediately shut down the burner when this test is performed. An ohmmeter can verify that the flow switch is operational and can also be used for troubleshooting purposes. For instance, with the boiler shut off, disconnect the wires to the

Universal Flow Switch, © Caleffi North America, Inc.

Figure 8-7. A paddle-type flow switch uses a blade suspended in the water piping. When water flows through the pipe, it pushes the paddle closed and engages a set of contacts.

Dwyer Instruments, Inc.

Figure 8-8. A differential pressure switch measures the flow across the circulating pump to prove flow through the boiler.

Functional Devices, Inc.

Figure 8-9. A current relay, or current sensing switch, measures the amperage through the circulating pump wiring to prove flow through the pump.

LWCO and measure the resistance between the contact terminals. If the LWCO is a paddle, differential pressure, or current relay-type sensor, there should be infinite resistance. If the contacts show they are closed, replace the sensor.

8.4 Backflow Preventer

A *backflow preventer* is required in the hydronic heating system to prevent the boiler water from flowing backward into a domestic water source—be it a well or municipal water system, **Figure 8-10**. For instance, some hydronic heating systems use an antifreeze or glycol solution to prevent pipes from freezing in the event of a flame failure, and the backflow preventer is used to prevent this solution from flowing backward and contaminating the drinking water supply. When the domestic or potable water source enters a building from the main water supply, it should only flow in one direction—into the building. However,

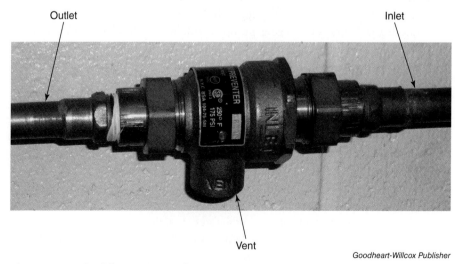

Goodheart-Willcox Publisher

Figure 8-10. A backflow preventer is used to prevent any boiler water from flowing backward into the domestic water supply.

conditions could change—such as a sudden pressure change due to a ruptured water supply line, allowing this water source to flow backward. When this situation occurs, the water pressure within the hydronic heating system is greater than the incoming or makeup water pressure. Because fluid will always flow from a higher to a lower pressure, the boiler water can actually flow backward and seep into the main water supply system.

CODE NOTE

Backflow Prevention

Backflow prevention is required on all hydronic systems using makeup water. It is addressed in the International Plumbing Code under Section 608.13. This section states that backflow preventers shall conform to ASSE (American Society of Safety Professionals) or CSA (Canadian Standards Association) standards, and that the relief opening shall discharge by an air gap.

The backflow preventer is made up of two separate check valves with a vent port in between them, **Figure 8-11**. When the inlet water pressure is greater than the boiler water system's pressure, the upstream check valve allows the makeup water to pass through and fill the hydronic system. If for some reason the hydronic system water pressure becomes greater than the supply-side makeup water pressure, the downstream side check valve will open, allowing the boiler's system water to safely drain through the vent port and preventing any back-siphoning.

The backflow preventer is installed between the water supply source and the pressure-reducing valve, **Figure 8-12**. It should always be mounted horizontally with the vent port pointed downward to allow any fluids to drain naturally from the hydronic system if a problem should occur. Always remember to check for the proper direction of flow on the device. To avoid any debris from interfering with the backflow preventer, flush the supply piping before installation. A strainer located on the inlet side will capture any inbound particulate and may need to be removed and cleaned if any problems occur with the inlet water

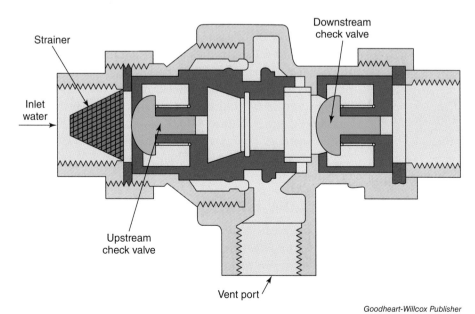

Goodheart-Willcox Publisher

Figure 8-11. This illustration shows the internal workings of the backflow preventer.

DID YOU KNOW?

Using a Backflow While Doing the Backstroke

Backflow preventers are also required on the supply water source for municipal swimming pools and spas. Any outdoor supply water source—such as a faucet or hose bib—must be protected from the potential siphoning of the pool or spa water back into the municipal water supply. Although these types of backflow preventers look different from the ones used for hydronic heating system applications, they function in the same manner.

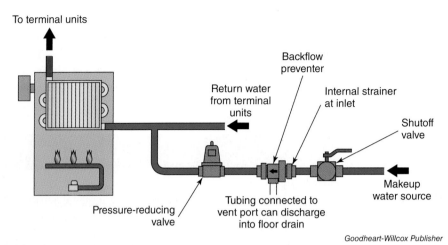

Goodheart-Willcox Publisher

Figure 8-12. This illustration shows that the backflow preventer should be installed between the boiler water source and the pressure-reducing valve.

pressure. Some water seepage from the vent port can take place during initial filling of the hydronic system, especially if the supply-side water pressure falls below atmospheric pressure. This is acceptable and should cease once the valve seats have properly sealed. Otherwise, any leaks during standard boiler operation are not normal and need attention. Repair kits for backflow preventer valves that include replacement valve seats are available from some manufacturers.

8.5 Aquastat Relay

An *aquastat relay* is a common control device used on many earlier residential and light commercial boilers, **Figure 8-13**. It is a temperature-controlled switch designed to cycle the boiler's burner controls based on the hot water temperature. Aquastats are, in fact, a combination controller for hot water boilers. Some aquastats have the capability of adjusting the high-limit, low-limit, and differential temperature settings. In addition to temperature control, aquastat relays are designed to cycle the hot water circulating pump as well as control the space temperature setting.

Aquastat relays have a copper sensing bulb attached to the back of the relay housing, **Figure 8-14**. This sensing bulb is not directly immersed in the boiler water, but rather housed in an immersion well inside the boiler's heat exchanger, thus eliminating the need to drain the boiler if the relay ever needs replacing. In addition, most aquastats include a built-in transformer for interfacing low-voltage thermostat controls as well as a relay to energize the circulating pump, **Figure 8-15**. Knowing how the aquastat relay controller operates will help technicians and building owners understand why their boiler is shutting down unexpectedly.

Goodheart-Willcox Publisher

Figure 8-13. The aquastat relay is a boiler device that controls the hot water temperature and the circulating pump.

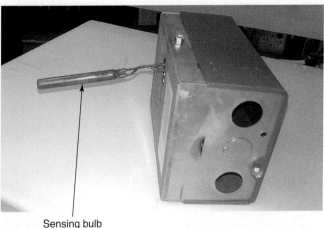

Goodheart-Willcox Publisher

Figure 8-14. The sensing bulb on the aquastat relay is placed in an immersion well, which is mounted inside the boiler heat exchanger. The bulb senses the hot water temperature.

8.5.1 Low-Limit Control Setting

Some aquastat models have low-limit settings that are preset at the factory. On models with adjustable low-limit temperature control settings, this set point is typically set at the lowest desired hot water temperature. This is based on outdoor climate conditions and the desired space temperature setting, but is usually no lower than 120°F. When

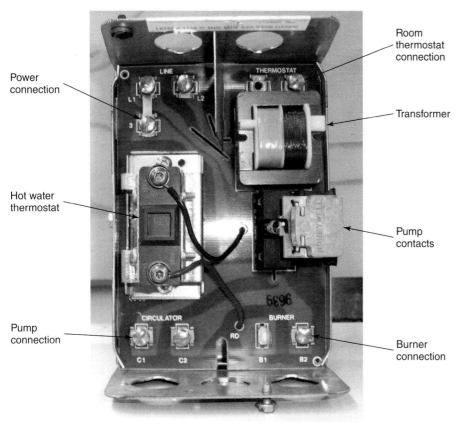

Goodheart-Willcox Publisher

Figure 8-15. This illustration shows the internal workings of the aquastat relay.

the boiler water temperature falls below the low-limit setting, the boiler's burner is energized and begins to heat the water to maintain the desired temperature set point.

8.5.2 High-Limit Control Setting

Most aquastat relays have an adjustable high-limit temperature control setting. This is the highest allowable water temperature setting before the controller de-energizes the boiler's burner circuit. If the controller has a fixed temperature differential set point, the burner will restart when the water temperature falls 10°F below the high-limit temperature setting. This scenario will maintain a relatively consistent hot water temperature throughout the hydronic heating system. In most cases, the high-limit control set point should be set between 180°F and 200°F. To prevent the boiler hot water from reaching its boiling point, it is highly recommended that the high-limit set point be limited to no more than 200°F.

8.5.3 Differential Temperature Control Setting

The differential control setting found on some models of the aquastat relay allows for setting the difference between cut-in and cut-out hot water temperatures, **Figure 8-16**. For instance, if the desired hot water temperature set point is 160°F and the differential is set at 20°F, the cut-in temperature set point would be 140°F. The lowest recommended differential temperature set point is usually 10°F. Anything lower than that can lead to short cycling of the heating circuit and circulating pump. When the boiler short cycles, it is turning on and off too quickly. This situation leads to an uncomfortable environment, causes the system to be inefficient, and can damage the equipment over time. Temperature

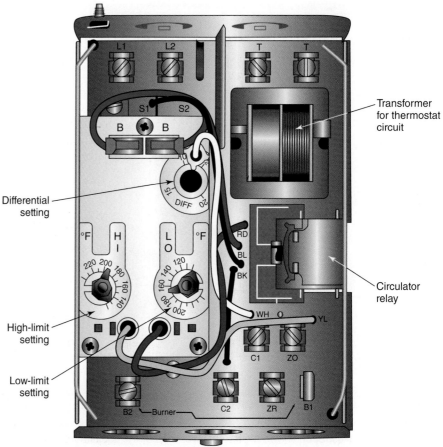

Figure 8-16. Some aquastat relays have adjustments for high-limit, low-limit, and differential temperature control settings.

differentials that are set too high will lead to wide space temperature swings and uncomfortable conditions.

8.5.4 Sequence of Operation

The sequence of operation for an aquastat relay is as follows:
1. Upon a call for heat from the conditioned space, the thermostat closes, sending a signal to the aquastat relay.
2. The aquastat relay energizes the circulating pump and the burner circuit simultaneously.
3. The burner operates until the hot water temperature set point is satisfied, then the aquastat relay de-energizes the heating circuit.
4. The circulating pump continues to operate until the space temperature set point is met.
5. When the space temperature has reached its set point, the thermostat opens, which sends a signal to the aquastat, opening the pump contacts and de-energizing the circulating pump.

It is important to remember that the thermostat is controlling the conditioned space, and the aquastat is controlling the boiler hot water temperature. Both devices operate together to maintain the proper space temperature conditions. Aquastat relays can be used in conjunction with multiple zone valves to control multiple spaces with one boiler. Zone valves and their function are covered in detail in Chapter 9, *Valves*.

8.6 Outdoor Reset Controller

One of the most popular temperature controls for hydronic systems is the ***outdoor reset controller***. Although it has existed for years, the outdoor reset controller is now a required control included with every boiler package purchased today. The outdoor reset controller aims to reduce operational costs without sacrificing comfort. To understand how the outdoor reset controller works and its benefits, we must first look at how a boiler heating system is designed.

In order to accurately design a functional hydronic heating system for residential or light commercial applications, we must first collect accurate information on the following two parameters:

- The proper outdoor winter design temperature
- The correct heat load on the building

The boiler's heating system is typically designed to deliver the highest water temperature output in order to match the building's heat loss on the coldest day of the year. However, there is one flaw with this concept: most areas in North America have fewer than 10 days per year when the outdoor temperature meets or falls below the winter design temperature. What this means is that for the remaining days of winter, the hydronic heating system is essentially oversized. In addition, most boiler controls operate at a single set point that does not allow for modulation of the hot water temperature. To provide precise indoor space temperatures, the heat supplied to the building should equal the proportional heat loss from the building. If too much heat is supplied to the building, the conditioned space will overheat and create an uncomfortable environment. If too little heat is supplied, the space can become drafty and uncomfortable. An outdoor reset controller corrects these situations by automatically adjusting the boiler's hot water temperature set point based on the outdoor air temperature. As the outdoor air temperature drops, the hot water supply temperature is increased. When the outdoor air temperature increases, the hot water supply temperature set point decreases. This is accomplished automatically by a controller that calculates a reset ratio, **Figure 8-17**. This ratio can be programmed by the installer and adjusted as needed.

Goodheart-Willcox Publisher

Figure 8-17. Outdoor reset controllers such as this will automatically control the hot water temperature based on the outdoor air temperature.

The reset ratio is the degrees of hot water temperature change for each degree of outdoor air temperature change. For instance, a ratio of 1:1 means that for every degree the outdoor air temperature falls, the boiler water temperature set point increases one degree. A ratio of 1:2 means that a 1-degree fall in outdoor air temperature results in a 2-degree rise in the hot water temperature. You can study these ratios in **Figure 8-18**. According to this graph, if the controller were set up for a 1:1 ratio when the outdoor temperature reached 0°F, the boiler water temperature set point would be 170°F. When the outdoor air temperature increased to 50°F, the hot water temperature would decrease to 120°F. The proper reset ratio setup is based on the type of terminal unit utilized. Conventional baseboard fin tube heaters typically require a 1:1 ratio. However, higher-mass terminal units such as PEX piping used in radiant floor heating will incorporate a higher offset, such as a 4:1 ratio. For proper ratio setup, technicians should refer to the operation manual provided by the manufacturer of the outdoor reset controllers. A simple outdoor

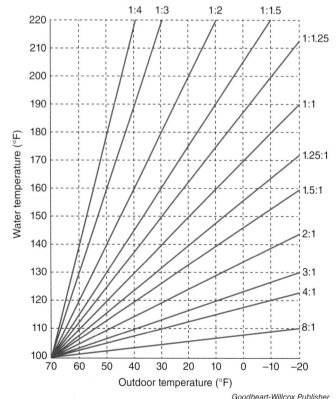

Goodheart-Willcox Publisher

Figure 8-18. The reset ratio programmed into the outdoor reset controller is determined by the types of terminal devices used.

tekmar Control Systems Ltd.

Figure 8-19. An outdoor reset control package is typically made up of a sensor for outside air temperature, a sensor for hot water temperature, and a microprocessor controller.

reset control package consists of sensors for outside air and hot water, plus a microprocessor controller. See **Figure 8-19**.

The use of an outdoor reset controller offers many benefits, including:

- **A substantial reduction in fuel usage.** Because the boiler operates at a lower supply water temperature, an energy cost savings of 10%–15% can be realized.
- **A more consistent indoor air temperature.** Because the hot water temperature tracks the outdoor air temperature, the rate of heat output from the boiler is closer to the rate of heat loss in the building. This results in a decrease of space temperatures "overshooting" or "undershooting."
- **Continuous water circulation.** Because of a more consistent hot water temperature, the circulating pump will operate for longer cycles. This results in a more uniform delivery of heat through terminal devices, which is especially important for radiant floor heating systems.
- **Noise reduction.** Because changes in the hot water temperature set point are gradual, there is less likelihood of piping rapidly expanding due to temperature fluctuation. This results in the reduction of "water hammer."

TECH TIP

Outdoor Reset Control with Cast-Iron Boilers

There may be some concern about applying an outdoor reset controller to an older cast-iron boiler, due to the fact that these types of boilers are subject to premature failure if the return water temperature falls below 140°F for an extended period of time. Fortunately, the manufacturers of outdoor reset controllers have built in a minimum water temperature setting for the protection of conventional boilers.

8.7 Electronic Controls

Newer boiler packages for residential and light commercial hydronic heating systems typically offer *solid-state circuitry* as standard equipment for the control of hot water temperatures. These electronic control modules offer many features not found with conventional heating controls. By providing additional features such as multipoint temperature monitoring, night setback control, outdoor air reset control, and even control of domestic hot water (DHW), these solid-state control modules can reduce fuel operating costs while providing a more comfortable environment.

GREEN TIP

Wireless Controls Save Energy—and Money!

Wireless hydronic controls overcome difficult installation problems because there is no intrusion into walls and ceilings, thus making them easier and less costly to install.

Efficiency gains are realized when data generated by wireless controls through the controller software can be sent to computers, handheld devices, or even building management systems to accurately assess the building's performance.

Another important benefit of wireless controls is the improved energy savings that customers can realize. Today's wireless controls offer such benefits as optimized demand calculations from the room thermostat to the boiler controller. This feature allows the system to "learn" how the building reacts to changes in room and outdoor air temperatures to ensure the equipment is energized for optimal periods of time, which not only helps the environment but also translates to energy consumption savings of up to 13% for the customer.

It is important that the installation and service technician reviews and understands the setup and programming instructions for these modern electronic control modules that accompany most boilers used today for residential and light commercial hydronic heating applications. The following sections examine several types of solid-state control modules found on boiler packages.

8.7.1 High Efficiency Two-Stage Control Module

This boiler offers a two-stage firing system providing high-fire and low-fire capabilities, **Figure 8-20**. The two-stage thermostat built into the solid-state control module offers "Low/High/Low" temperature controls, which automatically fire the boiler on its highest output capacity only when extra capacity is needed. This controller features a domestic hot water (DHW) priority function, available by connecting an aquastat to the control board. When the controller receives a DHW call for heat, the boiler's set point will ramp up to 180°F until the DHW demand is satisfied. It also has a programmable outdoor reset controller that allows the operator to enter a reset curve designed to meet the system's requirements by specifying the targeted boiler water temperature at both low and high outdoor temperatures. Additional temperature control features include adjustments for maximum and minimum water temperature settings as well as high-fire offset.

Goodheart-Willcox Publisher

Figure 8-20. An example of a boiler with a two-stage modulating controller.

8.7.2 Integrated Boiler Control with Built-In Diagnostics

The integrated control on this direct-vent gas boiler features a three-function design. It incorporates a temperature limit control, a low-water cutoff control, and a built-in boiler reset control, **Figure 8-21**. The built-in diagnostics provide protection against potentially dangerous low-water conditions. They also utilize "thermal targeting" technology to conserve fuel by monitoring heating demand and establishing target boiler temperatures below the high-limit setting.

Adjustable controls offer flexibility: the user can adjust the high-limit settings as well as a dial-in economy setting for maximum fuel economy. Lighted LED displays alert the user when there is a low-water condition, when the thermal targeting function is active, and when the boiler is purging latent heat from the system. There is also a low-water test button to ensure that the low-water cutoff function is working properly.

8.7.3 Fully Modulating Control Module

This controller is capable of producing up to three set point temperatures to meet different space heating demands, **Figure 8-22**. This means that up to three separate hot water loops can be controlled based on settings from three different space heating demands. Input signals from various temperature sensors activate the burner circuit and circulating pump on demand. The control module regulates the combustion inducer blower speed and gas valve output to control the boiler's firing rate. Parameters input by the user control the space temperature

Goodheart-Willcox Publisher

Figure 8-21. This boiler offers integrated controls with built-in diagnostics.

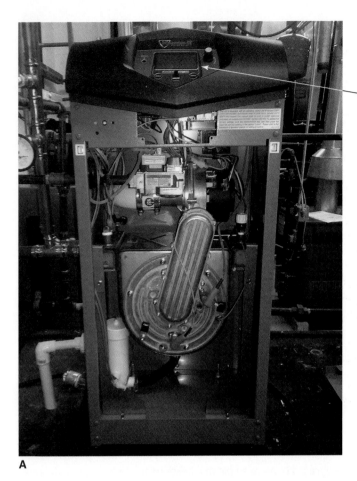
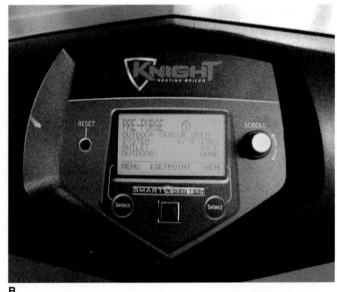

Solid-state modulating control board

Goodheart-Willcox Publisher

Figure 8-22. A—An example of a boiler with a solid-state fully modulating controller. B—A close-up view of the control board.

and domestic hot water temperature based on the boiler outlet water temperature, inlet water temperature, and system and outdoor temperatures.

The control module can be set up for DHW priority and space heating cycling. In addition to a modulating capability of 20%–100%, other features include burner antishort cycling, night setback, and freeze protection. System monitoring includes run time and alarm outputs, cycle counting, error logging, and a service reminder for regular maintenance schedules.

TROUBLESHOOTING

Diagnosing a Defective Control Board

Modern boilers with modulating controls and integrated system diagnostics can be relatively easy to troubleshoot if the technician knows what to look for. This procedure covers a generic approach to troubleshooting control boards. Be sure to have the electrical diagram handy and know the proper sequence of operation for the equipment.

1. Using a volt-ohm meter, check for 120-volt power to the boiler and for 24 volts to the control board.

(continued)

2. Be sure to determine if there is a low-voltage fuse protecting the board. If there is, check the fuse to see if it has blown. If it is blown, find the source of its cause before proceeding.
3. Most boards use a series of input and output signals to control the boiler. Begin by checking for voltage at the appropriate input terminal—such as from the thermostat to the board. Measure the voltage between this input terminal and to the common lead on the control transformer. There should be 24 volts.
4. If the proper voltage is read at the input terminal, determine which output terminal should be sending out a voltage signal in response—such as to the circulating pump control. If there is no reading at this output terminal, the control board may be defective.
5. Continue using this same procedure based on the sequence of operation to determine if there is a defective input or output signal on the control board (such as an input from the flow switch and an output to the ignition control).
6. Another method of troubleshooting is to use the LED diagnostics that most modern control boards possess. Follow the troubleshooting guide provided with the boiler to interpret the fault code based on the number of LED flashes or by a numeric display, and perform the appropriate service to the system.
7. One more diagnostic method is to observe the condition of the control board itself. If there are visual burn marks on the front or rear of the board, it may have suffered a short circuit or voltage surge. If there are white powdery marks on the terminals on either side of the board, the board may have suffered water damage. Both conditions typically require a board replacement.

CAREER CONNECTIONS

"I Have Never Seen That Before"

No matter how many years an HVAC technician spends troubleshooting systems, there will always come a day when that never-before-seen problem with a system or device will be encountered. Whether it stems from working on unfamiliar equipment, issues with new technology, or just dumb luck, sooner or later every technician will cross paths with that "one-of-a-kind" problem.

The key to success with these types of troubleshooting issues is a lesson in how the technician chooses to deal with them. There may be times when hours of investigation lead to 15 minutes' worth of repairs. Sometimes the answer seems out of reach, and the technician may feel like the only person in the world with this problem. Fortunately, the situation does not need to be an exercise in frustration if the person servicing the equipment keeps these valued principles in mind:

- Every problem is caused by something that has gone wrong, and every problem has a solution.
- Sometimes creative thinking—thinking outside the box—can lead to accelerated answers.
- It is not a bad thing to ask for help.

Remember, patience is the key to effective problem solving, and patience means not cutting corners on workmanship or taking chances by working unsafely. Effective problem solving takes time and many hours of experience. Sometimes reaching the top of one's profession involves working on a few of those "one-of-a-kind" problems.

Chapter Review

Summary

- Boilers incorporate both safety devices and control devices to protect the boiler and make it more energy efficient. Safety devices protect the boiler, hot water piping, mechanical room, and building occupants from dangers that can occur when controlling a pressurized vessel for comfort heating and hot water generation. Control devices increase system efficiency, reduce energy consumption, and make the controlled space more comfortable for the building occupants.
- Pressure relief valves protect the boiler and hydronic system piping in the event that the boiler's burner circuit fails to de-energize.
- Pressure relief valves should be tested annually for proper functionality. To test the valve, manually lift the lever on top and allow hot water to flush through the valve and down through the discharge tube. Then release the lever and allow it to close, making sure it shuts off and seals tight.
- A low-water cutoff switch is designed to de-energize the boiler's burner circuit if the water level within the boiler falls below a predetermined point.
- It is important to maintain the proper water flow through the boiler when it is operating.
- Water flow switches are used to prove proper flow through the boiler before allowing the burner circuit to be energized.
- There are several different types of flow switches available for protecting the boiler: paddle switches, differential pressure switches, and current relays.
- A backflow preventer is used to protect the domestic water source from contamination.
- Aquastat relays are commonly used on earlier boilers and are used for multiple functions: adjusting the high-limit, low-limit, and differential temperature settings, cycling the hot water circulating pump, and controlling the space temperature setting.
- Different types of aquastat relays have different control settings.
- Outdoor reset controllers are standard equipment on new boiler packages.
- The outdoor reset controller adjusts the boiler's hot water temperature set point based on the outdoor temperature.
- Outdoor reset controllers offer many benefits in addition to fuel savings.
- Most of today's boilers incorporate some type of electronic temperature controls.
- Electronic temperature controllers can vary between two-stage and fully modulating controls.
- Integrated solid-state controls can operate boilers for both heating and domestic hot water generation.

Know and Understand

1. Boiler safety devices protect _____.
   A. the boiler
   B. the hydronic piping
   C. the building occupants
   D. All of the above.

2. Most of the hydronic devices found on boiler systems are rated for at least _____.
   A. 30 psi
   B. 40 psi
   C. 50 psi
   D. 60 psi

3. Pressure relief valves are usually rated to open at _____.
   A. 30 psi
   B. 40 psi
   C. 50 psi
   D. 60 psi
4. What is the term used for operating a boiler below its proper water level?
   A. Low firing
   B. Water logging
   C. Dry firing
   D. Dry logging
5. *True or False?* Federal law requires that all boilers must incorporate a low-water cutoff switch.
6. A flow switch is used to _____.
   A. modulate the flow of water through the hydronic system
   B. limit the flow of water through the boiler
   C. re-direct the flow of water through the terminal units
   D. ensure that there is proper water flow through the boiler before the burner is activated
7. Which type of flow switch incorporates normally open contacts?
   A. Paddle switch
   B. Differential pressure switch
   C. Current relay
   D. All of the above.
8. A backflow preventer should be installed _____.
   A. between the water source and pressure-reducing valve
   B. between the water source and expansion tank
   C. between the pressure-reducing valve and pressure relief valve
   D. between the pressure-reducing valve and expansion tank
9. How do aquastat relays sense the boiler water temperature?
   A. By a sensor strapped to the boiler outlet piping
   B. By a copper sensing bulb attached to the back of the relay housing
   C. By a thermistor connected to the boiler return piping
   D. None of the above.
10. *True or False?* The outdoor reset controller adjusts the boiler water temperature downward as the outdoor air temperature falls.
11. What outdoor reset ratio would conventional baseboard fin tube heaters require?
    A. 1:1 ratio
    B. 1:2 ratio
    C. 1:3 ratio
    D. 1:4 ratio
12. *True or False?* A fully modulating controller regulates the combustion inducer blower speed and gas valve output to control the boiler's firing rate.

Apply and Analyze

1. Describe the difference between a boiler safety device and a control device.
2. Pressure relief valves contain a tag or rating plate that lists what information?
3. How often should pressure relief valves be inspected and tested?
4. How does a low-water cutoff switch function?
5. If the low-water cutoff switch cannot be mounted directly on the boiler, where is the next best location?
6. Describe the function of a differential pressure switch to prove flow through the boiler.
7. Explain how a backflow preventer is used to protect the domestic water source from contamination.
8. Should a backflow preventer be mounted horizontally or vertically in the piping system? Explain why.
9. Describe the multiple control features that an aquastat uses to control the boiler hot water temperature.
10. What is the maximum recommended high-limit control set point on an aquastat relay? What condition could occur if this set point is set at a higher temperature?
11. Describe the theory behind an outdoor reset controller and its functionality.
12. Explain why the reset ratio of an outdoor reset controller differs based on the types of terminal units that are being used.
13. What built-in precautions have the manufacturers of outdoor reset controllers incorporated into their product to allow their use on conventional boilers?
14. Describe the difference between two-stage electronic boiler controls and fully modulating controls.
15. What are some of the additional features that solid-state controllers offer for the control and monitoring of a hydronic heating system?

Critical Thinking

1. Explain why a current relay may be a more reliable device for proving water flow versus a paddle-type switch.
2. A circulating pump is connected to an aquastat relay and fails to energize when there is a call for heat. What troubleshooting steps should be taken to diagnose the problem?

9 Valves

Chapter Outline
9.1 Isolation Valves
9.2 Flow Control Valves
9.3 Temperature Control Valves
9.4 Specialty Valves

Dmitry Kalinovsky/Shutterstock.com

Learning Objectives

After completing this chapter, you will be able to:
- List the main categories of hydronic valves.
- Describe the purpose of an isolation valve.
- Explain the difference between a gate valve and a ball valve.
- List the different types of flow control valves.
- Explain how a check valve operates.
- Describe how a globe valve can affect pressure drop through a hydronic system.
- Summarize how a pressure-reducing valve operates.
- Explain how balancing valves are used with individual hydronic control loops.
- Differentiate mixing valves from diverting valves.
- List the different types of terminal units that incorporate zone valves.
- Describe how thermostatic control valves regulate space temperature on individual terminal units.
- Explain how purging valves remove air from a newly filled hydronic system.
- Summarize how a differential pressure bypass valve operates.

Technical Terms

balancing valve
ball valve
check valve
differential pressure bypass valve (DPBV)
diverting valve
flow coefficient (Cv)
flow control valve
gate valve
globe valve
handwheel
heat motor actuator
isolation valve
mixing valve
pressure drop
pressure-reducing valve
purging valve
spring-loaded check valve
swing check valve
thermostatic mixing valve (TMV)
thermostatic radiator valve (TRV)
zone valve

Hydronic heating systems utilize many different types of valves for a wide variety of applications, ranging from isolation purposes to temperature control. The system designer needs a good understanding of these different types of valves to ensure proper use. To clarify the different types of valves used in hot water heating systems, this chapter categorizes valves according to the following applications:
- Isolation valves
- Flow control valves
- Temperature control valves
- Specialty valves

9.1 Isolation Valves

Hydronic heating systems sometimes require service or repair to certain areas of the system or specific components. When this occurs, it is useful to have the capability of isolating these specific items from the rest of the system. This eliminates the need to drain the entire heating system just to work on one area. *Isolation valves* are used for this very purpose. Isolation valves can be beneficial at the inlet and outlet of the boiler water lines, both sides of a circulation pump, and at the inlet and outlet of a zone valve. The two most common types of isolation valves used in hydronic heating systems are:
- Gate valves
- Ball valves

9.1.1 Gate Valves

The *gate valve* gets its name from the mechanism inside the valve that enables the water to flow through it, **Figure 9-1**. This mechanism looks like a gate or disc that moves up and down when the handle of the valve is turned either open or closed. This gate is connected to a stem, which in turn is connected to the handle. The handle is also referred to as the *handwheel*.

wisawa222/Shutterstock.com

Figure 9-1. An example of a gate valve.

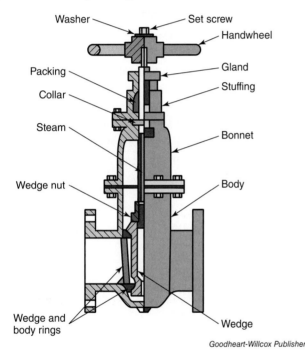

Goodheart-Willcox Publisher

Figure 9-2. This illustration shows the internal workings of a gate valve.

Fablok/Shutterstock.com

Figure 9-3. An example of a ball valve.

These types of valves offer good shutoff protection, sealing tightly when closed. They are bidirectional and because there is very little restriction to water flowing through the valve, they provide minimal pressure loss. However, as they require several rotations to go from fully open to fully closed, they cannot be opened or closed quickly, and they are not designed to regulate water flow. They are subject to leakage around the packing nut but can be repaired by either tightening the packing nut or replacing the packing gland, **Figure 9-2**. Gate valves are designed to be used in both high-pressure and high-temperature applications. Also, they are designed to be left in either a fully opened or fully closed position. When the gate valve is left in a partially opened or closed position, the valve will generate excessive noise, known as chatter, and can experience premature wear from the velocity of the heated water pushing against the gate. Most gate valves have a brass or bronze body, but they can also have a body made of PVC material or stainless steel.

9.1.2 Ball Valves

Ball valves serve the same purpose as gate valves but offer several advantages over conventional gate valves. The ball valve prevents the flow of water by means of a small sphere, or ball, within the body of the valve, **Figure 9-3**. This sphere actually has an opening running through the center of it, which either allows or stops the water flow depending on its opened or closed position. When the ball valve is open, the opening through the sphere is in line with the pipe, allowing water to pass through. When closed, the sphere opening is perpendicular to the flow of water through the piping, which stops the flow of water completely. The ball valve is controlled by a lever requiring only a quarter turn versus a handwheel. This valve is closed when the lever is perpendicular to the piping flow, **Figure 9-4**.

Because of the lever-type design, ball valves open and close more easily and quickly than gate valves. Also, the user or technician can tell at a glance whether the valve is open or closed. Ball valves are very durable and rarely freeze in cold environments, even after many years of service. In addition to being reliable, they are also quite versatile. However, the ball valve may not be best suited for limited space applications due to the fact that there needs to be enough room for the handle to be rotated 90 degrees. Just as with gate valves, ball valves can be constructed of bronze, brass, PVC, or stainless steel.

9.2 Flow Control Valves

Flow control valves differ from isolation valves in that they are designed to modulate, restrict, or prevent water from flowing in a certain direction. They are more specialized for use with hydronic heating systems and include the following types:

- Check valves
- Globe valves
- Pressure-reducing valves
- Balancing valves

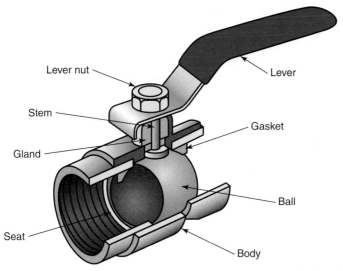

Figure 9-4. This illustration shows the internal workings of a ball valve.

9.2.1 Check Valves

Simply put, the *check valve* allows water to flow in one direction and automatically prevents backflow in the opposite direction through the hot water piping. Also known as a *one-way valve*, it is considered a self-automated valve in that it does not require assistance to open or close. Check valves continue to work even when there is a loss of power to the hydronic system. This is because the check valve relies on the pressure of the water in the system to open and close.

There are two popular types of check valves used with hot water applications:
- Swing check valve
- Spring-loaded check valve

The *swing check valve*, also known as the *tilting disc check valve*, uses a swinging disc attached to the top of the valve by a hinge to allow or prevent water flow through the valve, **Figure 9-5**. When water flows through this type of check valve in the desired direction, the disc swings up, allowing water to pass through. The force of the water flow keeps the disc open. If the water flow reverses, the change in motion along with gravity causes the disc to close, preventing water from flowing in the opposite direction, **Figure 9-6**. This type of check valve can be used on the outlet piping of a heating terminal unit to prevent water from backflowing

Figure 9-5. An example of a swing check valve.

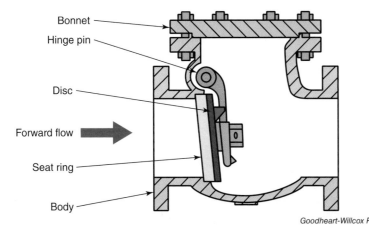

Figure 9-6. This illustration shows the internal workings of a swing check valve.

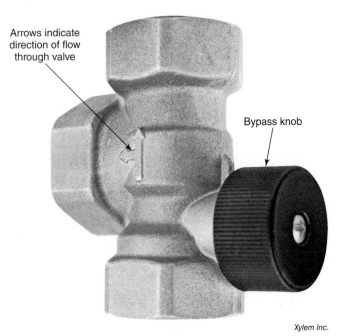

Figure 9-7. This flow control/check valve prevents gravity circulation of hot water out of the boiler.

through the unit, which could create an overheating situation. It should be installed in a horizontal position with the bonnet in the upright position. In addition, the basic swing check valve should be installed with a minimum of 12 pipe diameters of straight pipe connected upstream of the valve body. This will minimize the amount of water turbulence within the piping.

Another type of specialty check valve shown in **Figure 9-7** is designed to be piped into the hydronic system in a vertical position. Today's modern hydronic heating systems are designed so that hot water should only leave the boiler when there is a call for heat and the hot water circulating pump is energized. However, without the use of a special flow control/check valve, as illustrated in **Figure 9-7**, unchecked hot water would be allowed to circulate by gravity out of the boiler and into nearby heating zones whenever the circulating pump is off, thus causing those zones to become overheated.

Install this type of flow control/check valve on the discharge piping side of the boiler. It has a bronze weight that rides up on the valve stem whenever the circulating pump is operating. When the pump is de-energized, the bronze weight drops down onto the valve seat and keeps the valve shut, preventing off-cycle hot water gravity circulation.

TECH TIP

Check Valves and Emergency Heat Requirements

If gravity circulation is needed—say when the circulation pump is disabled for some reason, these types of vertically mounted, flow control/check valves come equipped with a bypass handle that allows the valve to circulate water in both directions. This can provide some temporary heat to the building until the proper repairs can be completed!

Figure 9-8. An example of a spring-loaded check valve.

Spring-loaded check valves can be installed in both horizontal and vertical piping, **Figure 9-8**. An arrow stamped on the side of the valve indicates the direction of flow. The spring-loaded check valve works by allowing a small spring inside the valve to close against the valve disc if the water is not flowing in the appropriate direction. As with the swing check valve, make sure that the spring-loaded check valve is installed downstream at least twelve pipe diameters of straight piping to minimize turbulence.

9.2.2 Globe Valves

The *globe valve* resembles the gate valve but has major differences. Named for their spherically shaped body, globe valves are designed for modulating the flow of water—they are not intended to be used for hydronic isolation, **Figure 9-9**. Although they can be used to start and stop the flow of water through the hydronic system, they should not be used specifically for on-off service. Here is why: the flow pattern through a globe valve involves an abrupt change in direction. Water enters the lower chamber, and then flows upward through a gap in the valve before it exits through the upper chamber. Although the flow through this gap allows for precise flow through the valve, it also contributes to a greater resistance in flow, which results in a considerable ***pressure drop*** through the system.

Globe valves must be installed in the proper direction, as indicated by a directional arrow marked on the valve's body. They are most often used for the control of fluids through terminal devices like fan coil units and large heating coils—such as those found on air handling units. Globe valves are sized according to their range of flow control, pressure drop, and the specific application for which they are being used. The factor used for sizing these valves is referred to as the ***Cv*** or ***flow coefficient.*** The *Cv* is the volume of water flow in gallons per minute (GPM) that passes through the valve at a pressure drop of 1 psi at 68°F. (There are online *Cv* calculators available that are based on water flow and pressure drop.) **Figure 9-10** shows a simplified illustration of a globe valve. These valves offer superior throttling capacity and flow control. They provide good shutoff capability and are available in multiple configurations such as straight through valves, tees, and angle patterns. However, because of their high pressure drop, globe valves require a greater closing force, which warrants a larger motor actuator in order to properly seat the valve when closed.

Goodheart-Willcox Publisher

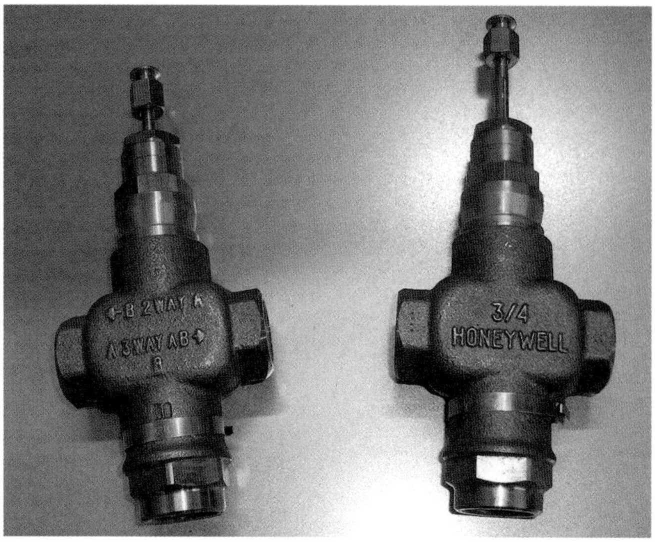

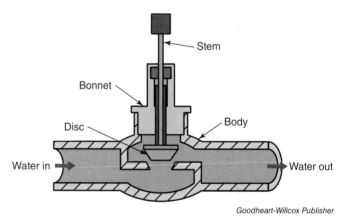

Figure 9-9. Globe valves come in many different sizes and types.

Goodheart-Willcox Publisher

Figure 9-10. This illustration shows the internal workings of a globe valve.

9.2.3 Pressure-Reducing Valves

A ***pressure-reducing valve*** is also referred to as a *water-regulating valve* or a *feed water valve.* The purpose of this valve is to deliver the correct amount of makeup water to the hydronic heating system at the correct pressure. Because most hydronic systems can lose water for a variety of reasons—such as leaks in the piping or from system servicing, an automatic makeup water source is necessary to replace lost water. This is why a constant supply of domestic water should always be available to the hydronic heating system. The domestic water source that feeds the hydronic system is typically at a much higher pressure (50–60 psi) than is needed (typically 12 psi). The pressure-reducing valve steps down the source inlet pressure to the proper system pressure, **Figure 9-11.**

Goodheart-Willcox Publisher

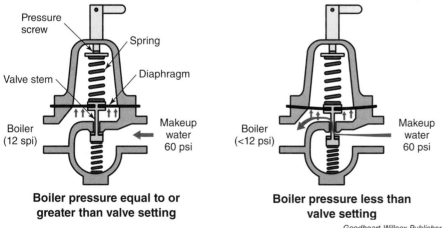

Figure 9-11. Pressure-reducing valves are used to step down the water pressure from the domestic source to the system operating pressure.

Various manufacturers offer a variety of pressure-reducing valves, but most operate in the same fashion. The valve itself consists of a spring-loaded diaphragm that operates much like the gas pressure regulator discussed in Chapter 4, *Gas Burners and Ignition Systems*. When the boiler water pressure is equal to or greater than the valve setting, the valve stem closes and no makeup water passes through. When the boiler water pressure falls below the reducing valve setting, the spring force becomes greater than the diaphragm force and makeup water is allowed to flow into the system until the pressures are equalized. This spring is adjustable to allow the operator to set the boiler system pressure as required. The correct system pressure setting will vary according to building height and system capacity, as discussed in Chapter 7, *Hydronic Piping Systems*. One feature shown on the pressure-reducing valve in **Figure 9-12** is called a "fast-fill" lever. This device increases the adjustment spring, allowing for full system pressure provided by the domestic water source when initially filling the boiler, thus saving time. It also can be used for purging the system to flush any contaminants. Most pressure-reducing valves include a removable strainer on the inlet side of the valve. This strainer prevents unwanted particulate from entering into the hot water system and can be removed for easy cleaning.

The pressure-reducing valve should be installed close to, or in line with, the expansion tank. This ensures that it is located at the "point of no pressure change." **Figure 9-13** illustrates where this point is located. Because the system pressure can change at any other point in the system due to various drops (such as on the outlet side of a terminal device), the system fill pressure can also change unless this connection point is adhered to. The point of no pressure change is discussed in Chapter 10, *Circulating Pumps*.

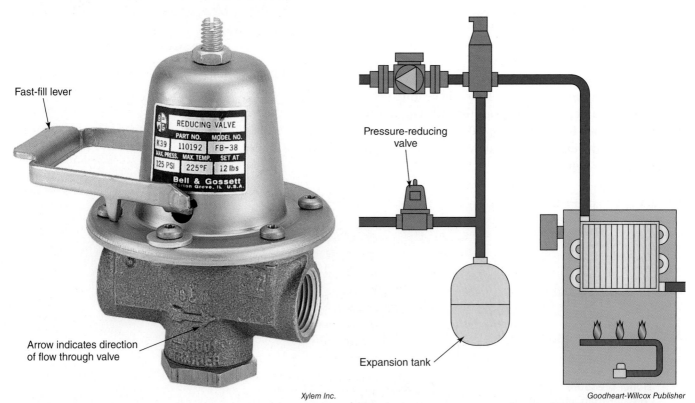

Figure 9-12. This pressure-reducing valve has a fast-fill lever for use when the boiler is initially filled.

Figure 9-13. The pressure-reducing valve should be installed close to the expansion tank.

TECH TIP

Keeping the Makeup Water On—or Not?

Should the water source flowing to the pressure-reducing valve remain open at all times? This question is still up for debate. Some manufacturers of pressure-reducing valves want you to keep the boiler water supply valve shut *except* when servicing the boiler. By doing so, there is a better chance of detecting a system leak, as evidenced by a reduction of water pressure within the system. Some companies also point out that replenishing makeup water to the boiler too frequently can increase the likelihood of system corrosion due to the high oxygen content in the makeup water. However, most building owners and maintenance personnel almost never think to check the water level or system pressure in hydronic heating systems until there is an indication of a problem—such as no heat. One method of getting around this argument is to install a low-water cutoff switch (see Chapter 8, *Boiler Control and Safety Devices*). This device will not only protect the boiler from "dry firing," but also act as an indicator if there is a potential problem with system leaks.

9.2.4 Balancing Valves

Every hydronic heating system must be designed to meet the proper flow rate in gallons per minute (GPM) through each respective piping loop. When this condition is met, the system is considered balanced. An unbalanced system can produce wide temperature variations between zones and increase energy consumption. Some specific hydronic piping configurations outlined in Chapter 7, *Hydronic Piping Systems*, are considered "self-balancing." However, not all piping configurations have the capability of balancing themselves, as is the case in **Figure 9-14**, which is a two-pipe direct return system. Without some assistance, this configuration will result in uneven temperatures and an inefficient system. Piping loops such as this can be assisted with the help of **balancing valves**.

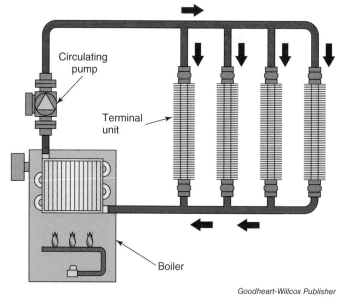

Goodheart-Willcox Publisher

Figure 9-14. Piping configurations such as this one need to incorporate balancing valves to ensure the proper flow through each loop.

There are several different types of balancing valves available. A simple ball valve could be used for balancing, but it will not offer the precision or accuracy of a valve that has the capability of measuring and dialing in the correct water flow for each heating loop. Most balancing valves, such as the one shown in **Figure 9-15**, have two pressure taps that allow for connecting a pressure reading device—such as a digital manometer. A manometer will read the differential pressure across the valve. The technician can then "dial in" the proper pressure drop to ensure the correct flow rate through the respective hydronic loop. This position can be locked in as a "full open" position. If the valve is closed during service, it will only open to the stop position that was set by the technician during commissioning. Other types of balancing valves, such as the one shown in **Figure 9-16**, have a built-in flowmeter housed in a bypass circuit on the valve body. This flowmeter provides fast and easy circuit balancing without the need for an additional pressure gauge, and it can be shut off during normal operation.

Most balancing valves should be installed on the discharge or return side piping of any terminal unit. A minimum length of three pipe diameters of unrestricted straight pipe upstream and one pipe diameter downstream should be maintained immediately adjacent to the balancing valve to maintain the correct flow through the valve.

Xylem Inc.

Figure 9-15. This balancing valve has built-in ports to connect a pressure gauge.

QuickSetter™ Balancing Valve, ©Caleffi North America, Inc.

Figure 9-16. This balancing valve has a built-in flowmeter for accurate calibration.

Goodheart-Willcox Publisher

Figure 9-17. This three-way mixing valve has two inlet ports and one outlet port.

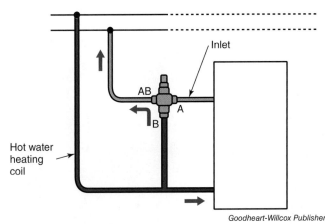

Goodheart-Willcox Publisher

Figure 9-18. An illustration showing how a mixing valve is used to control the temperature through a hot water coil.

9.3 Temperature Control Valves

Some valves used in hydronic heating systems specifically maintain the proper temperature of hot water through such devices as fan coil units, radiators, unit ventilators, and hot water coils found in air handling units. These temperature control valves are usually modulating by design and can be used with electronic actuators linked to a direct digital control system or building automation system. Some are simple open/closed valves for basic residential applications. Others are used as stand-alone temperature control devices and have internal thermostatic controls.

9.3.1 Mixing Valves

A ***mixing valve*** is a specialized valve that blends two sources of water to create a desired outlet water temperature. It can generate the proper water temperature to a hot water coil or terminal unit for space temperature control, or protect the boiler return water from dropping below a desired temperature.

Three-way mixing valves have two inlet ports and one outlet port, **Figure 9-17**. Remember the phrase "two in and one out" to distinguish them from diverting valves, discussed in the next section. A rotating spool inside the mixing valve alternately opens one port while simultaneously closing the opposite port. On three-way mixing valves, the valve seat or disc opens and closes the respective inlet ports by riding up and down on the valve stem. It is important to remember that technically the three-way valve is not considered "fully open" or "fully closed," but rather it is in "full flow" or in "bypass" mode with regard to the position of the inlet ports.

An application for a three-way mixing valve is illustrated in **Figure 9-18**. Here the valve controls the hot water temperature through a heating coil. Notice that the supply water from the boiler is entering both the bottom of the heating coil and also through the "B" port of the three-way valve. Water flows out of the heating coil and into the "A" port of the valve. Port "AB" is piped back to the return side of the boiler. A motorized actuator controls the valve position according to the discharge air temperature off the heating coil. When the discharge air temperature calls for heat, the valve will modulate—closing the "B" port and allowing full flow from the "A" port through the "AB" port. This allows 100% hot water flow through the coil. As the discharge air temperature off the coil reaches its set point, the actuator opens the "B" port while simultaneously closing the "A" port to maintain an accurate air temperature. Hot water now bypasses the coil and flows from the "B" port through the "AB" port and back to the boiler, **Figure 9-19**.

Another mixing valve is known as a ***thermostatic mixing valve (TMV)***, **Figure 9-20A**. This type of valve has a built-in sensing element that responds to changes in the hot water temperature at its outlet. This feature automatically adjusts the ratio of inlet water temperatures to the desired outlet temperature. When used as a return water installation, as shown in **Figure 9-20B**, the valve controls the return water temperature back to the boiler to keep this water temperature elevated on a conventional boiler. This prevents combustion gases from condensing. When used as a supply water installation, the thermostatic mixing valve

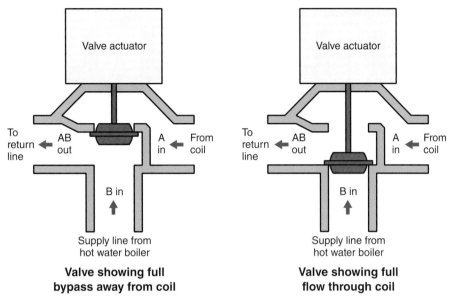

Figure 9-19. This illustration shows the internal ports of a three-way mixing valve.

recirculates a portion of the supply water back to the boiler until it has reached the desired temperature. Once the supply water is hot enough, the mixing valve opens to allow flow through the system.

A four-way mixing valve can also be used to connect low-temperature terminal units (such as radiant floor heating) to conventional boilers. **Figure 9-21** shows how these types of valves operate. The hot water supply from the boiler is allowed to flow either to the terminal unit or back to the boiler. Likewise, return

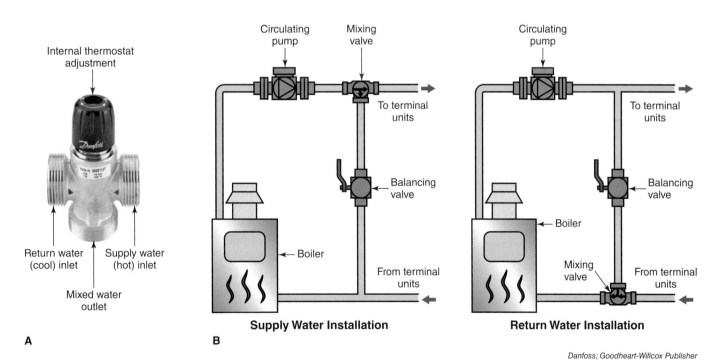

Figure 9-20. A—Shows an example of a thermostatic mixing valve. B—Shows how this valve can be used to control the boiler water temperature.

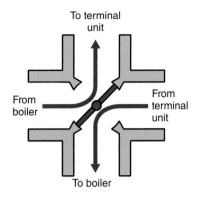 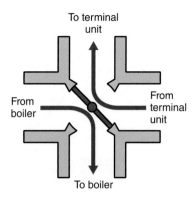 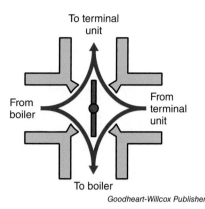

Figure 9-21. An illustration of a four-way mixing valve.

water from the terminal unit can flow either back toward the boiler or be bypassed back to the heating loop. This mixing effect ensures both consistent temperatures throughout the heating loop and a boiler return temperature elevated enough to prevent excessive condensation.

TECH TIP

Using Mixing Valves to Prevent Conventional Boiler Condensation

As stated in earlier chapters, conventional boilers require that their return water temperature be maintained above 140°F. Conventional boilers tend to condense the products of combustion when the return water temperature approaches 130°F. This can create major operational, installation, and safety issues. Flue gas condensation is very acidic, with a pH between 4 and 5. This acidic condensate can reduce efficient heat transfer, damage the heat exchanger, and corrode the venting system, causing premature failure. Mixing valves can effectively prevent harmful condensation from taking place.

Both three-way and four-way mixing valves can be manually operated by attaching a handle to the valve stem. This allows a constant fixed position for the valve setting, maintaining a constant balance between supply and return water temperature. However, most modern hydronic heating systems incorporate either a two-position or fully modulating actuator affixed to the valve for better water flow and greater temperature control.

9.3.2 Diverting Valves

A hot water *diverting valve* can be used when the water flow needs to take separate paths depending on the specific application for the hydronic system. When this is the case, a three-way diverting valve can be used. To remember the flow orientation through the three-way diverting valve, remember the phrase "one in and two out." **Figure 9-22** shows a graphic example of a diverting valve. This valve has the same ports as a mixing valve; however, on this particular valve, water flows into the "AB" port and exits through either the "A" or "B" port depending on the position of the valve actuator.

One application for the three-way diverter valve is when it is paired with a heating coil when full flow is not required because of partial load system conditions. In this instance, the hot water is diverted back to the boiler by repositioning the valve actuator when the discharge air temperature off the coil is near set

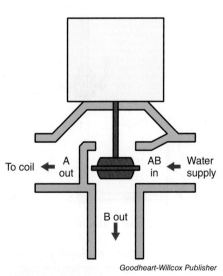

Figure 9-22. This illustration shows the configuration of a three-way diverting valve.

point. Another application is when the supply water from the boiler is being used for both space heating and for generating domestic hot water. In this instance, the system controller sends a signal to the valve actuator based on a call for heat from either the conditioned space or from the system water heater. The control strategy may signal that domestic hot water generation overrides space heating to allow the diverting valve to send more hot water to the storage tank until the set point is satisfied.

9.3.3 Zone Valves

Zone valves are commonly used in residential and light commercial hydronic heating applications to control the space temperature of individual zones. These are typically motorized valves that have 24-volt controls, **Figure 9-23**. Some zone valves have gear motors to actuate the valve open and closed, while others have *heat motor actuators*, **Figure 9-24**. Both types are two-way, two-position valves and consist of a separate motor and valve body. As the name implies, a gear motor zone valve uses a set of gears powered by a 24-volt motor to drive the valve open and closed. A heat motor valve has a bimetal strip that warps to open the valve when electrical current is applied. Gear-driven valves are quick opening and lend themselves to common terminal unit applications such as baseboard fin tube heaters, fan coil units, and radiators. Heat motor valves are better suited for radiant floor zones because their slow opening response time reduces the likelihood of water hammer and provides a more even hot water flow for better space temperature control.

Some zone valves come equipped with an end switch. This switch closes once the valve is fully open, which provides two functions: it can energize both the circulating pump and boiler control, and it also proves that the valve is open before these devices are allowed to energize.

Zone valves are also available in three-way configurations, **Figure 9-25**. A three-way zone valve can be used in systems with multiple heating zones, with each zone receiving its own valve. By incorporating this feature, water is allowed to bypass its respective zone if the valve is not calling for heat. This reduces the chance of deadheading the circulating pump if no zones are calling for heat and the pump is required to run.

Goodheart-Willcox Publisher

Figure 9-23. A typical zone valve used for residential and light commercial use.

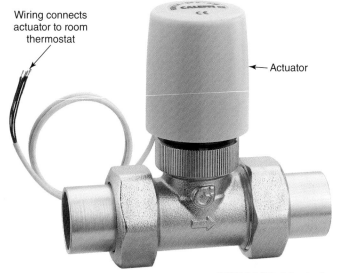

© 2012 Caleffi North America, Inc.

Figure 9-24. An example of a heat motor zone valve.

> ### DID YOU KNOW?
>
> **Three-Way Zone Valves Prevent Pump Deadheading**
> Pump deadheading occurs when too many zone valves are closed and there is no other flow path available to the circulating pump. When this situation occurs, the pump impeller continues to circulate the same volume of water as it rotates in the pump casing. The water temperature inside the pump housing increases as a result of friction—sometimes to the point where it turns to vapor. This vapor can result in excessive heat to the pump bearings and ultimately lead to premature failure. Properly installed three-way zone valves can prevent this situation from happening.

3-way Diverting Valve, ©Caleffi North America, Inc.

Figure 9-25. An example of a three-way zone valve.

9.3.4 Thermostatic Radiator Valves

Thermostatic radiator valves or *TRVs* are self-contained, spring-loaded valves mounted directly onto terminal units such hot water radiators and baseboard fin tube units, **Figure 9-26**. They control space temperature automatically by modulating the flow of hot water into the terminal unit. Because each terminal unit has its own control, the TRV provides an economical alternative to more complex control systems by allowing multiple zones to be controlled independently of each other in a "stand-alone" fashion. For example, a residential bedroom could have the capability of night setback by simply turning a dial, while the living room could maintain a higher set point simultaneously.

> **DID YOU KNOW?**
>
> **Set It and Forget It**
>
> Why do thermostatic radiator valves have numbers instead of degree settings? (See **Figure 9-27**.) The numbers on this valve do not correspond to a precise temperature, but rather a comfort level. It would be next to impossible to accurately control the space temperature according to a degrees setting on the TRV. This is because every application will vary depending on the size of the space being controlled, the temperature of the heating water, and how well the building is insulated. However, through trial and error the best control setting can be achieved. When the ideal set point has been established—"set it and forget it."

The valve itself consists of two main parts: the valve body and the valve head or operator. Inside the valve body, a disc or valve seat is connected to the valve stem. The valve stem is spring loaded, with the spring holding the valve in an open position when no operator is attached to it, **Figure 9-28**. The valve head or operator is filled with fluid and is temperature sensitive. With the valve head attached to the valve body, an increase in room temperature causes the fluid inside the valve head to expand, which pushes down onto the valve spring and closes the valve. Conversely, as the space temperature falls, the fluid contracts and the spring pressure overrides the fluid pressure inside the valve head, causing the valve to open. Some thermostatic radiator valves come equipped with operators that have remote set-point dials. These dials allow the temperature to be set at the user's eye level rather

Peter Gudella/Shutterstock.com

Figure 9-26. Thermostatic radiator valves are mounted directly onto terminal units such as hot water radiators.

Calek/Shutterstock.com

Figure 9-27. Thermostatic radiator valves have numbers instead of degrees to set the temperature set point.

than down near the inlet of the terminal unit. The remote dial is connected to the valve operator by a capillary tube filled with fluid. This fluid operates the valve in the same fashion as the valve-mounted operator, only remotely.

TROUBLESHOOTING

Freeing a Stuck TRV

One commonly occurring problem with TRVs is that they tend to stick open or closed. This issue can occur at the beginning of the heating season when the valve has not functioned over the summertime. When this happens, start by turning the adjustment knob to the highest temperature setting. Next, remove the valve operator from the valve body by unscrewing the thumbwheel, and then manually move the valve stem open and closed several times. This will normally free up the valve and allow it to operate effectively. If necessary, spray the stem with some penetrating oil to help it move freely.

TECH TIP

Using a TRV for Emergency Heating

The body of the TRV is held open by a spring, which makes it a "normally open" valve. This means that in an emergency, if the valve actuator fails, the valve head can be removed from the valve body and hot water will be allowed to flow freely through the valve. This can provide temporary emergency heat to the conditioned space until the valve head can be repaired or replaced.

9.4 Specialty Valves

Some hydronic valves do not fit into the previous categories. These types of valves can be considered specialty valves because they have specific functions.

9.4.1 Purging Valves

Because it is essential to remove any trapped air from the hydronic heating system, a *purging valve* can assist in air removal—specifically during the initial filling of the system. A purging valve consists of two ball valves joined together at 90° angles to each other, **Figure 9-29**. The main body of the valve is in line with the main hydronic piping. The secondary valve provides a means to flush the system and purge any air and debris when the system is first filled. This secondary valve has a threaded port for connecting a hose. The hose can then be routed into a pail or down a drain.

The key to effective air purging is fluid velocity. It is important to place the pressure-reducing valve in the "fast-fill" position as indicated on the lever when flushing the system, unless you are using a portable flush pump.

9.4.2 Differential Pressure Bypass Valves

Hydronic systems with four or more heating zones can experience a problem with excessive differential pressure when only one or two zones are open and calling for heat. This is especially

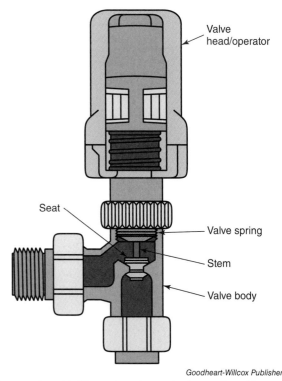

Goodheart-Willcox Publisher

Figure 9-28. This illustration shows the internal workings of a thermostatic radiator valve.

Figure 9-29. An example of a purging valve.

Labels: Inline ball valve; Outlet ball valve; Outlet port (hose thread with cap). Webstone/NIBCO

Differential Pressure By-pass Valve, ©Caleffi North America, Inc.

Figure 9-30. An example of a differential pressure bypass valve.

true when these zones incorporate two-way valves versus three-way valves, which allow the excess water to bypass the zone. Excessive differential pressure can cause increased velocity noise and a higher-than-normal flow rate through the zones calling for heat, resulting in poor space heating. When this situation arises, the problem can be alleviated by installing a *differential pressure bypass valve (DPBV)*, **Figure 9-30**. The DPBV's construction is similar to that of a pressure regulator. The DPBV consists of a disc or valve seat, a compression spring, and an adjustment knob. When the compression spring is adjusted to the proper pressure using the adjustment knob and the water force acting on the disc is low enough, the valve remains closed. However, when this differential pressure exceeds a force greater than the spring pressure, the disc opens and allows water to flow through the valve, balancing the pressure between the two points where the valve is installed.

This valve can be installed downstream at the end of the zone valve piping circuit or near the boiler outlet after the pump discharge—as long as it is piped between the supply and return piping, **Figure 9-31**. The bypass valve is manually adjusted to the correct pressure setting so that when only one or two zone valves are open and calling for heat, the DPBV will modulate open to relieve the excess pressure created by the circulating pump.

PROCEDURE

Installing a Differential Pressure Bypass Valve

To correctly install and set up a DPBV, use the following steps:

1. Be sure to install the valve with the flow arrow in the correct direction, flowing from the supply to the return piping.
2. Turn the adjustment knob in a clockwise direction until it stops.
3. Refill, purge, and restart the system.
4. When the heating system has reached its normal operating temperature, open all zone valves.
5. Wait approximately two minutes and begin adjusting the DPBV counterclockwise.
6. When an increase in temperature is measured on the outlet side of the bypass valve, stop the adjustment. This temperature increase indicates that there is now flow through the valve.
7. Once the flow is established through the valve with all zones opened, turn the adjustment knob approximately one full turn clockwise to stop flow through the valve.
8. The valve is now set to bypass excess flow as the zones close.

TECH TIP

The Advantage of a Variable Speed Pump

The DPBV is not needed when using a pressure-controlled variable speed pump. The controls for this type of pump will adjust the pump speed to maintain the proper system pressure automatically based on how many zones are calling for heat. Variable speed pumps are discussed in Chapter 10, *Circulating Pumps*.

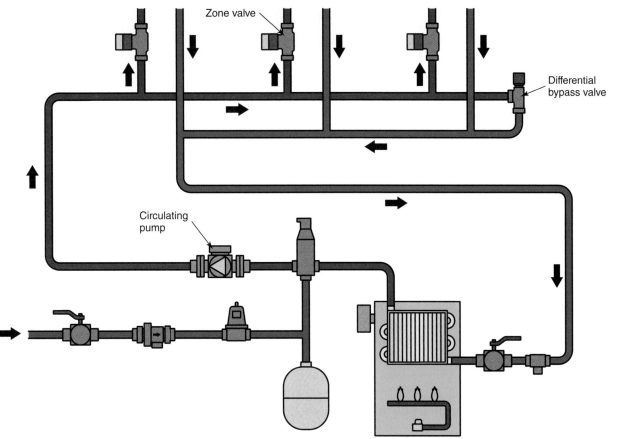

Figure 9-31. Differential pressure bypass valves can be installed downstream at the end of the zone valve piping circuit or near the boiler outlet after the pump discharge.

GREEN TIP

Using Modulating Valves and Variable Speed Pumps

Small changes to the hydronic system design can go a long way toward reducing energy consumption in both residential and commercial buildings. These changes can be as simple as incorporating modulating zone valves and variable speed circulation pumps.

When sized and installed correctly, modulating zone valves can provide optimal hot water return temperatures, lower energy losses in return piping, and more stable environmental temperatures. These features can result in lower energy costs.

Variable speed circulation pumps can provide an additional level of zone heating control and can be used to solve heating problems by either increasing water circulation (to get more heat out of a zone) or decreasing pumping speed (to provide more even heat). In addition, variable speed circulation pumps have been shown to operate nearly an entire year at less than 20% of design horsepower.

When combined, modulating zone valves and variable speed circulation pumps will result in a greener building!

Chapter Review

Summary

- Hydronic heating systems incorporate a number of different types of valves that are used for multiple applications. They fall into four main categories: isolation valves, flow control valves, temperature control valves, and specialty valves.
- Isolation valves are used to separate different areas of the hydronic system for service and repair.
- Both gate valves and ball valves can be used as effective isolation valves. Gate valves use a disc that moves up and down when the valve is turned open and closed. Ball valves use a sphere with an opening through it to allow or stop water flow.
- Flow control valves may be used for modulated, restricted, or prevented flow in a certain direction. They include check valves, globe valves, pressure-reducing valves, and balancing valves.
- Check valves limit the flow of water to one direction and may be either swing type or spring-loaded type.
- Globe valves are primarily used for flow control in temperature control applications, but can increase the pressure drop through a system. They are sized according to their flow coefficient or Cv.
- Pressure-reducing valves are also referred to as a water-regulating valves or feed water valves. They are used in conjunction with the makeup water system to limit the pressure on a hydronic system.
- Balancing valves are designed to regulate the proper flow in GPM through respective heating loops.
- Mixing valves are used to regulate the water temperature through hot water coils for effective temperature control. They may be three-way or four-way type valves and can regulate the return water temperature at the boiler.
- Diverting valves are used to divert water to separate control devices, including domestic hot water heating.
- Zone valves are two-way or three-way control valves used to control water flow to individual zones. Zone valves typically incorporate either gear motors or heat motor actuators. Gear-driven valves lend themselves to common terminal unit applications such as baseboard fin tube heaters, fan coil units, and radiators. Heat motor valves are better suited for radiant floor zones. Zone valves can include end switches to energize circulating pumps.
- Thermostatic radiator valves are used on individual terminal units to provide stand-alone temperature control. They can be equipped with remote set-point dials.
- Purging valves are used to eliminate air from a newly filled hydronic heating system.
- Differential pressure bypass valves are used to prevent multiple loop hydronic heating systems from becoming overpressurized. When differential pressure exceeds a force greater than the spring pressure in the DPBV, the disc opens and allows water to flow through the valve, balancing the pressure between the two points where the valve is installed.

Know and Understand

1. Which of the following is considered an isolation valve?
   A. Purging valve
   B. Globe valve
   C. Ball valve
   D. Zone valve
2. *True or False?* Gate valves offer both good isolation and effective flow control.
3. Which advantage does a ball valve offer over a gate valve?
   A. They can be opened and closed more quickly.
   B. The user can tell at a glance whether it is open or closed.
   C. They are more durable.
   D. All of the above.
4. *True or False?* A swing-type check valve can effectively be installed in both horizontal and vertical piping.
5. Which of the following valves should *not* be used specifically for on-off service?
   A. Gate valve
   B. Globe valve
   C. Ball valve
   D. Zone valve
6. The factor that is used for sizing globe valves is referred to as _____.
   A. the flow differential
   B. the flow coefficient
   C. the flow variable
   D. the flow speed
7. The pressure-reducing valve is used to reduce the pressure between _____.
   A. the supply and return side of the boiler
   B. the inlet and outlet of each heating zone
   C. the circulating pump and DPBV
   D. the domestic water source and the hydronic system
8. Balancing valves are used to specifically balance the _____ between each zone.
   A. temperature
   B. pressure
   C. water flow
   D. aeration
9. *True or False?* Mixing valves are offered in two-way, three-way, and four-way configurations.
10. To remember the flow orientation through a three-way diverting valve, remember the phrase _____.
    A. two in and two out
    B. two in and one out
    C. one in and two out
    D. three in and one out
11. What is the purpose of the end switch found on a zone valve?
    A. To energize the boiler
    B. To energize the circulating pump
    C. To energize the zone valve
    D. Both A and B are correct.
12. What is the purpose of a remote set-point dial on a thermostatic radiator valve?
    A. It allows the user an easier way to change the temperature set point.
    B. It can vary the flow of water through the radiator.
    C. It can be used for purging air from the system.
    D. It removes the valve head to provide emergency heat to the conditioned space.
13. *True or False?* A purging valve can be used to remove debris from the system as well as air.
14. A differential pressure bypass valve is best suited to which type of system?
    A. A system with a variable speed circulating pump
    B. A system with only one zone
    C. A system with multiple zones using two-way zone valves
    D. A radiant floor system

Apply and Analyze

1. Explain the difference between a gate valve and a ball valve.
2. What is the difference between an isolation valve and a flow control valve?
3. Describe how a check valve can prevent off-cycle hot water gravity circulation.
4. Why should a globe valve *not* be used for isolation purposes?
5. What other names are used for the pressure-reducing valve?
6. Describe the procedure for adjusting a balancing valve.
7. How is a three-way mixing valve used to control the hot water temperature through a heating coil?
8. Explain how a four-way mixing valve is used to keep the boiler return temperature elevated enough to prevent excessive condensation.
9. Can a diverting valve be used in conjunction with domestic hot water generation? Explain why or why not.
10. Describe the two different types of two-way zone valves used on individual heating loops.
11. Explain how thermostatic radiator valves are used to develop a "stand-alone" hydronic system.
12. Describe the purpose of a differential pressure bypass valve.

Critical Thinking

1. What would be the effect of using a differential pressure bypass valve on a system that incorporated three-way mixing valves on each terminal unit?
2. What would be the consequence of mistakenly using a three-way diverting valve on a heating coil that is intended for use with a three-way mixing valve?

10 Circulating Pumps

Chapter Outline

10.1 How a Circulator Works
10.2 Types of Circulating Pumps
10.3 Circulator Installation and Placement
10.4 Circulating Pump Performance
10.5 Pump Sizing and Selection
10.6 Service and Repair

NavinTar/Shutterstock.com

Learning Objectives

After completing this chapter, you will be able to:
- Describe the function of a hot water circulating pump.
- List the components that make up the circulating pump.
- Explain how energy is transferred through the pump components.
- Describe the use of different types of pump couplings.
- Explain how base-mounted pumps differ from inline pumps.
- State why pump placement in relation to the expansion tank is important.
- Describe how a pump performance curve is developed, explaining the relationship between flow and head pressure.
- Describe the benefits of both series and parallel pumping.
- List the steps used for proper pump sizing.
- Apply manufacturer's pump curves for pump selection.
- List the different service issues that can be encountered with hydronic pumps as well as their diagnosis and repair.

Technical Terms

aquastat
base-mounted circulator
centrifugal pump
close-coupled pump
coupling
delta-T circulator
delta-P circulator
electronically commutated motor (ECM)
end suction circulator
feet of head
flexible coupling
hot water circulating pump
impeller
inline circulator
parallel pumping
point of no pressure change
pump motor
pump performance curve
series pumping
split coupling
static pressure
three-piece circulator
variable frequency drive (VFD)
volute
wet rotor circulator

In the early days of hydronic heating, hot water systems relied on gravity as a means of water circulation. These systems utilized the physics of natural convection in order to circulate heat through the building's structure. In other words, they relied on the fact that heated air and fluid rises, **Figure 10-1**. Although systems such as these did in fact keep the building warm, they were far less efficient than the modern systems of today. The main reason we now enjoy a

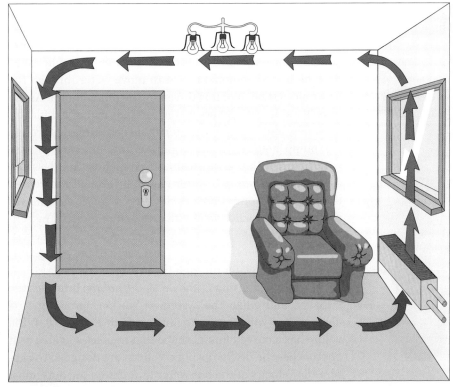

Sergey Merkulov/Shutterstock.com

Figure 10-1. Before the use of hot water circulating pumps, buildings relied on natural convection to keep occupants warm.

192 Hydronic Heating: Systems and Applications

Goodheart-Willcox Publisher

Figure 10-2. Hot water circulating pumps, or circulators, are used to efficiently move fluid through a hydronic system.

more efficient hot water system is because of the ***hot water circulating pump***. With the incorporation of the circulating pump, today's hydronic systems are much more efficient in transferring heat and, with smaller piping, can realize the same results as gravity systems. It is important to point out that this device used for circulating water through the hydronic heating system is technically not a pump, but rather a circulator. This is explained later in this chapter. Whether the industry labels these devices as circulators, pumps, or circulating pumps, they all perform the same function—to move hot water efficiently through the hydronic heating system, **Figure 10-2**.

This chapter discusses the function of the circulating pump used by today's industry and defines different types of circulators. In addition, this chapter covers the mechanics of pump curves and how they relate to pump sizing and performance. Emphasis is placed on proper pump selection as well as the service and repair of hydronic system circulators.

10.1 How a Circulator Works

Circulators or circulating pumps, **Figure 10-3**, consist of three main components:
- Motor
- Impeller
- Pump housing

The ***pump motor*** is used to drive the pump by rotating the pump's impeller. A circulating pump may incorporate several different types of motors, depending on the pump size and application. These motors may be fixed or variable speed, with smaller motors incorporating single-phase power and larger pumps using three-phase power.

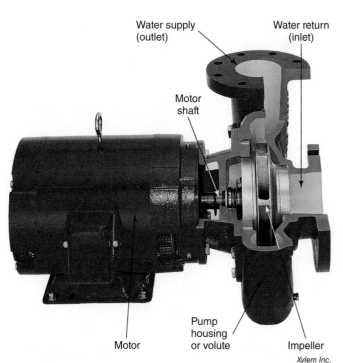

Xylem Inc.

Figure 10-3. This illustration shows a cutaway view of a circulating pump's components.

The circulating pump's ***impeller*** is a circular disc that contains turning vanes and is located in the pump housing. It is the main component used to move water through the hydronic system, and it is driven by the pump motor. As the impeller spins, it gathers water through its curved blades, or vanes, and slings the water by accelerating it through the pump housing. It accomplishes this function by centrifugal force. This is why hydronic circulators are classified as ***centrifugal pumps***. Water entering through the center, or eye, of the pump impeller develops velocity as it moves through the impeller vanes. This velocity produced by centrifugal force generates energy, which creates a differential pressure between the inlet and discharge of the pump. This pressure forces water through the hydronic system.

The pump's housing is known as the ***volute***. It houses the impeller, and when the system is working properly, it is always filled with water. As illustrated in **Figure 10-4**, the volute has a curved funnel shape that increases in area as it approaches the discharge port. Volutes are designed to capture the velocity of the hot water as it enters the outermost diameter of the impeller. This velocity is then converted into pressure. Volutes are designed with different shapes that contour to how they will be connected to the system's piping.

Copyright Goodheart-Willcox Co., Inc.

The body of the volute may be made of cast iron, bronze, brass, or stainless steel. The use of cast iron is not recommended with open-type hydronic systems that rely on excessive makeup water, due to the fact that they are subject to corrosion caused by oxidation.

TECH TIP

Hydronic Circulators versus Positive Displacement Pumps
It is important to point out that hydronic circulators use centrifugal force to move water and, unlike most refrigeration compressors, are not positive displacement pumps, **Figure 10-5**. Positive displacement pumps move fluid by repeatedly enclosing a fixed volume and moving it mechanically through the system. The name *circulator* is used because hydronic circulators do not add a large amount of pressure between the inlet and outlet ports. The amount of differential pressure generated by a circulator is enough to move water through the hydronic system. This is due to the fact that these types of hydronic systems are classified as "closed systems" and do not need a large amount of positive displacement to achieve their purpose. This topic is covered in detail under *Circulating Pump Performance*.

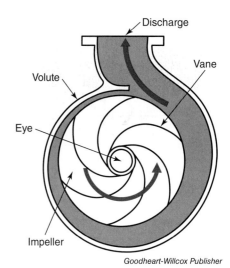

Figure 10-4. The pump housing, or volute, has a curved funnel shape that increases in area as it approaches the discharge port.

10.2 Types of Circulating Pumps

Hydronic circulators come in many shapes and sizes. They fall into two categories:
- Inline circulators
- Base-mounted circulators

Inline circulators are characterized by having their inlet and discharge ports in a straight line in relation to the hot water piping. *Base-mounted circulators* are typically larger circulators that are mounted on the floor on a rigid base.

10.2.1 Inline Circulators

Inline circulators are available as either three-piece circulators or wet rotor circulators, **Figure 10-6**. A *three-piece circulator* is a traditional type of circulating pump. This style of pump is designed for residential and light commercial applications. The three-piece circulating pump consists of the pump body (volute), the motor, and a bearing assembly—otherwise known as the *coupling*. With this type of circulator, the motor is isolated from the actual pump body, which allows for easy service or replacement of the motor. Three-piece circulator motors can be either inline or base-mounted and require periodic lubrication of the motor bearings. They can come with oil cups on the motor assembly that allow for easy lubrication on a regular basis or as maintenance-free assemblies, which have become an industry standard.

The motor on a three-piece circulating pump is connected to the pump assembly by a coupling. Pump couplings can be split, flexible, or close coupled. *Split couplings* consist of a solid metal device that connects the pump to the motor shaft. This type of coupling is made of steel and offers no flexibility, **Figure 10-7**. They are usually found on larger base-mounted circulators, **Figure 10-8**. The purpose of the split coupling is to connect a standard frame motor to the end of the pump. This can make the pump easier to service,

DID YOU KNOW?

The Classic Design of the Volute
The name *volute* was inspired by its resemblance to the scroll-like top of an Ionic-order column found in classical architecture. This top part of the column is called a volute.

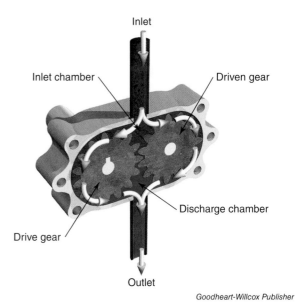

Figure 10-5. A circulator moves water through centrifugal force rather than "pumping" the water as a positive displacement pump does (shown here).

Figure 10-6. Two types of inline circulators are shown here. A—A three-piece circulator. B—A wet rotor circulator.

Figure 10-7. This image shows a split-type coupling.

Figure 10-8. An example of a base-mounted pump.

but if the motor shaft begins to vibrate due to bearing wear, the vibration will be transferred directly to the pump, potentially causing a premature failure.

Flexible couplings, or *couplers*, are found on smaller inline pumps and also on larger base-mounted pumps. These are the most popular types of couplings for three-piece circulators. Flexible couplings can consist of metal bars connected to stainless steel springs or two separate brass collars connected to a single, continuous helical spring, **Figure 10-9.** On larger pumps, they are typically made of a rubber material sandwiched between two metal hubs. These types of flexible couplings can connect a standard frame motor to a base-mounted pump to provide a certain amount of flexibility. Flexible couplings act as sacrificial connectors. If the pump bearings should seize up, flexible couplings are designed to break in order to protect the motor bearings and pump shaft. The idea behind this scenario is that it is much less expensive to replace a flexible coupling than a pump shaft or motor bearing (or even a motor). The flexible coupler also helps to compensate for any slight misalignment that may occur between motor shaft and pump shaft.

Direct-coupled or *close-coupled pumps* do not actually have couplings. They instead have a single shaft that extends from the motor to the pump body and are usually found on larger base-mounted pumps. The motor shaft and impeller are connected together so that the motor bearings must absorb the entire torsion load. This means that the motor bearings must be large enough to handle any vibration that may occur, which could require a specialized motor. While close-coupled pumps often take up less floor space, they can require specialized service.

Wet rotor circulators are popular for residential and light commercial hot water heating applications. These small- to medium-sized pumps combine the motor, pump shaft, and impeller into one assembly that is contained in a single housing, **Figure 10-10**. What makes the wet rotor circulating pump unique is that the motor is cooled and lubricated by the water enclosed in the hydronic heating system. Because of this configuration, these pump motors need no external lubrication, making them virtually maintenance free. In addition, their compact design allows the wet rotor circulating pump to be installed in tighter spaces where a conventional three-piece circulator would be too large. However, this type of pump is subject to premature failure if it is allowed to run for an extended period of time when there is no water present to help keep it lubricated—such as in the case of a major water leak.

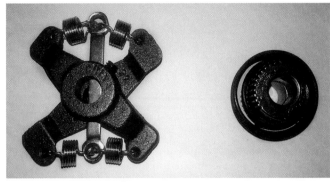

Goodheart-Willcox Publisher

Figure 10-9. These are two examples of flexible pump couplings.

10.2.2 Base-Mounted Circulators

Base-mounted circulators are mounted to the floor on a secured base and can be categorized as *end suction circulators* or *split-case pumps*. With end suction circulators, water enters one end and is discharged out of the top, **Figure 10-11**. Just as with the three-piece circulating pump, end suction circulators contain an impeller housed within a volute. When water passes through the end of the volute casing and through the impeller, it generates a high velocity, which in

A

Figure 10-10. An example of a wet rotor circulator.

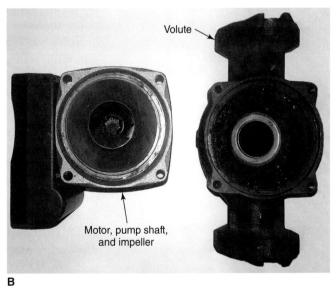

B

Goodheart-Willcox Publisher

Figure 10-11. An example of a close-coupled end suction circulator.

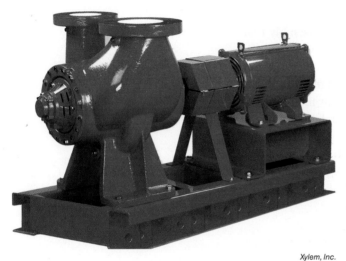

Figure 10-12. An example of a split-case pump.

turn, is converted into high pressure as it discharges out of the pump.

Split-case pumps have a double volute design, which increases the pump efficiency and lowers life-cycle costs, **Figure 10-12**. The impeller on base-mounted end suction pumps can be trimmed and balanced for customized requirements. The benefits of this pump include reduced noise, low vibration, and long bearing life.

Close-coupled end suction pumps are similar to wet rotor circulators in that the pump shaft and motor shaft are one and the same. With this type of pump, the impeller is directly attached to one end of the motor shaft. Close-coupled end suction pumps require less space and offer a compact construction design. A permanent, rigid alignment of the pump and motor eliminates the need for additional future alignment, which can prolong pump seals and bearing life.

10.3 Circulator Installation and Placement

One question that presents itself when designing the hydronic heating system is, Where should the circulating pump be placed—on the supply or return piping side of the boiler? Some designers feel that the pump works better if it is pumping away from the boiler, which means placing it on the supply side piping. Others have determined that locating it on the return piping—adjacent to the boiler—will prolong the life cycle of the pump because it is subjected to cooler water, thus reducing stress on the pump bearings. In general, either location can be acceptable—but be sure to follow manufacturers' certified instructions.

Whether the circulating pump is mounted on the supply or return side piping, one thing is certain: the location and placement of the circulating pump in relation to the expansion tank is very important. Because the system pump is circulating water throughout a closed loop, it can be considered a water mover, rather than a water lifter. This is due to the fact that the static pressure in the hydronic heating system is defined by the pressure reducing valve and the expansion tank. When the circulating pump is operating, it will create a differential pressure between its inlet and outlet and in some cases, this pressure may be less than the static fill pressure. This is important to understand because the circulator uses the point where the expansion tank is connected to the system as its ***point of no pressure change.*** This is the point in the system where the pressure always remains constant regardless of what the system is doing. If the circulator is pumping away from the expansion tank, it will add its differential pressure to the system's static fill pressure. If the circulator is pumping toward the expansion tank, it will remove its differential pressure from the system's static fill pressure. This latter situation can lessen the system's ability to remove air and can also lead to cavitation within the pump housing. Here is why: assume that the expansion tank is pressurized to 12 psi and the differential pressure across the pump is 15 psi. With the pump discharge pointed at the expansion tank, the pressure at the pump outlet cannot change, so the pressure on the inlet side of the pump will drop to account for the 15-psi differential. This will create a vacuum in the system when the pump starts, as shown in **Figure 10-13**. If a leak should occur in this situation, it will draw unwanted air into the system.

By pumping away from the expansion tank, the vacuum situation is eliminated. This is accomplished by relocating the circulating pump to the downstream side of the expansion tank. **Figure 10-14** shows that the inlet pressure of the circulator will now be 12 psi and the discharge pressure will be 27 psi (12 + 15). The system now has enough positive pressure for operative circulation, and any excess air can be effectively removed through the air vent.

Figure 10-15 shows another example of what could happen by installing the circulating pump to discharge into the inlet of the expansion tank. Notice how the heating loop experiences a negative pressure when the pump is positioned on the wrong side of the expansion tank. When the pump is located downstream of the expansion tank, as in **Figure 10-16**, the heating loop has pressure to spare. Many existing residential and commercial hydronic systems have been piped

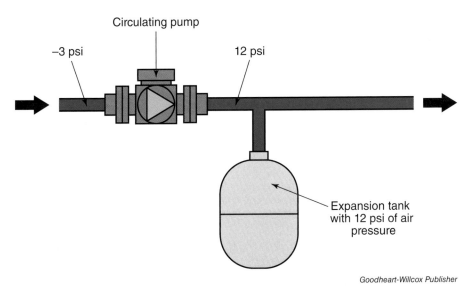

Goodheart-Willcox Publisher

Figure 10-13. If the pump is located before the expansion tank, a vacuum will be created in the circuit and flow will be lost.

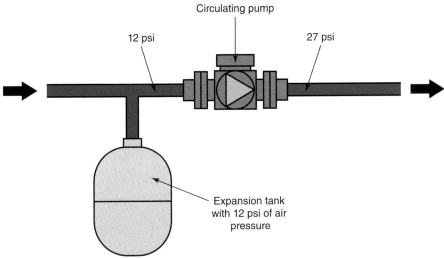

Goodheart-Willcox Publisher

Figure 10-14. By moving the pump to the outlet side of the expansion tank, the vacuum is eliminated and there is plenty of pressure for the system loop.

198 Hydronic Heating: Systems and Applications

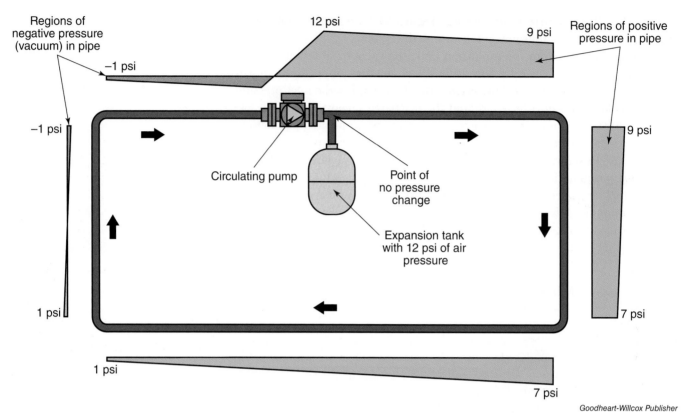

Figure 10-15. This illustration shows another example of how installing the circulating pump on the wrong side of the expansion tank can cause a system vacuum.

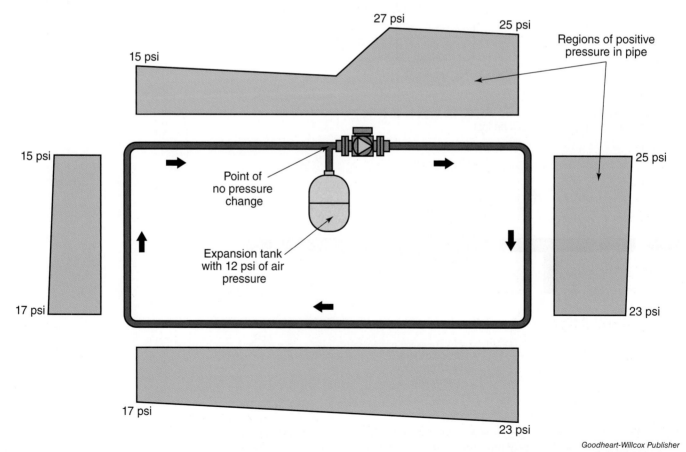

Figure 10-16. Installing the pump on the discharge side of the expansion tank will provide sufficient pressure in the heating loop.

incorrectly, similar to the scenario in **Figure 10-15**. Why do these systems seem to continue to operate effectively throughout the years? For a number of reasons: typically, these systems have only one or two floors, and there is not an extreme length of piping in the system. This limits the amount of pressure loss due to friction as well as low pressure at high points in the system. However, to maintain proper system flow and to minimize air from entering the system, it is highly recommended to always *"pump away from the point of no pressure change."*

10.4 Circulating Pump Performance

The task of the circulating pump is to move water from the boiler through the hydronic system and back to the boiler. It is not the responsibility of the pump to *"lift"* the water to the top of the heating loop—that task was already performed when the system was first filled. Because the system is completely filled with water (hopefully without any air), the only responsibility of the circulating pump is to move the water around. When thinking of a hydronic heating loop, visualize a Ferris wheel, **Figure 10-17**. When a Ferris wheel rotates, the weight that is elevated to the top of the loop is counterbalanced by the weight traveling downward. There is no "lifting," but rather just turning of the wheel. Because the wheel is balanced, all it has to do is overcome the friction from its bearings in order to set the wheel in motion. A circulator acts the same way: when it is energized, the weight of the water in the hydronic loop flowing upward is perfectly balanced with the weight of the water flowing downward. In the case of the circulating pump, the friction that it has to overcome is caused by the water passing through valves and terminal units, and by washing against the inside of the piping.

As outlined in Chapter 7, *Hydronic Piping Systems*, there is a major difference between static pressure and pump pressure in the hydronic system. **Static pressure** is caused by gravity and the weight of the column of water as determined by the building height, **Figure 10-18**. Pump pressure or head pressure is determined by the number of fittings, valves, terminal units, and the overall length of piping. It is this head pressure, or ***feet of head***, that the pump has to overcome to achieve the proper system flow in gallons per minute, **Figure 10-19**. In order to compare different types of circulating pumps and their performance, as well as choose the correct pump for a specific application, hydronic system designers rely on ***pump performance curves***.

10.4.1 Pump Curves

Centrifugal-type circulating pumps used in residential and light commercial hydronic heating systems convert energy provided by an electric motor into energy within the water being pumped. This energy is transferred from the impeller into the water and is considered a combination of velocity energy, pressure energy, and static elevation energy. To effectively analyze these forces, a performance curve can be plotted onto a coordinate system that compares the pressure to flow. Typically, the flow, measured in gallons per minute, is plotted on the x-axis, and the head pressure, measured in feet, is plotted on the y-axis.

Flat_Enot/Shutterstock.com

Figure 10-17. The hydronic loop can be compared to a Ferris wheel in that water at the top of the loop is counterbalanced by the weight of the water traveling downward.

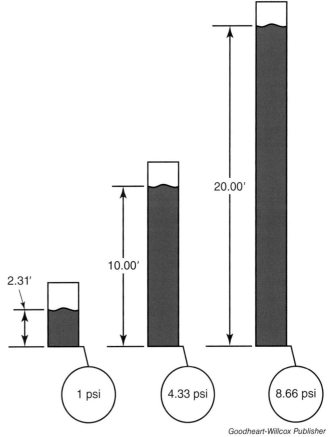

Goodheart-Willcox Publisher

Figure 10-18. Static pressure in a hydronic system is caused by gravity and the weight of the column of water as determined by the building height. As the height of the building increases, the amount of static pressure rises.

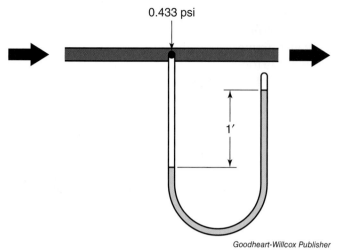

Figure 10-19. The feet of head in a hydronic system is created by the resistance to flow within the piping. The circulating pump must generate 1 psi of pressure for each 2.309 feet of head, which is created by the number of fittings, valves, terminal units, and overall length of piping.

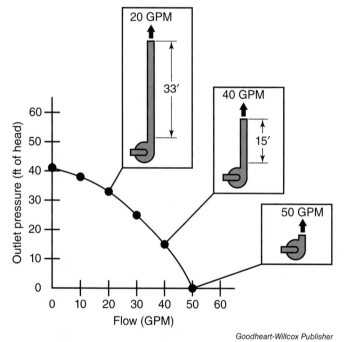

Figure 10-20. These graphs show an example of how a pump curve can be established by connecting the pump to a tank of water and increasing the height of piping, which the circulator must pump against.

To understand how a pump performance curve is developed, consider the following example in **Figure 10-20**:

- The inlet of a pump is connected to a tank of water and subjected to no resistance to flow. This scenario allows the maximum flow of water that the pump can produce. Say, for example, that this maximum flow is 50 gallons per minute (GPM).
- If a 15′ section of vertical pipe is attached to the pump discharge, the pump will experience resistance (known as feet of head) and the flow will be reduced to 40 GPM.
- If the pipe is extended, the pumping capacity will diminish proportionally until it cannot overcome the resistance.
- The data produced by this example can be plotted onto a performance curve to show the pump's flow capability.

Because it would be difficult for the pump manufacturer to simulate the previous scenario, they instead use valves to restrict the flow of water into the pump to simulate pressure in feet of head. Pressure gauges are installed on the inlet and discharge sides of the pump, and the pressure difference is then established in psig (pounds per square inch gauge). Next, this pressure is converted to feet of head. Remember from Chapter 7, *Hydronic Piping Systems*, that when there is a pressure difference of 1 psig, it is equivalent to 2.31 feet of head. When the valve located on the inlet of the pump is throttled closed, the pressure differential increases, thus decreasing the system flow in GPM. The flow can be determined by measuring the volume of water that is pumped in gallons and then dividing this number by the time elapsed in minutes to calculate gallons per minute. Curiously, you will notice that the pump amperage decreases as the flow decreases. This is because water recirculates in the pump rather than being discharged out of the pump. The pump does not have to work as hard, so amp draw decreases, **Figure 10-21**.

By systematically plotting the various points for water flow and feet of head resistance, a pump curve can be established. This process is repeated for different pump impeller sizes so that impellers in a variety of sizes can be used for the same pump volute.

The pump curve illustrated in **Figure 10-22** shows the various components found in a typical pump curve:

- **Flow.** Measured in gallons per minute (GPM).
- **Head.** Amount of resistance the pump must overcome to achieve GPM flow.
- **Impeller trim.** This is the size of the impeller used to develop its respective performance.
- **Horsepower.** This is the size of the motor required to achieve the respective performance.
- **Net positive suction head (NPSH).** This is the minimum amount of pressure on the suction side of the pump to overcome pump entrance losses. If sufficient NPSH is not met, the pump will cavitate, which will affect performance and pump life.

- **Efficiency.** When selecting the best pump for an application, efficiency can be an important factor. The higher the efficiency, the less energy required to operate the pump for a specific performance point.

10.4.2 Series and Parallel Pumping

Special applications sometimes require additional water flow or additional pumping capacity to overcome head pressure in order to accomplish the proper system design. When these cases arise, pumps can be configured to operate in a *series pumping* or *parallel pumping* design.

When additional head pressure is needed without additional flow capacity, pumps may be piped in a series configuration, as shown in **Figure 10-23**. An example of where a series configuration might be used is with a large radiant floor system. This design is more economical than trying to find a larger pump to fit the application. The pump curve shown in **Figure 10-24** illustrates this configuration's performance. The feet of head essentially double while the flow rate increases slightly.

Sample of Relation between Flow and Motor Current Draw		
Outlet flow (GPM)	Outlet pressure (psi)	Motor current (amps)
50	10	15
40	25	10
0	50	5

Goodheart-Willcox Publisher

Figure 10-21. This table shows a sample of pump performance and motor current draw at various throttling valve positions.

TECH TIP

Series Pumping Configuration
When configuring circulating pumps in series, they do not need to be coupled together. In fact, these pumps can be located on opposite ends of the boiler's inlet and discharge piping. When piped in this fashion, they create a "push/pull" arrangement. This can distribute the differential pressure more evenly and reduce the chance of pump cavitation, which can result from pumps coupled too close together.

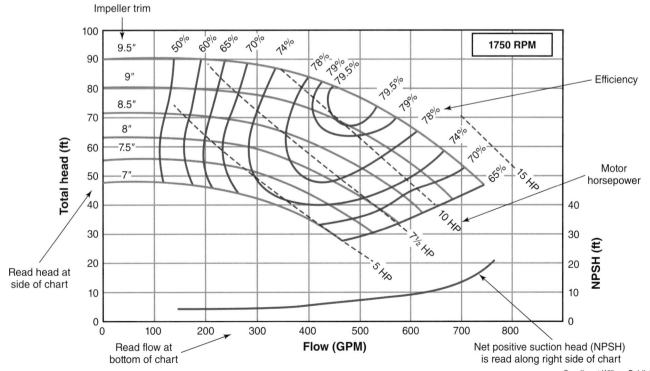

Figure 10-22. This graph shows the various components of a pump performance curve.

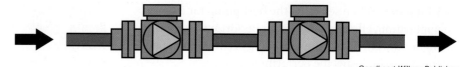

Figure 10-23. An example of circulating pumps piped in series.

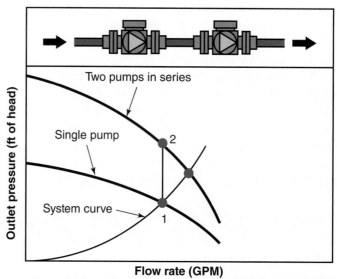

Figure 10-24. A performance curve showing two pumps piped in series. Notice that the feet of head essentially doubles while the flow rate increases slightly.

If the application calls for additional flow capacity such as where numerous terminal devices are being used on a short piping circuit, pumps may be installed in a parallel fashion—as shown in **Figure 10-25**. In this arrangement, the flow rate in GPM is essentially doubled while an increase in head pressure is kept minimal, **Figure 10-26**.

TECH TIP

Check Valves with Parallel Pumping
Always remember to install check valves on the outlet side of pumps that are piped in parallel. If this practice is not adhered to, there is a chance that the water flow can "short circuit" back through the adjacent pump if one of the pumps should be de-energized.

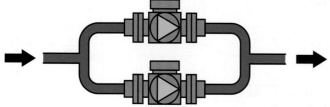

Figure 10-25. An example of circulating pumps piped in parallel.

10.5 Pump Sizing and Selection

Once the initial hydronic system layout has been designed, the next step is to properly size and select the best circulating pump for the application. Pump sizing is based on the required hydronic flow through the system against the measured system pump head. When these factors have been calculated, the selection process can begin.

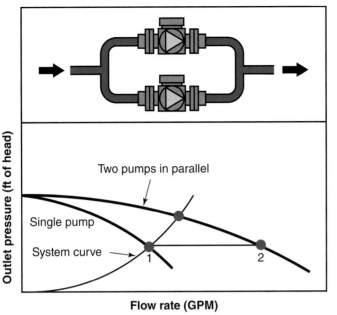

Figure 10-26. A performance curve showing two pumps piped in parallel. Notice that the flow rate essentially doubles while the feet of head increases slightly.

10.5.1 Pump Sizing

All circulating pumps are sized for the required maximum flow in GPM through the system while overcoming the system head loss for the longest zone.

PROCEDURE

Calculating Pump Size

To find the required pump size for a given system, follow these steps:

1. Calculate the required hot water flow in GPM through the system. This can be done by using the sensible heat formula for water as outlined in Chapter 7, *Hydronic Piping Systems*:

$$GPM = \frac{Btu}{500 \times \Delta T}$$

where
   GPM = gallons per minute of water flow
   500 = constant
   ΔT = temperature difference across the system

As a simplified example, if the boiler has an output of 100,000 Btu/hr and the normal design temperature difference is 20°F, the GPM flow requirement would be:

$$\frac{100,000}{(500 \times 20)} = 10 \text{ GPM}$$

2. Calculate the resistance to flow in feet of head. This is determined on the longest heating circuit flowing away from and back to the boiler. As discussed in Chapter 7, *Hydronic Piping Systems*, you will need to determine the resistance of this loop by analyzing the following:
   - Size of piping used
   - Equivalent length of pipe for each fitting used
   - Friction loss through the piping circuit
   - Pressure loss through the boiler and respective terminal units

Charts are available on the Internet for calculating equivalent length and friction losses. Manufacturer's data can be referenced to determine pressure losses through respective equipment.

One simplified calculation for determining feet of head loss is to multiply the length of piping through the longest heating circuit by 0.06. As an example, use this formula:

ft-head = longest pipe run × 0.06

If the longest run is 150′ in length, feet of head loss would be:

150 × 0.06 = 9.0 ft-head.

Once the flow in gallons per minute and the pressure loss in feet of head are known, these values can be used to research performance curves from various pump manufacturers to select the best circulating pump for the system application.

10.5.2 Pump Selection

With a variety of circulating pumps available in today's market, the best selection is based on the pump's optimal operating performance with the best efficiency, at a competitive price.

One method for selecting the best circulating pump is to first choose a performance curve and then select a range on that curve that is approximately in the midpoint of the curve itself. This typically covers the pump's intended usage and provides reliable operation. Selecting a pump to the left of this midpoint does not necessarily cause any significant problems; however, it does not provide a margin for error should the head pressure be augmented. Increases in head pressure can occur when control valves begin to close, which will restrict flow through any terminal units. Selection to the right of the midpoint of the curve is not recommended because operation in this area can cause cavitation, which can lead to pump damage and reduced flow to terminal units located at the end of the heating loop. Pump selection above the curve is highly inadvisable since the pump is not designed to work in this area. Fortunately, many pump manufacturers offer pump selection applications as well as flow and resistance calculators on their websites to aid in appropriate pump selection, **Figure 10-27**.

10.5.3 Variable Speed Circulators

Several pump manufacturers now offer variable speed circulators. These self-adjusting, high-efficiency circulating pumps utilize *electronically commutated motors (ECM)* for variable speed control, **Figure 10-28**. An ECM uses an internal

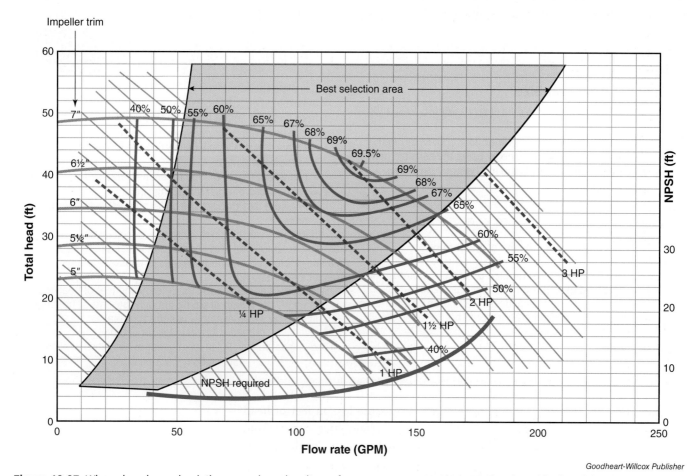

Figure 10-27. When choosing a circulating pump based on its performance curve, start by targeting the midpoint of the curve.

microprocessor control to vary the speed of the pump. Variable speed circulators can be categorized as:

Delta-T (ΔT) circulators

Delta-P (ΔP) circulators

Delta-T circulators control their pumping speed based on the temperature difference across a load. This feature can be used for several applications, such as to control the water flow through a boiler or hot water coil. Delta-T circulators work by monitoring the temperature across individual zones and adjusting the circulator's speed to a desired temperature differential setting. This can make them particularly useful for radiant floor systems.

Once installed in the hydronic system, delta-T circulators can be programmed for several different applications. These circulators are equipped with temperature sensors connected to the inlet and outlet of each zone or device to be controlled. In the "delta-T" mode, the temperature difference can first be programmed according to the user's requirements. In this mode, the circulator will start at its lowest speed. If the temperature difference is less than the desired set point, the circulator will gradually increase its speed until it reaches its target. If the temperature difference is greater than the set point, the circulator speed will decrease until it reaches its target, at which point it will vary its speed to maintain the desired temperature difference.

Goodheart-Willcox Publisher

Figure 10-28. An example of a delta-T circulator. These circulators operate off a differential in-system temperature.

Delta-T circulators may also be programmed for constant speed mode as well as "set-point" mode. In set-point mode, the circulator will vary its speed to maintain a fixed set point temperature. The circulator speed will increase to raise the temperature and lower its speed to decrease temperature. This feature can be used to maintain a constant temperature through the boiler.

Delta-P circulators are used to maintain a constant pressure through the system or an individual zone, **Figure 10-29**. These types of circulators can be used where the system pressure may be changed, such as with thermostatic radiator valves (TRV) used on radiation terminal devices. With this application, the circulator will automatically modulate its pumping speed to allow for differences in system pressure—when a zone valve or thermostatic radiator valve opens or closes, for example.

Xylem, Inc.

Figure 10-29. An example of a delta-P circulator. This circulator operates off a differential in-system pressure.

Delta-P circulators can be programmed to maintain a constant pressure or a proportional pressure based on the application. When programmed for constant pressure, the circulator maintains a constant pressure at any flow demand. With this application, the desired feet of head can be set on the user interface. When programmed for proportional pressure, the circulator is continuously modulating its pump speed based on flow demand.

Today's larger pumps utilize **variable frequency drives (VFD)**, which can be controlled using differential pressure sensors, **Figure 10-30.** A VFD is an adjustable-speed device used to control AC motor speeds and torque by varying motor input frequency and voltage. Hydronic systems that employ a large number of heating coils with modulating hot water valves are normally equipped with VFD-controlled circulating pumps. Here is how it works: A differential pressure sensor is placed toward the far end of the heating loop. As the building heating demand increases, multiple heating valves will begin to open. This creates a pressure drop across the system. The differential pressure sensor detects this pressure drop and sends a signal to the VFD, which in turn increases the speed of the pump to

Figure 10-30. This pump operates with a VFD—variable frequency drive. These types of pumps use pressure sensors to control pump speed.

Xylem, Inc.

maintain a desired system pressure set point. As the heating valves close, the VFD again signals the circulator, reducing the pump speed.

10.6 Service and Repair

Hydronic circulators can be operated as a single pump system for multiple zones (where zone valves are used for individual zone control), or they can be used to control individual zones in place of zone valves. In either case, there are service and repair issues that need to be addressed. This section classifies various service issues along with their proper diagnosis and repair.

Problem: Circulating pump will not run.

Solutions: First check for voltage to the pump motor, **Figure 10-31**. After ruling out any obvious causes related to connectivity, check to see if the pump is operating. This can be as simple as listening to the pump to determine whether it is running (listen carefully, because newer inline pumps can run very quietly) or feeling the pipes between the pump inlet and outlet—they both should be hot. If the pipes are cool to the touch and the pump is not operating, there may be other issues.

Other causes for a pump that will not run:
- If the thermostat is not calling for heat and the room temperature is below the thermostat set point, determine that the thermostat is not the issue by bypassing the contacts across the heating circuit. Also, check to see that the thermostat has fresh batteries (if applicable).
- Check to see that the boiler water temperature is above the cut-in set point temperature. This is typically controlled by the boiler *aquastat*. Check the boiler gauge temperature versus the aquastat set point, plus the high, low, and differential settings.
- Check the pump relay to see if it is calling for heat and that the contacts are closed. On aquastats, the relay is built in and should be energized by a call for heat from the thermostat. Check the aquastat by jumping the thermostat

Goodheart-Willcox Publisher

Figure 10-31. Technician checking for power to the pump motor.

terminals. The relay should energize and the contacts should close. Also, check for 120 V power to the pump terminals on the aquastat. The relay contacts on the aquastat can be manually closed to determine if there is a problem with the relay or thermostat. If the pump runs when the relay contacts are pushed closed, the relay is at fault and the aquastat needs to be replaced.
- Some systems have separate pump relays that may be controlled by the thermostat or by another external device. Use the same troubleshooting techniques as with the aquastat to determine whether the pump relay is working properly.
- If the system has zone valves with end switches, check to see that these end switches are closed. If these valves have not operated for a while, the internal linkage can stick, which will prevent the end switch from closing.

> **SAFETY FIRST**
>
> **Caution with Aquastats**
> There is a potential for exposed electrical circuits within the aquastat. Be careful when working inside the aquastat to prevent electrical shock. If you do not have experience working in this device, seek training or call for professional help!

Problem: Circulating pump has power but will not run.

Solutions: Check to see if the pump itself is damaged. If the circulator has power but is not pumping, it may have a frozen or damaged impeller.
- See if the pump casing is very hot. Wet-type circulators rely on water flow to keep the bearings lubricated. If the impeller is frozen up, the pump casing will be very hot and there will be no water flow. The pump will need to be disassembled to verify that there is a problem with the pump impeller.
- If it is a three-piece pump, there should be an exposed coupling visibly turning. Check that the coupling is not damaged, loose, or broken.
- If possible, measure the pump's current draw, **Figure 10-32**. If the current reading is well above the full load amperage (FLA) reading, the pump motor is failing due to a breakdown in the motor windings. If the current draw is well below the FLA, typical reasons for motor failure include a worn or broken impeller, an impeller that is loose, or a broken linkage.
- Other causes of low current readings are no water flow due to an air lock, a plugged strainer, or a closed valve in the circuit that should be open.
- Check the motor capacitor (if applicable). There may be a defective start or run capacitor. Use a capacitor tester, and if the capacitor is defective, replace it with the same microfarad rating or one that does not exceed a 10% higher rating.

Goodheart-Willcox Publisher

Figure 10-32. Technician checking the pump amperage. A higher-than-normal FLA will indicate failing motor windings.

Problem: Circulating pump is leaking.

Solutions: A common spot where most circulating pumps leak is at the mechanical seal due to failure. This simply could be because the flange mounting bolts were not securely tightened. Or with three-piece pumps, the flanges are supporting the entire weight of the pump and are beginning to wear.

A small leak may go unnoticed because the water is so hot, it evaporates too quickly. However, a buildup of white mineral salts and corrosion on the flange mounts is a telling sign that there is a flange leak. The best time to repair this type of leak is at the end of the heating season when it will be more convenient to shut down the boiler, let it cool down, and possibly drain it if necessary. If this type of leak is left unchecked, it could lead to more difficult and costly repairs.

Problem: Circulating pump will not stop running.

Solutions: When this situation occurs, first check the thermostat. Although thermostats seldom just quit working, there is the remote possibility that the

Goodheart-Willcox Publisher

Figure 10-33. Technician checking the end switch on a zone valve. The switch should be closed in order for the pump to run.

heating contacts on the thermostat are stuck in the closed position. Simply disconnecting the heating leads will determine whether the thermostat is the problem. If the circulating pump de-energizes when the thermostat wires are disconnected, the problem has been found. The next step is to check the relay contact at the aquastat or at the pump relay itself. There are occasions when either the contacts are fused closed or the relay coil will not de-energize. Again, disconnecting the wires to the aquastat relay or pump relay will pinpoint the problem. Then, check each zone valve (if applicable). Many zone valves have end switches that energize the pump when the valve proves open. These end switches can stick closed, especially at the beginning of the heating season when they have not operated over the summer months, **Figure 10-33.** However, if this is the case, a zone is likely overheating as well, which will also indicate a problem.

TECH TIP

Continuous Operation by Design

On some residential applications in very cold climates and especially on some commercial heating applications, the circulating pump is designed to operate continuously when the outdoor temperature falls below a certain set point. This is to keep water circulating throughout the system to avoid any freeze-up. Always remember to look for this control scenario when diagnosing a service call for a pump that runs continuously.

Problem: Circulating pump is noisy.

Solutions: A noisy circulating pump usually indicates a need for lubrication. Three-piece circulators should include oil cups on smaller pumps and grease fittings on larger pumps. Try to find out from previous maintenance reports the last time when the pump was lubricated. If there are no records to be found, chances are it is time for lubrication. On inline circulating pumps, excessive noise could be a sign of a failing impeller. Because a wet-type circulator relies on water for bearing lubrication, the problem could be a lack of water flowing through the pump.

If a banging noise is heard when the pump de-energizes or if the sound of trickling water is present when the pump is running, these sounds could indicate air in the system. The banging noise is a sign of water hammer and should be addressed quickly.

TECH TIP

Noisy Aquastat and Pump Relays

Noise from the aquastat or pump relay may simply be a noisy relay. This symptom does not necessarily mean that the aquastat or relay needs replacing, but the sound can be annoying. Remember that the relay coil acts as an electromagnet when the relay is energized. The buzzing noise is simply current passing through the coil. Sometimes the aquastat or relay can be repositioned, or some insulation can be placed between the relay and the device that it is connected to. If this problem persists, perhaps it is time to replace the device.

Problem: Circulating pump does not heat well.

Solution: This problem, and the previous issues with banging pipes and trickling noises, all indicate air in the hydronic system. Air removal can be a simple fix or a time-consuming chore, depending on where the air is located and how much air is in the system. Air blockage at a terminal unit can lead to a complete lack of water flow and a "no heat" service call. As mentioned in previous chapters, no air should be present in the hydronic heating system, and excessive amounts of air can lead to problems with noise, corrosion, and a failed heat exchanger.

To effectively remove unwanted air in the circulating system, begin at the highest terminal unit in the building. If there is a bleed screw or air vent on the unit, open the port to discharge the air. Wait until water discharges from the vent, then close the vent and proceed to the next terminal unit. Also check the air vent on the top of the air scoop. This is usually located at the top of the expansion tank.

Some hot water systems will have isolation valves on each zone at the boiler, with a drain cock or hose bib connected to each zone. Begin by closing all isolation valves except one. With the isolation valve closed on the first zone to be purged, open the hose bib connected to that zone until any trapped air is removed. Proceed to the next zone and repeat the process until all air is completely purged from the system.

Air removal can be time-consuming and sometimes frustrating to resolve, but it is necessary in order to maintain a properly working system.

CAREER CONNECTIONS

Test and Balance Technician

HVAC testing, adjusting, and balancing (TAB) technicians work as specialized contractors who perform testing and balancing services on commercial building construction projects.

TAB specialists perform hydronic measurements on HVAC systems and adjust the flows through hydronic devices to achieve optimum performance of the building's environmental equipment. This balancing is based on the design flow values set forth by the mechanical engineer for the project. Upon completion of the project, the TAB contractor submits a written report that summarizes the testing and balancing findings and any deficiencies found during the process. In some instances, facility managers will use the TAB contractor to identify any HVAC issues that might be found within the facility.

Many TAB contractors and technicians hold professional balancing certifications to validate their skill levels. Qualifications for work in this industry include a thorough understanding of various HVAC systems and their operations. TAB technicians must be proficient in using HVAC tools of the trade. They also need good written and verbal communication skills. Entry-level technicians will possess a high school diploma or higher, and advanced technicians may be required to supervise the work of other technicians.

The entry-level salary for TAB technicians ranges between $12–$14 per hour in the United States, with journeyman wages ranging from $25–$35 per hour.

Chapter Review

Summary

- Circulating pumps are used to transport hot water through the hydronic system.
- Circulators consist of a motor, impeller, and pump housing (known as a volute).
- Centrifugal-type circulating pumps convert energy provided by an electric motor into energy within the water being pumped. This energy is transferred from the impeller into the water and is considered a combination of velocity energy, pressure energy, and static elevation energy.
- Circulating pumps are categorized as inline circulators or base-mounted circulators. An inline circulator is a type of centrifugal pump that has the inlet and discharge ports in a straight line in relation to the hot water piping. A base-mounted circulator is a circulator pump that is mounted on the floor and secured to a rigid base.
- Inline circulating pumps are classified as either three-piece circulators or wet rotor circulators.
- Three-piece circulators use a variety of different couplings to connect the motor to the pump: split couplings, flexible couplings, or close-coupled pumps.
- The location where the expansion tank is connected to the piping system is known as the point of no pressure change. Circulator placement is critical in relation to the point of no pressure change.
- The purpose of the pump is to circulate the water through the piping system rather than lifting the water to the top of the loop.
- Pump curves are used to evaluate the performance of the circulator. Pump curves compare the circulator flow in gallons per minute (GPM) against the amount of head pressure that the circulator has to pump against.
- Circulating pumps may be piped in a series or parallel arrangement to overcome higher head pressure (series) or when greater flow capacity is needed (parallel).
- Water flow in GPM and pressure in feet of head are calculated when determining pump size.
- Circulating pump selection is based on the comparison and evaluation of various pump curves.
- There are numerous service issues that can be encountered with circulating pumps.

Know and Understand

1. Before the days of circulating pumps, what was the means of hydronic heating?
   A. Natural conduction
   B. Natural convection
   C. Natural radiation
   D. Natural condensing
2. What is the device that contains turning vanes and is located in the pump housing?
   A. Impeller
   B. Propeller
   C. Centrifugal disc
   D. Venturi
3. The pump's chamber is known as the:
   A. Volute
   B. Cavity
   C. Cylinder
   D. Venturi
4. On which type of circulator is the motor isolated from the actual pump body?
   A. Wet rotor
   B. Three-piece
   C. Reverse suction
   D. Base-mounted
5. *True or False?* Wet rotor circulators combine the motor, pump shaft, and impeller in one assembly contained in a single housing.
6. The point where the expansion tank is connected to the system is known as the _____.
   A. point of cross pressure
   B. point of permanent suction
   C. point of high pressure
   D. point of no pressure change
7. What type of pressure does the pump have to overcome to achieve the proper system flow?
   A. Static pressure
   B. Velocity pressure
   C. Head pressure
   D. Terminal pressure
8. *True or False?* On the pump performance curve, the GPM flow is plotted on the y-axis.
9. When additional head pressure is needed without additional flow capacity, pumps may be piped in a _____ configuration.
   A. series
   B. parallel
   C. cross-linked
   D. perpendicular
10. If a boiler has an output of 50,000 Btu/hr and the design temperature difference is 25°F, what would the GPM flow requirement be?
    A. 4 GPM
    B. 5 GPM
    C. 10 GPM
    D. 15 GPM

Apply and Analyze

1. Describe the function of the three main components of a circulating pump.
2. Explain how the impeller uses centrifugal force to move water.
3. What is the difference between an inline circulator and a base-mounted circulator?
4. Describe two different types of pump couplings.
5. What issue can happen if a wet rotor circulator is allowed to run with no water present?
6. Explain what is meant by the "point of no pressure change" and how it relates to circulating pump placement.
7. Describe how pump manufacturers configure their performance curves.
8. How is resistance to flow in feet of head calculated in a hydronic system?
9. Describe the selection criteria for choosing a pump based on its performance curve.
10. Explain what issues can create pump noise.

Critical Thinking

1. What is the purpose of listing the required horsepower on a pump performance curve?
2. What effect would a pump with variable speed capability have on its performance curve?

11 Terminal Devices

Chapter Outline

11.1 Choosing a Piping Configuration
11.2 Finned-Tube Baseboard Units
11.3 Radiators
11.4 Fan Coil Units
11.5 Unit Heaters
11.6 Other Types of Terminal Devices

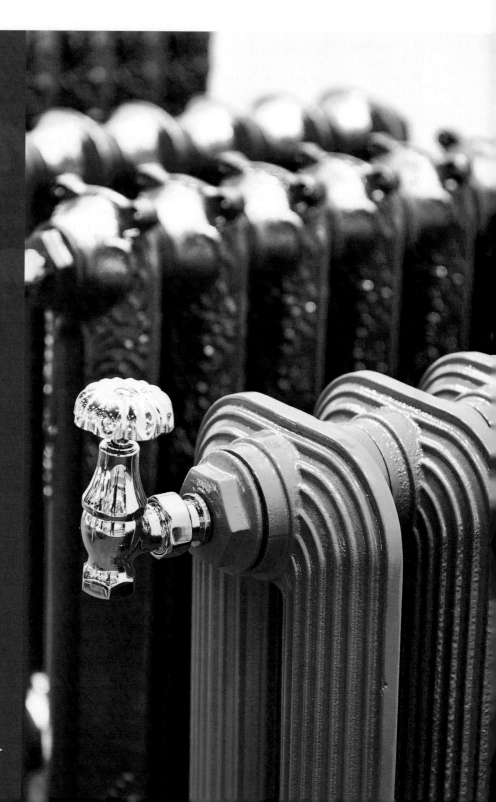

Sergey Ryzhov/Shutterstock.com

Learning Objectives

After completing this chapter, you will be able to:
- Describe the factors used in selecting a terminal device.
- Understand how finned-tube baseboard heaters are sized.
- Explain how finned-tube baseboard heaters are installed.
- Describe how a radiator-type terminal device operates.
- Differentiate between high-mass and low-mass terminal devices.
- Describe the procedure for installing radiators.
- Compare fan coil units to other types of terminal devices.
- Explain the procedures for sizing fan coil units.
- List the different types of temperature controls used with terminal devices.
- Describe how a unit ventilator uses outside air for ventilation.

Technical Terms

cabinet unit heater
escutcheon plate
fan coil unit
finned-tube baseboard heater
heat emitter
high-mass terminal unit
hydronic towel warmer
Institute of Boiler and Radiator Manufacturers
interpolate
low-mass terminal unit
radiant panel
radiator
terminal device
under-cabinet fan coil unit
unit heater
unit ventilator

One of the most important components in a hydronic heating system is the *terminal device*. Sometimes referred to as a *heat emitter*, the terminal device is the final component in the hydronic piping plans and is used to heat the conditioned space. The role of the terminal device is to transfer heat from the water located in the hydronic piping into the space to be heated, either by convection or radiation.

There are a wide variety of terminal devices the system designer can choose from. The choice of terminal units depends on several factors. These include:
- The amount of heat needed for the conditioned space
- The space constraints of the heated area
- The economic issues that play into the cost of the hydronic system
- The architectural aspects of the space to be heated

This chapter covers various types of terminal devices widely used in the industry today. It discusses their advantages and disadvantages and proper sizing and installation.

11.1 Choosing a Piping Configuration

Before a terminal unit can be chosen for a project, the type of piping layout must first be configured. There are several types of piping loops from which to choose, as discussed in detail in Chapter 7, *Hydronic Piping Systems*. As a reminder, here are four typical configurations that can be considered:
- **Series loop:** This is simplest and most economical piping arrangement. It is used when installation costs are an issue, and it is typically controlled by a single-zone thermostat, **Figure 11-1**.
- **Parallel or two-pipe loop:** This piping configuration can incorporate multiple zones, controlled with either zone valves or circulating pumps, **Figure 11-2**.
- **Split loop:** This is a variation of the series loop piping arrangement, typically found on larger systems, that divides the entire system into two separate loops, **Figure 11-3**.
- **One pipe loop with diverter tees:** This arrangement allows for automatic zone control using diverter tees. It can also be used with thermostatic control valves, **Figure 11-4**.

Once the piping configuration is established, the hydronic heating designer can choose the ideal terminal devices.

214 Hydronic Heating: Systems and Applications

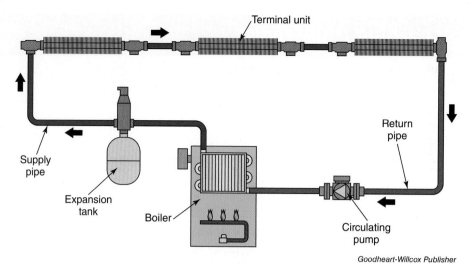

Figure 11-1. An illustration of a series-loop system.

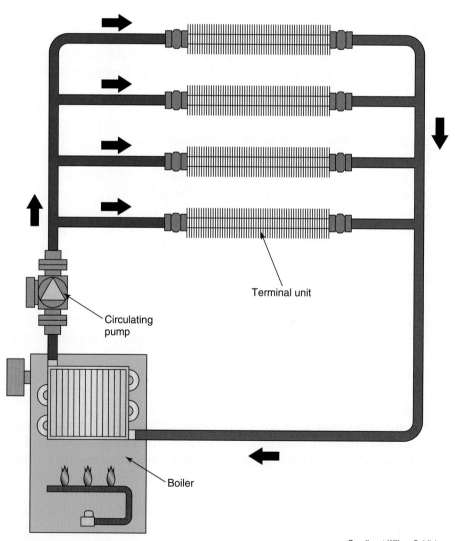

Figure 11-2. An illustration of a parallel or two-pipe system.

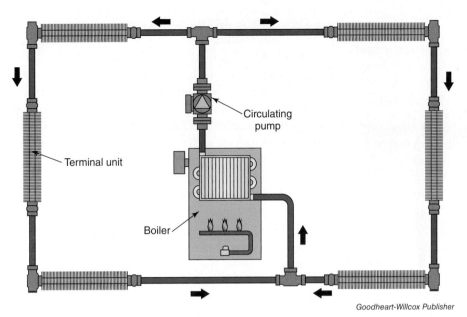

Figure 11-3. An illustration of a split-loop system.

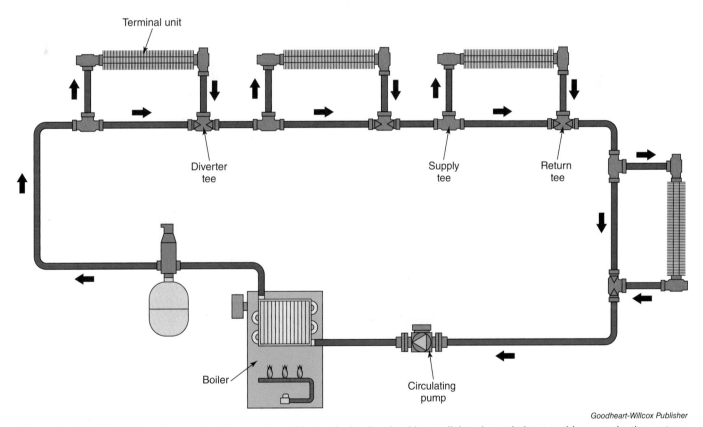

Figure 11-4. An illustration of a one-pipe loop system with terminal units piped in parallel to the main loop. In this example, the system would incorporate the use of diverter tees or thermostatic control valves.

11.2 Finned-Tube Baseboard Units

One of the most versatile and widely used types of terminal units is the *finned-tube baseboard heater*, **Figure 11-5**. Available in a variety of styles and heat outputs, finned-tube heaters have proven to be effective heat emitters for many

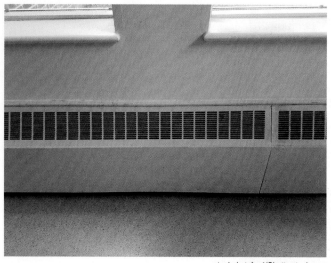

Figure 11-5. An example of a baseboard finned-tube heater.

years. These devices use convection as a means of heat transfer. Convection heating occurs when air currents circulate between warm and cool areas of a surrounding fluid. The theory of this principle is that when air is heated, it has less density and will rise—whereas colder air has greater density and will fall. In the case of finned-tube baseboard units, air enters the bottom of the unit and then passes over the heated tubing. The tubing used with this type of terminal unit is typically copper with aluminum "fins." These fins are fitted over the copper tubing and increase the amount of heat transfer between the fluid in the tubing and the surrounding air, **Figure 11-6**. The copper and aluminum heating elements found in finned-tube heaters are encased inside a metal enclosure with a fitted damper at the top of the enclosure that is used to direct the flow of air currents as they exit the terminal unit, **Figure 11-7**. Although most finned-tube baseboard heaters are copper tubing with aluminum fins, some may be made of steel piping with either aluminum or steel fins, **Figure 11-8**. This type of terminal unit is designed to transfer a greater amount of heat in Btu per foot for larger commercial applications. They are more robust and resistant to abuse than residential grade tubing.

Figure 11-6. Aluminum fins connected to copper tubing provide greater heat transfer in finned-tube baseboard heaters.

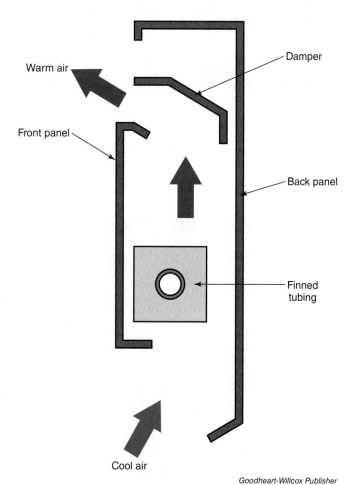

Figure 11-7. This illustration shows the components found in finned-tube baseboard heaters.

Goodheart-Willcox Publisher

Figure 11-8. Steel finned-tube heaters used in commercial applications are more rugged than residential types and can provide a higher heating output.

11.2.1 Sizing Finned-Tube Baseboard Units

Residential and light commercial finned-tube baseboard heating elements consist of either 1/2″ or 3/4″ copper tubing covered with aluminum fins that are approximately 2 1/2 in^2. They are typically purchased in lengths ranging from 2′ to 10′. Manufacturers of finned-tube heaters rate the heating output capacity of their units based on the number of Btu/hr per linear foot. This data is calculated using an entering air temperature of 65°F. As shown in **Figure 11-9**, these output ratings are determined by the entering hot water temperature and the water flow in gallons per minute through the piping.

To accurately size finned-tube baseboard units, first calculate the heat load of each room. Consult Chapter 13, *Building Heating Loads and Print Reading* for information on load calculations. Second, determine the differential temperature

> **DID YOU KNOW?**
>
> **I=B=R**
>
> I=B=R stands for the ***Institute of Boiler and Radiator Manufacturers***. It is a nationally recognized certification organization formerly known as the Hydronics Institute. IBR is nationally recognized as the leading organization for the certification of pressurized vessels and hydronic devices.

Hot Water Ratings														
Fine/Line 30—No. 30-75 baseboard with 3/4″ E-75 element														
Water Flow	Pressure Drop†	110°F	120°F	130°F	140°F	150°F	160°F	170°F	180°F	190°F	200°F	210°F	215°F	220°F
1 GPM	47	160*	210*	260*	320*	380	450	510	580	640	710	770	810	840
4 GPM	525	160*	220*	270*	340*	400	480	540	610	680	750	810	860	890

BTU/hr./ft. with 65° entering air †Millinches per foot. *Ratings at 140°F and lower temperatures determined by multiplying 150°F rating by the applicable factor specified in testing and rating standard for baseboard radiation.

NOTE: Ratings are for element installed as per drawing shown here, with damper open, with expansion cradles. Ratings are based on active finned length 5′ to 6′ less than overall length) and include 15% heating effect factor. Use 4 gpm ratings only when flow is known to be equal to or greater than 4 gpm; otherwise 1 gpm ratings must be used.

Slant/Fin (stlantfin.com)

Figure 11-9. This illustration shows Btu ratings for various finned-tube baseboard heaters.

through the terminal unit. Typically, a 20°F temperature differential can be used for sizing purposes. Next, calculate the flow rate in gallons per minute through the terminal unit. This figure can be calculated using the sensible heat formula for water:

$$\text{GPM} = \frac{\text{Btu/hr}}{(500 \times \Delta T)}$$

Using this formula, the flow rate for a terminal unit that has a heating capacity of 60,000 Btu/hr and a temperature drop of 20°F would be calculated as follows:

$$\text{GPM} = 60,000 / (500 \times 20)$$
$$\text{GPM} = 60,000 / 10,000$$
$$\text{Flow rate} = 6 \text{ GPM}$$

One method of sizing finned-tube baseboard units is based on IBR standards. This method takes into account the average water temperature in the hydronic circuit. The average water temperature is typically configured as being 10°F cooler than the average boiler output temperature. Once the boiler output temperature has been determined, the manufacturer's thermal rating table can be consulted in order to find the Btu/hr per linear foot of heating element (copper tubing, for example) that is best suited for that application.

PROCEDURE

Sizing a Finned-Tube Heater

Following is a step-by-step example for finned-tube heater sizing based on IBR standards:

1. **Calculate the heat loss of the building space.** Refer to ACCA Manual J for load sizing requirements or follow a similar procedure to determine the heat loss of the conditioned space. (A complete guide to load calculations is covered in Chapter 13, *Building Heating Loads and Print Reading*).

2. **Determine the total length of the required finned-tube baseboard heater.** Measure the length of wall space available, taking into account the space used for furniture and door swing. Subtract two feet from the total length of available outside walls to account for valves and piping connections, then round off to the nearest half foot. Divide this number into the Btu/hr heat loss that was calculated from step #1. This will provide the actual number of Btu per linear foot required.

3. **Determine the water flow in gallons per minute (GPM) through the finned-tube element.** The flow through the heating terminal unit can be calculated by referring to Chapter 7, *Hydronic Piping Systems*. This will be calculated after the boiler capacity and proper pipe sizing have been determined. Otherwise, use the sensible heat formula for water, as previously mentioned.

4. **Determine the design water temperature.** When using a two-pipe system, the average water supply inlet temperature to each terminal unit should be about the same. This temperature is equal to the design water temperature minus half the temperature drop. As an example, 20°F is the typical design temperature drop used across most terminal units. One-half of this temperature drop is equal to 10°F. The design boiler water temperature for sizing purposes can now be configured by subtracting 10°F from the boiler water outlet temperature.

   Note: Boiler outlet design temperatures may vary based on the type of boiler used and its application. Consult Chapter 3, *Boilers*, for the type of boiler being used for this exercise.

(continued)

5. **Select the finned-tube heater from the manufacturer's chart, Figure 11-9.** Using the design water temperature selected, choose the element model for each different condition from the manufacturer's catalog chart. *Interpolate* between the values if necessary. Interpolation is a method of estimating new data points within the range of two different known data points. Cross-check this selection with the calculation from step #2 for accuracy. Otherwise, divide the design heating load estimated in step #1 by the manufacturer chart selection rating and plan accordingly. Always round up to the nearest whole foot of baseboard length.

TECH TIP

Accounting for Heating Effect Factor

The chart in **Figure 11-9** has a "15% heating effect factor" as a footnote. This means that the values listed for the finned-tube heating outputs are 15% higher than those calculated by the manufacturer's testing results. The designer should consider this detail when selecting a terminal unit.

TECH TIP

One Row or Two?

Most manufacturers of finned-tube baseboard units offer both one-row and two-row heating elements. In some cases, a two-row heating unit may be the best choice when additional Btu output capacity is needed in a given space.

11.2.2 Installing Finned-Tube Baseboard Units

The installation of finned-tube baseboard heaters is not complex, and most manufacturers provide comprehensive step-by-step instructions to make the task relatively easy. The installer should be mindful of the following points to avoid any unplanned issues when installing baseboard units:

1. The first step is to locate where the hot water piping will penetrate the flooring and drill the appropriate-sized holes to provide for risers. These holes should be 1/2″ larger than the pipe size to allow for heat expansion. Install the risers close to the inside edge of the hole to allow room for the riser to move to the outside edge of the hole during pipe expansion, **Figure 11-10**.

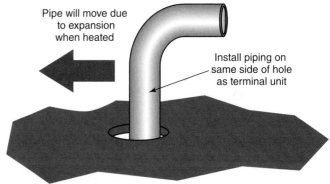

Goodheart-Willcox Publisher

Figure 11-10. The first step for installing finned-tube heating is to drill holes for the system risers.

TECH TIP

Drilling Holes for Risers

Be sure to locate any floor joists, plumbing pipes, or electrical wiring below the floor before drilling holes for risers. Use a common reference point above the floor, and measure from this point below the floor to locate any possible obstacles beneath the floor. Another tip is to drill a small pilot hole through the top of the floor where the riser is to be located, and fish a solid wire through this hole to determine where the hole will end up beneath the floor.

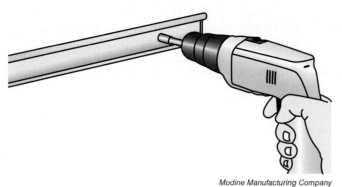

Modine Manufacturing Company

Figure 11-11. Attach the baseboard enclosures using the appropriate fasteners.

2. Attach the baseboard enclosure to the wall using the proper fasteners. For typical drywall, conventional nails work, **Figure 11-11**. Plaster or masonry walls may require the use of wall anchors or masonry nails. **Note:** Be sure to leave enough room at the ends of the baseboards for the later installation of accessories.
3. Attach the heating elements to the wall plate and solder the connections, **Figure 11-12**. Be sure to properly clean and flux all joints and flush the system when all soldering is completed. **Note:** It is recommended to install expansion compensators on longer piping runs to prevent straining and buckling of the heating elements at the soldered joints, **Figure 11-13**.

TECH TIP

Copper Press Fittings versus Soldering

As an alternative to soldering copper pipe, installers may now use copper press fittings for making piping connections. These flameless fittings are specially designed connectors that join traditional pipes using a hydraulic tool to press the fittings into place. Press fittings are also known as press-connect joining and crimping. Using press fittings instead of soldering can save time and reduces the risk of using an open flame—especially in hazardous areas.

4. Install the front cover over the enclosure. This front panel should lock into place to secure a tight fit, **Figure 11-14**. Once installed, the damper on the front of the enclosure can be adjusted for proper heat output.
5. Attach any accessories: these include any end caps, wall trims, splice plates, and *escutcheon plates* over the piping penetrations through the floor, **Figure 11-15**.

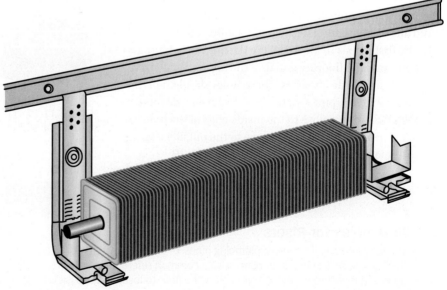

Modine Manufacturing Company

Figure 11-12. Next, attach the heating elements to the wall plates and solder in place.

TECH TIP

When Space Is an Issue

Sometimes it is not possible to drop the hot water return line through the floor at the end of the terminal unit because of space limitations. When this is the case, a 180° pipe elbow can be installed at the end of the finned-tube element and the return pipe can be run within the baseboard enclosure, **Figure 11-16**. If possible, install an air vent onto this elbow to provide a point for air elimination, **Figure 11-17**.

Metraflex

Figure 11-13. Expansion compensators should be used on longer piping runs to prevent straining and buckling of the heating elements at the soldered joints.

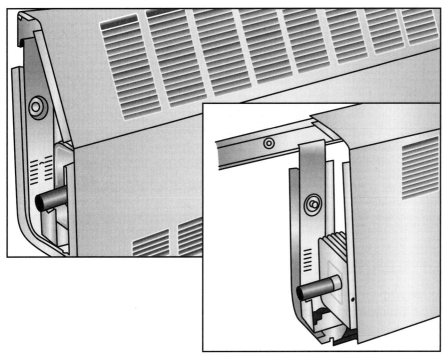

Modine Manufacturing Company

Figure 11-14. Once the heating elements are in place, secure the front cover of the baseboard unit.

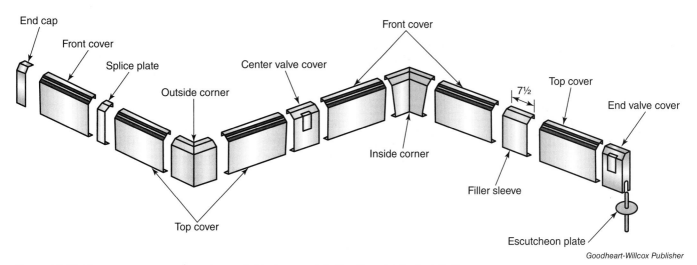

Goodheart-Willcox Publisher

Figure 11-15. There are many accessories available to complete the finned-tube installation.

Slant/Fin (stlantfin.com)

Figure 11-16. When it is not possible to drop the hot water return line through the floor at the end of the terminal unit because of space limitations, a 180° pipe elbow can be installed at the end of the finned-tube element and the return pipe can be run within the baseboard enclosure.

Goodheart-Willcox Publisher

Figure 11-17. Consider adding an air vent when installing finned-tube units with 180° elbows.

11.3 Radiators

As the name implies, ***radiators*** transfer heat primarily by means of radiation rather than convection. Both types of heat emitters—radiators and convectors—essentially utilize both types of heat transfer. However, radiators provide several advantages as far as effective and economical heat transfer when compared to convection units.

The term *radiator* conjures up images of old-style cast iron radiators that were used primarily with steam heating systems, **Figure 11-18**. These types of terminal units are still used today. However, modern radiators incorporate newer innovations that make them much more effective in terms of heat transfer. Radiators and ***radiant panels*** can be classified as either ***low-mass*** or ***high-mass units***. Low-mass radiators are typically made of lightweight materials such as aluminum and do not have the ability to retain heat for long periods of time. Consequently, low-mass radiators need to function constantly when there is a call for heat. High-mass radiant systems are made of heavier materials and can retain heat for longer periods of time, even after the heating source has cycled off, **Figure 11-19**. Low-mass radiant systems respond quickly to changes in space temperature and tend to heat up and cool down faster to reach their desired temperature. High-mass radiators are much slower to reach their desired temperature, but in turn will retain their heat over longer periods of time.

Aleksei Isachenko/Shutterstock.com

Figure 11-18. An illustration of a conventional cast-iron radiator.

11.3.1 Benefits of Radiators

Regardless of which type of radiator is chosen for a particular heating project, these types of terminal units have distinct advantages over conventional convection units (such as finned-tube baseboard heaters):

- **Radiators require less linear wall space.** This feature can be very attractive for limited spaces where extra heating output is needed, such as kitchens and bathrooms.
- **Radiators are designed to operate at lower water temperatures.** Because they possess a greater ability to retain heat, radiators are very functional at lower water temperatures as opposed to finned-tube baseboard heaters. This feature also means that boilers can operate at lower supply water temperatures, which can increase fuel efficiency.
- **Most radiators have greater durability.** Because most radiators are constructed of heavy gauge steel, they can withstand damage and overall wear and tear.
- **Most radiators operate at a greater temperature difference across the coil.** This makes the design and selection less critical than for other types of terminal units.

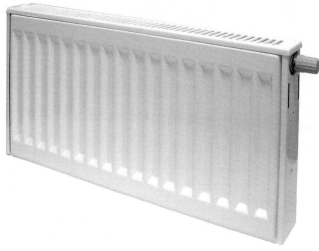

Bosch Thermotechnology

Figure 11-19. High-mass radiators such as this one can retain heat for longer periods of time—even after the heating source has cycled off.

11.3.2 Sizing Radiators

Radiator sizing is similar to finned-tube baseboard sizing. There are a few exceptions (and advantages) that the designer will want to take into consideration:

- Because of their unique design, radiators or radiant panels can be mounted under windows or on exposed walls without taking up as much linear footage as conventional convection-type heaters. This feature gives the designer greater flexibility.
- Because radiators can operate at lower temperatures than finned-tube heaters or other types of conventional convection units, they are usually sized and specified at a broader range of temperatures than conventional heaters. This allows for a greater margin of error when sizing.
- Radiators can be specified in capacities as either Btu per linear foot or as total Btu/hr output per heating unit. This allows the designer to choose a single unit rather than a total number of feet of finned-tube piping, **Figure 11-20**.
- The calculated average water temperature chosen for sizing the unit is based on the median point between the entering water temperature (EWT) and the leaving water temperature (LWT). For instance, if the chosen EWT is 170°F and the chosen LWT is 150°F, the average water temperature would be 160°F. Apart from these differences, the sizing procedure is similar to that of finned-tube baseboard units.

11.3.3 Installing Radiators

The procedure for installing radiant wall panels is also similar to that of finned-tube heaters. However, the installation procedure for radiators may allow for greater flexibility in certain instances. For example, piping may be run through

Panel Radiator Ratings Model SRP-1									
Model Type	BTU/FT/HR @ LISTED AWT & 65°F EAT								
	200°F	195°F	185°F	175°F	165°F	155°F	145°F	135°F	125°F
SRP-1 01	231	181	162	144	127	109	93	77	62
SRP-1 02	416	326	292	260	228	197	167	139	112
SRP-1 03	615	483	433	384	337	292	248	206	165
SRP-1 04	811	636	571	507	445	385	327	271	218
SRP-1 05	1012	794	712	632	555	480	408	338	272
SRP-1 06	1219	956	858	761	668	578	491	407	327
SRP-1 07	1429	1122	1006	893	784	678	576	478	384
SRP-1 08	1638	1285	1153	1023	898	777	660	547	440
SRP-1 09	1846	1449	1300	1154	1013	876	744	617	496
SRP-1 10	2057	1614	1448	1285	1128	976	829	688	552
Panel Radiator Ratings Model SRP-2									
SRP-2 01	480	374	331	293	254	216	192	158	125
SRP-2 02	862	672	595	526	457	388	345	284	224
SRP-2 03	1274	994	879	777	675	573	510	420	331
SRP-2 04	1680	1310	1159	1025	890	756	672	554	437
SRP-2 05	2096	1635	1446	1279	111	943	838	692	545
SRP-2 06	2524	1969	1742	1540	1338	1136	1010	833	656
SRP-2 07	2960	2309	2042	1806	1569	1332	1184	977	770
SRP-2 08	3392	2646	2340	2069	1798	1526	1357	1119	882
SRP-2 09	3820	2980	2636	2330	2025	1719	1528	1261	993
SRP-2 10	4260	3323	2939	2599	2258	1917	1704	1406	1108

Goodheart-Willcox Publisher

Figure 11-20. Radiators can be sized by Btu per linear foot or total Btu/hr output per heating unit.

walls as well as between floors. In another example, the area where the radiator is to be installed may lend itself to a more convenient location for running the supply and return piping. Just as with unit selection, there are differences in the installation of radiant panels compared to finned-tube units. These differences include:

- Radiators should not be used with series loop piping systems or with gravity-type systems.
- Mount the radiator a minimum of 4″ off the floor to provide for adequate air currents, **Figure 11-21**.
- Use good judgment when determining the location of the radiator to provide for proper air circulation across windows and exposed walls.
- In larger conditioned spaces, use several identically sized panels spaced evenly across the room as opposed to sizing one large heater.

In addition to providing flexibility to the heating requirements of the conditioned space, radiators offer more flexibility in terms of architectural aesthetics compared to conventional convection heaters. There are many different designs to choose from when it comes to radiant panels. These include flat-tube panel radiators, fluted-channel panel radiators, and even towel warmer radiant panels, **Figure 11-22**. All provide similar features that make radiant heating the popular choice for hydronic heating designs.

Gadelshina Dina/Shutterstock.com

Figure 11-21. When installing wall-mounted radiators, provide the proper spacing between the floor and heating element to allow for adequate convection currents across the heater.

Bilanol/Shutterstock.com

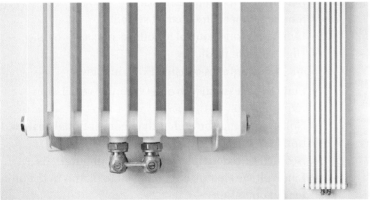

LightField Studios/Shutterstock.com

Symonenko Viktoriia/Shutterstock.com

Figure 11-22. Radiators are available in many different colors and configurations.

11.4 Fan Coil Units

A *fan coil unit*, or FCU, is simply a convector that uses an integrated fan to transfer heat throughout the conditioned space, either with or without the use of connected ductwork, **Figure 11-23**. Fan coil units consist of a cabinet that houses the fan, a heat exchanger coil (finned-tube coil), and the appropriate controls.

Figure 11-23. An example of a fan coil unit that can utilize both hot and chilled water coils.

Figure 11-24. This fan coil unit is referred to as a cassette unit.

Figure 11-25. This free-standing fan coil unit is used to heat an individual classroom.

They can utilize either or both hot water and chilled water for heating and cooling purposes. Some fan coil units can be recessed into a ceiling such as the one in **Figure 11-24**. This type of unit is sometimes referred to as a *cassette*. Fan coil units are usually installed in areas that experience high heat losses, such as building entrance areas or vestibules. However, they may be used as a primary heating source for individual spaces such as in classrooms, offices, or other occupied areas, **Figure 11-25**. One issue to consider when selecting a fan coil unit for heating applications is to ensure that noise from the fan does not become an issue. Always check with the potential space occupants to ensure that noise will not be a problem.

Fan coil units are available in horizontal or vertical configurations as well as ceiling, floor-mounted, and freestanding models. Fan coil units that are flush-mounted inside of a wall or mounted against a wall are referred to as **cabinet unit heaters**, **Figure 11-26**. Most fan coil units also include a filter for improved indoor air quality. The orientation of the fan and coil may be either "draw-through" or "blow-through." In the draw-through design, the fan is ahead of the coil so the air is drawn through the coil. In the blow-through design, the fan pushes the air through the coil. In some cases, a draw-through design may be more favorable since it offers a more even airflow across the coil. This configuration lends itself to better air and heat distribution where duct transition spaces tend to be more confined.

Advantages of using FCUs include:
- Relatively low equipment cost
- Simple temperature control system
- Individual zone control for offices or classrooms
- High levels of flexibility

Some disadvantages of using FCUs include:
- Limited accessibility to outside air for ventilation requirements
- Noise levels may be higher than conventional heating units
- Not well suited to open floor plan configurations
- Line-voltage electric circuits need to be run to the unit for fan operation
- Units located above ceilings may present maintenance problems

11.4.1 Sizing Fan Coil Units

The procedure for sizing fan coil units is as follows:
1. **Calculate the heat loss of the space where the fan coil is to be used.** Consult ACCA Manual J for residential and light commercial applications. Use either ACCA Manual N or the ASHRAE Handbook of Fundamentals for commercial spaces, and calculate the heat loss of the conditioned space in Btu/hr.

2. **Determine entering and leaving water temperatures through the FCU.** Most applications utilize a maximum entering water temperature of 180°F and leaving temperature of 160°F; however, check to see if these temperatures correspond to other devices being used within the hydronic heating system.
3. **Calculate the proper flow in gallons per minute (GPM).** Use the sensible heat formula for water to determine the proper GPM flow through the fan coil unit:

$$GPM = \frac{Btu/hr}{(500 \times \Delta T)}$$

where

   GPM = Water flow through the FCU
   500 = Constant
   ΔT = Temperature difference between entering water and leaving water

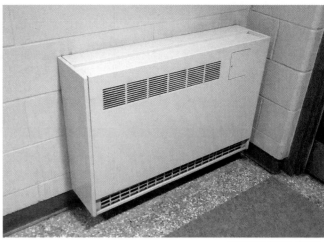

Figure 11-26. This type of fan coil unit is also known as a cabinet unit heater.

4. **Check for proper airflow through the unit.** Using the following formula, the airflow in cubic feet per minute (cfm) can be determined:

$$cfm = \frac{Btu/hr}{(1.08 \times \Delta T)}$$

where

   cfm = Airflow through the FCU
   1.08 = Constant
   ΔT = Temperature difference between entering airflow and leaving airflow

5. **Select a unit based on the calculated data.** Consult the manufacturer's specifications to choose a fan coil unit based on the required Btu/hr and airflow.
6. Other considerations to be aware of when selecting the appropriate fan coil unit include:
   - **Noise criteria:** Select a unit that is based on acceptable sound levels for the given space.
   - **Duct considerations:** If the unit is to be ducted, check for airflow rates based on the calculated static pressure through the ductwork. Ensure the fan is capable of delivering the proper airflow.
   - **Condensate drain pan:** If the fan coil unit is to be installed above a ceiling and is to be used in conjunction with a chilled water coil, install a condensate drain pan below the unit to prevent damage to the ceiling and objects below in the event of a leak.

11.4.2 Fan Coil Unit Temperature Controls

There are a number of temperature control strategies that can be incorporated with fan coil units. The simplest is a line-voltage thermostat that controls the fan and hot water valve operation, **Figure 11-27**. Other strategies for temperature controls include the following:
- **Variable fan speed control:** Fan speeds can be varied based on heating and cooling operations. These may be controlled by a selector switch or through a microprocessor control. A higher fan speed is more conducive for cooling control versus heating.

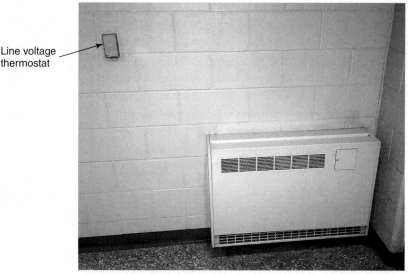

Goodheart-Willcox Publisher

Figure 11-27. Fan coil units can be controlled simply by a wall mounted line-voltage thermostat.

- **Modulating hot water valve:** More precise space temperature control can be achieved by modulating the hot water valve based on room temperature. This can be done through either a stand-alone digital thermostat or a microprocessor control, **Figure 11-28**.
- **Occupancy scheduling:** Greater efficiency can be achieved by scheduling the fan coil unit operation based on building occupancy. This feature can incorporate night setback control and morning warmup.

GREEN TIP

Night Setback and Lower Water Temperatures

Simple modifications to the hydronic system can result in increased efficiency and energy savings. Among these implementations include incorporating an occupancy schedule. One of the easiest ways to save energy is to simply shut off the equipment. This can be done if no one is in the building by scheduling the system to respond to building occupancy, which most digital controllers can easily achieve. When the building is coming out of a night setback mode, the equipment can be programmed to go into full heating mode (morning warmup) for a short period of time to quickly and efficiently bring the building up to the occupancy temperature set point.

Another energy-saving method is to lower the supply water temperature to the distribution system. Most modern hydronic systems such as radiant floor heating are already utilizing a lower supply water temperature for optimum performance. By lowering the hot water set point, the boiler fuel consumption is reduced and money is saved. Even though the equipment may need to run longer, the space temperature will be more comfortable—another added benefit. This is because the terminal unit's discharge temperature is closer to the room temperature. When this occurs, the two bodies of air will blend more evenly, which reduces stratification.

These simple implementations result in money saved and a more comfortable environment—a win/win result!

Goodheart-Willcox Publisher

Figure 11-28. For greater temperature control, fan coil units may use either option. A—A modulating hot water valve. B—A digital controller with built-in room temperature sensor, as seen here.

11.5 Unit Heaters

Another type of terminal unit used with hydronic heating is a *unit heater*. Unit heaters are small, compact heating devices that consist of a heating coil and propeller fan housed in a sheet metal cabinet, **Figure 11-29**. They often include directional-type louvers at the heater outlet, which are used to adjust the flow of heat either vertically or downward, **Figure 11-30**. Unit heaters are a good source for heating an unconditioned space such as a garage or shop. They are typically hung from the ceiling in the corner of the space to be heated, **Figure 11-31**. Unit heaters require a power source to energize the fan, a hot water source, and a temperature controller.

11.5.1 Applications for Unit Heaters

Unit heaters are most frequently installed for heating applications in the following locations:
- Warehouses
- Garages (both residential and commercial service)
- Workshops
- Factories
- Parking facilities
- Mechanical rooms

Hydronic unit heaters provide a good alternative where gas-fired appliances are not preferred due to fire hazards or potentially explosive atmospheres in the area to be heated. In cases such as these, the hydronic unit heater can be equipped with spark-proof or explosion-proof controls.

Other advantages of hydronic unit heaters include:
- Simple installation
- Wide range of heating outputs—from 25,000 to 400,000 Btu/hr
- Quick solution for temporary heating
- No duct distribution network needed
- Simple controls
- Easy maintenance

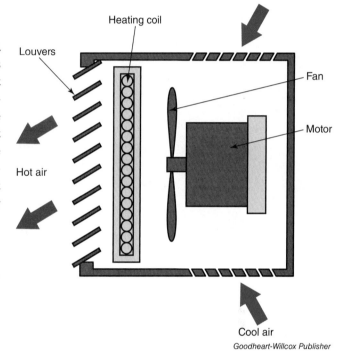

Goodheart-Willcox Publisher

Figure 11-29. This illustration shows the components that make up a unit heater.

Goodheart-Willcox Publisher

Figure 11-30. Unit heaters can be equipped with directional louvers to adjust the flow of heat either vertically or downward.

Goodheart-Willcox Publisher

Figure 11-31. Unit heaters are a great source of heat for a barn, garage, or shop.

Disadvantages of unit heaters include:
- Potentially excessive stratification, or layering of warm air from floor to ceiling
- Limited heating capability for larger areas
- Not conducive to heating multiple spaces simultaneously
- Potential noise issues

11.5.2 Controlling Unit Heaters

The control of a basic unit heater can be as simple as a space heating thermostat that cycles the fan on and off. This control strategy does not require a hot water valve. For better control, the following strategies can be implemented:

- **Space thermostat controlling a zone valve:** In this scenario, a line-voltage space thermostat energizes the zone valve. Once the valve is fully opened, a built-in end switch energizes the unit heater fan.
- **Low-voltage thermostat controls a hot water valve:** The thermostat can be electronic or mechanical. On a call for heat, the thermostat energizes a hot water valve to open and a relay to turn on the unit heater fan. This type of control needs a step-down transformer but does not require an additional electrical circuit for 120-volt controls (it can use the fan circuit).
- **Digital thermostat controls modulating hot water valve:** This is the best strategy for tight temperature control. A digital thermostat with built-in microprocessor modulates a hot water valve to control space temperature. The unit heater fan runs continuously, or is activated by an aquastat connected to the hot water return piping so that the fan is energized only when the unit heater coil is fully heated.

11.6 Other Types of Terminal Devices

There are variations to the previously discussed terminal units as well as some specialty terminal units. These types of terminal units include, but are not limited to:
- Unit ventilators
- Under-cabinet fan coils
- Hydronic towel warmers

11.6.1 Unit Ventilators

The *unit ventilator* has been around since the 1950s and is a primary source for the heating and ventilation of classrooms, **Figure 11-32**. Unit ventilators consist of a heating coil, fan or blower, dampers, and temperature controls. They are typically floor-mounted and available in both horizontal and vertical configurations, **Figure 11-33**. Unit ventilators provide a good heating source for individual spaces and offer the ability to incorporate outdoor air for ventilation purposes. In addition, they provide economical room-by-room zoning and can be controlled by individual room thermostats or by a building automation system.

Even with the advantages that unit ventilators provide, there are some disadvantages to using them. Because they are installed as floor-mounted units within the space, noise

Modine Manufacturing Company

Figure 11-32. Unit ventilators (such as the one shown here) are traditionally used for heating classrooms.

can be a factor. Their airflow is limited to about 2000 cfm, and they do take up floor space. This can be an issue to consider from an architectural standpoint. Also, if emergency repairs are needed, they will have to be scheduled around the space occupancy requirements.

Unit ventilators are controlled much like other types of terminal units used for individually heated spaces. They can incorporate a space temperature thermostat, which can control the fan, blower, heating valve, and outside and return air dampers. However, unlike other types of terminal units used for space heating purposes, unit ventilators can utilize outside air for ventilation and also for free cooling, **Figure 11-34**. Follow the steps mentioned earlier for sizing fan coil units to size unit ventilators.

> **CODE NOTE**
>
> **Outdoor Airflow for Ventilation**
> Proper outdoor airflow is essential for acceptable indoor air quality. The required amount of outdoor air for ventilation is often specified by local codes and is referenced by ASHRAE Standard 62 for indoor air quality standards.

11.6.2 Under-Cabinet Fan Coil Units

In areas where conventional fan coil units cannot be mounted—primarily where space considerations are a problem—the use of **under-cabinet fan coil units**, otherwise known as *kick-space* or *toe-kick heaters*, can be incorporated. These types of heaters are a space-saving alternative to conventional fan coil units or baseboard heaters, **Figure 11-35**. They are designed to be installed under cabinets, such as bathroom vanities—or they can fit inconspicuously

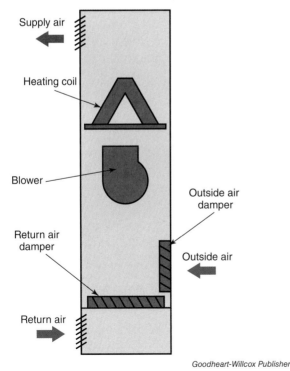

Figure 11-33. This illustration shows a vertical unit ventilator and components.

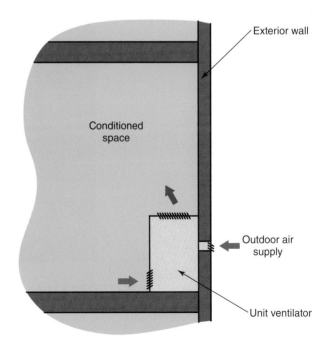

Figure 11-34. Unit ventilators can utilize outside air for ventilation and free cooling.

Myson

Figure 11-35. An example of an under-cabinet fan coil unit, also known as a kick-space heater.

Myson

Figure 11-36. Under-cabinet fan coil units draw air from their sides and back and push the air through a heating coil.

inside of a wall or floor. Kick-space heaters incorporate a centrifugal-type fan or blower that gently blows heated air into the space. They can be equipped with variable speed fans and can be controlled automatically according to the space heating demand. Air is drawn from the sides or back of the heater by the blower and pushed through the hot water coil, **Figure 11-36.** Kick-space heaters can be controlled with built-in thermostats or by wall-mounted space thermostats. Many of these heaters come equipped with aquastats that ensure the hot water coil temperature is above a certain set point before the blower is allowed to energize.

Considerations for installing a kick-space heater include being able to access the area around the cabinet or vanity for inlet and return hot water piping. Line-voltage wiring installation will also be necessary for powering the blower motor. In addition, openings will need to be made on the front and underside of the cabinet in order for the heating to be properly installed. An access panel should be installed in the event that the heater needs future servicing. The installer should take care not to damage the outside of the cabinet when installing the heater.

11.6.3 Hydronic Towel Warmers

For years, people in Europe have been using *hydronic towel warmers* as a way to improve the comfort level in their bathrooms and kitchens. These towel radiators not only provide warmth as a bathroom or kitchen heater but also provide the luxury of gently warming towels and bathrobes for personal use, **Figure 11-37.**

Olena Yakobchuk/Shutterstock.com

Figure 11-37. Hydronic towel warmers provide warmth for bathrooms and offer the luxury of gently warming towels and bathrobes.

Most hydronic towel warmers are available as flat-panel or round tube–type heaters. Both of these designs provide an expansive heating surface that warms towels completely—like they just came out of the clothes dryer. They can either be used as supplemental heat or as a sole source of heat for the designated room. They can also make a great complement to in-floor radiant heating. Most towel warmers are available in heating output ranges of 800 to 8000 Btu/hr and can provide the necessary space heating requirement for any size bathroom.

Towel radiators can provide an attractive means of heat for not only bathrooms and kitchens but also laundry rooms, foyers, spas, saunas, pool rooms, and hot tub areas. There are many different styles, colors, and sizes to choose from that can complement any decor, **Figure 11-38**.

Andrei Kobylko/Shutterstock.com

Figure 11-38. Towel radiators come in many different styles, colors, and sizes to choose from.

Chapter Review

Summary

- Terminal devices are the final component in the hydronic piping plans that are used as a means to heat the conditioned space by transferring heat from the water located in the hydronic piping into the space to be heated.
- There are several factors that go into determining the proper type of terminal device to be used. When considering which type of terminal unit to use, the proper piping layout will first need to be configured.
- Finned-tube baseboard units use convection as a means of heat transfer. They are made up of copper tubing with aluminum fins.
- Calculations used in sizing finned-tube baseboards include space heat loss, water flow, and design water temperature.
- Considerations for installing finned-tube baseboard heaters include the type of wall fasteners to hang the enclosure, size of the holes to drill for the risers, and underfloor obstacles.
- Radiators transfer heat primarily by means of radiation versus convection.
- There are several benefits of using radiators versus convectors. They require less linear wall space, operate at lower water temperatures, have greater durability, and operate at a greater temperature difference across the coil. Also, the installation of radiators provides more flexibility compared to the installation of convectors.
- Radiator sizing is similar to finned-tube baseboard sizing.
- Fan coil units are convectors that incorporate integral fans as a means of heat transfer. They are available in various configurations and can also be known as cabinet unit heaters. They can be used for both hot and chilled water applications.
- When sizing fan coil units, the heat loss of the space is required along with the proper air and water flow.
- There are a number of temperature control strategies that can be incorporated with fan coil units.
- Unit heaters are small, compact heating devices that consist of a heating coil and propeller fan housed in a sheet metal cabinet. They can be used for various heating applications.
- Unit heaters typically use wall thermostats to control their fan and hot water valves.
- Unit ventilators consist of a heating coil, fan or blower, dampers, and temperature controls. They are traditionally used for heating classrooms.
- Unit ventilators can bring in outside air for ventilation and free cooling.
- Under-cabinet fan coil heaters are a space-saving alternative to conventional fan coil units or baseboard heaters.
- Hydronic towel warmers can be used for space heating as well as for the heating of towels and bathrobes.

Know and Understand

1. The method of heat transfer used by most terminal units is either _____.
   A. convection or radiation
   B. convection or conduction
   C. radiation or conduction
   D. radiation or condensation

2. Why do some finned-tube baseboard heaters use steel piping and fins rather than copper tubing and aluminum fins?
   A. Steel is designed to transfer a greater amount of Btu per foot.
   B. Steel is designed to withstand greater abuse.
   C. Both A and B are correct.
   D. None of the above.

3. The IBR method of sizing finned-tube baseboard heaters takes into account the _____ of the hydronic circuit.
   A. overall length
   B. average piping size
   C. average water temperature
   D. number of terminal units
4. To interpolate between two different values on a manufacturer's finned-tube sizing chart means to _____.
   A. choose between either one value or the other
   B. estimate a new value halfway between the two known values
   C. add the two values together
   D. subtract the two values
5. When drilling holes through flooring for finned-tube risers, why should the holes be sized 1/2″ larger than the pipe size?
   A. To provide for future piping
   B. To provide for miscellaneous fittings
   C. To allow the installer to see what is below the floor
   D. To allow for thermal expansion
6. High-mass radiant panels are made from _____ materials than low-mass panels.
   A. heavier
   B. lighter
   C. the same
   D. more expensive
7. *True or False?* Radiators can be specified in capacities as either Btu per linear foot or as total Btu/hr output per heating unit.
8. Fan coil units that are flush-mounted inside of a wall or are mounted against a wall are referred to as _____.
   A. unit heaters
   B. unit ventilators
   C. cabinet unit heaters
   D. wall-mounted ventilators
9. *True or False?* One disadvantage of using a fan coil unit is that the noise level may be higher than for conventional heating units.
10. What would be the approximate cfm output of a fan coil unit that has a capacity of 6000 Btu/hr and a 15°F temperature rise?
    A. 400 cfm
    B. 370 cfm
    C. 550 cfm
    D. 600 cfm
11. What is the purpose of connecting an aquastat to the return piping of a unit heater?
    A. To prevent the coil from overheating
    B. To prevent the space from overheating
    C. To energize the fan only when the coil is fully heated
    D. To raise the temperature of the return water
12. What main advantage do unit ventilators have over other conventional types of terminal devices?
    A. Their coils can be used for chilled water as well as hot water.
    B. They can utilize outside air for ventilation and free cooling.
    C. They can be both floor-mounted and ceiling-mounted.
    D. They are compact and can be concealed easily in a room.

Apply and Analyze

1. Explain some of the factors that go into selecting a proper terminal unit.
2. Describe the steps that go into sizing a finned-tube baseboard heater.
3. What is the "15% heating effect factor" listed at the bottom of manufacturer's sizing data and how does this factor affect sizing finned-tube baseboard heaters?
4. When installing finned-tube heaters, what is a possible solution when the hot water return line cannot be routed through the floor at the end of the terminal unit because of space limitations?
5. List and describe the differences between low-mass and high-mass thermal radiators.
6. Explain why radiators have significant advantages over conventional convection heaters.
7. What are some of the considerations to be aware of when selecting fan coil units?
8. List and describe the control strategies that can be used with fan coil units.
9. List some of the areas where unit heaters would *not* be a good fit for heating applications.
10. Explain how a unit ventilator could be used for both heating and cooling applications.
11. Describe the location and control of a "toe-kick" heater.
12. When and where would a hydronic towel warmer be used for space heating applications?

Critical Thinking

1. Describe some various building applications where different types of terminal units would be used in the same hydronic piping system.
2. Would it be a good idea to use both low-mass and high-mass terminal units in the same hydronic piping system? Why or why not?

12 Radiant Heating Systems

Chapter Outline

12.1 Principles of Using Radiant Heating
12.2 Heating Sources for Radiant Systems
12.3 Radiant Heat Piping Systems
12.4 Radiant Floor Piping Configurations
12.5 Radiant Wall and Ceiling Panels
12.6 Designing Radiant Heating Systems
12.7 Radiant Heating Controls
12.8 Radiant Heating for Snow and Ice Melt Systems

ArtMari/Shutterstock.com

Learning Objectives

After completing this chapter, you will be able to:
- Compare radiant heating systems to conventional heating systems and explain how they are different.
- Describe how radiant heating relates to human comfort.
- Summarize the difference between high-mass and low-mass radiation.
- Compare various heating sources for radiant systems.
- Describe the differences between various types of radiant piping materials, including the different types of PEX, PEX-AL-PEX, and PE-RT tubing.
- Explain where PEX tubing with an oxygen barrier is required.
- Compare the various piping configurations that are available for radiant floor heating.
- Explain the procedure for installing slab-on-grade radiant floor systems.
- Identify how concrete or gypsum thin-slab floor systems differ from slab-on-grade systems.
- Describe the process for installing a thin-slab floor heating system.
- List the types of radiant flooring systems that are available for both above- and below-floor applications and explain how they are designed and installed.
- Describe how radiant wall and ceiling panels function and list their installation steps.
- Demonstrate design concepts for radiant heating systems.
- Explain how radiant heating systems are controlled.
- List the different applications where snow and ice melt systems may be used and their installation steps.

Technical Terms

below-floor suspended tubing
buffer tank
coefficient of performance
diffusion
dry radiant hydronic system
expanded foam board (EPS)
extruded foam board (XPS)
geothermal heat pump
high-mass radiant system
ideal heating curve
low-mass radiant system
manifold station
oxygen barrier
PE-RT tubing
PEX-AL-PEX
PEX (crosslinked polyethylene)
prefabricated radiant floor panel
radiant heating system
radiant ceiling panel
radiant wall panel
slab-on-grade radiant piping
socket fusion
solar thermal storage
snow and ice melt system
thermal memory
thin-slab radiant flooring system
water-to-water heat pump
wet radiant hydronic system
vapor barrier

Up to this point, our discussions have focused on conventional hydronic heating systems and the terminal devices typically used with them. In this chapter, we discuss *radiant heating systems* in detail and how they compare to conventional hydronic heating systems. A ***radiant heating system*** is one in which thermal radiation is used to transfer heat from the piping to objects in the space to be heated, rather than the air. We will cover the main principles of radiant heating, different methods of installation, types of piping and controls, and various applications for radiant systems.

12.1 Principles of Using Radiant Heating

In previous chapters, we discussed how heat travels from warm to cold. Furthermore, heat travels in three different ways: through convection, conduction, and radiation. With most conventional hydronic heating systems, heat actually is transferred by all three methods. For instance, heat from the burner inside the boiler comes in contact with the water-filled heat exchanger, which is a form of heat transfer by conduction. When the heat from the hot water inside the hydronic piping gives up its heat into the air, there is heat transfer by convection. However, with radiant heat transfer, heat is conveyed in a direct physical path and heats the objects in the conditioned space, not the air. This is the main difference with heating by radiation rather than by conduction or convection. Another main difference between conventional hydronic heating systems and radiant heating systems is that in radiant systems the piping is considered the terminal device and is usually enclosed within walls, floors, and sometimes ceilings—completely out of sight. One additional principal difference is that radiant heating systems typically circulate the heated water at lower temperatures than conventional systems. With a conventional hydronic heating system, the hot water supply temperature typically ranges between 140°F and 180°F. With radiant heating, this temperature range is usually between 100°F and 140°F.

12.1.1 How Radiant Heating Relates to Human Comfort

By definition, comfort is controlling the rate at which a body loses heat. The human body operates at a core temperature of approximately 98.6°F and generates 300–400 Btu/hr depending on the level of activity. By controlling the rate at which our bodies absorb or reject heat, we can better control our individual comfort level. As stated in previous chapters, most individuals are comfortable at a room temperature of 70°F for heating applications. By maintaining a heating output source close to 70°F, there is less fluctuation in space temperature and less chance for heat stratification between floors and ceilings. Radiant heating has the capability to deliver this temperature into the conditioned space more evenly—and at a lower supply temperature—than a forced-air system. It accomplishes this by first delivering the heat at a lower supply temperature than forced air, and then by using heat from radiation to warm objects rather than air—thus keeping the space at a more comfortable temperature. **Figure 12-1** illustrates how this is accomplished. Notice the *ideal heating curve*. This curve represents the optimum comfort level in degrees Fahrenheit from floor to ceiling for an average person at rest. Compare this curve to the radiant floor heating curve. Notice that there is minimal difference between the two curves. **Figure 12-2**

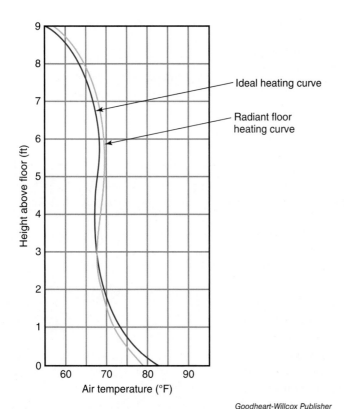

Goodheart-Willcox Publisher

Figure 12-1. This illustration compares the radiant floor heating curve to that of the ideal heating curve based on human body temperature.

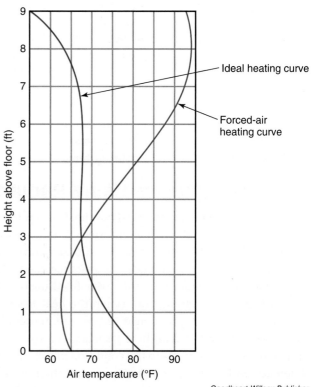

Goodheart-Willcox Publisher

Figure 12-2. This illustration compares a heating curve for a forced-air heating system to that of the ideal heating curve based on human body temperature.

represents a forced-air heating system. Notice that there is a large discrepancy between the ideal heating curve and the forced air heating curve. When heating with a forced air system versus a radiant floor heating system, not only are there sizable increases in stratification from floor to ceiling but also the temperatures at floor level are significantly lower, **Figure 12-3**. Considering the fact that a person's feet will typically lose heat faster than his or her head, one can see that this configuration is less than desirable.

> **DID YOU KNOW?**
>
> **Radiant Heating—A Brief History**
> Radiant heating dates back to ancient times, when the Romans warmed rooms by running the flues of slave-tended, wood-burning fires under elevated marble floors. This kept the occupants' toes and togas nice and toasty.
>
> Many centuries later in the United States, architect Frank Lloyd Wright embedded copper pipes in the concrete floors of his Usonian homes (in Lakeland, Florida) and warmed them with hot water. A few postwar subdivisions, including Levittown, New York, followed suit, but when the copper pipes eventually corroded most homeowners abandoned their radiant heating systems rather than jackhammer through their floors to get rid of them.

12.1.2 High-Mass versus Low-Mass Radiation

The difference between high-mass and low-mass radiation is an important concept to understand because it can significantly affect the best design for a hydronic heating project. The difference between these two types of radiation systems mainly involves their capacities and abilities to store heat energy. *Low-mass radiant systems* possess a very limited capacity to store heat energy, while *high-mass radiant systems* have a much higher capacity for energy storage. An example of a low-mass system is finned tube baseboard radiation, **Figure 12-4**. An example of high-mass radiation is embedding radiant heating tubes in

Room Air Temperatures

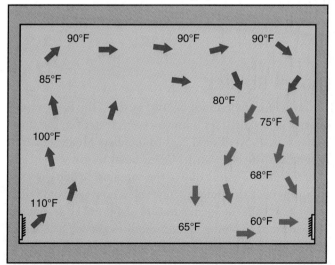
Forced-air heating system

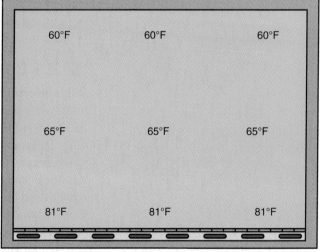
Hydronic heating system

Goodheart-Willcox Publisher

Figure 12-3. This illustration shows the effects of temperature stratification using a forced-air heating system.

Figure 12-4. An example of a low-mass heating system.

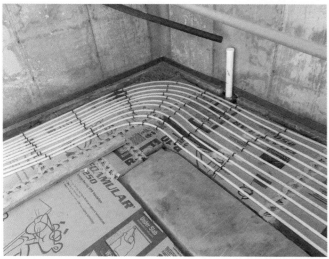

Figure 12-5. An example of a high-mass heating system.

concrete, **Figure 12-5**. There are three major differences between low- and high-mass radiant heating systems:

- **Energy retention:** Low-mass radiant panels made of lightweight materials (such as aluminum) cannot retain heat energy for long periods of time and need to be constantly functional whenever heating is required. Conversely, high-mass radiant systems that use materials such as concrete or gypsum tend to retain heat energy for extended periods and maintain levels of heat transfer long after the heating source is de-energized.
- **Response speed:** Low-mass radiant systems will respond quicker to required changes in the conditioned space. Once a low-mass system is set to the desired heat output temperature, it will heat up and cool down much faster to reach the temperature required. On the other hand, high-mass systems tend to take longer to reach their desired temperature.
- **Material type:** Because low-mass heating systems typically have higher performance requirements than high-mass systems, they usually consist of higher conductivity materials such as copper or aluminum.

12.2 Heating Sources for Radiant Systems

There are many different methods of generating hot water to supply a radiant heating system. The system designer needs to be aware of the advantages and disadvantages of the various types of systems. Many different factors must be evaluated before settling on the right source for generating hot water to the system. These factors include the type of heating fuel available, whether a renewable energy source should be used versus conventional fossil fuels, the size and scope of the project, the overall cost of the project, and the projected payback. Here we discuss four different types of hot water generating systems:

- Conventional boilers
- Condensing boilers
- Geothermal heat pumps
- Solar thermal storage

12.2.1 Conventional Boilers

When selecting the right boiler for the radiant heating system, the first consideration is whether the boiler is designed for a lower return water temperature. As discussed in earlier chapters, conventional cast-iron boilers are designed to allow for a return water temperature of at least 140°F. Colder return water temperatures can cause premature boiler failure due to the condensing flue gas temperatures impinging upon the heat exchanger. However, a conventional boiler can be used on a radiant system with the incorporation of a primary-secondary loop system. The primary-secondary piping system requires the installation of two separate piping circuits. The first or primary circuit simply circulates water throughout the boiler and provides a constant water temperature through the boiler's heat exchanger. The secondary circuit is used to supply the proper water temperature through the individual radiant heating circuits, with each circuit acting independently of the others, **Figure 12-6**. Primary-secondary circuit piping is discussed in greater detail in Chapter 7, *Hydronic Piping Systems*.

12.2.2 Condensing Boilers

A condensing boiler is a high-efficiency boiler with output efficiency ratings usually above 90%, **Figure 12-7**. These boilers achieve a higher efficiency rating by "squeezing" more heat out of the fuel. This simply means that more heat from combustion is transferred into the boiler water, and less heat passes through the chimney. This is accomplished by incorporating a secondary heat exchanger into the boiler. When flue gases from combustion pass through this secondary heat exchanger, water vapor condenses out of the gases. This secondary heat exchanger is made of a noncorrosive stainless steel, which resists the effects of flue gas condensation. Condensing boilers do not use conventional metal for their flue vents, but rather plastic piping such as PVC or PP that resists corrosion caused by acidic condensate from the products of combustion. Condensing

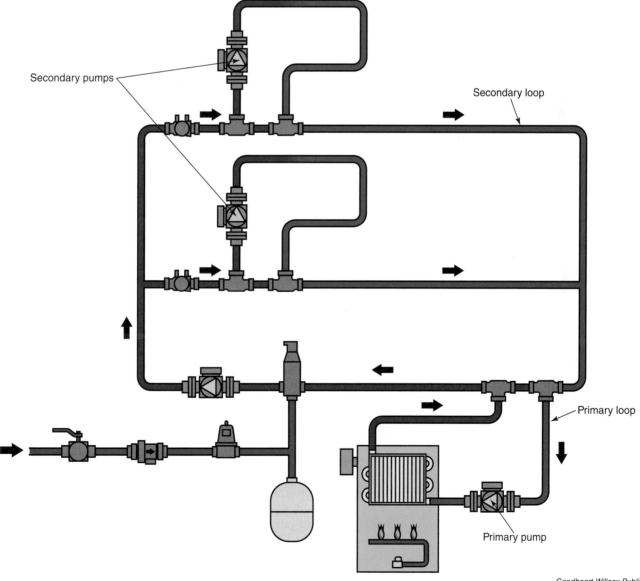

Goodheart-Willcox Publisher

Figure 12-6. An example of a boiler piped with a primary-secondary loop system.

boilers can be piped directly to individual radiant heating circuits. Boiler types and their configurations are covered in more detail in Chapter 3, *Boilers*.

12.2.3 Geothermal Heat Pumps

Geothermal heat pumps can be used for forced air heating systems as well as for hydronic heating applications. A ***geothermal heat pump*** uses the refrigeration cycle to transfer heat from the ground into the conditioned space. Pipes filled with water or antifreeze that are buried in the ground move heat into the heat pump's refrigerant and then, using air or water, transfer this heat into the building. When used with hydronic heating, a geothermal heat pump incorporates two coaxial heat exchangers and the use of the refrigeration cycle for heat transfer, **Figure 12-8**. The refrigeration compressor extracts heat from the primary heat exchanger water source and transfers this heat into the secondary heat exchanger's water, where it is then circulated through the heating system. This use of a ***water-to-water geothermal heat pump*** is a viable option for radiant heating applications for the following reasons:

- **There is no gas-fired heat exchanger.** With geothermal heat pumps, there is no need for venting of flue gases and no worry of heat exchanger failure due to flue gas condensation.
- **A higher level of energy efficiency can be achieved.** Geothermal heat pumps use the refrigeration cycle as a means of heat transfer rather than the combustion of fossil fuels. These types of heat pumps are rated according to their ***coefficient of performance*** and can achieve higher energy efficiency than conventional boilers.
- **The earth's crust has an abundant amount of heat available.** With geothermal heat pumps, heat is absorbed from the Earth. This means that when sized properly, geos can deliver the required amount of heat needed—no matter what the weather is like outside.
- In addition to these advantages, geothermal heat pumps deliver hot water to individual radiant heating loops at an ideal lower temperature application—typically 90°F to 120°F.

When installing a heat pump for radiant applications, a ***buffer tank*** should be incorporated. A buffer tank serves as a storage vessel for the heated water after it leaves the heat pump, **Figure 12-9**. When a buffer tank is applied, the heat pump will cycle on and off to maintain the proper temperature within the tank by circulating water through a primary loop between the heat pump and tank. A separate pump can then be installed between the tank and the individual heating loops. This configuration will ensure that an abundance of hot water is always available no matter what load is placed on the building, and it will prevent short cycling of the heat pump during periods of high heating loads.

Goodheart-Willcox Publisher

Figure 12-7. An example of a high-efficiency condensing boiler.

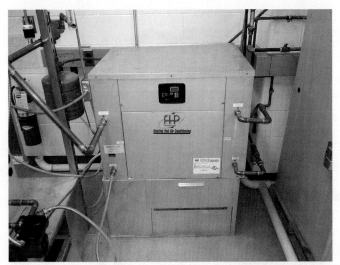

Goodheart-Willcox Publisher

Figure 12-8. A water-to-water geothermal heat pump is a viable alternative as a heating source for radiant hydronic systems.

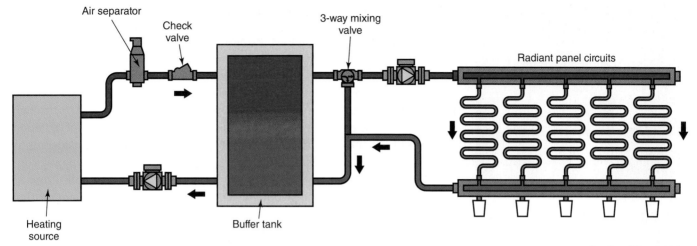

Figure 12-9. A buffer tank should be used when incorporating a heat pump for hot water generation.

GREEN TIP

Clean and Efficient Geothermal Energy

Geothermal heat pumps are one of the most sustainable sources of renewable energy available. They can provide a lower operating cost than any conventional type of residential or commercial heating system available. A geothermal system differs from a conventional furnace or boiler because of its ability to transfer heat via the refrigeration cycle versus the standard method of producing heat by burning fossil fuels.

Super-efficient geothermal heat pumps provide quiet, clean heating while reducing energy bills by up to 70%. Compared to solar or wind energy, geothermal systems can perform even if the sun does not shine or the wind does not blow.

12.2.4 Solar Thermal Storage

A renewable energy alternative to conventional fossil fuels is the use of *solar thermal storage* for hot water generation. Our sun provides an abundant level of thermal energy every day. It is estimated that two weeks of the sun's energy radiating onto the Earth is equivalent to all of the known amounts of oil, gas, and coal available. The challenge has been to harness this energy and transfer it into hot water to heat our homes and buildings.

Solar thermal storage is achieved by first using solar collectors to capture the sun's energy, **Figure 12-10**. These collectors, or solar panels, are typically flat plate or vacuum tube collectors filled with water or glycol and mounted in the path of the sun's rays. As water in the collector is heated, it is circulated through a storage tank and then pumped into radiant heating circuits, **Figure 12-11**. Solar thermal storage systems have the capability of generating supply water temperatures in excess of 150°F, depending on their size and efficiency. They are ideal as a supplement to fossil fuel or geothermal heating systems and can increase system and fuel efficiencies dramatically. Chapter 19, *Solar Thermal Storage* covers these systems in greater detail.

Figure 12-10. An example of a vacuum tube solar collector.

Goodheart-Willcox Publisher

Figure 12-11. A solar collector storage tank is used to circulate water to the heating loops.

12.3 Radiant Heat Piping Systems

Radiant hydronic piping systems use several different types of plastic piping to transport heated water through the "terminal device." In this case, the terminal device is typically the floors, walls, or ceiling of a heated space. Regardless of whether the tubing is run through a floor, wall, or ceiling, the intent is still the same: using the heated water from the tubing to warm the thermal mass through which it is running. This thermal mass then conveys heat by means of radiation to warm the people and objects within the conditioned space. Considerations must be made when choosing the type of piping to be used and the piping configuration. This section covers different types of tubing and their applications as well as several different piping configurations typically used with radiant heating systems.

12.3.1 Radiant Piping Materials

When selecting the type of piping material for a radiant heating system, several factors need to be considered. The selected piping should adhere to the following characteristics:

- It must have the ability to readily transfer heat.
- It must be flexible and expandable.
- It should be easy to join together and connectable with other types of piping materials.
- It must have longevity.
- It should be resistant to freezing temperatures.
- It should be reasonably priced.

The two types of tubing that will be discussed in this section are the most widely used in the radiant heating industry today: PEX and PE-RT tubing.

Pawel G/Shutterstock.com

Figure 12-12. PEX tubing is crosslinked polyethylene made from a high-density polyethylene polymer.

12.3.2 PEX Tubing

PEX stands for *crosslinked polyethylene*. It is made from a high-density polyethylene polymer (HDPE), which is melted and continuously extruded into tubing. PEX is flexible, resistant to scaling and corrosion, and easier to install than metallic or rigid piping, **Figure 12-12**. It has been used in Europe since the early 1970s and was introduced to the United States around 1980. PEX pipe usage has increased ever since, and today it replaces copper piping for many applications, including domestic hot and cold water plumbing as well as radiant heating systems. Most PEX tubing is available in pipe sizes from 3/8″ up to 1″ in diameter.

PEX tubing or piping is classified as type A, B, or C. The different classifications are not different *grades* of PEX; rather the letters are used to identify the manufacturing process.

- **PEX type A** is the highest quality. It is more flexible and has *thermal memory*—in other words, it can be heated and will return back to its original shape if it becomes kinked.
- **PEX type B** is less flexible than PEX type A. It is more susceptible to freezing under extreme conditions; however, it is typically more competitively priced than type A or C.
- **PEX type C** is softer than type B, but it is more environmentally friendly to manufacture. It has little or no coil memory and is more prone to cracking.

Another consideration when choosing the proper type of PEX tubing is whether an *oxygen barrier* is required. If the tubing has to come in contact with any valves, pumps, boilers, or any other fittings or devices that contain ferrous (iron) materials, then an oxygen barrier is required, **Figure 12-13**. With conventional PEX, oxygen molecules will diffuse into the water through the tubing walls and can corrode any ferrous-based materials. PEX tubing with a built-in oxygen barrier has a special external coating that prevents the *diffusion* of oxygen into the water. The external coating on oxygen barrier PEX has a shinier finish compared to conventional PEX.

Another type of PEX tubing is called **PEX-AL-PEX**. The "AL" stands for aluminum. PEX-AL-PEX has an internal layer of aluminum sandwiched between two layers of conventional PEX tubing. The aluminum layer adds a coil memory feature that allows it to retain its shape when the tubing is bent, **Figure 12-14**. This feature can be very useful during a one-person installation. In addition, the aluminum layer provides an oxygen diffusion barrier, eliminating the need for any additional external coatings. The pressure ratings for PEX-AL-PEX are also higher than that of conventional PEX.

Viega, LLC

Figure 12-13. PEX tubing with an oxygen barrier is required if the tubing has to come in contact with any ferrous (iron) materials.

TECH TIP

The Color of PEX

PEX piping comes in several different colors. Oxygen barrier PEX is normally either red or white. PEX without an oxygen barrier comes in red, blue, or white. This makes it easy to use with domestic plumbing to designate hot- and cold-water lines (red for hot, blue for cold). PEX-AL-PEX is generally orange in color. The color of PEX does not affect any of the tubing's ratings.

12.3.3 PE-RT Tubing

PE-RT tubing is made up of a polyethylene resin (PE) that has been designed to operate at raised temperatures (RT). It is constructed with five layers of materials that give it significant strength and chemical resistance, **Figure 12-15**. The core and outermost layers are made of polyethylene. The middle layer is an ethylene vinyl alcohol oxygen barrier, sandwiched between two layers of adhesive. PE-RT tubing is more flexible than other types of tubing and retains its memory better than conventional PEX after it is bent. PE-RT tubing has been used in Europe for the past 35 years and was first introduced to North America in 2003. The applications for PE-RT tubing include plumbing, snow and ice melting, and geothermal and hydronic radiant heating systems. It is approved for both plumbing and mechanical codes throughout the United States and Canada. PE-RT tubing is available in sizes ranging from 1/4″ to 6″ and can withstand pressures of up to 100 psi at 180°F. One key difference between PE-RT and PEX tubing is that PE-RT can be heated and melted to be joined using *socket fusion*, similar to HDPE piping used in geothermal loop fields, **Figure 12-16**.

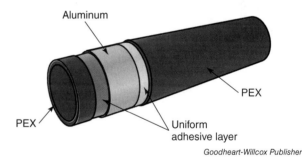

Goodheart-Willcox Publisher

Figure 12-14. PEX-AL-PEX tubing has an internal layer of aluminum, which adds a coil memory feature that allows it to retain its shape when bent.

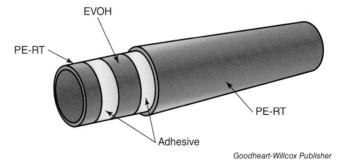

Goodheart-Willcox Publisher

Figure 12-15. PE-RT tubing is more flexible than other types of tubing and retains its memory better than conventional PEX after it is bent.

DID YOU KNOW?

HDPE for Geothermal Ground Loops

High-density polyethylene piping (HDPE) is commonly used for geothermal ground loops and can be connected using heat fusion. This process melts the piping and joins it together using special tools. HDPE piping can be joined together using butt splicing or with special fittings or sockets of the same material.

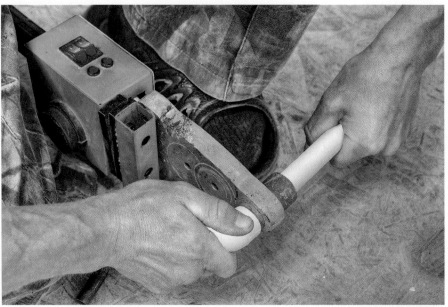

Grigvovan/Shutterstock.com

Figure 12-16. Socket fusion is a method of joining piping such as PE-RT by heating the pipe and fusing it together.

CODE NOTE

Mechanical Code Standards for Radiant Piping Materials

Both crosslinked polyethylene (PEX) tubing and raised temperature polyethylene (PE-RT) tubing are recognized in all major building codes for use with hot-water and cold-water distribution applications, water service lines, and for radiant floor heating systems.

Code standards such as the International Residential Code (IRC), International Plumbing Code (IPC), International Mechanical Code (IMC), as well as the Canadian Standards Association (CSA), outline the use of these radiant piping materials through tables based on ASME, ASTM, and CSA standards.

The installer must determine which codes are applicable to the specific project and must ensure compliance with all local, state, and federal code regulations and standards. Codes are constantly reviewed and updated. The use of PEX tubing has been adopted in the model codes since 1993.

12.4 Radiant Floor Piping Configurations

There are many piping configurations available for radiant heating installations. The most common method is radiant floor heating, which can be classified as either ***wet radiant hydronic systems*** or ***dry radiant hydronic systems***.

Wet systems include:
- Slab on grade
- Concrete or gypsum thin slab

Dry systems include:
- Above-floor prefabricated panels
- Below-floor staple-up or suspended tube systems

12.4.1 Slab-on-Grade Radiant Floors

The ***slab-on-grade radiant piping*** configuration is very popular and is used primarily with new construction. The term *slab-on-grade* means that a layer of concrete is poured over an existing grade (the ground floor of a basement, garage, shop, etc.) with the radiant tubing embedded into the concrete, **Figure 12-17**. Because this practice is typically done during the initial construction phase, it makes for an easier installation of the piping and accompanying controls compared to installing this type of system as a retrofit project.

Items to consider when choosing a slab-on-grade radiant floor system:

- **Always start with a plan.** Regardless of which type of radiant floor system is selected, always start by developing a floor plan for which rooms will receive heating and what the tubing layout will look like. Include the number of circuits per room, overall length of piping per room, and where each circuit will be terminated.
- **Consider tubing depth within the slab.** Depending on the overall thickness of the slab, the optimal depth of the tubing should be between 1″ to 2″ deep. Tubing should not be deeper than 4′ below the surface of the slab because tubing that is too deep in the slab increases the heating response time, meaning it will take longer for the floor to reach the desired temperature. It will also require more energy to bring the slab up to temperature and may require tubing with a larger diameter.
- **Determine tubing spacing within the slab.** There should be an equal distance between the individual tubes within a loop. The typical spacing of PEX and PE-RT tubing is 8″ on center. Closer spacing can be applied to areas with higher heat loss or in areas where a more comfortable floor is desired—such as in bathrooms. Closely spaced areas are usually no less than 4″ apart. Tubing spacing should not exceed 12″ regardless of the application. Wider spaced tubing may require a higher than average inlet water temperature, which could affect the performance and efficiency of heat-generating equipment such as geothermal heat pumps or solar thermal storage systems.

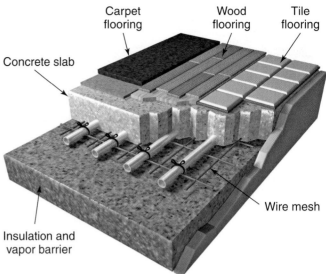

Uponor, Inc.

Figure 12-17. A slab-on-grade radiant floor system is a layer of concrete poured over an existing grade with the radiant tubing embedded into the concrete.

12.4.1.1 Installing Slab-on-Grade Radiant Floor Systems

Once the decision has been made to install a slab-on-grade radiant floor system, the following installation procedure should be followed:

1. **Start by installing a vapor barrier.** The first step is to install a layer of polyethylene sheeting over a well-compacted and properly leveled base. This sheeting will act as a *vapor barrier*, which is necessary to prevent the migration of water vapor from the soil into the concrete slab. The recommended thickness of the sheeting is between 10 and 15 mils—with 6 mils as an absolute minimum. Be sure to properly tape all seams and overlap the edges for maximum protection.
2. **Install under-slab insulation.** This insulation will cover the vapor barrier, **Figure 12-18**. This is an absolutely essential step—as much as 70% of the heat generated by the boiler can be lost through the Earth

Michael Helsel

Figure 12-18. Insulate over the vapor barrier with a minimum R-5 insulation board, as seen here.

nikkytok/Shutterstock.com

Figure 12-19. Once the insulation is installed, reinforce the cement slab with either rebar or welded wire fabric (shown here).

if the concrete slab is not properly insulated. Two types of insulation are available: ***expanded foam board (EPS)*** and ***extruded foam board (XPS)***. Both are popular choices for thick slab insulation. However, extruded foam board will lose up to 50% of its R-value over its lifetime; therefore, expanded foam board is a better choice. Typically, a 2″ layer will be sufficient, but be sure that the insulation used has a minimum R-value of 5 and check with local building codes for the proper amount of insulation for your area. Be sure to properly insulate the perimeter with edge insulation and cover any exposed foundation walls. **Do not insulate areas that will be under support columns or bearing walls.**

3. **Reinforce the slab.** Lay down either welded wire fabric (WWF) or steel reinforcement bar (rebar). The tubing can be attached to the WWF or rebar to hold it in place while the concrete is poured, **Figure 12-19**.

TECH TIP

WWF: Sheets versus Rolled Fabric

If using welded wire fabric, sheets are recommended over rolled fabric. Sheets are noticeably easier to install and provide a flatter surface. The only downside is that the sheets will need to be tied together.

4. **Roll out the tubing.** Uncoil the rolls of tubing and lay out the individual loops onto the WWF or rebar. There are several different tubing layout patterns to choose from, depending on the amount of heat loss in a given zone, **Figure 12-20**. Zones with the greatest heat loss will have the tightest layout pattern. It is a good idea to mark the tubing layout patterns along the floor with spray paint before rolling out the tubing. Also, be sure to mark which ends of the tubing are supply and return so there is no confusion when connecting these ends to the manifold station.

5. **Fasten the tubing.** Connect the tubing to the reinforcement fabric using nylon cable ties, metal twist ties, or plastic clips, **Figure 12-21**. Space the fasteners a minimum of 3′ apart along straight runs, and every 12″ along bends and arcs. Be careful not to puncture the tubing with any sharp edges along the slab's reinforced wire fabric or rebar.

A

Rigamondis/Shutterstock.com

B

tchara/Shutterstock.com

C

Wolfgang Filser/Shutterstock.com

Figure 12-20. Tubing patterns will vary according to the amount of heat loss in a given zone.

Copyright Goodheart-Willcox Co., Inc.

6. **Join the tubing to a manifold station.** The *manifold station* is where the supply and return ends of each individual loop are joined together and piped back to the boiler or heating device, **Figure 12-22.** These are typically housed within a finished wall, so it is important to determine and mark their exact location. The manifold station is the central distribution point for all the tubing circuits for each zone. The manifold size must correspond to the number of circuits within each radiant zone. The manifold station should also provide individual shutoff valves for each supply and return connection.

7. **Pressure test the entire system.** Once the piping is secured in place to the woven wire fabric or rebar, make sure that all valves are open to each circuit except for one of the main shutoff valves on the manifold station. Connect a device such as a compression hose adapter to the manifold station for pressure testing purposes. Pressurize the system with up to three times the normal operating pressure (typically 90 psi) of compressed air and check for leaks. The minimum test for each zone should be at least 24 hours. Remember to allow for slight pressure deviations based on the weather—extremely warm weather may create a small rise in pressure, extreme cold may reduce the pressure slightly. If a noticeable loss in air pressure is detected, test for leaks in any suspected areas by using a soapy bubble solution. Repair any leaks that are found and then pressure test again. **The use of water for leak testing is not recommended—there is the risk of freezing in cold weather, which could burst the tubing if unattended.**

nikkytok/Shutterstock.com

Figure 12-21. Fasten the tubing to the reinforcement fabric using nylon cable ties, metal twist ties, or plastic clips (as shown here).

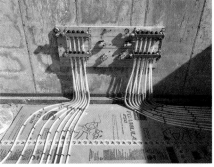

Michael Helsel

Figure 12-22. The manifold station is where the supply and return ends of each individual loop are joined together and piped back to the boiler or heating device.

TROUBLESHOOTING

Repairing Radiant Tubing Leaks

Leaks found in radiant floor tubing can be easily repaired without having to replace the entire line of tubing. Once the leak has been found, cut out the damaged tubing and repair it using a PEX coupling to bind the piping together and secure the fitting with two cinch clamps.

For larger damaged areas, cut a new piece of PEX tubing and repair the damaged area with two couplings. "Push fit" couplings may also be used instead of standard couplings.

TECH TIP

Prepare Tubing for Control Joints

Concrete floors usually have saw-cut joints called control joints cut into the floor surface after it has been poured and before it is fully cured. These joints provide for expansion of the concrete to prevent cracks in the floor. Determine where the control joints will be cut and wrap the tubing in these locations with a thin plastic sleeve to prevent it from bonding with the concrete. Secure the tubing under the wire fabric so that it is not damaged by the saw.

12.4.2 Concrete or Gypsum Thin-Slab Floors

There are several methods of installing a hydronic radiant heating system over an existing surface such as a conventional wood-framed floor. When the installation of a slab-on-grade radiant system is not possible, one alternative is to install a ***thin-slab radiant flooring system***. This system is a more practical application for existing homes and buildings where radiant heating is desired. Thin-slab systems incorporate either a formulated concrete or poured gypsum underlayment material over the radiant tubing, **Figure 12-23**. The installation procedure for thin-slab floor heating is similar to a slab-on-grade radiant system, with a few exceptions and requirements.

One issue that must be considered is that thin-slab installations typically add 1 1/4″ to 1 1/2″ to the floor height. This requires adjustments to the rough-in openings of doors and the riser heights of stairs. Another issue to consider is that the thin-slab system will add substantial weight to an existing floor. Typically, a poured gypsum floor will add 13 to 15 pounds per square foot to the dead load, or constant weight, of an existing floor structure. Standard weight concrete thin slabs add about 18 pounds per square foot per 1 1/2″ thickness. This is typically not an issue for basement floors or slab on grade, but should be considered for wood-framed floors on upper levels. Do not assume that the floor structure can support the additional weight of either type of thin-slab application. When in doubt, have a qualified structural engineer or designer verify that the proposed changes to the building can be supported by the existing framing. Otherwise, determine what changes are necessary to support the additional load.

12.4.2.1 Installing Thin-Slab Floor Systems

The installation of a thin-slab system begins by stapling or securing the tubing to the existing subfloor. If the tubing is being installed over an existing concrete floor, install a vapor barrier and insulation as described for slab-on-grade systems. Consult the tubing manufacturer for the acceptable method of fastening the tubing to the floor. Once all tubing circuits are in place, pressure test each

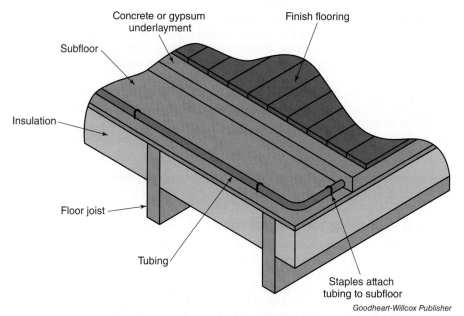

Goodheart-Willcox Publisher

Figure 12-23. Another method of radiant floor heating is the thin-slab system. It incorporates either a formulated concrete or poured gypsum underlayment material over the radiant tubing.

zone as outlined for slab-on-grade systems. When all tubing circuits are determined to be leak-free, spray the floor with a layer of sealant and bond coating to minimize water absorption into the subfloor. This will also strengthen the bond between the subfloor and concrete.

ANDREY-SHA74/Shutterstock.com

Figure 12-24. Some manufacturers offer a floor panel system that provides an easily assembled modular board that includes a holding grid, as seen here.

> **TECH TIP**
>
> **Using Insulated Floor Panels**
> Some manufacturers offer an insulated floor panel system that provides an easily assembled, insulated modular board that includes a vapor barrier and radiant tube holding grid. This system simplifies installation of hydronic radiant floor heating systems in both slab-on-grade and above-grade installations, **Figure 12-24**.

The concrete or gypsum underlayment material is now prepared and mixed outside of the building, then poured over the tubing circuits. The poured mixture is typically quite fluid, and when installed it becomes self-leveling with minimum floating necessary, **Figure 12-25**. Once cured, the finished floor product resembles plaster and is very hard and durable. It can be covered with almost any flooring finish including carpeting, vinyl sheeting, ceramic tile, or glued-down wood flooring. Always follow the finished flooring manufacturer's instructions for proper installation over a thin-slab floor.

12.4.3 Above-Floor Prefabricated Panels

Prefabricated radiant floor panels provide another alternative to slab-on-grade radiant floor installations. Several manufacturers fabricate these types of panels, which are designed to fit over an existing wooden subfloor or over slab-on-grade concrete—such as in a basement, **Figure 12-26**. Prefabricated panels can be used where the existing floor cannot support the additional weight of a thin-slab installation or where the finished flooring requires a large number of fasteners for completion. Prefabricated radiant floor panels are essentially heat transfer plates. They consist of wooden or plywood panels with notched grooves that are designed to hold 5/16″ to 1/2″ PEX or PE-RT tubing in place. Sheets of aluminum on the underside of these panels enhance the heat transfer between the tubing and the flooring.

brizmaker/Shutterstock.com

Figure 12-25. Once the tubing is in place, the concrete or gypsum underlayment material is now poured and finished.

12.4.3.1 Installing Above-Floor Prefabricated Panel Systems

The installation of a prefabricated floor panel system begins with designing and laying out the zones to be heated. This will reduce installation time and the amount of materials used. Once the floor surface is thoroughly cleaned, begin by snapping chalk lines along the edge of the room to align the first row of panels. Lay the panels in place and cut around any obstacles such as cabinets, posts, or irregular walls, **Figure 12-27**. When all panels are properly placed, secure them using staples or screws according to the manufacturer's instructions. In some cases, panels must be secured with adhesive. Use blank pieces of plywood to fill any gaps in floor areas that are not being heated, such as under cabinets. Stagger end panels to avoid creating any "dead-end" circuits

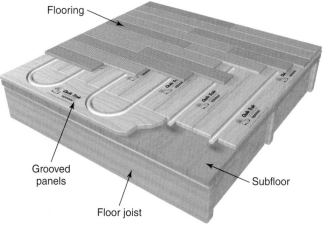

Uponor, Inc.

Figure 12-26. Prefabricated radiant floor panels are designed to fit over an existing wooden subfloor.

Dmitry Melnikov/Shutterstock.com

Figure 12-27. Once the prefabricated floor panel system is designed, set the panels in place and secure them firmly.

Viega, LLC

Figure 12-28. Be sure that end panels are staggered to avoid creating any dead-end circuits where the tubing is looped around each row.

where the tubing is looped around each row, **Figure 12-28**. When all panels are secured, vacuum the grooves where the tubing is to be inserted and apply a bead of silicone into each groove. Lay the tubing into the grooves before the silicone dries. Protect the tubing where it penetrates any floor openings by installing plastic sleeves through the floor penetrations. Connect the individual zoned circuits to the manifold station and pressure test each circuit for leaks. Once all panels are installed and circuits tested, the room is ready for its finished floor covering, **Figure 12-29**.

12.4.4 Below-Floor Suspended Tubing Systems

An alternative to prefabricated radiant floor panels is to install the radiant system below the existing flooring using ***below-floor suspended tubing***. This installation is similar to the procedure for installing above-floor radiant panels and can be accomplished with new construction projects or as a retrofit. In this type of installation, heat is transferred from the bottom of the flooring and through the tubing to the floor above, **Figure 12-30**.

12.4.4.1 Installing Below-Floor Suspended Tubing Systems

One installation method for the below-floor radiant heating system is to install the PEX or PE-RT tubing within the underfloor joist cavity using pipe hangers fastened either to the sides of the floor joists or to the underside of the subflooring. The floor joists are drilled on opposite ends of the room and tubing is pulled through the joist space, looping the tubing from one joist bay to the next as necessary. After installing the tubing through the last joist cavity, it is run back to the floor opening and connected to the manifold station. The tubing can also be attached to the floor joists or to the underside of the subflooring using PEX clips spaced 8″ on center for 16″ joist bays and spaced 3′ apart. Avoid placing the tubing directly onto the underside of the subflooring as this can produce hot

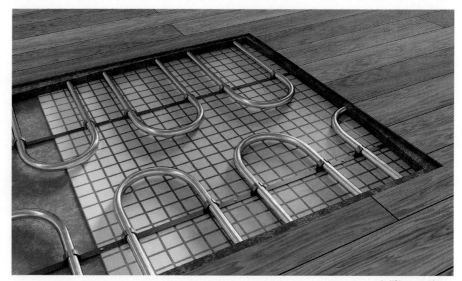

vchal/Shutterstock.com

Figure 12-29. Once all panels are installed and circuits tested, the room is ready for its finished floor covering.

spots over the tubing, particularly when the floor covering has a low resistance factor—the pipe hangers will prevent the tubing from making direct contact.

Another method of installing suspended tubing is to use aluminum heat transfer plates fastened to the tubing or directly to the subflooring and spaced evenly throughout the floor joist, **Figure 12-31**. This may be a better method because the aluminum heat transfer plates spread the heat energy from the tubing more evenly throughout the underside of the floor, compared to tubing only. This procedure eliminates any hot spots and provides a more consistent temperature at the floor surface. The installation procedure is identical to that of suspended tubing; however, space the aluminum plates 3″ to 6″ apart for the best heat transfer coverage. The plates can be secured to the subflooring using staples, nails, or screws; or they can be suspended just below the flooring using PEX-AL-PEX or PE-RT tubing.

Note: Whether using aluminum heat transfer plates or suspending the tubing using pipe hangers, this installation requires that insulation be installed underneath the tubing to allow the heat energy to be transferred to the subflooring and not downward below the tubing. A minimum of R-11 fiberglass insulation is required—even if the tubing is installed over a heated space. If the tubing is installed over a crawlspace, use insulation with a minimum R-19 value. The use of standard, unfaced insulation is adequate for either installation.

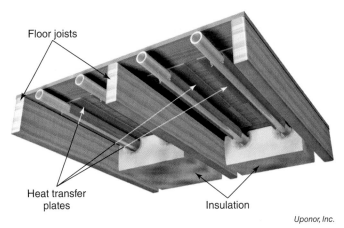

Uponor, Inc.

Figure 12-30. With below-floor suspended tubing, heat is transferred from the bottom of the flooring and through the tubing to the floor above.

12.5 Radiant Wall and Ceiling Panels

When altering the flooring is not an option for a radiant heating system, an alternative is to install either ***radiant wall panels*** or ***radiant ceiling panels***. Because the principle theory behind radiant heating is to heat the objects in a room—not the air—both radiant wall and ceiling panels can achieve the same level of comfort as that of a radiant flooring system.

Radiant wall and ceiling panels are considered low-mass systems due to the fact that the wall covering (such as drywall) has a lower thermal mass than that of concrete used in radiant flooring systems. Because of this fact, radiant wall and ceiling panels can respond much more quickly to changes in heat load requirements and can heat up much faster, which makes them more attractive—especially when considering using a temperature setback option. In addition, radiant wall and ceiling panels are much easier to retrofit than radiant flooring systems.

Ultra-Fin Radiant Floor Heating System

Figure 12-31. Aluminum heat transfer plates that are fastened to the tubing or directly to the subflooring can also be used for underfloor radiant heating.

12.5.1 Installation of Radiant Wall and Ceiling Panels

The installation procedure for radiant wall and ceiling panels is similar to that of radiant flooring systems.

1. The tubing is placed within the wall or ceiling cavity of an interior or exterior partitioned wall. Begin by drilling holes in the partition studs to route the supply and return tubing, **Figure 12-32**.

JuneJ/Shutterstock.com

Figure 12-32. Begin the radiant wall or ceiling panel installation by first drilling holes in the partition studs to route the supply and return tubing.

2. Be sure to install a minimum of R-5 insulation behind the tubing and against the wall. If installing tubing in an exterior wall, a higher level of insulation is required. Foam insulation boards with reflective facing are recommended.
3. Once the holes are drilled and insulation backing is in place, run the tubing within the partitioned wall or ceiling. Be careful not to kink the tubes, **Figure 12-33**.
4. Install aluminum plates between the wall and tubing to increase the amount of heat transfer, **Figure 12-34**.
5. Run the tubing to the manifold station and terminate in the same manner as with radiant floor systems, **Figure 12-35**.
6. The wall covering or ceiling is now ready to be installed.

TECH TIP

Balancing the Radiant Floor System
Once the radiant heating system is installed, the final step is to balance the system. To accomplish this, the installer simply adjusts the valves on the manifold station for each heating loop. Pressure gauges or flowmeters need to be installed to accomplish this task. The valves on shorter heating loops will typically need to be closed further, as they will have less resistance and a lower pressure drop.

12.6 Designing Radiant Heating Systems

As discussed in previous chapters, an accurate design of the heating system—whether for the boiler or a terminal device—depends on an accurate heating load calculation of the building. Review Chapter 13, *Building Heating Loads and Print Reading*, which discusses how to calculate heating loads, before attempting to design a system. For radiant heating systems, a room-by-room heat load

J.J. Gouin/Shutterstock.com

Figure 12-33. Once the holes are drilled through the wall or ceiling studs and insulation backing is in place, run the tubing within the partitioned wall or ceiling. Be careful not to kink the tubes.

Photo by Warren Gretz, NREL 11480

Figure 12-34. When installing radiant wall panels, use aluminum plates between the wall and tubing to increase the amount of heat transfer.

Alexander Raths/Shutterstock.com

Figure 12-35. The final step when installing radiant wall panels is to run the tubing to the manifold station and terminate in the same manner as with radiant floor systems.

calculation is necessary. Once this is accomplished, determine the following parameters:

- Btu/hr heat loss per square foot for each zone
- Radiant floor surface temperature
- Tubing size
- R-value of the finished flooring
- Supply and return water temperature differential
- Tubing spacing
- Required supply water temperature
- Total loop length
- Fluid flow in GPM
- Feet of head loss

Btu/hr heat loss per square foot for each zone: Divide the heat loss of the room or zone to be heated by the total square foot area of that zone. Remember to subtract those areas where radiant tubing will not be installed—such as under counters or vanities.

Radiant floor surface temperature: Once the Btu per square foot is calculated, the designer can determine the floor surface temperature. Use the chart shown in **Figure 12-36** to estimate this factor.

Example: A calculation of 25 Btu per square foot and a room temperature set point of 68°F would result in a radiant floor surface temperature of 80.5°F. This estimate helps determine the comfort level for the customer and can be adjusted by moving the room temperature set point either up or down.

Tubing size: Most radiant flooring systems use either 3/8″ or 1/2″ PEX or PE-RT tubing. Both sizes are fairly equal with regard to heat output per square foot when used in a high-mass radiant application. However, tubing size does affect pressure loss and GPM flow rates. Smaller tubing results in higher head loss than larger tubing. On the other hand, 1/2″ tubing provides nearly twice the volume of fluid flow per cubic inch compared to 3/8″ tubing. Both factors can affect circulator sizing. Smaller diameter tubing is suggested for shorter loop lengths, and larger diameter tubing is recommended for larger areas.

R-value of the finished flooring: The R-value of the finished floor covering will differ depending on the type of material. Knowing the R-value of the finished

Radiant Floor Surface Temperatures (°F)											
	Heat Loss (Btu/hr per ft²)										
		10	15	20	25	30	35	40	45	50	55	60
Set Point Temperature (°F)	75	80.0	82.5	85.0	87.5	90.0	92.5	95.0	97.5	100.0	102.5	105.0
	74	79.0	81.5	84.0	86.5	89.0	91.5	94.0	96.5	99.0	101.5	104.0
	73	78.0	80.5	83.0	85.5	88.0	90.5	93.0	95.5	98.0	100.5	103.0
	72	77.0	79.5	82.0	84.5	87.0	89.5	92.0	94.5	97.0	99.5	102.0
	71	76.0	78.5	81.0	83.5	86.0	88.5	91.0	93.5	96.0	98.5	101.0
	70	75.0	77.5	80.0	82.5	85.0	87.5	90.0	92.5	95.0	97.5	100.0
	69	74.0	76.5	79.0	81.5	84.5	86.5	89.0	91.5	94.0	96.5	99.0
	68	73.0	75.5	78.0	80.5	83.0	85.5	88.0	90.5	93.0	95.5	98.0
	67	72.0	74.5	77.0	79.5	82.0	84.5	87.0	89.5	92.0	94.5	97.0
	66	71.0	73.5	76.0	78.5	81.0	83.5	86.0	88.5	91.0	93.5	96.0
	65	70.0	72.5	75.0	77.5	80.0	82.5	85.0	87.5	90.0	92.5	95.0
	64	69.0	71.5	74.0	76.5	79.0	81.5	84.0	86.5	89.0	91.5	94.0
	63	68.0	70.5	73.0	75.5	78.0	80.5	83.0	85.5	88.0	90.5	93.0
	62	67.0	69.5	72.0	74.5	77.0	79.5	82.0	84.5	87.0	89.5	92.0
	61	66.0	68.5	71.0	73.5	76.0	78.5	81.0	83.5	86.0	88.5	91.0
	60	65.0	67.5	70.0	72.5	75.0	77.5	80.0	82.5	85.0	87.5	90.0

Goodheart-Willcox Publisher

Figure 12-36. Radiant floor surface temperatures can be estimated by knowing the Btu per square foot of heated space and the desired room temperature set point.

flooring is necessary to select the appropriate supply water temperature. The chart shown in **Figure 12-37** displays the correlation between the type of floor covering and R-value. For instance, nylon saxony carpeting that is 1/4″ thick will have an approximate R-value of 0.88.

TECH TIP

Floor Covering Temperature Limitations

Certain floor coverings have temperature limitations. For instance, solid hardwood flooring can shrink and expand with fluctuating temperatures, leaving unsightly gaps and cracks. For this reason, special considerations are necessary when using various types of flooring with radiant heating systems. Most hardwood floors have a maximum floor surface temperature of 80°F. All other flooring types have a maximum floor surface temperature of approximately 87.5°F. Be sure to consult the flooring manufacturer for their recommendations when choosing a particular floor covering.

Supply and return water temperature differential: This is the temperature drop between the inlet and outlet of the manifold station. A differential temperature of 20°F is common for most radiant floor projects, although a 10°F difference can be used. This calculation is also used when sizing the circulator. Keep in mind that a 10°F temperature differential will double the fluid flow requirement in GPM, compared to a 20°F temperature differential.

R-Value for Select Carpets and Carpet Pads					
Thickness (in.)					
Carpeting					
	1/4	3/8	1/2	5/8	3/4
Commercial glue down	0.60	0.90	1.20	1.50	1.80
Acrylic level loop	1.04	1.56	2.08	2.60	3.12
Acrylic plush	0.83	1.25	1.66	2.08	2.49
Polyester plush	0.96	1.44	1.92	2.40	2.88
Nylon saxony	0.88	1.32	1.76	2.20	2.64
Nylon shag	0.54	0.81	1.08	1.35	1.62
Wool plush	1.10	1.65	2.20	2.75	3.30
Carpet Pads					
Rubber (solid)	0.31	0.47	0.62	0.78	0.93
Rubber (waffled)	0.62	0.93	1.24	1.55	1.86
Hair and jute	0.98	1.47	1.96	2.45	2.94
Prime urethane (2-lb. density)	1.08	1.62	2.16	2.70	3.24
Bonded urethane (4-lb. density)	1.04	1.56	2.08	2.60	3.12
Bonded urethane (8-lb. density)	1.10	1.65	2.20	2.75	3.30

Goodheart-Willcox Publisher

Figure 12-37. This chart shows the R-values for various types of finished floor coverings.

Tubing spacing: The tubing spacing is a function of the type of floor construction, water temperature, and overall desired comfort. Decreasing the tubing spacing lowers the required supply water temperature and produces a more even surface temperature, but it increases the amount of tubing required. For slab-on-grade and thin-slab applications, the maximum tubing spacing is 9″ on center. Anything wider than this will cause uneven floor temperatures. Spacing tends to narrow along perimeter walls—such as in basements—as seen in **Figure 12-38**.

Required supply water temperature: The factors that affect the supply water temperature are:
- Installation method
- Btu per square foot load
- Room temperature set point
- R-value of the floor covering
- Temperature differential between supply and return water

The chart shown in **Figure 12-39** outlines these parameters and is one method to determine the required supply water temperature for the heating loop.

Example: A zone has a heat loss of 22 Btu per square foot, a desired room temperature set point of 68°F, a 20°F temperature differential, and no appreciable R-value for the floor covering. By using these parameters and the chart in **Figure 12-39**, the required supply water temperature for this zone would be approximately 98°F.

Michael Helsel

Figure 12-38. Tube spacing will be narrower along basement perimeter walls.

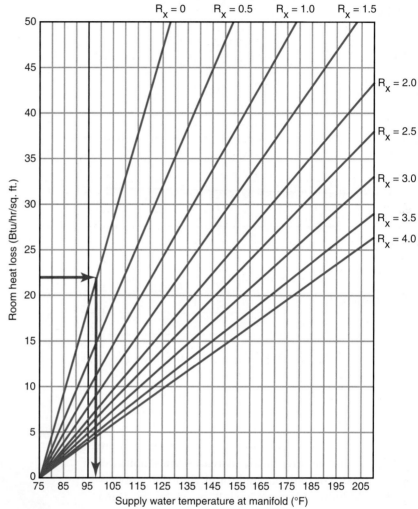

Figure 12-39. This table is used to determine the proper supply water temperature for the heating loop.

Total loop length: The length of tubing for each zone is measured from the outlet of the manifold station to its inlet. To determine the amount of tubing that should be installed for each individual zone, use the following calculations:
- For tubing to be spaced 12″ on center, multiply the square footage of the zone by 1.0.
- For tubing to be spaced 10″ on center, multiply the square footage of the zone by 1.2.
- For tubing to be spaced 9″ on center, multiply the square footage of the zone by 1.33.
- For tubing to be spaced 8″ on center, multiply the square footage of the zone by 1.5.
- For tubing to be spaced 6″ on center, multiply the square footage of the zone by 2.0.

TIMEOUT FOR MATH

Calculating Total Loop Length
A room measures 12′ × 10′ with the tubing installed at 9″ on center. What is the total loop length for this zone?
Answer: 12 × 10 = 120 ft^2 × 1.33 = 159.6 total linear feet
Note: Be sure to include any additional tubing footage measuring from the inlet of the manifold station to the zone as well as any footage from the zone back to the manifold station.

Fluid flow in GPM: Once the previous factors are determined, the next step is to select a circulator for the radiant flooring system. In order to meet the calculated heat load requirement, the proper water flow through each heating loop must be maintained. The proper flow rate per zone is based on the total heating load, the total loop length, and the supply and return temperature differential. When these parameters have been calculated, the water flow through an individual zone can be determined by referencing the chart in **Figure 12-40** and by using the following information:

- Btu per square foot of floor space
- On-center distance between tubing loops
- Area in square feet being heated

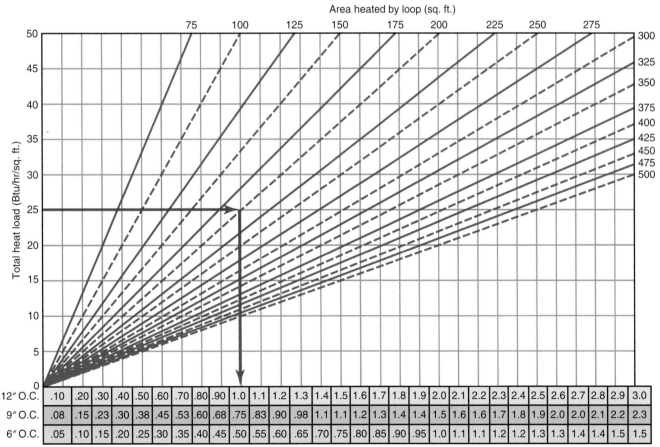

Figure 12-40. Use this chart for calculating the GPM flow for circulating pump sizing.

Example: A room measures 200 ft² in area and has a heat load of 25 Btu/ft². The tubing is spaced 9″ on center. Using **Figure 12-40**, the minimum flow for this area will be 0.75 GPM.

Feet of head loss: The final determination when selecting a circulating pump is to calculate the total pressure loss in feet of head. To do this, the following information must be known:

- Water flow per loop in gallons per minute (GPM)
- Total loop length
- Tubing size
- Type of tubing used
- Supply water temperature

The table in **Figure 12-41** can be used to determine the head loss per foot of pipe using 3/8″ tubing and 100% water (versus a water and glycol mix). If glycol is required for freeze protection, a different table should be referenced. These tables can be found on the Internet or by consulting the PEX fitting manufacturer.

Example: The flow has been calculated through a particular zone as 0.45 GPM. The total loop length is 160′ and the tubing being used is 3/8″ PEX with 100% water (no glycol). The supply water temperature is 110°F. Using the table in **Figure 12-41**, the feet of head per foot of tubing is 0.03692. When this number is multiplied by the total loop length of 160′, the total feet of head for this loop is equal to 5.9.

TECH TIP

Adjusting Loop Lengths and Tubing Size

If the feet of head loss is higher than desired once all pressure loss calculations are complete, the loop lengths may need to decrease or the tubing size may need to increase. Once these factors have been changed, recalculate the pressure drops using the new loop length and tubing size.

Head Loss for 3/8″ Hydronic Tubing, 100% Water (ft head loss per ft of pipe)

Velocity (ft./sec.)	GPM	Water Temperature (°F)												
		80	90	100	110	120	130	140	150	160	170	180	190	200
1.0	0.30	0.01912	0.01845	0.01785	0.01732	0.01685	0.01642	0.01603	0.01568	0.01535	0.01505	0.01478	0.01453	0.01430
1.1	0.33	0.02286	0.02207	0.02136	0.02074	0.02017	0.01967	0.01921	0.01879	0.01841	0.01805	0.01773	0.01744	0.01716
1.2	0.36	0.02688	0.02595	0.02513	0.02441	0.02375	0.02316	0.02263	0.02214	0.02170	0.02128	0.02091	0.02056	0.02024
1.3	0.39	0.03117	0.03011	0.02917	0.02833	0.02758	0.02690	0.02629	0.02573	0.02522	0.02474	0.02431	0.02392	0.02355
1.4	0.42	0.03572	0.03452	0.03345	0.03251	0.03165	0.03088	0.03019	0.02955	0.02897	0.02843	0.02794	0.02749	0.02706
1.5	0.45	0.04054	0.03919	0.03799	0.03692	0.03596	0.03510	0.03431	0.03359	0.03294	0.03233	0.03178	0.03127	0.03079
1.6	0.48	0.04562	0.04411	0.04277	0.04158	0.04051	0.03954	0.03867	0.03786	0.03713	0.03645	0.03583	0.03526	0.03473
1.7	0.51	0.05095	0.04928	0.04780	0.04648	0.04529	0.04421	0.04324	0.04235	0.04154	0.04078	0.04010	0.03947	0.03888
1.8	0.54	0.05653	0.05469	0.05306	0.05161	0.05027	0.04911	0.04804	0.04706	0.04616	0.04533	0.04457	0.04387	0.04322
1.9	0.57	0.06237	0.06035	0.05856	0.05697	0.05553	0.05423	0.05306	0.05198	0.05100	0.05008	0.04925	0.04848	0.04777
2.0	0.60	0.06844	0.06624	0.06427	0.06255	0.06098	0.05957	0.05829	0.05711	0.05604	0.05504	0.05413	0.05330	0.05252

Goodheart-Willcox Publisher

Figure 12-41. Use this chart for calculating the feet of head loss for circulating pump sizing.

When sizing the circulating pump for the entire system, add the total flow of all loops combined. However, when calculating the feet of head for pump sizing, simply select the loop with the highest pressure drop. Remember to add the supply and return piping between manifold stations and any other devices that the circulator will be serving when calculating feet of head.

12.7 Radiant Heating Controls

Controls for radiant floor heating systems can be simple or complex, depending on the size and application of the system and the desires of the heating designer for increased efficiency and performance. Some components for controlling a radiant floor heating system include:

- Space thermostats
- Electric or electronic zone valves
- Flowmeters
- Temperature gauges
- Floor or slab sensors
- Outdoor reset controllers
- Microprocessor controllers

Space thermostats: Today's space thermostats used for radiant heating control provide many features not found on conventional thermostats, including touchscreen design, seven-day programmable scheduling, and auxiliary sensor inputs for monitoring floor temperature and outdoor air temperature. In addition, space thermostats can directly control zone valves or be connected to a microprocessor panel for more precise temperature control, **Figure 12-42**.

Electric or electronic zone valves: Low-voltage electric zone valves operate on a 24 V circuit and have been used widely in the hydronic heating industry for years. These are normally two-position zone valves designed to operate directly from a thermostat or through a relay control board, **Figure 12-43**. Electronic zone valves are microprocessor-driven valves that offer better temperature control through modulation, **Figure 12-44**. These valves can open and close anywhere between 0% and 100% of their total range to provide optimal control based on space heating demand. In addition, electronic zone valves do not rapidly spring-return closed, which lessens the chance for water hammer within the circuit.

Photo courtesy of Watts

Figure 12-42. An example of a modern thermostat used with radiant heating applications.

Goodheart-Willcox Publisher

Figure 12-43. This is an example of a two-position zone valve that can be controlled directly from a thermostat or control panel.

Goodheart-Willcox Publisher

Figure 12-44. This is an example of an electronic zone valve that has modulating capability.

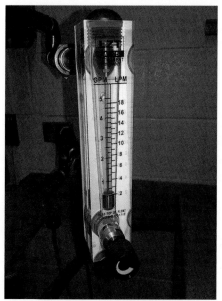

Figure 12-45. Flowmeters are used to monitor proper fluid flow rates through heating loops.

Flowmeters: Flowmeters provide monitoring of the flow rates of water through individual heating loops or through the entire hydronic heating system. Monitoring the flow rate can determine the proper amount of Btu delivered to the space and can also be an indicator of circulating pump performance. These devices can be incorporated into the system along with temperature gauges and can also be used for system analysis and troubleshooting, **Figure 12-45**.

Temperature gauges: Temperature gauges are located at various points in the hydronic heating circuit, including at the inlet and outlet of the manifold station. They provide a monitoring point for determining proper water temperature at a glance. Temperature gauges can be used to regulate the temperature drop across a circuit and can be instrumental as a troubleshooting tool to decide if the system is operating at peak efficiency, **Figure 12-46**.

Floor or slab sensors: Floor and slab sensors are used not only to assist in controlling the space temperature but also as a limit switch when incorporating radiant heat with hardwood floors. These sensors can be connected to a microprocessor control to reset the space temperature, or they can be used to shut off the zone valve in the event that the water temperature within the heating loop exceeds the desired set point.

Outdoor reset controllers: These controllers should be considered standard equipment on any hydronic heating system. The outdoor reset controller is used to regulate the hot water temperature set point based on the outdoor air temperature. It functions by lowering the supply water temperature set point as the outdoor air temperature rises. Conversely, it raises the hot water temperature set point proportionally as the outdoor air temperature falls. The outdoor reset controller provides maximum fuel efficiency for any hydronic heating system and can be used to control the entire system or individual heating loops, **Figure 12-47**.

Microprocessor controllers: The solid-state microprocessor controllers found in many of today's radiant heating systems provide a logic-based control panel for multiple heating loops. Most of these controllers can command multiple zone valves to open and close by either two-position control or full modulation, **Figure 12-48**. They provide inputs for solid-state room sensors and

Figure 12-46. Temperature gauges provide a monitoring point for determining proper water temperature and can be used to regulate the temperature drop across a circuit.

Figure 12-47. Outdoor reset controllers provide maximum efficiency in a radiant heating system by adjusting the hot water set point based on the outdoor air temperature.

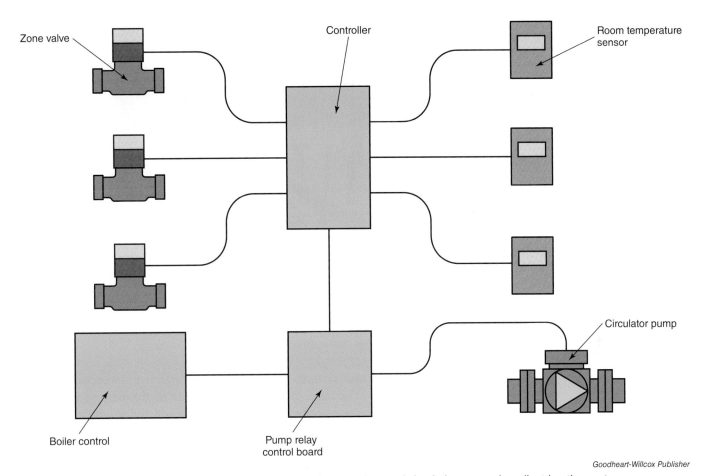

Figure 12-48. Microprocessor controllers are used to control zone valves and circulating pumps in radiant heating systems.

offer web-based access for remote monitoring and control through the Internet. These controllers typically incorporate solid-state switching relays for the control of individual zone valves and system circulating pumps, **Figure 12-49**.

By combining several or all of the above-mentioned controls, the radiant heating designer can develop a system that is effective and efficient while also being user-friendly. More resources on creating and designing radiant heating systems are available from the many manufacturers that offer these types of controls.

12.8 Radiant Heating for Snow and Ice Melt Systems

Applications for radiant heating are not limited to wall, ceiling, and flooring systems. A very practical use for radiant heating systems is for *snow and ice melt*. These types of systems have become quite popular for today's homes and businesses as a cost-effective and convenient method of automatically clearing snow and ice from many types of outdoor surfaces, **Figure 12-50**. With snow melt systems, shoveling is virtually eliminated. In addition, the use of chemicals for snow and ice melting is reduced or eradicated. Liability issues resulting from slip-and-fall accidents and personal injury due to slippery or snow-covered walks and driveways are also minimized.

Figure 12-49. An example of a microprocessor controller.

City of Holland

Figure 12-50. Snow and ice melt systems provide an effective method of keeping sidewalks, driveways, and even roads clear.

Snow and ice melt systems work as follows: An effective system detects snow and ice buildup on outdoor surfaces by means of an in-ground sensor. This sensor sends a signal back to the controller, which opens hot water valves located at the radiant system manifold. The heated fluid from the heating source warms the surface using PEX or PE-RT piping embedded in the concrete, sand, or asphalt. This in turn melts the snow and ice, keeping the area clear at all times.

When installed properly, snow and ice melt controls should protect the heating equipment at all times from damage due to extreme temperature fluctuations as well as protect the outdoor surfaces from repetitive freezing and thawing cycles. The in-ground sensors should detect the actual conditions on snow and ice melt surfaces to ensure that melting begins when the snow starts to fall. These sensors should also activate the system to cycle off as soon as they detect that the outdoor surface is dry, **Figure 12-51**. This will optimize the system's energy efficiency by not allowing the boiler or heat source to run longer than necessary.

ND700/Shutterstock.com

Figure 12-51. A good snow melt system will cycle off as soon as it detects that the outdoor surface is dry.

12.8.1 Applications for Snow and Ice Melt Systems

Snow and ice melt systems are often requested by customers looking to meet various health, environmental, and safety standards, such as the requirements set forth by the Americans with Disabilities Act (ADA). Customers also want to install these types of systems to protect themselves from lawsuits and liability resulting from slip and fall injuries.

Snow and ice melt systems can be effectively used in a wide selection of applications:
- Sidewalks
- Entryway steps

- Driveways
- Entry ramps
- Parking garages
- Aircraft hangars
- Pool decks

When combined with an efficient heating source system—such as geothermal or solar thermal storage, snow and ice melt systems can provide an innovative and effective system that can work smarter, faster, and easier than conventional methods of ice and snow removal.

> **GREEN TIP**
>
> **Using Waste Heat to Keep the Sidewalks Clear**
> In 1988, the city of Holland, Michigan, became home to the largest municipally run snowmelt system in North America. By using waste heat from a nearby power generating plant, water is heated and circulated through 120 miles of tubing underneath city streets and sidewalks. With water that is heated up to as high as 95°F, the system can melt snow at the rate of up to 1″ per hour—even at outdoor air temperatures of 20°F. By utilizing waste heat as its heating source, this sustainable system essentially provides free energy to the snow melt system.

12.8.2 Designing Snow and Ice Melt Systems

The design process for sizing snow and ice melt systems for both commercial and residential use can be planned by professional engineering groups or with help from system manufacturers. Software programs are also available to assist in system design. The process begins with an accurate heat-loss calculation to create optimum and efficient system performance. Designing these systems is a similar process to that of conventional radiant heating arrangements; however, there are several differences:

- The designer must first determine the amount of snow that must be removed per hour based on the average snowfall for the given area. In snow and ice melting, there must be a change in state (solid to liquid). Therefore, the designer must calculate the Btu/hr output required to melt the proper amount of snow or ice at the required rate.
- The designer must determine the lowest outdoor design temperature at which the system will operate during the snow or ice melt phase.
- Once these criteria are determined, the designer must calculate the total Btu/hr required to accomplish the amount of snow and ice melt needed at design conditions.
- Snow and ice melt sensors must be strategically placed within the slab. When dealing with sloping pavement, areas at higher elevations will dry out faster, and the melted snow and ice will drain to lower areas. If the drier areas continue to be supplied with heat, the result will be large amounts of wasted energy and high operating costs.
- The designer should plan for moisture runoff and design a drainage system that will prevent melted snow and ice from accumulating on lower areas of the slab.

In addition to these design points, take time to calculate the properly sized heating system (boiler) when determining radiant heating for snow and ice melting. Often, a dedicated boiler will be used for this type of application. Also, consult system manufacturers for the proper controls and peripheral devices that accompany this type of system. Most manufacturers of snow and ice melt systems provide information on such design criteria as type and size of piping to use, as well as the proper spacing of the tubing and correct zoning techniques.

12.8.3 Snow and Ice Melt Installation

Although there are similarities between the installation of a snow and ice melt system to that of a radiant flooring system, there are differences that must be adhered to. Because the system is located outdoors, the installation criteria differ as outlined in the following steps:

1. Remove existing walkway or driveway surface materials.
2. Be sure the earth grade is firmly packed—use sand or gravel as a base.
3. Install a minimum of R-5 insulation onto the finished grade.
4. Use rebar or woven wire fabric to strengthen the concrete base.
5. Staple the tubing to the insulation or use clips or ties to fasten the tubing to the rebar or wire fabric, **Figure 12-52**.
6. It is recommended that the tubing be placed 2″–3″ below the concrete surface for a faster response time.
7. If using interlocking pavers for a finished surface (such as for a sidewalk), encase the tubing in a 1 1/2″ sand bed. Pavers can then be installed over the sand bed.
8. If tubing is to be installed under an asphalt finish, it should be encased in a 3″ layer of stone dust or sand media and compacted. **Note:** Flush the tubing with cold water during asphalt installation and keep doing so until the asphalt has cooled.

When designed and installed properly, snow and ice melt systems can provide years of trouble-free service. These systems can reassure homeowners and business owners that their sidewalks and driveways will remain free of snow and ice, making a safe and secure environment for their families, employees, and customers.

City of Holland

Figure 12-52. Just as with slab-type radiant floor systems, the tubing for snow melt systems must be securely fastened to the woven wire fabric.

Chapter Review

Summary

- In radiant heating systems, heat is conveyed in a direct physical path and heats the objects in the conditioned space, not the air.
- The human body responds more favorably to radiant heating sources than to forced-air heating sources.
- The difference between low-mass and high-mass radiant heating systems is the capacity and the ability to store heat energy. Low-mass radiant heating systems have a limited capacity to store heat as compared to high-mass heating systems.
- There are many different types of systems available for generating hot water for radiant heating, including conventional boilers, condensing boilers, geothermal heat pumps, and solar thermal storage systems.
- Radiant heating systems primarily use PEX and PE-RT tubing as a means of heat transfer. PEX tubing is classified as type A, B, or C. When used with ferrous types of devices and fittings, PEX tubing must include an oxygen barrier. PE-RT tubing is designed to operate at raised temperatures and is constructed with five layers of materials that give it significant strength and chemical resistance.
- There are two classifications for radiant floor heating systems: they can be wet radiant hydronic systems or dry radiant hydronic systems.
- A slab-on-grade radiant floor system is a layer of concrete poured over an existing grade with the radiant tubing embedded into the concrete.
- Steps for installing a slab-on-grade flooring system include laying down a vapor barrier over the existing grade, followed by a layer of insulation board.
- Radiant heating systems should be pressure tested for tubing leaks before the tubing is covered over.
- Another method of installing radiant flooring is the use of thin-slab concrete or gypsum over an existing floor.
- Radiant flooring can also be installed by means of prefabricated above-floor panels or underfloor suspended radiant tubing.
- When radiant flooring is not an option, radiant panels are also available for installation in walls and ceilings.
- The steps that go into designing a radiant heating system include performing an accurate load calculation, properly sizing and spacing the tubing, and properly sizing the circulation pump based on system flow and feet of head loss.
- Controls for radiant heating systems can be simple or complex depending on the desires of the designer and the needs of the customer.
- Modern digital controls for radiant heating systems offer advanced efficiency and user-friendly interfaces compared to conventional controls.
- A very practical application for radiant heating is snow and ice melting. Snow and ice melt systems can be used for residential and commercial applications. Applications for snow and ice melt systems include driveways, sidewalks, and parking garages.
- The design process for snow and ice melt systems compared to conventional systems is quite similar except for some key differences for snow and ice melt systems, which include determining the proper design outdoor air temperature and proper placement of the in-ground slab sensors. Slab or in-ground sensors can detect when a snow and ice melt system shall be energized and de-energized.
- Because snow and ice melt systems are located outdoors, their installation criteria differ from that of conventional systems.

Know and Understand

1. With radiant heating, _____.
   A. heat is conveyed in a direct physical path
   B. objects are heated, rather than the air
   C. Both A and B are correct.
   D. Neither A nor B are correct.
2. The hot water supply temperature range used with radiant heating is usually between _____.
   A. 50°F–100°F
   B. 100°F–140°F
   C. 140°F–180°F
   D. 180°F–200°F
3. A heating system that has a very limited capacity to store heat energy is known as a _____.
   A. low-mass heating system
   B. high-mass heating system
   C. equal mass heating system
   D. radiant heating system
4. Which type of radiant tubing can be heated and returned back to its original shape if it becomes kinked?
   A. PEX type A
   B. PEX type B
   C. PEX type C
   D. None of the above.
5. *True or False?* If PEX tubing has to come in contact with any fittings or devices that contain ferrous materials, then an oxygen barrier is required.
6. PE-RT tubing is constructed with _____ layers of material that give it significant strength and chemical resistance.
   A. three
   B. four
   C. five
   D. seven
7. When installing a slab-on-grade radiant system, as much as _____ of the heat generated by the boiler can be lost through the Earth if the concrete slab is not properly insulated.
   A. 40%
   B. 50%
   C. 60%
   D. 70%
8. Radiant floor systems should be pressure tested with up to _____ times the normal operating pressure and for a minimum of _____ hours.
   A. 2, 12
   B. 3, 24
   C. 4, 48
   D. 5, 72
9. Typically, a poured gypsum floor will add _____ pounds per square foot to the dead load of an existing floor structure.
   A. 5 to 10
   B. 13 to 15
   C. 15 to 20
   D. 20 to 23
10. One type of radiant floor heating that can be used when the existing floor cannot support the additional weight of a conventional installation is known as _____.
    A. slab on grade
    B. thin-slab flooring
    C. above-floor prefabricated panels
    D. slab on slab
11. Radiant wall and ceiling panels are considered _____.
    A. high-mass systems
    B. low-mass systems
    C. equal mass systems
    D. standard mass systems
12. *True or False?* Smaller radiant tubing will result in a lower head loss than larger tubing but will provide nearly twice the fluid flow.
13. *True or False?* Decreasing the radiant tubing spacing will lower the required supply water temperature and produce a more even surface temperature, but it increases the amount of tubing required.
14. What is an advantage of using a modulating zone valve compared to a two-position valve?
    A. Modulating zone valves provide optimal heating control.
    B. Modulating zone valves decrease the chance of water hammer.
    C. Both A and B are correct.
    D. None of the above.
15. What device is used to detect snow and ice buildup on outdoor surfaces when using a snow melt system?
    A. An in-ground sensor
    B. An outdoor air sensor
    C. A tubing temperature sensor
    D. A microprocessor sensor

Apply and Analyze

1. Explain how radiant heat transfer differs from conductive and convective heat transfer.
2. Describe how radiant heating relates to human comfort.
3. How does high-mass radiant heating compare to low-mass radiant heating?
4. What are three different heat sources for radiant systems, and how do they compare?
5. Describe how different types of piping materials are used in radiant heating systems.
6. What are some of the differences between PEX tubing and PE-RT tubing?
7. List the installation procedures for installing slab-on-grade and thin-slab flooring systems.
8. Describe where above-floor prefabricated panels would be used instead of a thin-slab flooring system.
9. How does a below-floor suspended tubing system compare to prefabricated above-floor panels?
10. What are some advantages of using wall or ceiling radiant panels?
11. List the parameters that need to be determined when designing a radiant heating system.
12. What are the two factors used to select a circulating pump for a radiant heating system?
13. How do radiant heating controls compare to the controls used with a conventional heating system?
14. List the components that make up a snow and ice melt system.
15. How does the installation of a snow and ice melt system compare to other types of radiant heating systems?

Critical Thinking

1. Describe how radiant heating systems might be used for various commercial applications.
2. What would be the process for retrofitting a radiant system into a residential or commercial building that has a conventional heating system?

13 Building Heating Loads and Print Reading

Chapter Outline

13.1 Performing Building Heat Load Calculations
13.2 Heat Loss through Conduction
13.3 Heat Loss through Infiltration
13.4 Print Reading
13.5 Types of Blueprints

megaflopp/Shutterstock.com

Learning Objectives

After completing this chapter, you will be able to:
- Explain why performing accurate load calculations is essential to the success of a hydronic heating project.
- Explain the meaning of a building material's R-value and how this relates to thermal density.
- List the steps taken to calculate the building's design temperature difference.
- Describe how the U-value of a building material compares to its R-value.
- Compare how heat losses due to infiltration differ from heat losses due to conduction.
- List the methods used for finding the infiltration cfm of a building.
- Calculate heat loss due to infiltration using the sensible heat formula.
- Explain why a blower door test is a more advantageous method for finding heat loss due to infiltration.
- List the various types of prints used in residential and commercial boiler projects.
- Describe why the ability to read and interpret symbols and abbreviations is necessary to understanding blueprints and drawings.

Technical Terms

above grade
air change method
architect's scale
below grade
blower door test
built-up wall
crack method
design temperature difference
detail drawing
elevation drawing
floor plan
heat load calculation
infiltration
Manual J
Manual N
reciprocal
R-value
scale
slab-on-grade construction
sustainable design
thermal density
thermal resistance
thermal transmittance
transmission losses
U-value

Following the proper procedures for calculating heating loads and sizing equipment is essential to the success of a hydronic heating project. Many times, a hydronic system simply does not perform well after installation because the designer failed to use prudent sizing practices. In order to achieve a successful heating project, designers should master the following areas of expertise:
- Understand the proper procedure for accurate heat load calculations.
- Know how to interpret these calculated results.
- Utilize the information that has been gathered to correctly size and select the proper equipment.

With an ever-increasing demand for higher energy efficiencies and fuel savings, building designers and engineers are constantly striving to improve the thermal integrity of homes and buildings. Part of this process involves the implementation of *sustainable design*. This discipline involves the incorporation of materials and processes that reduce energy costs, improve productivity, and decrease the amount of environmental waste. Sustainable design seeks to reduce the amount of negative impact on the environment while improving the health and comfort of the building occupants. Along with sustainable design, there are a number of steps that go into the sizing and selection of a well-planned hydronic system. This chapter covers these steps along with the topic of reading and interpreting blueprints for boilers and hydronic systems.

GREEN TIP

Green Building Organizations

Several organizations promote green building design and construction—and designers, installers, and service technicians can become members. Among these, the US Green Building Council (USGBC) promotes the practice of LEED certification for homes and buildings. LEED stands for *Leadership in Energy and Environmental Design*, which is the most widely used green building rating system in the world.

13.1 Performing Building Heat Load Calculations

The first step in designing a successful hydronic heating system is to perform an accurate *heat load calculation* on the residential or commercial structure. An accurate heat load calculation determines the structure's heat loss based on the geographic location, the type of building materials used, and how resistant the building is to the infiltration of unwanted and uncontrolled outdoor air. An accurate heat load calculation helps accomplish the following goals:

- Ensure the proper sizing of the boiler and terminal units
- Design the piping system to deliver the correct amount of heat to each room
- Provide the most comfortable conditioned space at the lowest operating cost

The procedure for calculating heat losses on residential and commercial buildings primarily falls into two categories:

- Calculating heat losses resulting from conduction (transmission) through the building envelope
- Calculating the heat loss from convection as a result of uncontrolled infiltration around doors and windows (drafts)

13.2 Heat Loss through Conduction

Heat will migrate from a warm area to a cold area. Because of this naturally occurring phenomenon, buildings lose heat whenever the outdoor air temperature is colder than the indoor air temperature—no matter how well the building is insulated. This explains why homes and other buildings get cold in the wintertime. This heat transfer between the warm indoor air and the cold outdoor air occurs primarily by conduction—otherwise known as heat loss through transmission.

Transmission losses are due to the thermal transfer of heat energy through the building's envelope, as illustrated in **Figure 13-1**. These losses are a result of heat conduction naturally occurring between the indoors and outdoors through walls, windows, doors, ceilings, and even through the basement floor. If the building materials consist of a greater mass, the heat tends to transfer at a slower rate. The amount of mass found in the building's structural components is referred to as its *thermal resistance to heat transfer* and is expressed as the building material's *R-value*. The R-value of a building material is the rating system used to grade insulation products or the material's insulating properties. For instance, a 4″ common masonry brick has an R-value of 0.80 compared to 1/2″ gypsum wall board, which has an R-value of 0.45. Because the brick is made up of denser material, it loses heat more slowly than does the wall board.

In order to accurately calculate the building heat loss through conduction or transmission, the designer needs to know three components:

- The building's design temperature difference
- The area of the building component being analyzed (wall, door, window, ceiling)
- The R-value of the building material(s)

The building's *design temperature difference* is the difference between the indoor temperature and outdoor design air temperature. The typical indoor winter design temperature for most buildings is 70°F. This may vary slightly depending on the type of activity being performed in the building and the desired comfort level of the building occupants.

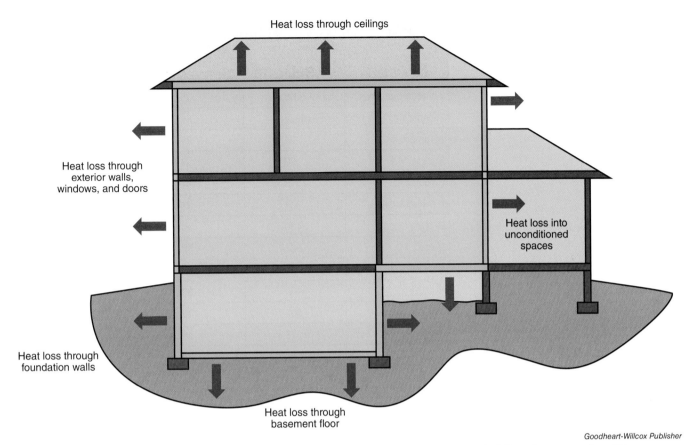

Figure 13-1. This illustration shows how heat losses can occur at various areas of the home or building. The areas colored blue represent the conditioned spaces, the areas colored yellow represent unconditioned spaces, and the building envelope is shown in green.

Outdoor winter design temperatures are usually the coldest average temperatures for a geographic region and can be translated into the "worst case scenario" for winter temperatures. The data for outdoor winter design temperatures is available from a variety of sources, including the following:

- Local airports, weather stations, or the National Weather Service
- The American Society of Heating, Air Conditioning, and Refrigeration Engineers (ASHRAE)
- The Air Conditioning Contractors of America (ACCA) *Manual J*

Figure 13-2 shows the outdoor design conditions for various cities across the United States. Notice that the winter design heating temperature is expressed as 99%. This means that the outdoor temperature for the listed city will be above this number 99% of the year, based on a 30-year average. Conversely, the outdoor temperature will on average be below this temperature for only 1% of the year, or approximately 88 hours. US states that are located geographically farther north will have colder outdoor design temperatures.

The second factor in calculating the building heat loss is the area of the building components being analyzed. This measurement is expressed in square feet. These components include the building's walls, doors, windows, ceiling, and the basement floor or slab on grade.

The third factor used for calculating heat loss is in determining the proper R-values of the building materials being used.

Heating Design Temperature for Select Cities			
City, State	Elevation (ft)	Latitude (degrees north)	Heating design temperature (°F)
Albuquerque, NM	5318	35	21
Atlanta, GA	1027	34	27
Baltimore, MD	24	39	17
Boston, MA	30	42	12
Chicago, IL	593	41	3
Cleveland, OH	804	41	9
Denver, CO	5430	40	7
Detroit, MI	627	42	9
Honolulu, HI	16	21	63
Houston, TX	105	30	33
Indianapolis, IN	807	40	6
Juneau, AK	23	58	8
Las Vegas, NV	2182	36	33
Miami, FL	30	26	52
New Orleans, LA	20	30	35
New York, NY	7	40	15
Oakland, CA	89	38	40
Oklahoma City, OK	1306	35	17
Orlando, FL	105	28	42
Seattle, WA	14	47	27
St. Louis, MO	580	38	8
Tucson, AZ	2556	32	34
Washington, DC	325	39	16

Adapted from ACCA Manual J, Residential Load Calculation

Figure 13-2. This table shows outdoor design temperatures for various cities throughout the United States.

13.2.1 Understanding R-Values and U-Values

As mentioned earlier, the resistance to heat transfer of the building construction materials is referred to as the material's R-value. By definition, an R-value is a measurement or rating that corresponds to a material's ***thermal density*** and how it resists heat transfer through conduction. The higher the R-value, the greater the resistance to heat transfer. **Figure 13-3** lists common building materials and their respective R-values. R-values can relate to a single type of building material or can be expressed as the R-value per inch of material.

For instance, in **Figure 13-3** a 1/2″ sheet of plywood has an R-value of 0.62, whereas particle board will have a variable R-value depending on the density or thickness of each sheet. All building materials have an R-value. Even an air space can be expressed as an R-value.

The Air Conditioning Contractors of America (ACCA) publishes data and tables on R-values for various types of building materials as well as the outdoor

R-Values of Common Building Materials

The R-value of a material determines how quickly heat is conducted across it. The values below are some of the more common R-factors for surfaces found on buildings in the U.S.

Exterior Walls with Siding

Exterior Walls with Siding Concrete block (8″)	R-Value	Wooden Frame	R-Value
Concrete block (8″)	2	Uninsulated with 2″ × 4″ construction	4.6
with foam insulated cores	20	with 1 1/2″ fiberglass	9
with 4″ on unisulated stud wall	4.3	with 3 1/2″ fiberglass; studs 16‴ o.c.	12
with 4″ insulated stud wall	14	with 3 1/2″ fiberglass and 1″ foam	20
Brick (4″)	**R-Value**	**StrongGreen™ Structural Panel**	**R-Value**
with 4″ uninsulated stud wall	4	3 1/2″ thick polyurethane panel × 24″ Wide	27
with 4″ insulated stud wall	14	6″ thick polyurethane panel	45
		3 1/2″ thick polyurethane panel w/ ceramic coating	47
		6″ thick polyurethane panel w/ ceramic coating	65

Floor

Floor Over unheated basement or crawl space vented to outside	R-Value	Concrete Slab	R-Value
Uninsulated floor	4.3	No insulation	11
6″ fiberglass floor insulation	25	1″ foam perimeter insulation	46
Over sealed, unheated, completely underground basement	**R-Value**	2″ foam perimeter insulation	65
Uninsulated floor	8	**StrongGreen™ Structural Panel**	**R-Value**
with 1″ foam on basement walls	19	3 1/2″ thick polyurethane panel × 24″ Wide	27
with 3 1/2″ fiberglass on basement walls	20	6″ thick polyurethane panel	45
Insulated floor, 6″ fiberglass	43	3 1/2″ thick polyurethane panel w/ ceramic coating	47
		6″ thick polyurethane panel w/ ceramic coating	65

Exterior Doors (Excluding sliding glass doors)

Calculate glass area of door as window

Wood Door	R-Value	Steel with Foam Core Door	R-Value
1 1/2″ no storm door	2.7	1 3/4″ Pella	13
1 1/2″ with 1″ storm door	4.3	3/4″ Therma-Tru	16
1 2/3″ solid core door	3.1		

Roof / Ceiling

Material	R-Value	Material	R-Value
No Insulation	3.3	6″ thick polyurethane panel × 24″ Wide	45
3 1/2″ fiberglass	13	12″ cellulose	46
6″ fiberglass	20	3 1/2″ thick polyurethane panel w/ ceramic coating	47
6″ cellulose	23	14″ cellulose	54
3 1/2″ thick polyurethane panel × 24″ Wide	27	6″ thick polyurethane panel w/ ceramic coating	65
12″ fiberglass	43		

Glass	R-Value	Low Emissivity	Drapes	Quilts
Single pane	0.9	1.1	1.4	3.2
Single w/ storm window	2		2.5	4.2
Double pane, 1/4″ air space	1.7		2.2	4
1/2″ air space	2	2.99	2.5	4.3
Triple pane, 1/4″ air space	2.6		3	4.8

Air Conditioning Contractors of America (ACCA)

Figure 13-3. This table shows R-values for common building materials.

winter design temperatures for various US cities. This information is available in the following publications: **Manual J** for residential load calculations (**Figure 13-4**) and **Manual N** for commercial applications (**Figure 13-5**). Extensive information on R-values is also available from the American Society of Heating, Air Conditioning, and Refrigeration Engineers (ASHRAE). This work is published through the *ASHRAE Handbook of Fundamentals* and includes extensive data on transmission values for various types of building materials as well as requirements for the correct amount of residential ventilation.

When all three of the above factors are known, the following formula can be used for calculating the heat loss on a particular area of the building:

$$Q = A \times TD/R$$

where

- Q = Heat loss in Btu/hr
- A = Area of the building component in square feet
- TD = Temperature difference between the design indoor and outdoor air in degrees Fahrenheit
- R = R-value of the building material

Using the R-value for this heat transfer formula can be clumsy and prone to error. Therefore, it is more common to utilize the material's U-value when calculating heating loads. Whereas R-value defines the **thermal resistance** of a building material, **U-value** measures the **thermal transmittance** of a building material. By definition, the U-value measures the quantity of heat in Btu per hour that will flow through 1 square foot of material in 1 hour at a temperature difference of 1 degree between the indoors and outdoors. The U-value of a given building material is the reciprocal of the material's R-value. A **reciprocal** is the inverse of a given value, which can be found by dividing 1 by the value. For instance, the U-value of a building material that has a value of R-5 would be 0.2 (1/5 = 0.2). Whereas a building material with a higher R-value results in a lower amount of heat loss, so too does a material with a lower U-value. **Figure 13-6** lists both the U-values and R-values for various construction materials.

Air Conditioning Contractors of America (ACCA)

Figure 13-4. ACCA Manual J is used to calculate building loads for residential structures.

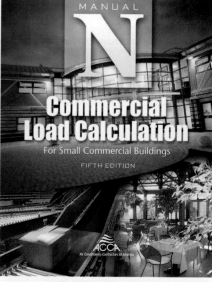

Air Conditioning Contractors of America (ACCA)

Figure 13-5. ACCA Manual N is used to calculate heating and cooling loads for commercial structures.

Constants for Heat Transmission (U- and R-Values)

Expressed in Btu per hour per square foot per degree temperature difference, based on 15 mph wind velocity.

General Wall Classification	Masonry Thickness 8"	
Masonry Construction		
	U	R
Brick–plain		
1/2" wallboard	0.29	3.51
1" polystyrene, 1/2" wallboard	0.13	7.54
Frame Construction		
	U	R
Wood siding on 1" wood sheathing, studs, gypsum board 1/2"	0.23	4.40
Wood siding, sheathing, studs, gypsum board 1/2", 1" polystyrene	0.07	14.43
Wood siding, sheathing, studs, 1/2" flexible insulation, gypsum board 1/2"	0.15	6.70
Wood siding, sheathing, studs, rock wool fill, lath and plaster	0.072	13.9
Note: Frame walls with single exterior finish same as walls with wood siding		
Stucco, wood siding, studs, gypsum board	0.30	3.32
Stucco on 25/32" rigid insulation, studs, 1/2" rigid insulation and plaster	0.20	5.00
Stucco on 1/2" rigid insulation, studs, rock wool fill, lath and plaster	0.074	13.50
Concrete Floors and Ceilings		
	U	R
4" thick concrete, no finish	0.65	1.54
4" concrete, suspended plaster ceiling	0.37	2.70
4" concrete, metal lath and plaster ceiling, hardwood floor on pine subflooring	0.23	4.35
4" concrete, hardwood and pine floor, no ceiling	0.31	3.23
Pitched Roofs		
	U	R
Asphalt shingles on wood sheathing	0.44	2.30
Asphalt shingles, 1" flexible insulation	0.136	7.93
Asphalt shingles, polystyrene 1"	0.132	7.56
Windows and Skylights		
	U	R
Single glass	1.04	0.96
Single glass and storm window	0.56	1.79
Double glass, intermediate air space 1/2"	0.58	1.73
Hollow glass tile wall 6" × 6" × 4" blocks	0.60	1.67
Brick Veneer on Frame Construction		
	U	R
Brick veneer, 1" wood siding, studs, gypsum board 1/2"	0.27	3.71
Rigid insulation, studs, gypsum board 1/2"	0.25	4.00
Brick veneer, 1" wood siding, studs, 1" polystyrene insulation, gypsum board 1/2"	0.07	14.43
Brick veneer, 1" wood siding, studs, rock wool fill, lath and plaster	0.074	13.5

Goodheart-Willcox Publisher

Figure 13-6. This table shows both R-values and U-values for various building materials and for built-up walls.

Once the U-value of the building material is known, it can be inserted into the following revised heat transfer formula:

$$Q = A \times U \times TD$$

where

- Q = Heat loss in Btu/hr
- A = Area of the building component in square feet
- U = U-value of the building material
- TD = Temperature difference between the design indoor and outdoor air in degrees Fahrenheit

Just as with R-values, published data is available through ASHRAE and ACCA for referencing the U-values of various building materials.

TECH TIP

Heat Transmission Multipliers

Manual J and Manual N will also combine the effects of U-values and design temperature differences into factors known as heat transmission multipliers (HTM). These HTM values are also listed in various tables and can help reduce the vast amount of data needed to perform accurate load calculations.

For instance, a wall has a U-value of 0.05 and a temperature difference of 70 degrees. The HTM for this wall is: $0.05 \times 70 = 3.5$. If the wall area equals 200 square feet, the heat loss calculation would be $200 \times 3.5 = 700$ Btu/hr.

In some cases, it is easier to calculate the total U-value for combined building materials that make up a component (such as a framed wall) rather than try to calculate each individual piece. A framed wall may also be known as a ***built-up wall***, which is a wall that contains the various building components used to construct the wall. Notice the illustration of a built-up wall in **Figure 13-7**. Even the inside and outside boundary air films are given an R-value. Once all of the R-values are totaled, the reciprocal is calculated to find the total U-value.

TIMEOUT FOR MATH

Calculating Heat Loss

Following is an example of how the R-value of a built-up wall can be calculated. This particular wall is constructed with:

- 1/2″ drywall: R-value = 0.63
- 4″ of fiberglass insulation: R-value = 4 per inch of insulation
- 3/4″ OSB sheeting: R-value = 0.51

The first step is to find the total R-value for the fiberglass insulation. This is done by simply multiplying 4″ of insulation by the R-value of 4 to get a total R-value of 16. The next step is to add all of the R-values together:

$$0.63 + 16 + 0.5 = 17.14 \text{ total R-value}$$

(continued)

Once the total R-value of the above wall is known, we may now incorporate it into the following problem: What will be the total heat loss of a house's wall measuring 10′ × 50′ that is located in Decatur, Alabama, and has a desired indoor temperature of 70°F?

Answer: The first step is to calculate the design temperature difference. By referencing the Manual J Outdoor Design Conditions for Decatur, Alabama, as shown in **Figure 13-2**, we can see that the winter design temperature is 16°F. Now subtract: 70°F – 16°F. The design temperature difference equals 54°F.

Next, calculate the wall area: 10′ × 50′ = 500 square feet of wall area.

Lastly, convert 17.14 R-value to a U-value.

Answer: 1/17.14 = 0.058 U-value

We may now solve for the following:

$$Q = A \times U \times TD$$
$$Q = 500 \times 0.058 \times 54$$
$$Q = 1566 \text{ Btu/hr}$$

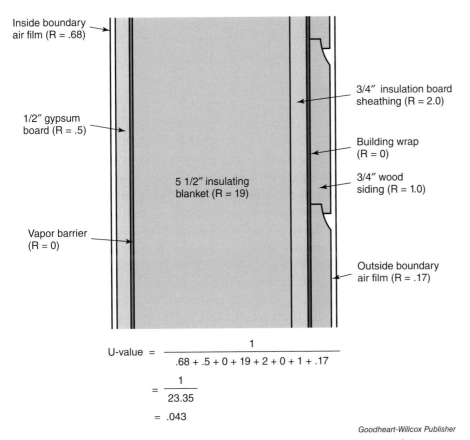

Goodheart-Willcox Publisher

Figure 13-7. An illustration of an exterior wall section showing the R-values of the building materials. Note that the U-value of this wall is the reciprocal of the combined R-values.

> **TECH TIP**
>
> **Converting R-Values to U-Values**
> When converting R-values to U-values, always add all R-values first and then find the reciprocal. Do not convert R-values to U-values and then add U-values individually.

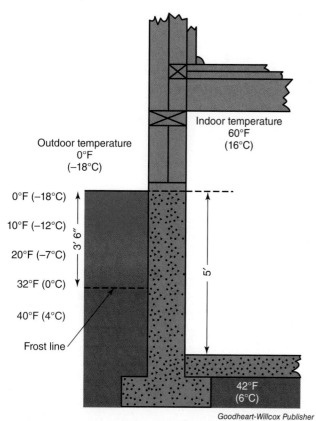

Figure 13-8. This drawing shows the difference in soil temperature based on soil depth and the corresponding frost line.

The same formula may be used for other building components such as windows, doors, and ceilings. One area of consideration when calculating building heat losses is the basement—especially if it is being used as a living space. Although concrete basement walls *below grade* should have minimal heat loss, they still need to be taken into consideration. However, keep in mind that basement walls constructed *above grade* can have significant heat losses, depending on geographic location. Notice the gradient temperature change of the basement in **Figure 13-8** corresponding to the height of the basement wall. The depth of the frost line will vary depending on the local climate and will affect the rate of heat transfer. The frost line—also known as frost depth or freezing depth—is the most common depth to which the groundwater in soil is expected to freeze. Another consideration is the calculation of heat loss through the basement floor. This calculation needs to be determined whether the floor is located in a basement or the building is *slab-on-grade construction*. The heat loss calculation through the concrete slab differs from previous formulas and is as follows:

1. Measure the feet of perimeter of building slab or foundation.
2. Calculate the total feet of perimeter length and multiply by one of the following U-values:
    a. No perimeter insulation: U-value = 0.80
    b. R-5 insulation (1″): U-value = 0.60
    c. R-11 insulation (2″): U-value = 0.50
3. Add this total heat loss to the other above-ground heat losses.

To minimize the heat loss of basement walls and floor slabs, a common building practice is to insulate the exterior foundation wall and the underside of the basement slab as shown in **Figure 13-9**.

Once the transmission loss calculations have been performed for all building components, these figures are then recorded on a spreadsheet. The form shown in **Figure 13-10** is an example of a Manual J spreadsheet that is used to total heat losses from the home being analyzed. After all heat loss calculations from transmission have been tabulated, the designer can then begin calculating heat losses as a result of infiltration.

13.3 Heat Loss through Infiltration

Building heat losses can also occur because of *infiltration*, also known as convection or the transfer of heat energy through a gas or liquid. In any given building, this transfer occurs as a result of uncontrolled cold air entering and exiting the

Chapter 13 Building Heating Loads and Print Reading 281

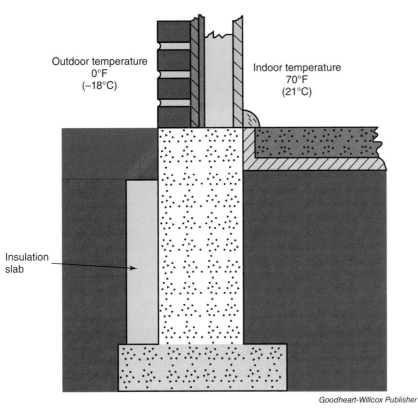

Goodheart-Willcox Publisher

Figure 13-9. By insulating the basement walls below grade as well as below the basement slab, heat losses can be reduced significantly.

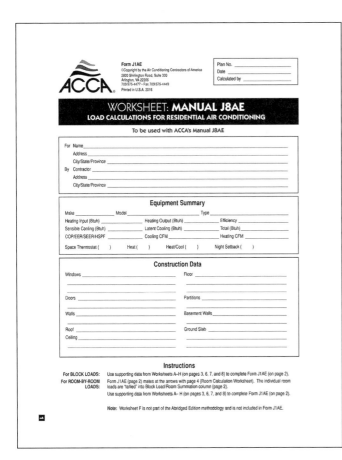
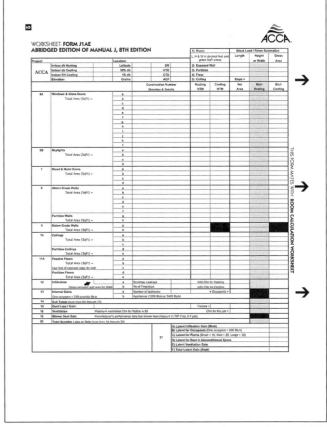

Air Conditioning Contractors of America (ACCA)

Figure 13-10. This illustration shows an example of a Manual J spreadsheet used to compile data on heat losses for a building.

structure. Simply put, heat loss from infiltration is a result of unwanted air that migrates into the building through cracks around doors and windows as well as drafts that penetrate walls—such as through the holes in electrical receptacles. Even if a structure is classified as "air tight," infiltration can still occur every time an outside door is opened. Losses from infiltration can be a substantial percentage of the total heat loss of the structure, which is why it is important to measure this variable carefully, ensuring its accuracy. The formula used to calculate the amount of heat loss as a result of infiltration is known as the sensible heat formula:

$$Q = \text{cfm} \times 1.08 \times TD$$

where

- Q = Heat loss in Btu/hr
- cfm = Cubic feet per minute of infiltration from outdoor air
- 1.08 = A constant that converts the specific heat and density of standard air into Btu per degree Fahrenheit per hour
- TD = Temperature difference between the design indoor and outdoor air in degrees Fahrenheit

13.3.1 Understanding the Crack Method, Air Change Method, and Blower Door Method

The variable that is sometimes elusive and vague in this formula—but still needs to be accurately calculated—is cfm. This is because heat loss through infiltration can be interpreted differently by various designers. Traditionally, there are two methods used for calculating the cfm of unwanted infiltration air:

- The crack method
- The air change method

The first method for calculating infiltration is the ***crack method***. This method is used to find the air leakage through cracks around doors and windows. The designer first measures the linear feet of crack around doors and windows, and then multiplies this figure by the estimated cubic feet of air per hour entering each foot of crack. Tables are available through the National Oceanic and Atmospheric Administration (NOAA) that can help determine the wind velocity for a given geographic area. This data can be used to calculate the cubic feet per hour (cfh) of air entering the building and to find the type of door or window being analyzed. Once the total cfh of infiltration is determined, this figure is divided by 60 to find the cubic feet per minute (cfm). Once the cfm is calculated, it is factored into the sensible heat formula to find the total heat loss due to infiltration.

Example:

Find the infiltration heat loss per hour of a 3′ × 6′ average non-weather stripped, wood frame, double-hung window based on a wind velocity of 20 mph:

Based on look-up tables from NOAA, the air leakage around this window at an average wind velocity of 20 mph is 59 cubic feet per hour per foot of crack. The linear foot of crack is:

$$(2 \times 6') + (2 \times 3') = 8 \text{ total feet}$$

The cubic feet of air per hour (cfh) for this window is:

18 total feet of crack × 59 cfh = 1062 cubic feet per hour of infiltration

The cubic feet of air per minute (cfm) for this window is:

1062 cfh ÷ 60 = 17.7 cubic feet per minute of infiltration

If this window were in a home that has a 70°F design temperature difference, the total Btu/hr heat loss due to infiltration would be:

17.7 cfm × 1.08 × 70°F = 338.12 Btu/hr

The *air change method* for calculating infiltration is used to estimate the number of times the total volume of air contained within the structure is displaced every hour. Tables are available through ASHRAE to assist in estimating this factor. An example of a very well-constructed or "tight" building would be 0.1 air changes per hour. Conversely, a "very leaky" structure may have up to 1.5 air changes per hour. Once this factor is determined, the cfm can be calculated by multiplying the total volume of the space above grade by the air change per hour, and dividing this number by 60. (Dividing by 60 converts cubic feet per hour into cubic feet per minute).

One advantage of using the air change method is that the entire heat loss of the building can be calculated as one formula, rather than having to itemize individual doors and windows.

Example:

Find the infiltration heat loss of a home that has an above-grade volume of 12,450 cubic feet and an estimated infiltration rate of 0.75 air changes per hour (ACH) based on look-up tables:

First, find the total air change rate of the building in cfh:

12,450 cubic feet × 0.75 ACH = 9337.5 cubic feet per hour

Next, convert cubic feet per hour (cfh) to cubic feet per minute (cfm):

9337.5 cfh ÷ 60 = 155.63 cubic feet per minute

Now, input this figure into the sensible heat formula. If this home has a 75°F design temperature difference, the total Btu/hr heat loss due to infiltration would be:

155.63 cfm × 1.08 × 75°F = 12,606 Btu/hr

A third method called a ***blower door test*** can effectively determine heat loss due to infiltration, **Figure 13-11**. A blower door consists of a frame with a flexible panel that fits over an exterior doorway. A powerful blower mounted onto the frame draws air out of the building, reducing the air pressure inside of the building. This causes outside air to flow inward through unsealed cracks and openings around the building, **Figure 13-12**. A pressure gauge then measures the differential pressure between the inside and outside of the building over a fixed period of time, **Figure 13-13**. This pressure is converted to an airflow measured in cfm. The results of this testing procedure are then used to determine the infiltration in cfm of the structure, which in turn can be used to calculate the heat loss due to infiltration.

TEC (The Energy Conservatory)

Figure 13-11. A blower door test is an accurate way to determine heat losses due to infiltration.

TECH TIP

Accounting for Other Infiltration Sources

When calculating heat losses from infiltration, remember that there are a number of factors that can contribute to infiltration besides cracks around doors and windows. These include such things as: fireplaces, flue vents on furnaces and water heaters with no dampers, vents on bathroom and kitchen hood exhaust fans, and ventilation or make-up air ducts that may be connected to the return duct on the furnace.

13.4 Print Reading

The ability to read and understand blueprints is an important skill in the design and installation of boilers and hydronic systems. Whether for a single-family

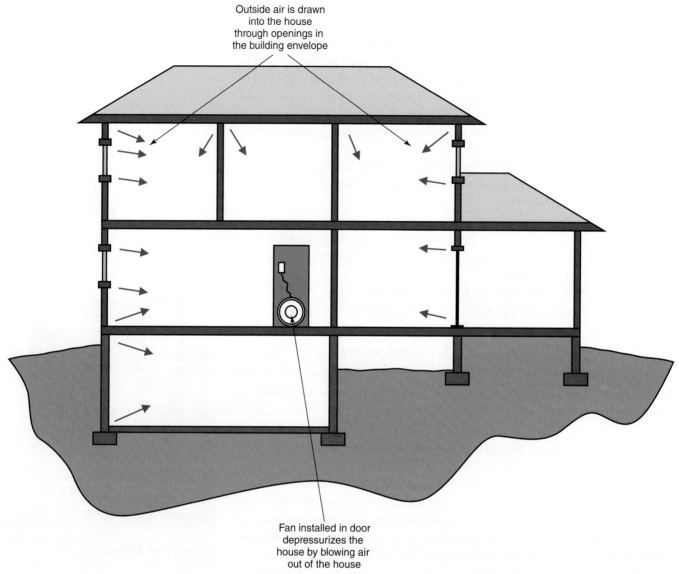

Figure 13-12. This illustration shows how air is drawn in through cracks in the house during a blower door test.

> **DID YOU KNOW?**
>
> **Load Calculation Software Programs**
>
> The heating and cooling loads on a given structure can be calculated using a pencil, a calculator, a spreadsheet, and the proper formulas. However, there are many software programs that make this task much easier and more accurate. These programs include Right-J, Rhvac by Elite Software, and Cool Calc.

home or a more complex commercial building, most construction projects include a comprehensive set of drawings or blueprints.

Construction prints contain the necessary information required to properly and efficiently design and install a successful hydronic project. Print reading skills, along with a solid background in understanding boilers and hydronic heating systems, ensure that installers and trade workers effectively carry out their duties on the jobsite.

This section focuses on the types of prints typically used for the installation of boilers and hydronic systems, along with the symbols and abbreviations used with blueprints.

13.5 Types of Blueprints

A complete set of prints for a construction project may contain several different types of drawings as they relate to the boiler, terminal units, and hydronic piping. These types of drawings may consist of the following:

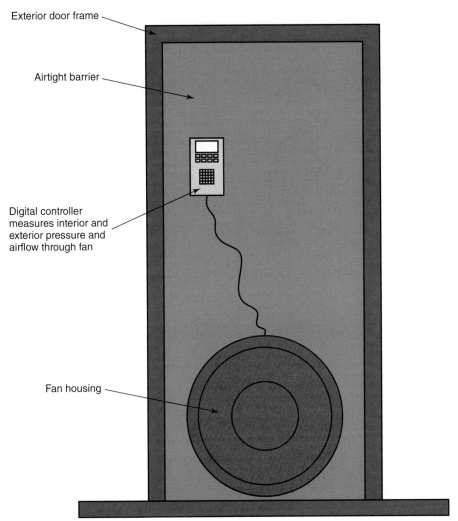

Goodheart-Willcox Publisher

Figure 13-13. The blower door uses pressure gauges to measure differential pressure inside and outside of the house to determine leakage based on cfm.

- **Isometric drawings.** These types of drawings are commonly used with piping diagrams to show the supply and return lines from the boiler. What is distinctive about isometric drawings is that they show the piping on three principle axes drawn 120° apart from each other, **Figure 13-14**.
- **Plot plans.** These drawings are scaled working prints showing the size and shape of the building located on the structure's property. These can be useful for boiler installations because plot plans usually show where the utilities—such as gas, electric, and water—enter the building.
- **Perspective drawings.** Otherwise known as artist renderings, perspective drawings are commonly used to allow the potential building owner to visualize the overall appearance of a new structure. They are commonly drawn to show the building from a single point.

The most common types of prints or drawings used in the construction and installation of a complete hydronic heating system are:
- Floor plans
- Elevation drawings
- Detail drawings

DID YOU KNOW?

The *Blue* in Blueprints

The development of blueprints dates back to 1842. John Herschel, a chemist, developed the process by taking an image drawn on semi-transparent paper and weighing it down on top of a sheet of paper that was pre-coated with a chemical mixture of potassium ferricyanide and ferric ammonium citrate. Once the drawing was exposed to light, the exposed parts of the print became blue, while the drawn lines remained white.

With today's modern computer-aided design programs, as well as modern drafting and printing methods, blueprints are not really blue anymore—but the name remains the same.

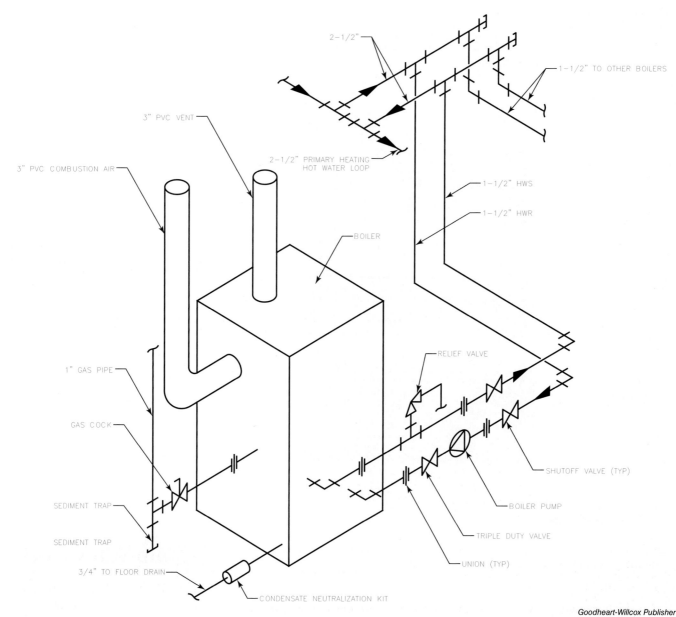

Figure 13-14. This is an example of an isometric drawing. With an isometric drawing, the axes are drawn 120° apart from each other.

13.5.1 Floor Plans

As the name implies, a *floor plan* is a plan view that shows the layout of equipment and piping when looking down from above. Floor plans illustrate the locations of heating equipment as well as walls, doors, windows, and other building components. These components are shown as they appear in an imaginary horizontal section five feet above the floor level. In other words, they are shown as if a plane were cut horizontally through the building five feet above the floor.

Floor plans are drawn to a *scale*, such as a 1/8″ scale. This means that 1/8″ is equal to one foot. By scaling blueprints, estimators can calculate the total length of piping needed for a particular project by measuring the total feet of pipe shown on the plans using an *architect's scale*. This measuring device is simply a ruler that has multiple scales to allow for easier measuring when estimating projects, **Figure 13-15. Figure 13-16** illustrates a typical floor plan drawing showing the locations of the piping and mechanical room.

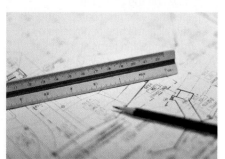

Nattakit Jeerapatmaitree /Shutterstock.com

Figure 13-15. An architect's scale is used for measuring piping and other components on scaled blueprints and drawings.

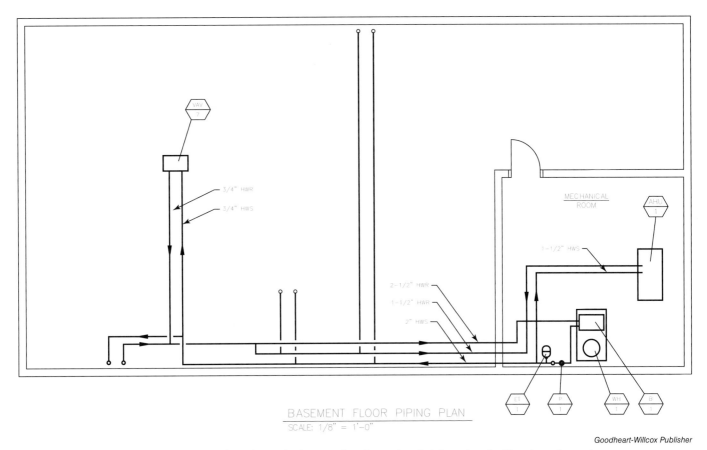

Figure 13-16. A floor plan is a plan view that shows the layout of equipment and piping when looking down from above.

When greater detail is needed, a portion of a floor plan can be expanded and shown as a detailed floor plan. A detailed floor plan is typically drawn on a scale of 1/4″ = 1 foot.

13.5.2 Elevation Drawings

An *elevation drawing* is a scaled view that looks directly at a vertical surface. Unlike a floor plan that has an imaginary cutting plane, elevations show views from floor to ceiling. These drawings display overhead piping and equipment as well as allow the viewer a different perspective of the orientation of the boiler and respective piping.

Figure 13-17 is an example of an elevation drawing showing the boiler, circulating pump, and associated equipment.

13.5.3 Detail Drawings

When special features of a particular piece of equipment are needed, a *detail drawing* is used. Details may be shown using plan views, elevations, or sectional drawings. Standard-sized drawings of floor plans or elevations are typically created using smaller scales and therefore do not always clearly show smaller details. By creating specific detail drawings, designers can show a specific part of a piece of equipment at a larger scale. However, keep in mind that some detail drawings are shown with the abbreviation *NTS*, meaning not to scale, **Figure 13-18**.

13.5.4 Understanding Symbols and Abbreviations

Blueprints and other drawings use a variety of symbols to represent various hydronic components, including fittings, valves, and temperature controls. Most

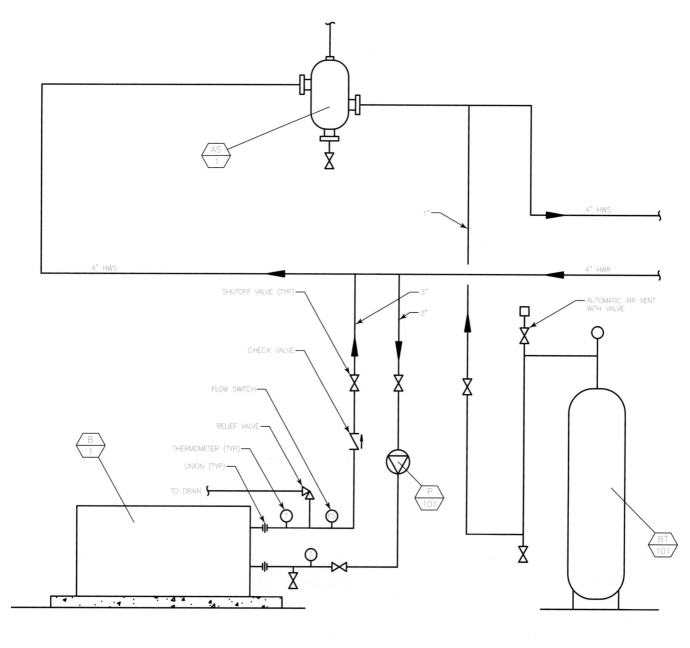

Figure 13-17. Elevation drawings show views of vertical surfaces from floor to ceiling.

symbols that appear on blueprints are based on ANSI Standards. The acronym ANSI stands for the American National Standards Institute, which is an organization that oversees the creation and endorsement of thousands of industry guidelines that directly impact businesses in nearly every sector of the economy. The use of symbols to represent hydronic components usually eliminates the need for descriptive notations. Without the use of symbols, the drawing of blueprint components would be a tiresome and redundant task. Therefore, the boiler installer

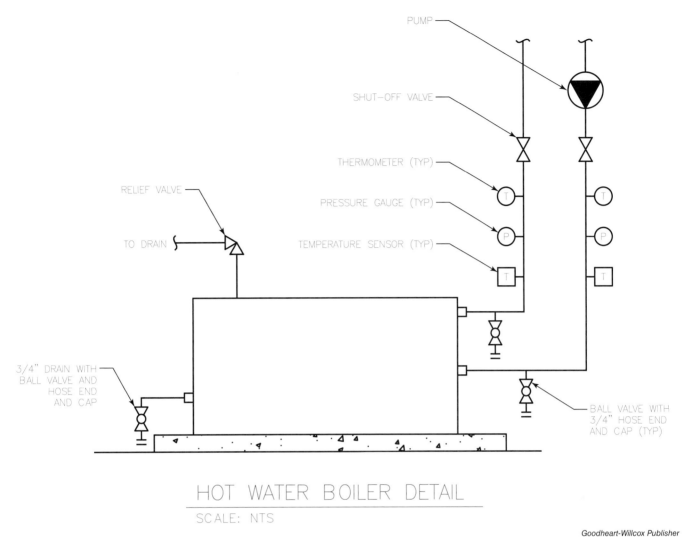

Figure 13-18. Detail drawings can show a specific part of a piece of equipment at a larger scale; however, they may not be drawn to a specific scale.

must become familiar with symbols and their proper usage in representing the devices and other components used in the construction of hydronic heating systems. A comprehensive list of hydronic symbols is shown in **Figure 13-19**.

Abbreviations are also used throughout print sets for boilers and hydronic heating systems to describe components and devices in an effort to conserve space within the drawings. Abbreviations save time and prevent confusion by conveying the proper information in a consistent manner. Uppercase letters are used to represent abbreviations, and they usually end with a period to distinguish them from actual words. **Figure 13-20** lists abbreviations used for a particular boiler room. Standard abbreviations eliminate confusion and misinterpretation of the working blueprints; therefore, only abbreviations that are utilized consistently in the hydronic industry should be used. Because some abbreviations can be used to represent different terms, a legend should be included with the construction drawing set showing the definitions of the abbreviations to prevent any confusion.

Common Piping Symbols			
Symbol	Description	Symbol	Description
	Gate valve		Check valve
	Globe valve		90° elbow up
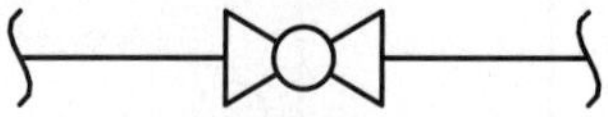	Ball valve	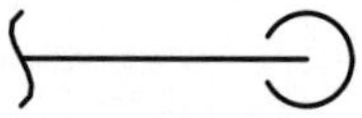	90° elbow down
	Manually operated		Tee down
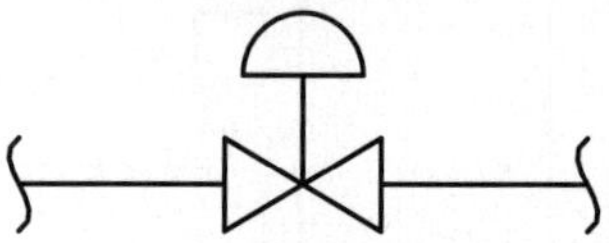	Pneumatically operated	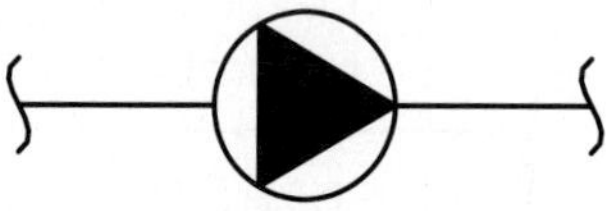	Inline pump
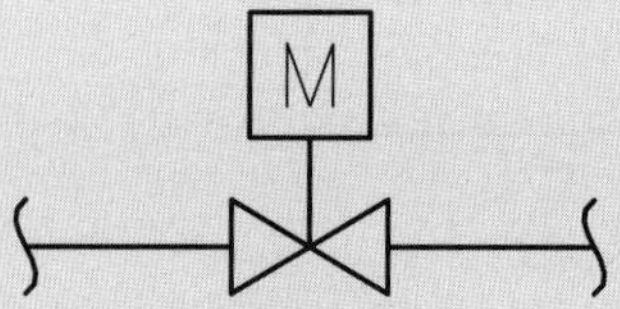	Motor-operated	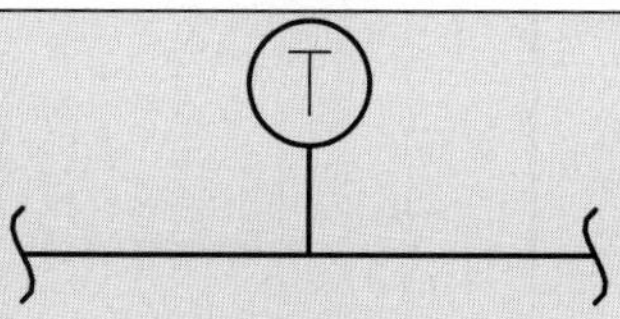	Temperature sensor
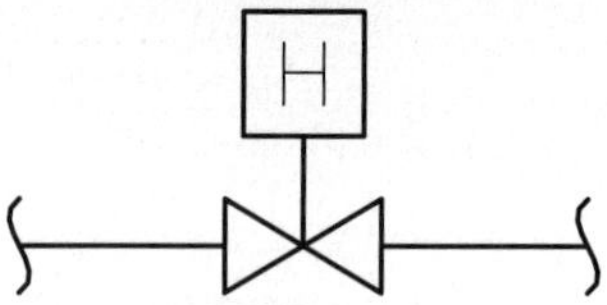	Hydraulically operated	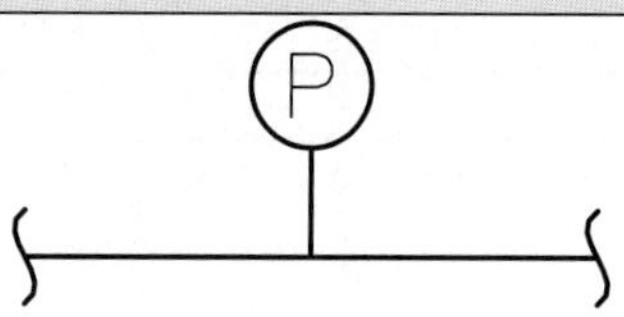	Pressure sensor
	Three-way valve	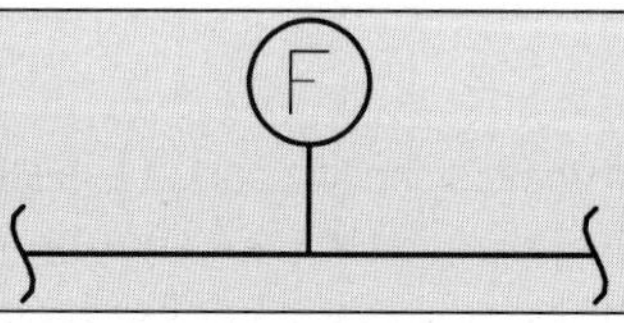	Flow sensor
	Butterfly valve		

Goodheart-Willcox Publisher

Figure 13-19. Blueprints use a variety of symbols to represent various hydronic components including fittings, valves, and temperature controls.

Common Abbreviations Used on Hydronic Drawings

Abbreviation	Description	Abbreviation	Description	Abbreviation	Description
AFF	Above finished floor	(E)	Existing	PRV	Pressure-reducing valve
ADJ	Adjustable	ESP	External static pressure	RP	Radiant panel
APD	Air pressure drop	FRL	Filter, regulator, lubricator	SOV	Shutoff valve
CUH	Cabinet unit heater	F&T	Float and thermostatic	SFW	Soft water
CHWR	Chilled water return	FCO	Floor cleanout	SF	Square feet
CHWS	Chilled water supply	GPM	Gallons per minute	STM	Storm
CO	Cleanout	HP	Horsepower	SA	Supply air
CW	Cold water	HW	Hot water	TW	Tempered water
CA	Compressed air	HWR	Hot water return	TWR	Tempered water return
CR	Condensate return	HHW	Heating hot water	MBH	Thousands BTU per hour
CFM	Cubic feet per minute	HHWS	Heating hot water supply	THR	Total heat rejection
DIA	Diameter	HHWR	Heating hot water return	TYP	Typical
DN	Down	IN WC	Inches water column	UNO	Unless otherwise noted
DB	Dry bulb	LWT	Leaving water temperature	WCO	Wall cleanout
EFF	Efficiency	NTS	Not to scale	WHA	Water hammer arrester
EWT	Entering water temperature	OA	Outdoor air	WB	Wet bulb
EA	Exhaust air	PD	Pressure drop		

Goodheart-Willcox Publisher

Figure 13-20. Abbreviations are used on drawings to eliminate the confusion and misinterpretation of various components and equipment.

CAREER CONNECTIONS

North American Technical Excellence (NATE): Certification and Training for Hydronic Heating Specialists

NATE is the nation's largest nonprofit certification organization for heating technicians. NATE exams represent real-world working knowledge of hydronic systems and validate the professional competency of installation and service technicians.

By successfully passing NATE exams in hydronic service and installation, technicians have the potential to advance within the companies through which they are employed. Advancement can take the form of higher wages or even positions of supervision or service manager.

Other successful NATE technicians may become building superintendents, cost estimators, or system test and balance specialists. Still others may advance into positions of sales and marketing, or even into areas of education.

Those technicians who possess the proper managerial skills and adequate funding can even develop their own contracting businesses.

Chapter Review

Summary

- In order to achieve a successful hydronic heating project, an accurate building heat load calculation first needs to be performed.
- Heat losses on residential and commercial buildings are primarily due to conduction and infiltration losses.
- Conduction, or transmission heat loss, is a result of heat traveling from warm to cold through the building envelope.
- The building's design temperature difference is the difference between the indoor temperature and outdoor design air temperature. The outdoor design temperature is based on historic weather data and will vary depending on geographic locations.
- The R-value of a building material is the rating system used to grade insulation products or the material's insulating properties. A building material's R-value corresponds to its thermal density. Various R-values of a built-up wall can be added together to determine the total R-value of that wall.
- U-values measure the thermal transmittance of a building material. The U-value of a built-up wall can be determined by calculating the reciprocal of the total R-value of that wall.
- Basement slab-on-grade floors have a different calculation for finding their heat loss compared to other areas of the building.
- Heat loss due to infiltration occurs because of uncontrolled cold air that enters and exits the structure.
- The crack method for calculating heat loss due to infiltration measures the perimeter of windows and doors and assigns a numerical value for air flow based on cubic feet per hour.
- The air change method for calculating infiltration is used to estimate the number of times the total volume of air contained within the structure is displaced every hour.
- Once the amount of infiltration air of a building is determined, this figure is used with the sensible heat formula to calculate the Btu heat loss.
- A blower door test is sometimes a more accurate method of calculating the amount of infiltration air found in a building.
- The ability to read and understand blueprints is an important skill in the design and installation of boilers and hydronic systems.
- There are several different types of prints used for hydronic heating projects, the most common of which are floor plans, elevation drawings, and detail drawings.
- Blueprints are drawn to specific scales.
- Symbols are used on blueprints to eliminate the need for descriptive notations.
- Abbreviations are used on drawings and blueprints to save time and prevent confusion by conveying the proper information in a consistent manner.

Know and Understand

1. In order to achieve a successful heating project, designers should master the following areas of expertise:
   A. Understand the proper procedure to develop accurate heat load calculations
   B. Know how to interpret both heating and cooling load results
   C. Know how to draft a full set of construction drawings
   D. Be able to teach customers how to calculate fuel efficiency
2. *True or False?* In order to perform an accurate heat load calculation, the designer must know the outdoor winter design temperature and the types of building materials being used.
3. Heat losses through the building envelope are also referred to as _____.
   A. losses through conduction
   B. losses through convection
   C. losses through radiation
   D. losses through evaporation
4. By definition, an R-value is a rating that corresponds to a building material's _____.
   A. thermal transmittance
   B. thermal density
   C. thermal reliance
   D. thermal energy
5. *True or False?* The U-value of a given building material is the reciprocal of the material's R-value.
6. What is the Btu heat loss of an 800 square foot wall with an R-value of 15 and a design temperature difference of 65°F?
   A. 78,000 Btu/hr
   B. 34,660 Btu/hr
   C. 7800 Btu/hr
   D. 3466 Btu/hr
7. Which type of wall will have a greater Btu heat loss?
   A. Uninsulated basement wall below grade
   B. Uninsulated basement wall above grade
   C. They will both be the same.
   D. Basement walls do not experience heat loss.
8. Which ACCA Manual is used for residential load calculations?
   A. Manual N
   B. Manual D
   C. Manual J
   D. Manual Z
9. Building heat losses due to infiltration are also known as _____.
   A. conduction losses
   B. convection losses
   C. radiation losses
   D. building envelope losses
10. Which formula is used to calculate heat loss due to infiltration?
    A. Latent heat formula
    B. Sensible heat formula
    C. Total heat formula
    D. Heat transmission multiplier
11. *True or False?* The crack method used to calculate infiltration losses takes into account the building's air changes per hour.
12. What would be the infiltration heat loss of a building with a total volume of 12,000 cubic feet, a design temperature difference of 70°F, and an air change rate of 0.5 per hour?
    A. 7560 Btu/hr
    B. 75,600 Btu/hr
    C. 420,000 Btu/hr
    D. 756,000 Btu/hr
13. Which type of blueprint will show views from floor to ceiling?
    A. Floor plan
    B. Elevation drawing
    C. Isometric drawing
    D. Architect's scale
14. *True or False?* Most symbols that appear on blueprints are based on AFUE Standards.

Apply and Analyze

1. Explain why sustainable design is important to the success of the hydronic heating project.
2. Describe the two major ways that a building loses heat.
3. Explain how a building material's R-value plays a role in reducing a building's heat loss.
4. What are the three major components used to accurately calculate heat loss through convection?
5. How is the total U-value for a building wall determined once the R-values for the individual wall materials are known?
6. Explain how the heat loss of a basement floor slab is calculated as compared to a basement wall.
7. Describe the two main methods used for calculating heat losses due to infiltration.
8. Describe the different types of blueprints used for hydronic heating systems.
9. How can an architect's scale be a useful tool when laying out a hydronic system?
10. Explain the difference between symbols and abbreviations used in blueprints.

Critical Thinking

1. Why might a blower door test be a more accurate way of calculating heat losses due to infiltration compared to other methods?
2. What are some of the problems that might arise if the designer does not perform an accurate building heat load calculation?

14 Boiler System Design Considerations

Chapter Outline

14.1 Boiler Selection
14.2 Combustion and Ventilation Air Requirements
14.3 Boiler Venting
14.4 Gas Piping

NavinTar/Shutterstock.com

Learning Objectives

After completing this chapter, you will be able to:
- Describe the steps taken when selecting a boiler based on fuel type, efficiency, and application.
- Explain why combustion and ventilation air requirements are important.
- Differentiate between combustion air requirements for a confined space and an unconfined space.
- Describe what steps must be taken to provide combustion air when the boiler is located in an unconfined space.
- Describe the four categories of boiler venting.
- Differentiate between venting requirements for a natural draft boiler and an induced draft boiler.
- Identify the conditions that merit Category B-vents, and explain why they are used for venting the products of combustion from a boiler.
- Contrast venting in a condensing boiler and a noncondensing boiler.
- Identify the steps required to convert a natural gas boiler to propane.
- Describe the steps taken when sizing gas piping for boiler applications.

Technical Terms

Category B-vent
confined space
direct vent boiler
draft hood
flue liner
free area
gas orifice
induced draft boiler
longest length method
negative draft
unconfined space

Once the heat load calculations for the building have been determined, the next step is to select the boiler for the hydronic system. Along with the boiler selection, it is also important to determine the proper combustion air and venting requirements for the system and follow the correct procedure for sizing the gas piping. This chapter covers the various factors that go into selecting the best possible boiler for the heating application as well as the proper sizing procedure for the combustion air requirements, venting, and gas piping.

14.1 Boiler Selection

When making a boiler selection, several factors should be considered to achieve the desired design intent. These considerations include:
- Fuel availability
- Maximizing efficiency
- System flexibility
- Budget constraints

In addition to these items, the boiler room must be large enough to house the boiler and its associated piping. Regardless of any additional considerations, the primary objective when selecting a boiler is to provide a system that meets the heating requirements of the building at the best efficiency while maintaining the most economical price.

14.1.1 Selecting a Boiler Based on Different Types of Fuels and Efficiencies

One criterion for selecting a boiler is to compare various fuel efficiencies. Most boilers have an AFUE classification or rating that may be anywhere from 80% efficiency to over 90% efficiency. AFUE is an acronym for Annual Fuel Utilization Efficiency, which measures how well the boiler utilizes the heating fuel. The higher the AFUE, the lower the user's heating bills could be.

For instance, a boiler that has an AFUE of 85% means that for every dollar spent on fuel, 85 cents is being used to heat the building and 15 cents is going up the chimney. Keep in mind that the fuel efficiency rating only compares boilers using the same type of heating fuel. It should not be used to compare boilers using different fuel types. For instance, a boiler using electricity as a fuel source will have an efficiency rating of 100% (no flue or chimney required, so 100% of the fuel is being utilized). However, the cost of electricity may be much more expensive than other types of fossil fuels.

The only way to make a comparison of heating costs for individual fuels is to take into account both the heating value for the fuel and the current cost for a unit of that fuel. As a review, following are the heating values for the most common types of fuels:

Natural gas = 1050 Btu per cubic foot or 105,000 Btu per hundred cubic feet
Liquified petroleum (LP) = 92,000 Btu per gallon
Fuel oil = 140,000 Btu per gallon
Electricity = 3413 Btu per kilowatt

When determining which type of fuel will be most cost effective, the great equalizer is to calculate the cost per million Btu for the various types of heating fuels being considered. By using the current cost of a given fuel and the heating value per unit of that same fuel in the following formula, we can calculate the cost for a million Btu of that fuel:

$$\frac{1{,}000{,}000 \times \text{Cost of the Fuel Unit}}{\text{Heating Value of Fuel Unit in Btu of that Fuel}} = \text{Cost per 1 million Btu}$$

Figure 14-1 shows a comparison of the cost per million Btu for various heating fuels. **Figure 14-2** compares these fuels and their prices when used with various boiler efficiencies.

Once the fuel type has been selected, the fuel efficiency ratings can be compared. Keep in mind that a higher efficiency boiler may cost more than one with a lower AFUE rating. Also, selecting a boiler with an efficiency rating of higher than 90% (condensing boiler) will come with special venting requirements compared to a lower AFUE model. This can be an important consideration when upgrading an existing system.

Another consideration when choosing a boiler based on efficiency is the long-term goals for the home or building being heated. If the home or building is an older structure, or the owner is considering upgrading to a newer space, a lower efficiency model may be a more economical decision. However, if the building is new construction, a higher efficiency model may be the right choice based on the equipment's life-cycle costs. Once these questions have been answered, comparisons can be made based on the payback rate of the new equipment. As

Cost Comparison of Common Heating Fuels				
Fuel	Unit	Btu per Unit	Unit Cost	Cost per Million Btu
Natural gas	100 cubic feet	105,000	$0.52	$4.95
Liquid petroleum	Gallon	92,000	$2.25	$24.46
Fuel oil	Gallon	140,000	$3.55	$25.36
Electricity	KWH	3412	$0.15	$43.96

Goodheart-Willcox Publisher

Figure 14-1. This table shows examples of various fuel costs per million Btu.

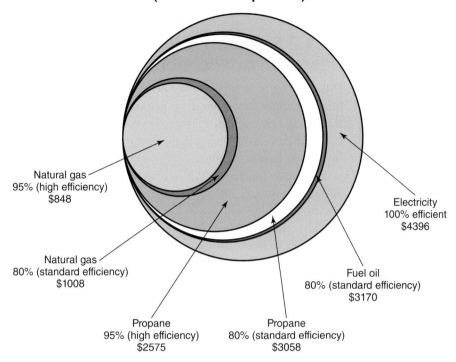

Figure 14-2. Calculating the cost per million Btu of various heating fuels can help the owner or designer choose which boiler is best for their application.

an example: A homeowner has an older boiler with an AFUE rating of 65%. They plan on selling the house within five years. Here are the costs of a new boiler:
1. AFUE rating: 95%; cost to install: $6800
2. AFUE rating: 90%; cost to install: $4500
3. AFUE rating: 85%; cost to install: $2200

If the average monthly heating bill is $150 per month, here are the average monthly savings based on efficiency:
1. 95% rated boiler − 65% current boiler = 30% potential savings per month
   $150 heating bill × 30% savings = $45.00 per month
2. 90% rated boiler − 65% current boiler = 25% potential savings per month
   $150 heating bill × 25% savings = $37.50 per month
3. 85% rated boiler − 65% current boiler = 20% potential savings per month
   $150 heating bill × 20% savings = $30.00 per month

Now we can calculate the annual fuel savings and divide these savings by the cost of the boiler to predict the estimated return on investment:
1. 95% rated boiler: $45.00 per month × 12 months = $540.00 per year savings
   Cost to install: $6800 ÷ $540 = 12.6 years payback period
2. 90% rated boiler: $37.50 per month × 12 months = $450.00 per year savings
   Cost to install: $4500 ÷ $450 = 10 years payback period
3. 85% rated boiler: $30.00 per month × 12 months = $360.00 per year savings
   Cost to install: $2200 ÷ $360 = 6.1 years payback period

Based on the above figures, a lower efficiency boiler may be a more economical choice for a homeowner who plans to move within six to seven years. However, a home or building owner who is planning to build new would probably want to go with a higher efficiency model that will provide savings over a longer period of time. Local building code requirements may also come into play. Another consideration is the additional value that the new boiler will add to the building when it comes time to sell.

14.1.2 Selecting a Boiler Based on Application

Different types of boilers may be more suited to a specific heating application. For instance, some heating applications may favor a high-mass boiler over a low-mass boiler. In Chapter 3, *Boilers*, we discussed the differences between high-mass heating applications and low-mass applications.

Low-mass boilers are typically lower cost, high efficiency, and space-saving. They also have a faster recovery time. A low-mass boiler will reach the desired temperature very quickly compared to a high-mass boiler because there is less water to heat. They are more desirable as replacement boilers where baseboard finned-tube heaters are used or when heating a single zone. Low-mass boilers typically have copper or stainless-steel heat exchangers. An example of a low-mass boiler is illustrated in **Figure 14-3**.

High-mass boilers should be considered for use with multiple zones or for the replacement of a conventional cast-iron boiler. A high-mass boiler can take advantage of an outdoor reset controller and handle multiple zones. It can also generate domestic hot water. High-mass boilers often have cast-iron heat exchangers and are quite durable. An example of a high-mass boiler is illustrated in **Figure 14-4**.

Another consideration when choosing a boiler based on the application is the type of terminal units to be used. As discussed in Chapter 12, *Radiant Heating Systems*, radiant floor heating systems typically use lower water temperatures compared to other types of terminal units. Therefore, selecting a higher efficiency condensing boiler may be preferred with this type of application.

Other considerations include:
- **Venting.** Not all boilers are vented equally. When utilizing a masonry chimney, the medium-efficiency boiler will require the installation of a flue liner. High-efficiency condensing boilers cannot be directly vented into a masonry chimney. They require the use of PVC or polypropylene (PP) material for proper venting.
- **Sizing.** Selecting a boiler that is oversized for the application will lower the efficiency, waste fuel, and cause it to short cycle (turning on and off too quickly). If the desired boiler is only offered by the manufacturer in a larger size, consider installing a 2-stage boiler or perhaps one with modulating controls.
- **Combining space heating with generating potable hot water.** Some of today's modern boilers offer the added feature of domestic hot water generation. This feature allows the building owner to virtually eliminate a conventional water heater. This topic is covered at length in Chapter 18, *Domestic Hot Water Production*.

Goodheart-Willcox Publisher

Figure 14-3. This image shows a low-mass boiler with a copper heat exchanger.

Bosch Thermotechnology

Figure 14-4. Shown here is an example of a high-mass boiler.

14.2 Combustion and Ventilation Air Requirements

As outlined in previous chapters, air is required for the combustion process. And with any gas-fired appliance, the correct amount of combustion air is essential to the proper functionality of the equipment. Always consult the boiler installation and maintenance guide for combustion air requirements and procedures for the introduction of combustion air into the mechanical room. Questions to consider when determining the proper amount of combustion air for the new boiler include:
- Is the boiler a single unit, or will multiple boilers be installed in the building?
- Are there other gas-fired appliances within the mechanical room?
- Is the boiler installation location considered an unconfined space or a confined space?

> **CODE NOTE**
>
> **Proper Combustion and Ventilation Air Requirements**
> Provisions for the proper amount of combustion and ventilation air are governed by Section 5.3, "Air for Combustion and Ventilation," of the National Fuel Gas Code, ANSI Z223.1. In Canada, CGA Standard B149 Installation Code for Gas Burning Appliances and Equipment or applicable provisions of the local building codes must be adhered to.

The first step in calculating the proper amount of combustion and ventilation air is to add the total Btu input of all appliances. Often, a residential boiler will be installed in the same equipment room as a domestic water heater. When this is the case, the Btu input of the water heater must be added to the Btu input of the boiler to calculate the necessary amount of combustion air. Next, determine whether the boiler room is considered an unconfined or confined space.

14.2.1 Unconfined Space Requirements

In some instances, the mechanical or boiler room may have enough space to satisfy the combustion air requirements for the gas-fired appliances installed in that room. According to NFPA 31, 54, and 58, an *unconfined space* is defined as a space that has at least 50 cubic feet of open area for every 1000 Btu of input. For example, this means that a 100,000 Btu/hr boiler must be located in a space that contains 5000 cubic feet to be considered unconfined.

> **TIMEOUT FOR MATH**
>
> **Confined or Unconfined?**
> A mechanical room houses a 50,000 Btu/hr boiler and a 40,000 Btu/hr water heater, **Figure 14-5**. What size would the mechanical room need to be for it to be considered an *unconfined space*?
> Answer:
> $$50{,}000 + 40{,}000 = 90{,}000 \text{ Btu/hr}$$
> $$90{,}000 \div 1000 = 90$$
> $$90 \times 50 = 4500$$
> The mechanical room would need to be at least 4500 cubic feet in size to be considered an unconfined space.

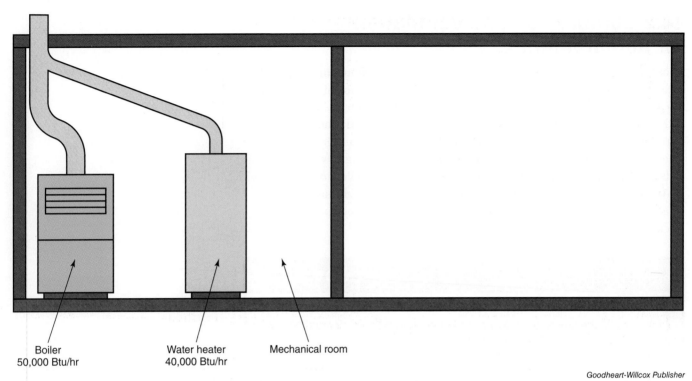

Figure 14-5. This mechanical room would need to be at least 4500 cubic feet in size to be considered an unconfined space based on the size of the appliances.

14.2.2 Confined Space Requirements

If the mechanical room is not large enough to be considered an unconfined space, it is a ***confined space*** and additional combustion air must be made available to the room. Combustion air can be introduced in several ways:

- From an interior space
- Directly from outside
- Through ducts

To take combustion air from another interior space that is adequately ventilated, create one opening at the ceiling level and one near the floor. Both openings should be at least one square inch per 1000 Btu of *free area*, but not less than 100 square inches total, **Figure 14-6**.

Note: Free area is the sum of the areas of all the spaces between the bars or fins of a grille opening measured in square inches. The grille manufacturer provides this information.

To take combustion air directly from outside the building with no ducting, create two permanent openings, each with a minimum free area of one square inch per 4000 Btu/hr. These openings must be located within 12″ from the bottom and 12″ from the top of the mechanical room, **Figure 14-7**.

To take the combustion air from the outdoors using a duct into the boiler room, create two openings on the outside of the building sized a minimum of one square inch per 2000 Btu/hr input of free area, with both openings spaced 12″ from the top and bottom of the room, **Figure 14-8**.

Note: If only a single opening can be provided to bring combustion air directly from the outdoors, make sure that the ducted opening size is a minimum of one square inch per 3000 Btu/hr input of free area. This opening must be located within 12″ of the top of the equipment room.

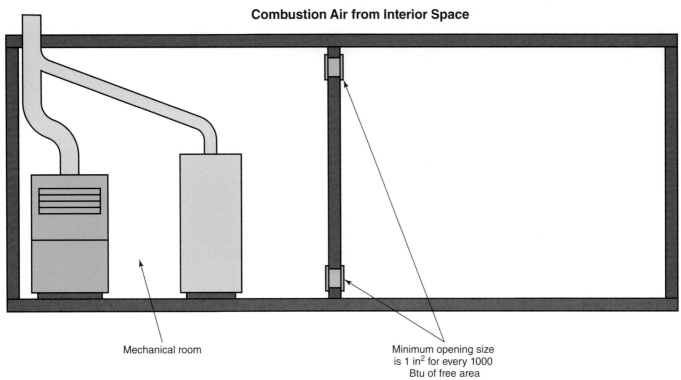

Figure 14-6. If the combustion air is taken from another interior space that is adequately ventilated, provide one opening at the ceiling level and one near the floor.

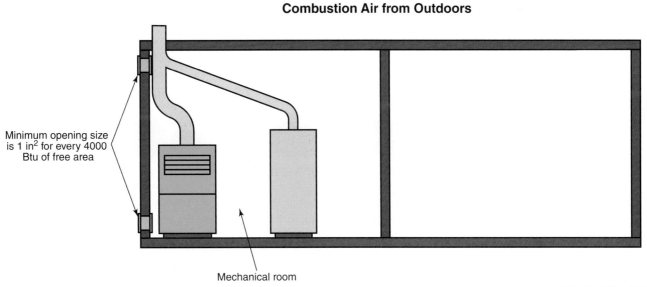

Figure 14-7. If combustion air is taken directly from outside of the building with no ducting, openings must be located within 12″ from the bottom of the mechanical room and 12″ from the top of the room.

On larger commercial applications, combustion air for the boiler is usually drawn in from outside using an opening through the boiler room wall. This opening must be sized based on the size of the boiler and any other gas-fired appliances within the space. Such an opening is usually closed off by a vent damper that is activated whenever the boiler calls for heat. It is crucial to provide an

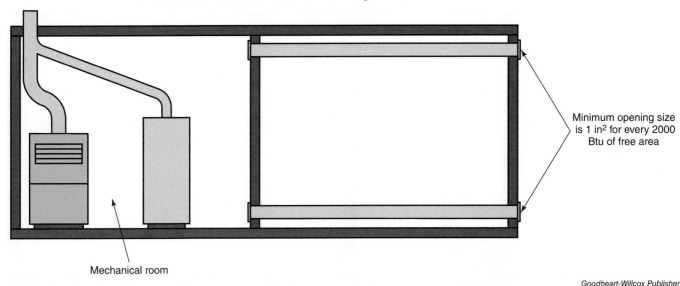

Figure 14-8. If the combustion and ventilation air is taken from the outdoors using a duct into the boiler room, provide two openings on the outside of the building.

interlocking end switch with this damper to ensure that the damper is fully open before allowing the boiler to energize, **Figure 14-9**.

Since combustion air openings are a requirement for conventional boilers with efficiency ratings below 90% AFUE, it is understandable why modern sealed combustion or ***direct vent boilers*** are so advantageous. Not only do these high-efficiency condensing boilers offer more fuel savings, they also receive all of their combustion air from the outdoors through a dedicated sealed opening, **Figure 14-10**. Keep in mind that most of these types of boilers require approved ventilation material, usually stainless steel, PVC, or polypropylene piping.

Note: Never mix different venting materials. Doing so could result in venting system failure, causing leakage of flue products into the building.

CODE NOTE

Combustion Air Venting Installation

Installation of combustion air venting must comply with local requirements and with the National Fuel Gas Code, ANSI Z223.1 for US installations and CSA B149.1 for Canadian installations. Inspect finished vent and air piping thoroughly to ensure all are airtight and comply with the instructions provided with the boiler as well as all requirements of applicable codes.

GREEN TIP

High-Efficiency Boilers and Sealed Combustion

Direct vented, high-efficiency boilers offer an additional advantage beyond increased fuel savings: sealed combustion. Because combustion air is drawn directly into the burner chamber from outdoors, these boilers prevent negative pressures that can occur in the home or building during burner operation of low- and medium-efficiency boilers. This reduces thermal losses directly associated with placing vents in walls and ceilings.

Goodheart-Willcox Publisher

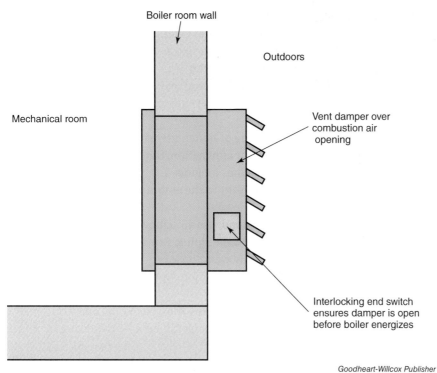

Figure 14-9. On larger commercial applications, combustion air for the boiler is usually drawn in from outside using an opening through the boiler room wall.

Goodheart-Willcox Publisher

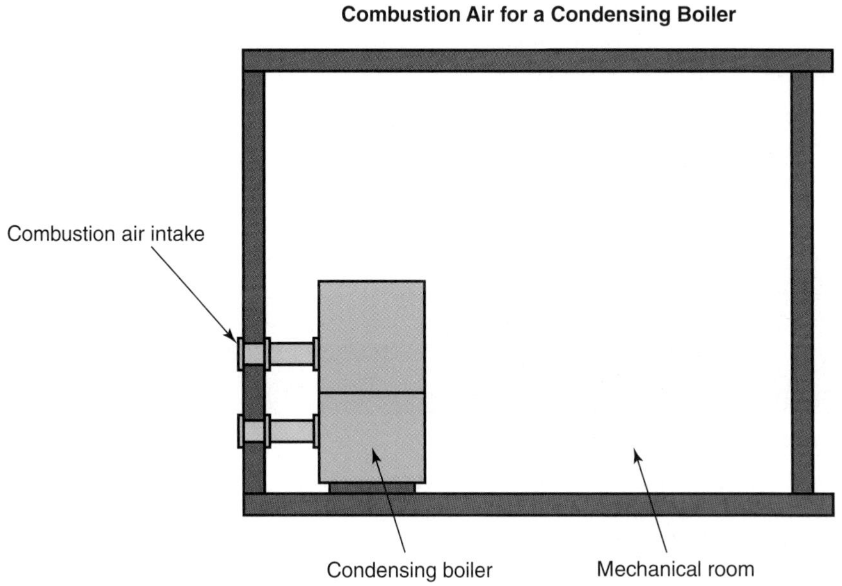

Figure 14-10. Condensing boilers have sealed combustion with separate intake and vent connections.

14.3 Boiler Venting

Proper venting practices are among the most critical elements of effective boiler installation. Products of combustion contain carbon monoxide, which can be lethal to humans if exposed over a prolonged period of time. The acceptable method of venting the boiler is based on its configuration and efficiency. Other factors that influence boiler venting include whether the installation is in conjunction with new construction or if the project involves retrofitting an old boiler.

The vertical venting of boilers involves the vent pipe or chimney passing through the roof of the building. Horizontal venting consists of a vent pipe passing through a side wall of the building. The size, location, and type of vent piping material varies depending on several factors:

- Whether the boiler is being retrofitted and using the existing venting system
- Whether the boiler uses a natural draft venting system or a power draft venting system
- Whether the boiler is condensing or noncondensing

Regardless of the type of boiler or application, the general guidelines for venting must be followed. These guidelines include:

- The connection from the boiler vent to the common vent or chimney should be as short as possible.
- Avoid the use of 90° elbows. Instead, use 45° elbows wherever possible.
- Properly support any horizontal vent piping. Consult the manufacturer's installation guide for support intervals.
- Horizontal vent runs must have a minimum upward slope of 1/4″ per foot of run.

Boiler venting is divided into four categories based on operating pressure inside the vent and whether the boiler is classified as condensing or noncondensing. These categories are as follows:

- **Category I boilers** are noncondensing appliances that operate with a nonpositive vent pressure. They may be natural or induced draft type appliances.
- **Category II boilers** are condensing appliances that operate under a negative pressure. These types of boilers are obsolete and no longer manufactured.
- **Category III boilers** are positive pressure, noncondensing appliances. These are direct sidewall vented without any additional apparatus. Flue pipes that operate under positive pressure require sealed connections and construction from noncorrosive materials.
- **Category IV boilers** are positive pressure, condensing appliances. This category applies to high-efficiency 90+ AFUE boilers. Because they are classified as condensing appliances, noncorrosive material such as PVC or PP piping must be used and all venting joints must be sealed.

> **CODE NOTE**
>
> **Venting Requirements**
> Vent installations for connection to gas vents or chimneys must be in accordance with Part 7, "Venting of Equipment," of the National Fuel Gas Code, ANSI Z223.1 in the United States and CGA Standard B149 Installation Code for Gas Burning Appliances and Equipment or applicable provisions of the local building codes in Canada.

14.3.1 Venting Requirements for Natural Draft Boilers

Natural draft boilers typically have an AFUE rating between 70% and 75%. These types of boilers use the natural buoyancy of heated air to exhaust the products of combustion from the boiler heat exchanger through the chimney rather than any mechanical means of venting. They usually include a *draft hood* as part of the venting system to entrain room air into the venting system, which mixes with the products of combustion, **Figure 14-11**.

Natural draft boilers can typically be vented directly into a masonry chimney without the need for any flue liner, as long as the chimney has been inspected and deemed structurally sound. A single-wall, galvanized vent pipe can be used from the boiler into the chimney as long as it meets gauge size requirements. However, most of these types of boilers use a double-wall **Category B-vent** instead of a masonry chimney for venting the products of combustion, **Figure 14-12**. Type B vents are factory-built, double-wall vent pipes that are used only for venting gas. They are always made with a galvanized exterior and an aluminum interior. Whenever natural draft venting must pass through a wall, ceiling, or roof, a Category B-vent is required. Additionally, the vent must adhere to the proper clearance from combustible material—a minimum of one inch per code.

Natural draft boilers are categorized as **negative draft** appliances and fall under Category I venting. This means the flue stack must have a negative differential pressure in relation to the boiler room in order for proper venting to take place. In a typical installation, the negative draft must be within a range of –0.02 to –0.05 in. WC to ensure proper operation. The draft reading should be made when the boiler is in a steady state of operation, usually 2 to 5 minutes after startup. If other gas-fired appliances—such as a water heater—are to be vented together, follow the approved venting tables for sizing the vent as outlined in the National Fuel Gas Code or the CGA Standard B149 Installation Code for Gas Burning Appliances and Equipment.

Goodheart-Willcox Publisher

Figure 14-11. This is an example of a natural draft boiler showing the draft hood.

Goodheart-Willcox Publisher

Figure 14-12. Shown here is an example of double-wall Category B-vent piping.

14.3.2 Venting Requirements for Induced Draft Boilers

Induced draft boilers include an inducer fan provided by the manufacturer to vent the products of combustion. By definition, an inducer fan draws or pulls the products of combustion through the heat exchanger, as opposed to a combustion blower, which pushes the flue gases through the boiler's heat exchanger. Induced draft boilers typically have an AFUE rating between 80% and 85% and are considered Category I appliances. They use metal venting with the proper metal gauge thickness and are sized according to the Btu output of the boiler. The same rules apply to induced draft boilers as for natural draft boilers with regard to single-wall and double-wall venting materials, meaning a Category B-vent is required whenever the vent passes through a wall, ceiling, or roof. One major difference is that induced draft boilers are not allowed to be vented directly into a masonry chimney without the use of a *flue liner*, **Figure 14-13**. These liners are made of corrugated stainless steel or aluminum and are connected directly to the boiler. They are routed through the existing chimney and vented directly outdoors, **Figure 14-14**. Flue liners are needed due to flue gas temperatures. Conventional natural draft boilers have flue gas temperatures as high as 500°F, whereas the flue gas temperature of induced draft boilers ranges between 300°F and 400°F. As flue gas temperatures fall, the potential for moisture condensing out of these gases rises. This condensation is highly acidic and can damage the chimney as well as the vent piping and heat exchanger—shortening the life of the boiler.

mipan/Shutterstock.com

Figure 14-13. A flue liner is made of corrugated stainless steel or aluminum.

14.3.3 Venting Requirements for Condensing Boilers

By definition, a condensing boiler has an AFUE rating above 90% and is considered a Category IV appliance. Because they have such high efficiencies, these types of boilers have the ability to condense water out of the flue gases. Since this condensate is very acidic, the use of metal venting for condensing boilers

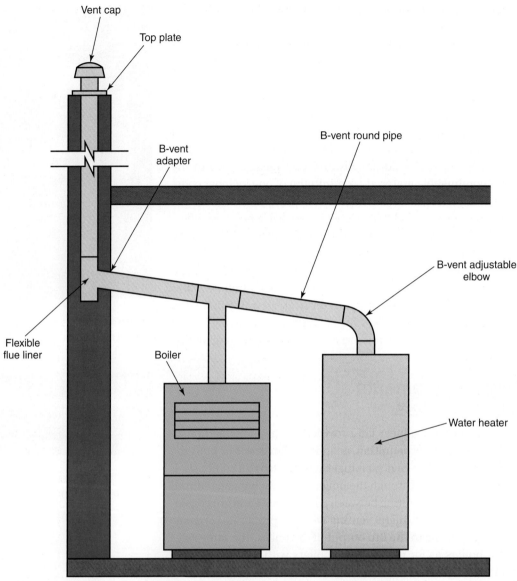

Figure 14-14. Shown here is an example of how a flue liner is routed through a masonry chimney.

is prohibited because it is prone to corrosion. Approved venting materials for condensing boilers include PVC, CPVC, polypropylene, and stainless steel. Condensing boilers provide sealed combustion chambers that allow combustion air to be piped directly into the boiler, as shown in **Figure 14-10**. Follow the manufacturer's installation instructions for the proper methods of cutting and joining these types of materials.

14.4 Gas Piping

The first step in properly piping gas is to verify whether the boiler fuel type is natural gas or propane. Most boilers ship from the factory designed for use with natural gas. When propane is required, the following steps must be followed to convert the boiler from natural gas to propane:

1. The burner *gas orifices* (spuds) must be changed to match the boiler capacity, **Figure 14-15**.

2. The gas pressure regulating spring located inside the gas valve must be changed.
3. The inlet gas pressure must be adjusted at startup to meet the propane requirements.

Goodheart-Willcox Publisher

Figure 14-15. An example of gas burner spuds showing the orifice. These must be changed when converting the boiler from natural gas to propane.

SAFETY FIRST

What Could Possibly Happen?
It is extremely important that the proper steps be performed when converting a natural gas boiler to propane. Because propane has a higher Btu content and operates at a much higher inlet pressure, the burner orifice sizes are smaller than those used for natural gas. If the natural gas spuds are not converted and propane is used on a natural gas boiler, the higher gas pressure would result in excess soot buildup in the heat exchanger and eventually cause boiler failure.

14.4.1 Gas Pipe Sizing

Ideally, the boiler would have a dedicated gas line from the meter. However, in most homes and commercial buildings, other gas-fired appliances are piped along with the boiler. When this is the case, the gas line to the boiler must be sized properly. Sizing tables are typically available from the boiler manufacturer. Other sources for determining gas pipe sizing are the International Fuel Gas Code (IFGC) or B149.1 for Canadian installations.

Gas pipe sizing tables are based on the rated input of the boiler and list capacities in cubic feet of gas per hour. To obtain the cubic feet per hour for a particular boiler, divide the rated input by 1000. For instance, a boiler that has an input rating of 150,000 Btu/hr would be sized based on 150 cubic feet per hour of natural gas (150,000 ÷ 1000). Other factors that need to be known when using gas pipe sizing tables are:

- The inlet gas pressure in pounds per square inch (psi)
- The gas pressure drop in inches water column (in. WC)
- The specific gravity of the gas
- The type of piping material

This information can be obtained from the gas supplier if it is not immediately known. There are also tables available for sizing propane boilers. The same factors listed above apply if using propane instead of natural gas.

One method of sizing gas piping is known as the ***longest length method***. It is considered a conservative approach to sizing, but it can be very effective. The longest length method applies the maximum operating conditions by setting the length of pipe to size any given part of the system to the maximum value. The following example outlines how this method is used. For this example, we will reference **Figure 14-16** and Table 402.4(1) from the International Fuel Gas Code (2009) in **Figure 14-17**.

Steps for Calculations:
1. Divide the system into appropriate sections.
2. Measure the length in feet between each section.
3. Determine the Btu input for each appliance. In this example, the boiler has a Btu/hr input of 120,000.
4. Determine the total Btu requirement for the overall length (add the amount of each appliance). In this example, the total Btu requirement is 240,000 Btu/hr.

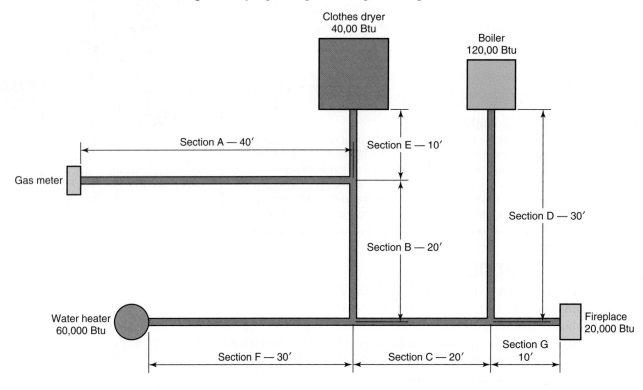

Figure 14-16. This illustration shows how to size gas piping based on appliance size in Btu.

5. Calculate the cubic feet per hour (CFH) to determine demand. In this example, total demand is 240,000 Btu ÷ 1000 = 240 CFH.
6. Use Table 402.4(1) (as shown in **Figure 14-17**) to determine the proper pipe size for each appliance.
   a. Begin by calculating the proper pipe size for Section A in **Figure 14-16**. Section A takes into account all gas fired appliances and extends from the gas meter to where the clothes dryer branches off. It must be capable of delivering 240 CFH of gas over 110′.
   b. In the left-hand column of In Table 402.4(1), locate lengths of 100′ and 125′ within the section Length of Pipe in Straight Feet. Always choose the larger amount—in this case 125′.
   c. Read down until this line intersects with the closest value to 240 or higher. (In this instance, use 269 CFH.)
   d. See left-hand column Pipe Size (Inches) to determine proper pipe size (269 CFH at 125′ requires a pipe size of 1 1/4″). This is the size of gas pipe that will be used from the meter to the intersection of the first appliance (the clothes dryer) **(Section A)**.
   e. Now work from left to right, continuing from the intersection of the clothes dryer (Section A) to the intersection of the water heater (Section B) in **Figure 14-16**, using the same method.
   f. The next calculation will be used to size Section B of pipe. (This will be the total Btu requirement minus the clothes dryer—40,000 Btu.)
   g. Continue to use the 125′ line for each calculation. Add the total Btu of each group of remaining appliances, round up the total to the highest figure referenced in the chart, and determine the required pipe size.

In this example, the boiler will require an inlet gas pipe size of 1″ **(Section D)**.

Pipe Capacity (Cubic feet per hour)								
Pipe Size (inches)	Length of Pipe (feet)							
	10	20	30	40	50	100	125	150
1/2	131	90	72	62	55	—	—	—
3/4	273	188	151	129	114	79	70	63
1	514	353	284	243	215	148	131	119
1 1/4	1060	726	583	499	442	304	269	244
1 1/2	1580	1090	873	747	662	455	403	366
2	3050	2090	1680	1440	1280	877	777	704

Goodheart-Willcox Publisher

Figure 14-17. This piping table—Table 402.4(1) from the International Fuel Gas Code (2009)—is used to calculate gas pipe sizing.

Chapter Review

Summary

- Boiler selection can be made based on fuel type, efficiency, and applications.
- All fossil fuel boilers require the correct amount of combustion air for proper functionality.
- Determine whether the boiler will be located in a confined space or unconfined space. If the boiler is to be installed in a confined space, the correct amount of combustion air must be introduced either from an adjacent space or outdoors.
- Proper venting practices are critical elements to an effective boiler installation.
- Boiler venting is divided into four categories based on the operating pressure inside the vent and whether the boiler is condensing or noncondensing.
- Natural draft boilers are considered negative-draft, Category I appliances. They are noncondensing and can be vented directly into a masonry chimney.
- Induced draft boilers use a combustion blower to assist in venting the products of combustion. They are also considered negative-draft, Category I appliances.
- When venting natural draft and induced draft boilers through a wall or roof penetration, double-wall Category B-venting material must be used.
- A condensing boiler has an AFUE rating above 90% and is considered a Category IV appliance. High-efficiency condensing boilers must use PVC or similar venting material because of the acidic nature of the condensate. It can be vented through a side wall or the roof as long as the proper code requirements are followed.
- The proper steps must be followed when converting a natural gas boiler to propane.
- Gas pipe sizing takes into account the number of appliances serviced, the Btu input of each appliance, and the overall length of the piping system.
- An effective method of sizing gas piping is known as the longest length method.

Know and Understand

1. Which type of fuel is sold in hundred cubic foot increments?
   A. Fuel oil
   B. Natural gas
   C. Liquified petroleum
   D. Electricity
2. What does the acronym AFUE stand for?
   A. Annual Fuel Utilization Economy
   B. Annual Fuel Usefulness Efficiency
   C. Annual Fuel Utilization Efficiency
   D. American Fuel Usefulness Economy
3. *True or False?* One method to select a boiler based on efficiency is to perform a return on investment calculation.
4. One good method of comparing the use of different types of heating fuels is to calculate their cost _____.
   A. per million Btu
   B. per thousand Btu
   C. per hundred Btu
   D. per gallon
5. By code, an unconfined space is defined as a space that has at least _____ cubic feet of open area for every _____ Btu of input.
   A. 55, 100
   B. 50, 1000
   C. 500, 1000
   D. 500, 100

6. What size would a mechanical room need to be for it to be considered an unconfined space if it contained a 40,000 Btu water heater and a 100,000 Btu boiler?
   A. 7000 cubic feet
   B. 8000 cubic feet
   C. 9000 cubic feet
   D. 12,000 cubic feet
7. If boiler combustion air is taken from openings to another interior space, both openings should be at least _____ square inch(es) per _____ Btu of free area.
   A. 2, 250
   B. 3, 500
   C. 1, 1000
   D. 4, 1200
8. Horizontal boiler venting runs must have a minimum upward slope of _____ per foot of run.
   A. 1/4″
   B. 1/2″
   C. 3/4″
   D. 1″
9. Which category is a positive pressure, noncondensing type of boiler?
   A. Category I
   B. Category II
   C. Category III
   D. Category IV
10. *True or False?* Condensing boilers can be vented directly into a masonry chimney without the need for any flue liner, as long as the chimney has been inspected and deemed structurally sound.
11. *True or False?* Approved venting materials for condensing boilers include PVC, CPVC, polypropylene, and galvanized steel.
12. When should a high-mass boiler be selected?
    A. When using baseboard finned-tube heaters
    B. When heating a single zone
    C. When replacing a conventional cast-iron boiler
    D. When needing to save space

Apply and Analyze

1. Explain the difference between a confined and an unconfined space with regard to combustion air requirements.
2. Describe how combustion air can be introduced into the boiler room if it is considered a confined space.
3. List and describe the four boiler venting categories.
4. Explain the difference between a natural draft and an induced draft boiler.
5. Under what conditions must a Category B-vent be used for natural draft and induced draft boilers?
6. What is a flue liner, and when must it be used for venting induced draft boilers?
7. Explain how a condensing boiler differs from a natural draft and induced draft boiler.
8. What venting category does a condensing boiler fall under and why?
9. Describe the steps that must be taken to convert a natural gas boiler to propane.
10. Why is venting an important issue when it comes to boiler selection?

Critical Thinking

1. What circumstances would justify installing a low-efficiency boiler over a high-efficiency boiler?
2. What problems arise when a boiler lacks adequate combustion air?

15 Boiler Installation

Chapter Outline
15.1 Boiler Installation
15.2 Boiler Preparation
15.3 Hydronic Piping
15.4 Gas Pipe Installation
15.5 Condensate Disposal
15.6 Field Wiring
15.7 Cascade Boiler Operation

Roman Zaiets/Shutterstock.com

Learning Objectives

After completing this chapter, you will be able to:
- Outline the steps for proper boiler installation, including codes and safety procedures.
- Describe what steps should be taken when preparing a boiler for installation, including boiler locations and proper clearances.
- Describe what the 10-2 rule is and how it is applied to vent terminations.
- Explain the advantages of using a primary-secondary piping arrangement over conventional piping arrangements.
- Describe the proper procedures for various pipe joining methods.
- Identify code-compliant piping materials for the gas supply to the boiler and explain how they should be properly joined together.
- Describe the proper method of condensate disposal for condensing boilers.
- Compare and contrast how to wire the various low-voltage components to the boiler.
- Identify functions a building management system can provide when interfaced with the boiler.
- Describe the advantages of using a cascade boiler system versus a larger individual boiler.

Technical Terms

10-2 rule
authority having jurisdiction
building management system (BMS)
capillary action
cascade boiler system
combustible floor base kit
concentric vent kit
condensate
condensate removal pump
hose bib
housekeeping pad
LonWorks
ModBus
near-boiler piping
neutralizing filter

The first rule when installing a boiler—whether for residential or commercial use—is to read the manufacturer's installation and maintenance manual. This guides the safe and accurate installation of a hot water boiler and its usage. The boiler installation should only be performed by a qualified professional installer. This may require the installer to have the proper licensing and certification depending on state and local codes. Many problems arise as a result of improper boiler and boiler component installation. Therefore, the installer should be aware of all federal, state, and local piping and mechanical codes in order to perform a safe and legal hydronic system installation.

15.1 Boiler Installation

Boiler installations fall into two main categories:
- New construction
- Boiler retrofits

With the construction of a new residence or commercial building, there are no previously installed components to remove or replace. This allows for easier accessibility and flexibility of the boiler installation and its corresponding piping. When retrofitting the replacement of an aging boiler, the installer needs to be aware of the configuration and condition of existing piping, **Figure 15-1**. In addition, the existing hydronic control and safety components should be examined to see if they need to be replaced. Because boilers vary in size, venting requirements, and temperature control requirements, it is important for the installer to perform a preliminary review of the location where the boiler is to be installed. Safety, codes, and the warranty should also be reviewed by the installation technician before proceeding with the hydronic system installation.

314 Hydronic Heating: Systems and Applications

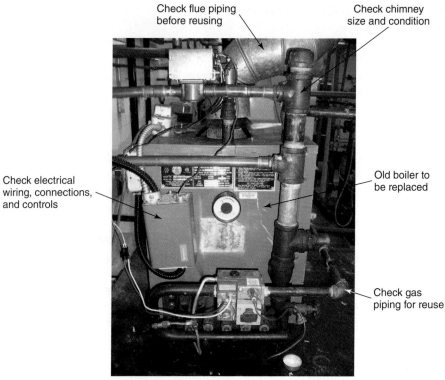

Figure 15-1. With new construction, there are no previously installed boiler components to remove or replace. When retrofitting a boiler, the installer needs to be aware of the configuration and condition of existing piping and electrical wiring.

15.1.1 Safety

Safety practices are discussed in depth in Chapter 2, *Safety*. When installing a new boiler, the manufacturer's installation and maintenance guide will outline the necessary safety steps, **Figure 15-2**. These guidelines include, but are not limited to, the following:

- Consult and abide by local building and fire regulations and other applicable safety codes.
- Contact the local gas utility company to inspect and authorize all gas and flue connections.
- Do not install the boiler outdoors or anywhere it will be exposed to freezing temperatures or to temperatures that exceed 100°F.
- To avoid property damage and personal injury, do not store materials against the appliance or the ventilation air intake system. Never cover the boiler, lean anything against it, store trash or debris near it, stand on it, or block the flow of combustion air to it.
- Under no circumstances must flammable materials such as gasoline be used or stored in the vicinity of the boiler or its corresponding combustion air intake.

15.1.2 Codes

Because of the complexity of boiler installations, it is important that the installer review and follow all local, state, and national codes. In this chapter, the appropriate code requirements are highlighted according to the installation procedure being covered. With most boiler installations, the latest editions of the following recognized code requirements shall be adhered to:

- The International Fuel Gas Code (IFGC). See **Figure 15-3**.

Figure 15-2. When installing a new boiler, the manufacturer's installation and maintenance guide will outline the proper safety steps that need to be followed.

- The International Mechanical Code (IMC)
- ANSI Standards; and for Canada, the CAN/CGA-B149
- ASME Boiler and Pressure Vessel Code and ASME CSD-1

In some cases, it may be necessary to consult the local *authority having jurisdiction* before proceeding with the boiler installation to ensure that local codes are adhered to.

15.1.3 Warranties

Before the boiler is installed, the technician should first inspect all components and equipment to ensure there is no damage. If there are signs of damage, the manufacturer or shipping company should be contacted before proceeding with the installation. When the installation is complete, be sure to fill out and send in the manufacturer's warranty application, **Figure 15-4**. This is a critical part of the overall boiler installation. The boiler's warranty may be voided if the appliance was installed improperly. In addition, the following items can affect a boiler's warranty:

- Excessive water hardness, which causes the buildup of scale
- Erosion of heat exchanger tubes caused by excessive water velocity through the boiler
- Damage due to contamination of the combustion air by such things as dust, dirt, lint, or corrosive chemicals

15.2 Boiler Preparation

Whether in a residential or commercial building, one of the first considerations when installing a boiler is the utility connections. If the project involves a retrofit, inform the building owner that the building may be without electricity or gas for an extended period of time. This consideration can be even more crucial depending on the season of the year; therefore, advanced planning is needed before the existing boiler is removed. However, this may not be as much of a critical issue for new construction, as installation planning typically is done well in advance.

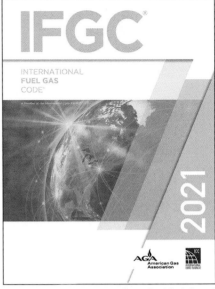

International Code Council

Figure 15-3. Code books such as the International Fuel Gas Code need to be reviewed by the installer to ensure proper boiler installation.

Fischer Boiler Warranty Registration Card

Name: _____

Street: _____

City: _____ State: _____ ZIP: _____

Phone: _____ Email: _____

Model #: _____ Serial #: _____

Installer Name: _____

Installer Phone: _____ Installation Date: _____

Mail this warranty card to

**Fischer Boiler Company, Inc.
206 Cornell Drive
Rueth, IL 60323**
or register online at the Fischer Boiler website

Goodheart-Willcox Publisher

Figure 15-4. Be sure to review the boiler manufacturer's warranty to identify any items that may not be covered.

Housekeeping pad

Goodheart-Willcox Publisher

Figure 15-5. A concrete pad, also known as a housekeeping pad, is used to keep debris and water away from the boiler.

Alhim/Shutterstock.com

Figure 15-6. When choosing the boiler location, considerations must be made for the water piping, gas piping, vent connections, and electrical connections.

Another consideration after selecting the placement of the new boiler is the installation of a *housekeeping pad*, **Figure 15-5**. This item is usually a concrete pad on which the boiler is placed, typically at least 4 inches high. The housekeeping pad is used to keep debris and water away from the boiler—an important addition if the boiler room has the potential for flooding. If this is the case, the installer should consider connecting water level detection switches around the boiler in the event the room should become flooded with water. These switches not only can automatically shut down the boiler, but also alert the building owner or maintenance personnel if there is an emergency.

15.2.1 Boiler Location

The most obvious location for a residential or commercial boiler is in the designated mechanical room. However, even in this location, issues such as piping layout and utility connections need to be considered, **Figure 15-6**. If the project involves retrofitting an existing boiler, the installer should consider how to position the boiler so as to take advantage of the existing water piping, gas piping, vent connections, and electrical connections. Other replacement boiler recommendations include:

- Check for and repair any system leaks in the hot water piping. Such leaks can lead to oxygen corrosion and heat exchanger cracks from hard water deposits.
- Determine whether the expansion tank is properly sized. If there is any doubt, or if the condition of the expansion tank is less than desirable, replace it.
- Check for any debris left inside the existing piping. If found, make sure the system is properly cleaned and flushed before making any new connections.
- Determine if any system freeze protection is required and whether the existing system needs to be flushed and refilled with the appropriate amount of antifreeze.

If the project involves replacing an existing larger boiler with multiple smaller boilers, floor space may be a consideration as well as the connecting of existing water and gas piping. Other items that need to be considered when determining the boiler location include the following:

- The boiler should be installed in an area that will prevent water damage should a leak occur in the water piping. If there is any doubt about the selected location, it is recommended that a suitable drain pan be installed under the boiler. This pan should be piped to an appropriate floor drain.
- The area around the boiler must be free of combustible materials. This includes gasoline, flammable paints, or any other flammable liquids. Vapors from these types of materials can ignite and cause a fire, resulting in substantial property damage, personal injury, or even death.

- The boiler must be installed so that the gas and electrical control components are protected from any dripping or spraying water during its operation and service.
- Ensure that combustion and ventilation openings are not obstructed. Check around the boiler area for any potential air contaminants that could cause corrosion to the boiler or the combustion air supply.

15.2.2 Clearances

When determining the appropriate location for the new boiler, the proper clearances must be adhered to. The boiler's clearances take into account two major issues:
- Clearances from combustible materials
- Clearances for service access

To ensure that proper clearances are maintained for combustible materials, always consult the installation guide that accompanies the boiler. Most boilers require the following clearances around the boiler:
- 6″ of clearance on the right side, left side, and rear of the boiler
- 24″ of clearance on the front of the boiler for service access
- The top clearance can vary based on the Btu output of the boiler but is typically between 15″ and 30″, **Figure 15-7**.

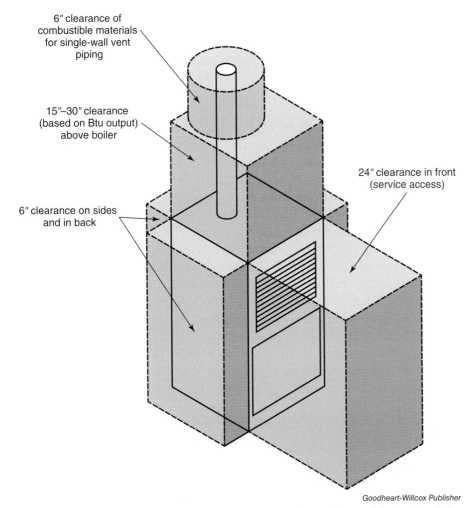

Goodheart-Willcox Publisher

Figure 15-7. Proper clearances from combustible materials and around the boiler will be clarified by the installation instructions.

In addition to boiler clearances, hot water piping and vent piping have their own rules regarding proper clearances:
- Hot water piping requires a minimum of 1/4″ clearance from combustible materials.
- Single-wall vent piping requires a minimum of 6″ from combustible materials.
- Double-wall vent piping requires a minimum of 1″ from combustible materials.

Other rules for clearances include the following:
- Boilers shall not be installed on carpeted floors or floors made up of combustible materials. For installation on a combustible floor surface, a special *combustible floor base kit* must be used. This kit consists of a durable fire-resistant material on which the boiler will be mounted.
- When installing a boiler in a residential garage or in adjacent spaces that open to a garage and are not part of the living space, all burners and burner ignition devices shall have a clearance above the floor of at least 18″.
- With garage installations, ensure the boiler is protected from any damage from motor vehicles.

CODE NOTE

Installation in Garages
If the boiler is to be located in a residential garage, it should be installed in compliance with the latest edition of the National Fuel Gas Code, ANSI Z223.1 and/or CAN/CGA-B149 Installation Code.

15.2.3 Venting Termination for Natural Draft and Induced Draft Boilers

On natural draft boilers, the termination of the venting on the outside of the building must be made with Category B-vent (double wall) or equivalent vent connectors, and there must be no reduction in vent diameter. This venting and all accessories, such as roof jacks, firestop spacers, vent caps, etc., must be installed in accordance with the manufacturer's instructions. When the vent or chimney passes through a flat roof, the minimum termination height is 3′. When the vent or chimney passes through a peaked roof, the *10-2 rule* must be followed: this means that the vent shall be terminated at least 2′ above the highest point of the roof within a 10′ radius of the termination. This vertical termination must be a minimum of 3′ above the point of exit, **Figure 15-8**. If the installation of a natural draft boiler requires a sidewall vent termination, an induced draft fan must be used. This fan must be properly sized and shall be listed by a nationally recognized testing agency. Furthermore, it must meet any local code requirements. Check with the boiler manufacturer for the proper use and installation of an induced draft fan for sidewall applications.:

TECH TIP

Examining the Venting System
Examine the venting system at least once a year. Check all joints and vent pipe connections for tightness. Also check for corrosion or deterioration. Immediately correct any problems observed in the venting system.

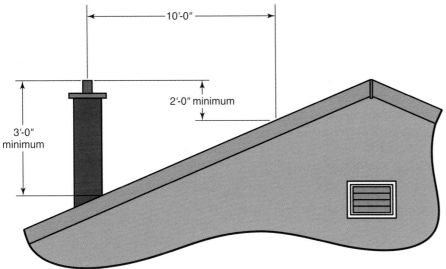

Figure 15-8. The 10-2 rule must be followed when venting a boiler through a pitched roof.

15.2.4 Venting Termination for Condensing Boilers

Condensing boilers can be vented through the roof or sidewall of a structure. Vertical venting can pass through ceilings and roofs as long as the proper clearances are adhered to. In addition, a masonry chimney can be used as a chase or passage for vent piping on a condensing boiler. Follow the same practices as with natural draft and induced draft boilers regarding proper terminations, or review the installation instructions, **Figure 15-9**.

A more common practice for the venting of condensing boilers is a sidewall termination. This practice tends to be less invasive and usually results in shorter pipe runs compared to venting through the roof. There are codes that apply to sidewall venting that the installer needs to be aware of. These include, but are not limited to, the following:

- Allow at least 6′ clearance from adjacent walls.
- Allow at least 3′ clearance above any forced air intake within 10′.

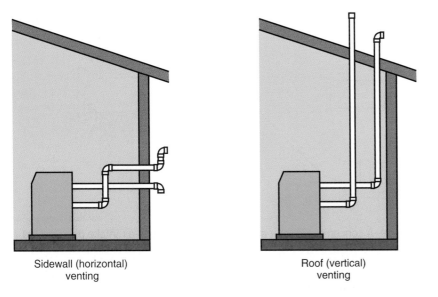
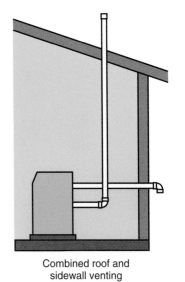

Figure 15-9. Shown here are some examples of how a condensing boiler can be vented.

- Terminate no closer than 12″ horizontally from any door or window or any other gravity air inlet.
- The combustion air inlet must terminate at least 12″ above the grade or snow line.
- Do not terminate the venting closer than 4′ horizontally from any electric meter, gas meter, or any other equipment.

Whether the boiler is natural draft, induced draft, or condensing, the 10-2 rule must be followed when venting through a roof penetration.

> **TECH TIP**
>
> **Installing a Concentric Vent Kit**
>
> When installing a two-pipe system on a high-efficiency boiler, consider using a *concentric vent kit*. This kit allows for the combustion air intake pipe and the exhaust vent to pass through a standard roof or sidewall, as an alternative to the standard two-pipe termination. This means that only one hole is required through the wall or roof where the pipes terminate, **Figure 15-10**. Without the use of these vents, the installation would require cutting two holes through the building, one for each pipe. They save time and money, reducing the amount of work required.

15.3 Hydronic Piping

Most residential and light commercial hot water boilers are designed to function in a closed loop pressurized system of not less than 12 pounds per square inch (psi). When designing the hydronic piping for the boiler, the installer must first determine what type of piping material will be used. Boiler manufacturers may allow for a variety hot water piping materials, including Schedule 40 black pipe—however, the most common type of piping material used is copper.

Begin the hydronic piping installation by locating the inlet and outlet ports of the boiler. The connection marked "inlet" should be used as the return port from

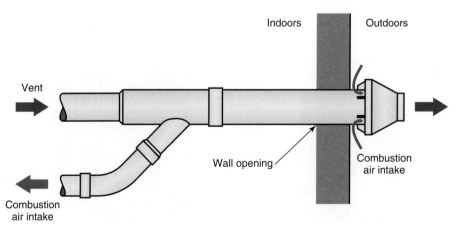

Goodheart-Willcox Publisher

Figure 15-10. A concentric vent kit allows for the combustion air intake pipe and the exhaust vent to pass through a roof or sidewall as an alternative to the standard two-pipe termination.

the system. The connection marked "outlet" should be connected to the supply side of the system. For ease of future service and maintenance, it is suggested that unions be piped to the inlet and outlet ports on the boiler.

The boiler's use and application will determine which piping configuration should be considered. Applications can vary between single-zone usage, multi-zone usage, radiant floor heating, and whether the boiler will be used for domestic hot water production. Many of today's boilers incorporate a primary-secondary piping arrangement, **Figure 15-11**. This type of configuration provides several advantages over conventional piping arrangements, including:

- Protecting the boiler against a lower than desired return water temperature. In noncondensing boilers, a return water temperature below 140°F can result in flue gas condensation that can damage the boiler's heat exchanger.
- The individual secondary circuits can be utilized for different types of heating loads, such as domestic hot water and radiant floor heating.
- Each secondary branch circuit can be individually controlled with a zone thermostat or circulating pump.

Chapter 7, *Hydronic Piping Systems,* describes various piping arrangements and their individual advantages. Always consult the manufacturer's installation and service instructions for choosing the best piping arrangement for your particular application.

15.3.1 Pipe Joining Practices

When using copper tubing for the boiler's hydronic piping, it is important that the technician use the proper tools for cutting and joining. Copper can be joined by soldering or by using press-type fittings.

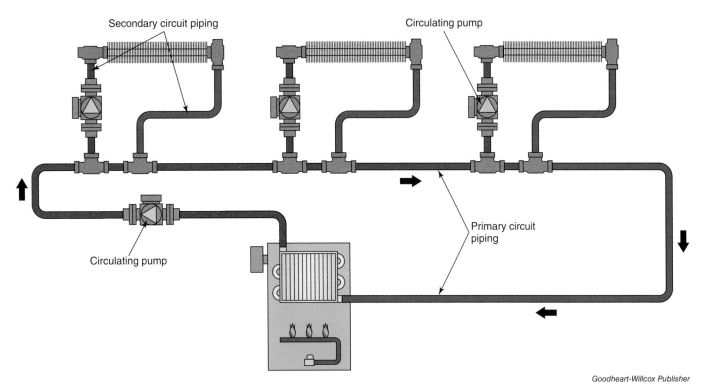

Goodheart-Willcox Publisher

Figure 15-11. This illustration shows an example of a primary-secondary piping arrangement, which is used in most of today's boiler applications.

PROCEDURE

Soldering Copper Tubing

The following are the proper steps for soldering copper tubing:

Cutting: Tubing cutters are used to ensure a square and clean cut, **Figure 15-12**. Avoid using a hacksaw, as it tends to leave a rough edge and cannot always provide a smooth 90° cut.

1. Measure and mark the location where the tubing is to be cut.
2. Align the cutting wheel with the mark.
3. Rotate the feed screw knob while turning the tubing cutter around the mark. Do not apply too much pressure to the feed screw knob while rotating, as this can pinch the tubing and ruin the cut, **Figure 15-13**.
4. After the cut has been made, remove any burrs from the cut using a deburring tool.

Preparing the joint for soldering:

1. Use an abrasive cloth to clean the male end of the tubing to be joined. With a wire brush, clean the female joint. Wipe each joint with a clean cloth.
2. Apply an approved flux to the male end of the tubing and inside the female end of the fitting using a clean brush—do not use your fingers. Avoid using too much flux.
3. Fit the ends together and inspect the joint to confirm that it fits snugly with no gaps between the tubing. Slightly twist the tubing when inserting the ends together, **Figure 15-14**.

Applying heat to the joint:

1. Use an approved gas for heating the joint. Gases that may be used include propane, MAPP gas, and acetylene, **Figure 15-15**. Take care not to overheat the joint!
2. Begin by heating the male end of the joint. Slowly draw the heat toward the female end of the joint and hold the torch steady until the joint is properly heated.
3. Apply solder to the joint, allowing *capillary action* to draw the solder into the joint. This means that the solder will be drawn to the heat source. Do not apply too much solder, **Figure 15-16**.
4. Remove the heat source and wipe the joint with a clean cloth—this will remove any excess solder.
5. Inspect the joint to ensure the solder has completely filled the gap.

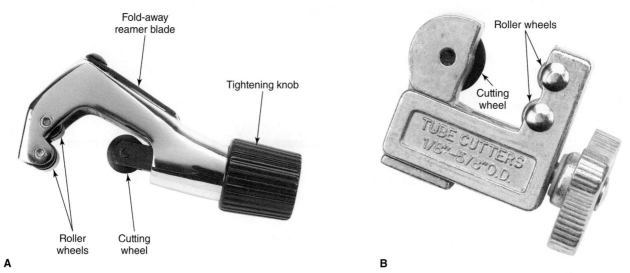

Goodheart-Willcox Publisher

Figure 15-12. These are examples of tubing cutters for copper tubing: A—A wheel-type tubing cutter. B—A mini-tubing cutter.

Aligning the cutting wheel

Tightening the tubing cutter
Uniweld Products, Inc.

Figure 15-13. Rotate the feed screw knob while cutting the tubing by turning the tubing cutter around the mark. Do not apply too much pressure to the feed screw knob while rotating.

Applying flux paste

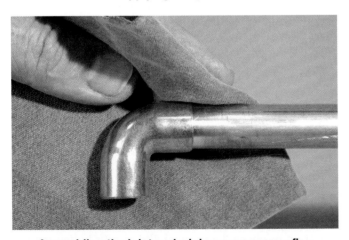
Assembling the joint and wiping away excess flux
Goodheart-Willcox Publisher

Figure 15-14. Prepare the joint to be soldered by cleaning and fluxing the fittings. Make sure that they fit snugly together.

BernzOmatic

Figure 15-15. Gases such as propane and MAPP gas are commonly used for soldering.

Heating the Joint

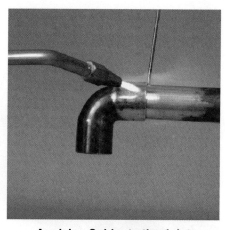
Applying Solder to the Joint

Goodheart-Willcox Publisher

Figure 15-16. Heat the female fitting and apply solder to the male fitting, allowing it to be drawn into the joint by capillary action.

Today, the use of press-type fittings for joining hydronic copper piping has become very popular, **Figure 15-17**. Although the initial cost of installation tools and fittings may be more than conventional methods of joining copper tubing, some installers prefer press fittings for several reasons:

- Press fittings can be joined faster, thus saving time.
- Press fitting connections can be very secure.
- There is no need for an open flame.
- Repairs can be made faster.
- Residual water in the piping system does not affect the process.

> **DID YOU KNOW?**
>
> **MAPP Gas**
>
> MAPP gas is a fuel gas based on a stabilized mixture of methylacetylene (propyne) and propadiene. It is safer and easier to use for soldering than acetylene. The name comes from the original chemical composition, methylacetylene-propadiene propane. MAPP gas was originally produced by Dow Chemical Company, then the Linde Group, and is now a generic product name. True MAPP gas production ended in North America in 2008. Many current products labeled "MAPP" are MAPP substitutes. These versions are composed almost entirely of propylene with small amounts of propane impurities.

PROCEDURE

Using Press-Type Fittings

Following are the required steps for using press-type fittings:

1. Cut the copper tubing at right angles using a quality tubing cutter.
2. Remove any burrs from the inside and outside of the tubing to prevent cutting the sealing rings.
3. Check the seal for the correct fit. Do not use any oils or lubricants.
4. Mark the proper insertion depth of the fitting. An improper insertion depth may result in an improper seal.
5. While turning slightly, slide the press fitting onto the tubing to the marked depth.
6. Insert the compression tool jaws onto the fitting and hold in place, **Figure 15-18**.
7. Start the pressing process by holding the trigger until the jaws have fully engaged the fitting.
8. After pressing, the jaws can be opened again.

Viega, LLC

Figure 15-17. Press-type fittings such as these have become very popular for joining hydronic piping.

15.3.2 Piping and Installation of Boiler Components

Once the piping configuration has been determined, the various boiler components can be installed. Many of these devices are collectively known as ***near-boiler piping*** because they are located on or near the boiler, **Figure 15-19**. Other devices such as zone valves, balancing valves, and pressure differential bypass valves may be located greater distances away from the boiler. However, these devices need to be taken into consideration when planning the overall hydronic piping design, along with the terminal devices and temperature control components. Following is a list of the common boiler components that will be piped into the system and their suggested locations. A detailed description of these components along with their functionality can be found in Chapter 6, *Boiler Fittings and Air Removal Devices*, Chapter 8, *Boiler Control and Safety Devices*, Chapter 9, *Valves*, and Chapter 10, *Circulating Pumps*.

Viega, LLC

Figure 15-18. Shown here is an example of how the compression tool is used to complete a press-type fitting connection.

DID YOU KNOW?

Boiler Trim
Another collective term for the boiler components is boiler trim. This expression refers to the standard safety and control devices installed on hydronic boilers. These devices include low-water cutoffs, relief valves, pressure and temperature gauges, and temperature controls.

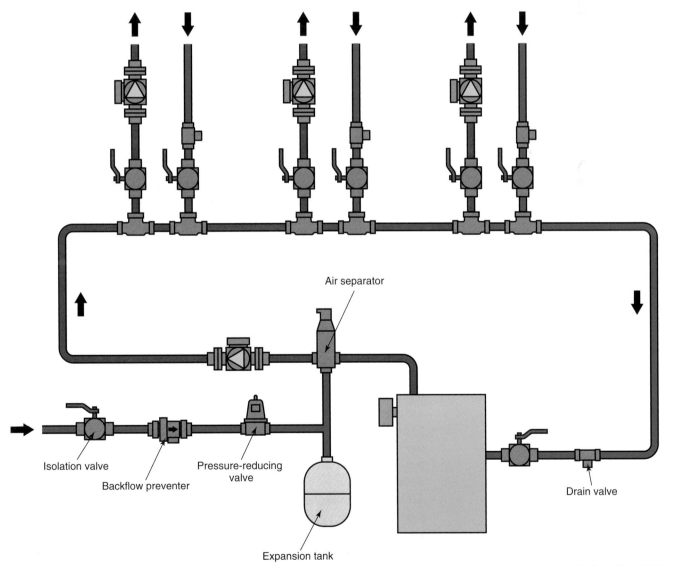

Goodheart-Willcox Publisher

Figure 15-19. This illustration shows an example of the devices that make up boiler piping.

Makeup water components: Starting at the municipal or domestic water source, the following components are combined to develop the makeup water piping system:

- **Isolation valves.** These are usually ball valves and can be located at various points in the piping system. They allow for isolating various components of the hydronic system when servicing the boiler. An isolation valve typically is connected at the inlet of the boiler's makeup water source.
- **Backflow preventer.** A backflow preventer is used to prevent the boiler water from flowing backward into a domestic or municipal water system. It is piped at the source of the makeup water—before the pressure-reducing valve.
- **Pressure-reducing valve.** Also referred to as a *water-regulating valve* or a *feed valve*, the pressure-reducing valve delivers the correct amount of makeup water to the hydronic heating system at the appropriate pressure—typically 12 to 18 psi. This device is installed between the backflow preventer and the boiler's primary piping loop.

Boiler safety components: These devices are used to protect the boiler and the building occupants from hazardous conditions:

- **Pressure-relief valve.** This device is usually mounted on top of the boiler. Codes require that every boiler have at least one properly rated pressure-relief valve set to open at or below the maximum allowable working pressure for each low-pressure boiler. When used for residential and light commercial boiler systems, these devices are usually rated to open at 30 psi.
- **Low-water cutoff.** A low-water cutoff is designed to de-energize the boiler's burner circuit if the water level within the boiler falls below a predetermined point—such as if there is an unseen leak in the system. Most boiler manufacturers require that the low-water cutoff switch be mounted at or near the outlet piping of the boiler.
- **Water flow switch.** This device ensures that there is proper water flow through the boiler before the burner is activated. The flow switch is wired in series with the heating circuitry. When the circulating pump is first activated, water flow through the piping is sensed by the flow switch. When the correct amount of flow is generated, the switch contacts are closed and the burner controls are energized. Water flow switches should be located near the outlet of the boiler.

Boiler control devices: Along with safety components, boiler control devices ensure that the hydronic system operates at optimum efficiency and provides accurate performance:

- **Circulating pump.** The circulating pump can be installed at various locations in the hydronic system, but it must be sized to meet the specific minimum flow requirements of the hydronic system. Most often it is piped near either the inlet or discharge port of the boiler.
- **Expansion tank.** Size the expansion tank for the proper water volume of the system and install it before the circulating pump. Always pump away from the expansion tank. Check with the boiler manufacturer for their exact desired location.
- **Air separation and eliminating devices.** The air separator is used to separate air from the water as it flows through the system. The eliminator or air vent removes the air from the system and, once it is separated, releases it into the atmosphere. These devices are sometimes mounted on top of the expansion tank but may be piped on the top of the boiler or at the highest

point in the hydronic system. Some systems may have more than one air eliminator device.
- **Pressure and temperature gauges.** These gauges may come readily installed on the boiler, or shipped loose to be installed separately. When installing the pressure/temperature gauge, be sure to mount it near the outlet or on the hot water supply line leading out of the boiler. These gauges must be easily within sight of the operator so that the proper temperature and pressure may be monitored.

Miscellaneous devices: The complete hydronic piping system may include, but is not limited to, the following devices:
- **Purge and drain valves.** If the boiler does not include a drain valve when shipped from the factory, one should be installed as part of the piping process. Furthermore, purge valves should be located at each heating zone to assist in purging air from individual zones at startup and during maintenance. These valves are usually *hose bib* type valves, **Figure 15-20**.
- **Balancing valves.** These valves are located in each individual heating zone and are used to ensure proper water flow through the zone.
- **Check valves.** These valves allow water to flow in one direction and automatically prevent backflow in the opposite direction through the hot water piping. Also known as a one-way valve, they are considered a self-automated valve in that they do not require assistance to open or close. Many circulating pumps are now equipped with check valves, eliminating the need for the additional valve.

Warren Price Photography/Shutterstock.com

Figure 15-20. This is an example of a hose bib, which is used as a drain valve for the boiler or to purge air from individual heating zones.

15.4 Gas Pipe Installation

Once the proper gas pipe size has been determined, as outlined in Chapter 14, *Boiler System Design Considerations*, follow the boiler manufacturer's installation instructions for proper installation. The first step is to properly support the gas piping. Use the correct pipe hangers for support, **Figure 15-21**. Refer to local code requirements for proper support distances, or consult the International Fuel Gas Code. Do not use the boiler or any of its accessories to support the gas pipe. Piping materials and joining methods must adhere to local codes or the authority having jurisdiction. In the absence of such requirements, refer to the following:

In the United States:
- The International Fuel Gas Code (IFGC)
- ANSI Z223.1
- NFPA 54

In Canada:
- The Natural Gas and Propane Installation Code
- CAN/CSA B149.1

The most common material used for gas piping is Schedule 40 black steel. Other materials such as copper, brass, or corrugated stainless steel tubing are also acceptable in some areas. However, there are some utilities that specifically prohibit the use of copper or galvanized tubing for gas piping—know what is acceptable in the area where the boiler is to be installed. Use approved pipe connectors such as elbows, couplings, unions, and reducers that are compatible with the type of piping being used. Do not mix different types of materials. After any cuts are made, the gas piping should be deburred and cleaned before threading and connecting. When cutting black pipe, make sure the ends are cut square.

7th Son Studio/Shutterstock.com

Figure 15-21. The first step in installing gas piping is properly supporting the pipe.

> **CODE NOTE**
>
> **Reducing Bushings**
> The use of reducing bushings on Schedule 40 black steel for gas piping is *not* allowed according to the 2009 IFGC 403.10.4 (5) (5.2).

When joining gas piping, use pipe joint compound suitable for natural gas and propane. Some areas allow for the use of Teflon™ tape for joining piping and fittings, as long as it is approved for use with natural gas and propane. Either pipe joint compound or Teflon tape should be applied to the male threads only of the gas pipe, **Figure 15-22**. Flared connections do not require pipe joint compound or Teflon tape.

The following three devices must be installed near the boiler where the gas pipe is to be connected to the gas valve:

- A manual shutoff valve connected in the vertical pipe approximately 5′ above the floor.
- A drip leg or sediment trap piped upstream of the gas controls.
- A ground joint union piped after the manual shutoff valve. See **Figure 15-23**.

When connecting the gas piping to the gas valve, use two wrenches: one wrench to tighten the piping and a second backup wrench connected to the valve. Failure to prevent the gas valve from turning could damage the gas line components. Once the piping is complete, always check for gas leaks on every fitting using an approved leak detector solution or leak detection device.

15.5 Condensate Disposal

High-efficiency boilers produce **condensate** from the products of combustion as a result of reduced flue gas temperatures. With these types of boilers, piping material such as PVC or PP must be used to drain the condensate from the boiler's heat exchanger. All piping materials must be approved by the authority having jurisdiction. If the condensate line has a long horizontal run, a vent connector may be necessary and the tubing size may need to be enlarged. The condensate line must be unobstructed and protected from freezing.

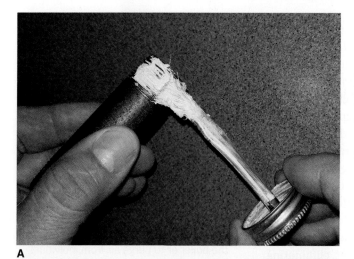
A

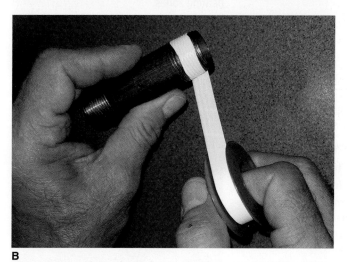
B

Goodheart-Willcox Publisher

Figure 15-22. Sealants used when joining gas piping. A—Pipe joint compound. B—Teflon™ tape.

If the boiler is located below the drain or the drain is a long distance away, a **condensate removal pump** is required, **Figure 15-24**. Select a pump approved for boiler condensate removal that contains an overflow switch to prevent property damage in the event of spillage. Boiler condensate is slightly acidic, typically with a pH between 3 and 5. For this reason, a **neutralizing filter** may need to be installed on the condensate line if required by local codes, **Figure 15-25**. *Avoid dumping acidic condensate into a septic system!*

> **CODE NOTE**
>
> **Condensate Piping Compliance**
> PVC and CPVC piping used for boiler condensate removal must comply with ASTM D1785 or D2845. Cement and primer must comply with ASME D2564 or F493. For Canada, use CSA or ULC certified PVC or CPVC pipe, fittings, and cement.

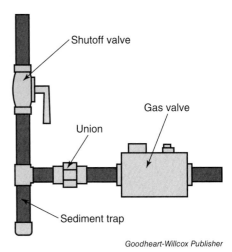

Figure 15-23. A gas piping layout must include a manual shutoff valve, a drip leg, and a union.

15.6 Field Wiring

Most boilers are designed to be field wired for a 120-volt service. When making any connection to the boiler or its electrical components, all local and national codes must be followed. In most cases, a qualified electrician should make the electrical connections to the boiler. If retrofitting a previous boiler, the existing electrical wiring may be used if it is in good condition and meets code requirements. In any event, electrical installations must comply with the following:

In the United States:
- The *National Electrical Code (NEC)* and any other local, state, and national codes or regulations

In Canada:
- CSA C22.1 Canadian Electrical Code Part 1 and any local codes

Many boiler manufacturers will provide terminal strips for connecting both line-voltage and low-voltage wiring. Line voltage is the standard voltage—120 volts—found in receptacles in the United States and Canada. Low voltage is typically 24 volts and requires a transformer to lower the line voltage from 120 volts down to 24. Be sure to review the installation guide provided with the boiler to determine where electrical connections should be terminated. Any additional wiring components, either line voltage or low voltage, must be compatible with the existing circuits, and the amperage draw of any additional components must be added to the boiler's total current draw to determine the proper wire size and overload protection for the entire boiler circuit.

Figure 15-24. A condensate pump must be used if the boiler is positioned below or a considerable distance from the condensate drain.

15.6.1 Line-Voltage Connections

Field-wiring connections begin at the breaker or service panel. Provide the properly sized overload protection based on the manufacturer's installation instructions or on the total amperage draw of the entire circuit. All line-voltage wiring must be enclosed in an approved conduit or an approved metal-clad cable. Install a fused disconnect or service switch at the boiler as required by code that is accessible to the operator. In most cases, this fused disconnect should be rated for 15 amps. The boiler must also be properly electrically grounded as required by the latest edition of the National Electrical Code ANSI/NFPA 70. Most boilers will provide a junction box or terminal strip to terminate any line-voltage connections as well as grounding wires.

Most boiler manufacturers require that the heating system circulating pump be purchased separately from the boiler. Both the boiler manufacturer and the wholesaler or dealer should be able to assist in the proper sizing of this pump. Review the installation instructions to determine if there is an existing pump

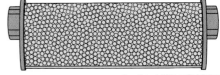

Figure 15-25. This is an example of a neutralizing filter used to reduce the acidity of the boiler condensate.

control relay or terminal strip connection for terminating the circulating pump control wiring, **Figure 15-26**. Also check to see that the maximum current draw of the pump does not exceed the full load amperage requirement of the boiler wiring circuit. Any field-installed pump circuit amperage must be added to the boiler circuit load to determine the minimum wire size and overload protection required. This includes any additional circulating pump that may be used in conjunction with the boiler for domestic hot water production.

15.6.2 Low-Voltage Connections

The boiler's low-voltage wiring originates at the manufacturer's provided step-down transformer, and the boiler's temperature control begins with the room thermostat. The thermostat should be installed on an interior wall away from the effects of drafts, sunlight, internal lighting fixtures, or any sources of heat—such as a computer or television. From this point, a variety of control strategies can be incorporated depending on the complexity of the boiler controls, the number of zones being serviced, and the desired operating efficiency. The boiler installation instructions outline the required low-voltage controls, and the manufacturer may provide a terminal strip for low-voltage wiring terminations, **Figure 15-27**.

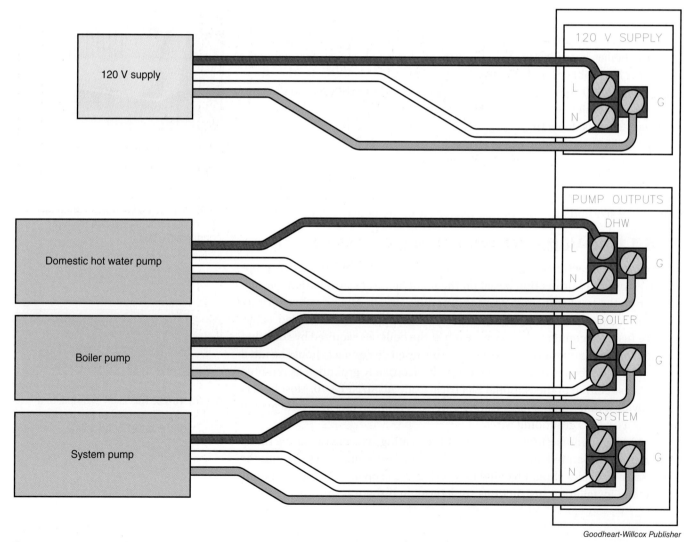

Goodheart-Willcox Publisher

Figure 15-26. Some boiler manufacturers may provide a terminal strip for making line-voltage connections.

Keep in mind that there are usually two separate temperature control circuits: one circuit to control the boiler water supply temperature and one circuit to control the space temperature. The control components for each of these circuits can include:

- **Hot water supply sensor or thermostat.** This sensor or thermostat is used to control the boiler water temperature and can be as simple as setting a fixed set point to control the cycling of the burner circuit. More advanced electronic hot water controls can include such items as temperature differential switches and high-fire offsets, **Figure 15-28**.

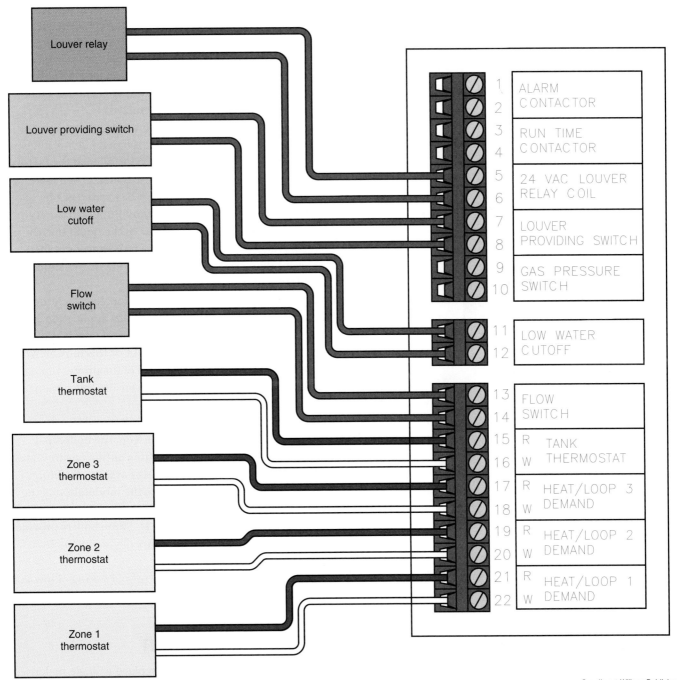

Goodheart-Willcox Publisher

Figure 15-27. This is an example of a low-voltage terminal strip for connecting control devices to the boiler.

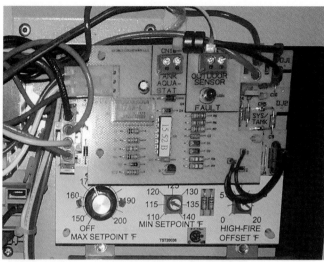

Goodheart-Willcox Publisher

Figure 15-28. This is an example of a boiler electronic temperature controller, including minimum and maximum set points and high-fire offset.

Goodheart-Willcox Publisher

Figure 15-29. Shown here is an example of an outdoor reset controller.

Universal Flow Switch, ©Caleffi North America, Inc.

Figure 15-30. A flow switch is used to shut down the burner in the event a system leak causes a low-water situation.

- **Outdoor temperature sensor.** An outdoor air temperature sensor may be provided with the boiler or added as an auxiliary item, **Figure 15-29**. It is used to reset the hot water supply temperature set point based on the outdoor air temperature, and it enhances fuel economy. A thorough explanation of outdoor reset controls is included in Chapter 8, *Boiler Control and Safety Devices*.
- **Combustion air damper relay and proving switch.** Larger boiler applications, or those located in a confined space, may incorporate a damper to provide outside air for proper combustion. When wiring this component into the boiler's low-voltage circuitry, it must be sequenced so that the damper is first commanded open when there is a call for heat, then a switch must prove that the damper is opened before allowing the burner to be energized.
- **Flow switch.** Boilers use a flow switch to guarantee water flow through the heat exchanger before allowing the burner to fire. They may be installed at the factory or shipped loose along with the boiler. When field installed, the flow switch must be located at the boiler hot water outlet. This is an important safety device and is wired in series with the boiler burner controls. It is used to shut down the burner in the event that a system leak causes a low-water situation, **Figure 15-30**.
- **Domestic hot water controls.** Some boilers are used for space heating plus the production of domestic hot water. When used for this application, a separate pump is installed to circulate water through the domestic hot water storage tank. Additional controls include a tank sensor or thermostat, which sends a signal to the boiler to be energized. Be aware of the proper piping arrangement provided by the boiler manufacturer to ensure this option works effectively.
- **Building management system.** Higher-efficiency boilers and larger commercial boilers may provide a control interface from the boiler to a digitally controlled building management system. This option ensures peak efficiency and performance from the hydronic heating system. ***Building management systems*** or ***BMS*** offer options such as the modulation of the circulating pump and burner control, the sequencing of multiple boilers, run time contacts, and alarm contacts, **Figure 15-31**. Some boilers provide interface modules that will communicate with protocol systems such as ***ModBus*** and ***LonWorks***.

Consult the boiler installation instructions for the proper wiring of low-voltage components and their applications.

15.7 Cascade Boiler Operation

In some instances, it may be more advantageous to replace a single boiler with multiple boilers. This strategy can offer higher operating efficiencies by sequencing multiple boilers as well as provide built-in redundancy should one boiler fail. This operation is known as cascading boilers, **Figure 15-32**. A *cascade boiler system* is comprised of multiple boilers linked together and controlled by a single temperature control system. This system can respond more efficiently to changes in heating loads and is easier to size for specific building heating needs. Some advantages of using a cascade boiler system include:

- Higher turndown capability when only one boiler is required.
- Flexibility with the installation footprint, allowing boilers to fit into irregular spaces.
- Increased reliability when heating needs to be provided by several boilers.
- Easier service and maintenance.
- Smaller boilers can be maintained by a single site operator.

genkur/Shutterstock.com

Figure 15-31. Some high-efficiency boilers and larger commercial boilers may provide a control interface from the boiler to a digitally controlled building management system.

PROCEDURE

Wiring a Cascade Boiler System

When wiring boilers for cascade operation, the following guidelines apply to wiring:

1. Select one unit as the "Lead" boiler. The remaining boilers will be designated as "Members."
2. Connect the system supply sensor and outdoor air sensor to the Lead boiler. This will be used to control the cascade system.
3. The system supply sensor should be located downstream of the boiler connections in the main system loop.
4. A high-limit and low-water cutoff switch is required for each boiler. They should be wired in series with each other.
5. If two or more boilers are used in the cascade system, their communication controls should be daisy-chained together.

Other requirements for boiler cascade systems should be outlined in the manufacturer's installation instructions.

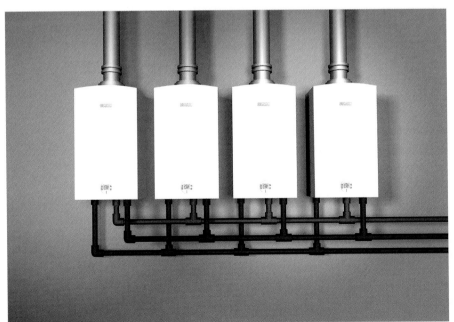

Bosch Thermotechnology

Figure 15-32. Here is an example of cascading multiple boilers together.

Chapter Review

Summary

- Before installing a boiler, it is recommended that the installer read the manufacturer's installation and maintenance manual.
- Boiler installations should only be performed by qualified professional installers.
- Understand the differences in the installation procedure between a retrofit boiler and new construction.
- When installing a new boiler, all safety procedures and code requirements must be followed.
- The installer of a new boiler should be aware of the issues that could void the warranty.
- When preparing the area where the new boiler is to be installed, be aware of issues that could lead to damage from such things as flooding, freezing, and corrosion.
- Be aware of the proper clearances around the boiler for service and from such things as combustible materials.
- High-efficiency condensing boilers must use PVC or similar venting material because of the acidic nature of the condensate.
- Whether the boiler is natural draft, induced draft, or condensing, the 10-2 rule must be followed when venting through a roof penetration.
- Various materials may be used for hydronic piping, but copper is the most common.
- There are several advantages to incorporating a primary-secondary piping arrangement when designing the hydronic piping system.
- The two main methods for joining copper tubing are soldering and using press-type fittings.
- Many of the devices that are piped into the hydronic system are collectively known as near-boiler piping or boiler trim.
- Hydronic piping components typically are used for temperature control or safety purposes.
- The three main categories for piping hydronic boiler components are makeup water devices, boiler safety devices, and boiler control devices.
- Gas pipe installation codes require that the piping be properly supported and the proper type of pipe sealant be used when joining pipe fittings.
- All gas pipe installations require that a shutoff valve, drip leg, and union be installed near the appliance.
- High-efficiency condensing boilers require that their condensate be disposed of properly.
- If the condensing boiler is located below the condensate drain, a condensate removal pump is required.
- Proper electrical codes must be followed when installing field wiring to boilers.
- Line-voltage wiring to the boiler must be of the proper wire gauge, run in the appropriate conduit, and have the correct overload protection.
- The boiler circulating pump typically is field provided; therefore, it must be sized, installed, and wired according to the manufacturer's instructions.
- Low-voltage devices are powered by a step-down transformer provided by the boiler manufacturer.
- The boiler low-voltage components are used to control temperature and for safety purposes.
- Some boilers can be interfaced with a building management system to allow for greater functionality.
- A cascade boiler system is used for a multiple boiler installation in which several smaller boilers are used to replace one large boiler.

Know and Understand

1. *True or False?* A retrofit installation allows for easier accessibility and flexibility of the boiler piping.
2. In some cases, it may be necessary to consult the local _____ before proceeding with the boiler installation to ensure that local codes are adhered to.
   A. law enforcement officials
   B. authority having jurisdiction
   C. health department
   D. building inspector
3. *True or False?* Excessive water hardness may affect the boiler's warranty.
4. What is the typical clearance on the front of the boiler for service access?
   A. 6″
   B. 10″
   C. 24″
   D. 36″
5. When installing a boiler in a residential garage, all burners shall have a clearance above the floor of at least _____.
   A. 6″
   B. 10″
   C. 12″
   D. 18″
6. The vertical termination of a boiler vent exiting a roof must be a minimum of _____ above the point of exit.
   A. 6″
   B. 12″
   C. 24″
   D. 36″
7. *True or False?* Approved venting materials for condensing boilers include PVC, CPVC, polypropylene, and galvanized steel.
8. As an alternative to the standard two-pipe termination on a condensing boiler, a _____ may be used.
   A. concentric vent kit
   B. eccentric vent kit
   C. geocentric vent kit
   D. convertible vent kit
9. Which type of piping configuration provides several advantages over conventional piping arrangements on a modern boiler installation?
   A. One-pipe direct return
   B. Two-pipe direct return
   C. Two-pipe reverse return
   D. Primary-secondary piping arrangement
10. Which of the following is *not* considered part of the makeup water piping assembly?
    A. Backflow preventer
    B. Pressure-reducing valve
    C. Pressure-relief valve
    D. Isolation valve
11. Where do most boiler manufacturers require that the low-water cutoff switch be mounted?
    A. At or near the inlet piping of the boiler
    B. At or near the outlet piping of the boiler
    C. At or near the expansion tank
    D. At or near the makeup water inlet
12. Which device is *not* required near the boiler where the gas pipe is to be connected to the gas valve?
    A. Gas pipe coupling
    B. Gas pipe union
    C. Manual shutoff valve
    D. Drip leg
13. When is a condensate pump required on the boiler?
    A. When the boiler is over 100,000 Btu input
    B. When the boiler uses propane rather than natural gas
    C. When the boiler requires an extended condensate line
    D. When the boiler is located below the condensate drain
14. *True or False?* Most boilers are designed to be field wired for 240 VAC service.
15. Where should the system supply sensor be located in a cascade boiler system?
    A. Downstream of the boiler connections in the main system loop
    B. At the outlet of the lead boiler
    C. Upstream of the last boiler located on the system loop
    D. At the farthest terminal device on the loop

Apply and Analyze

1. Explain why installing a boiler in a newly constructed building offers more flexibility than one being retrofitted.
2. List some of the safety guidelines that need to be followed when installing a new boiler.
3. What are some of the issues to consider when choosing the boiler location?
4. List the clearances that must be adhered to when installing a boiler.
5. Explain how the 10-2 rule is applied to vent terminations through a roof.
6. Define *near-boiler piping* and list the devices that fall into this category.
7. What three devices must be installed near the boiler where the gas pipe is to be connected to the gas valve?
8. Why should a neutralizing filter be used with condensate removal?
9. What devices are included in the low-voltage boiler circuit?
10. Describe how a cascade boiler system operates.

Critical Thinking

1. What issues could occur if boiler condensate were to be emptied into a septic system without it being neutralized?
2. What additional installation procedures might be necessary if a replacement boiler is smaller than the original?

16 Boiler Startup

Chapter Outline
16.1 Water Quality
16.2 Filling and Purging the System
16.3 Prestart Checklist
16.4 Igniting the Burner and Sequence of Operation
16.5 Combustion Testing
16.6 Testing Safety Devices
16.7 Freeze Protection

Alexander Raths/Shutterstock.com

Learning Objectives

After completing this chapter, you will be able to:
- Describe the required steps for proper boiler startup.
- List the boiler water substances to test for and the desired levels of each.
- Outline the required steps to initially fill and purge the hydronic system.
- Confirm the steps taken to ensure that the boiler is installed correctly and fully functional before starting it up.
- Identify the sequence of operation for most boiler startups.
- Explain how to properly set the manifold gas pressure on a boiler before startup.
- Explain why combustion testing is an important part of a boiler startup.
- Identify the elements tested for during combustion testing.
- Outline the process for testing safety devices during boiler startup.
- Explain why freeze protection is required for the hydronic fluid in certain climates.

Technical Terms

boiler drain valve
combustion analysis
condensate trap
fast-fill lever
high-fire offset
manometer
nitric oxide (NO)
pH
propylene glycol
purging valve
stack temperature
total dissolved solids (TDS)
trisodium phosphate
water hardness
zone isolation valve

Once the boiler has been installed, it is ready to be commissioned. The technician must have a working knowledge of the proper startup procedures covered in this chapter.

Boiler startup begins after initially filling the boiler. But before it is filled, it is good practice to first flush the boiler and the hydronic system to remove any contaminants. Contamination can occur from such things as soldering flux and pipe joint compound left over from the boiler installation. Also, debris such as sand, dirt, bits of solder, and slivers from the pipe threading can clog the system and ruin the pump seals.

Begin the flushing process by filling the boiler with water and adding about one pound of *trisodium phosphate* per every 50 gallons of water; or use an approved commercial boiler cleaning solution. Allow this solution to circulate through the system for about four hours, then drain the boiler. Clean any water filtering devices, then refill the boiler and the complete hydronic system as outlined in the following sections.

TECH TIP

Avoid Petroleum-Based Cleaners
Do not use petroleum-based cleaning or sealing compounds within the hydronic heating system. These can result in damage to gaskets and seals within the system.

16.1 Water Quality

Before initially filling the boiler, conduct a water testing procedure to verify the water is of good quality, **Figure 16-1**. Many different types of test kits are available for analyzing boiler water quality. Some of these kits contain ingredients known as reagents. A reagent is a substance or compound added to the boiler water sample to cause a chemical reaction or added to the sample if a reaction occurs. Once the reagent is added, the sampled water color is typically compared to a

alejandro dans neergaard/Shutterstock.com

Figure 16-1. Before initially filling the boiler, conduct a water testing procedure to verify the water is of good quality.

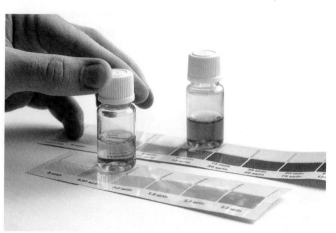

tetiana_u/Shutterstock.com

Figure 16-2. The pH level of the boiler water is an indication of whether the water is acidic or alkaline.

test strip to interpret the results. Other kits contain test strips that are dipped into the water sample to change color. The result is then compared to a color chart for analysis. Always follow the manufacturer's instructions when sampling boiler water, or have the water sample analyzed by a local chemical supplier. The proper water chemistry will help extend the life of the boiler by reducing the effects of scale buildup and corrosion within the closed-loop system. Test for the following components to ensure the desired levels of each:

Water Hardness: *Water hardness* is caused by dissolved calcium and magnesium. It is predominant in most areas of the United States, but it is higher in certain areas. Water hardness leads to scale, which if left unchecked can cause a reduction in boiler efficiency and obstructions in the hot water tubing. Most boiler manufacturers recommend that the fill water have a hardness rate between 5 and 12 grains per gallon. Water that registers above 12 grains per gallon may need to be softened before filling the system.

pH Levels: The *pH* of the boiler water is an indication of whether the water is acidic or alkaline. A pH level above 7 is considered alkaline or base, and a level below 7 indicates acidity. Boiler manufacturers recommend a pH level as close to 7 as possible. Levels below 6.5 can cause an increase in the rate of corrosion, whereas a pH of 8.5 or higher can potentially cause scale buildup, **Figure 16-2**.

Chlorine: Most municipal water supplies contain a certain amount of chlorine. Do not fill the boiler or operate it when the chlorine level is in excess of 150 ppm. If necessary, use fresh drinking water to fill the boiler.

Total Dissolved Solids (TDS): *Total dissolved solids (TDS)* are minerals, salts, metals, and charged particles that dissolve in water. The higher the amounts of TDS present, the greater the potential for corrosion due to increased conductivity within the water. Most boiler manufacturers require a TDS level below 350 ppm.

Monitoring the water chemistry within the boiler on an annual basis will prolong the life of the boiler and associated equipment.

16.2 Filling and Purging the System

Once the boiler fill water has met the correct water chemistry standards, it is time to fill the boiler. Following the correct procedure for filling and purging air from the boiler saves time and prevents unnecessary service calls later, **Figure 16-3**. The purging process for service and troubleshooting is covered in Chapter 17, *Boiler Maintenance and Service*. The procedure discussed here is primarily for initially purging air from a new or replacement system and may differ from the purging process during service for service reasons.

1. Begin by closing all automatic and manual air vents and **boiler drain valves**.
2. Fill the system by first opening the shutoff valve to the makeup water connection and engaging the **fast-fill lever** on the pressure reducing valve to increase the inlet water pressure (reference Chapter 9, *Valves*). Continue filling the system until it reaches the correct operating pressure according to the pressure gauge. (This may vary with each application.)
3. Once the system is entirely full, check for any leaks. Repair all leaks before proceeding.

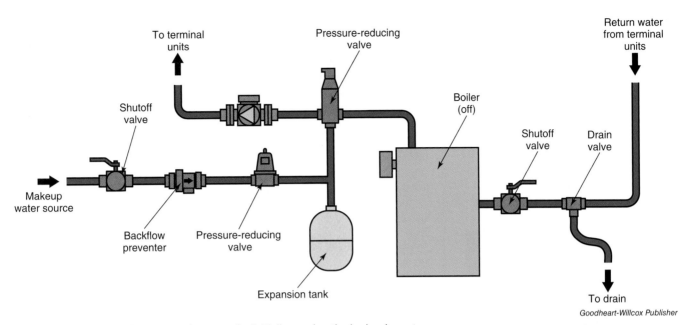

Figure 16-3. This illustration shows the setup for initially purging the hydronic system.

The next step is to purge any air from the hydronic system:
1. Begin by connecting a hose to the boiler drain valve and routing the hose to a proper drain where the water can be seen.
2. Ensure that any main *purging valves* and the pressure-reducing valve are closed.
3. Close all *zone isolation valves* and electric zone valves. If there is an isolation valve between the main water supply header and the individual zone branches, close this valve as well.
4. Slowly open the main purging valve to remove any air from the main piping header.
5. Close the main purging valve once any air is removed.
6. Begin purging the individual zones by opening the isolation valves and electric zone valves one zone at a time. Allow water to circulate through each zone. This will push any air out.
7. Continue to run water through each zone until no noticeable air is present. Close the isolation valve and zone valve and proceed with the next zone. Follow the same procedure until all zones have been purged.
8. Close the fast-fill lever on the pressure-reducing valve and the boiler drain valve. Remove the hose from the boiler drain valve and open all zone valves and isolation valves. Observe the system pressure rise to the correct cold-fill pressure.
9. Run the circulating pump for an extended period of time without firing the boiler. Depending on the size of the system, this may take several hours. There is no set period of time for initially running the pump; however, the longer the pump can run, the better the chances are for removing any excess air.
10. After the pump has circulated water for an extended length of time, open any strainers within the system and remove any accumulated debris. If an unusually large amount of debris is present, it may be a good idea to flush the system again.
11. Eliminate any residual air by opening any manual air vents or individual zone purging valves.

12. If the building is a multistory structure, begin on the lowest floor and open all air vents one at a time until only water emerges out of the vent. Repeat this procedure on each floor.
13. Refill the system with makeup water to the correct operating pressure.

TECH TIP

Let That Pump Run!
Oftentimes, other installation work still needs to be done after the boiler has been filled with water and the initial air purge process is complete.

The circulating pump can be temporarily wired into a separate circuit or extension cord and allowed to run while other installation tasks are being completed such as gas piping, venting, or low voltage wiring.

While the installer is finishing the installation processes, the pump can do its job by purging any unwanted air from the system and removing oxygen from the boiler water. Just be sure the circulating pump is correctly wired into the boiler panel before leaving the job!

16.3 Prestart Checklist

Before firing the boiler for the first time, follow these steps to ensure that the boiler was installed correctly and is fully functional:

1. **Purge any air from the gas lines.** After the gas piping to the boiler has been installed, there is usually excess air within the lines that must be purged. The time to purge all air from the lines will vary depending on the size and lengths of the gas piping runs. Begin by ensuring that the gas valves at the meter and the boiler are open. Start by cracking open a fitting near the boiler. This may be the union fitting or the cap located on the drip leg, **Figure 16-4**. Allow the gas to push the excess air out of the piping until there is evidence of a strong gas odor.

Goodheart-Willcox Publisher
Figure 16-4. Begin the boiler pre-start checklist by purging air from the gas lines.

SAFETY FIRST

Wait after Purging
Wait at least five minutes after purging air from the boiler's gas piping before attempting to light the boiler. This will give adequate time for the gas to disperse so there is minimal chance of a combustion hazard.

2. **Check for gas leaks.** Check for gas leaks by swabbing every gas fitting and connection with a soapy solution. Leaks will be evident by the solution generating bubbles, **Figure 16-5**. Wait several minutes to verify that there are no leaks. Another method for leak checking is to use an approved combustion gas electronic leak detector. Repair any leaks before proceeding.

3. **Inspect and fill the condensate system.** If the boiler is a high-efficiency 90+ model, inspect the condensate drain lines, pipe fittings, and *condensate trap*. The condensate trap is typically a P-type trap that may or may not be internal to the boiler, **Figure 16-6**. Be sure to fill the condensate trap with water before firing the boiler by following the manufacturer's directions. Failure to do so may allow the products of combustion to be vented into the boiler room.

Goodheart-Willcox Publisher
Figure 16-5. Always check for gas leaks using a soapy solution before starting up the boiler. If leaks are present, they will show up as bubbles.

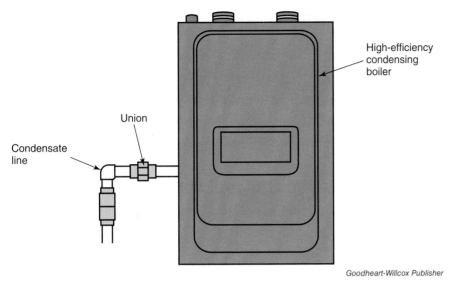

Figure 16-6. Part of the boiler pre-start checklist includes inspecting the condensate line connections and the condensate trap.

> **SAFETY FIRST**
>
> **LP Gas Conversion**
>
> If the boiler is designed for LP usage, verify that the proper gas conversion has been done correctly per the manufacturer's installation instructions before firing the boiler. (Reference Chapter 14, *Boiler System Design Considerations*.) Failure to perform this step may result in irreparable boiler damage.

4. **Check the electrical wiring.** Verify that all electrical connections are correct and securely fastened.
5. **Check the venting.** Inspect the vent piping and any combustion air piping for signs of deterioration, corrosion, sagging, or physical damage. Verify that this piping is installed correctly and secured properly at each connection, **Figure 16-7**.

16.4 Igniting the Burner and Sequence of Operation

Before firing the boiler for the first time, the startup technician must fully understand the proper sequence of operation. Review the installation instructions provided with the boiler for this step. Typically, the sequence of operation for most boilers begins by setting the space thermostat to call for heat. Once there is a call for heat, the sequence of operation includes, but is not limited to, the following steps:

1. If the boiler has a standing pilot for burner ignition, the pilot must first be lit. Follow the manufacturer's instructions for this procedure.
2. Upon a call for heat, a zone valve or zone pump is first energized before burner ignition takes place. Most zone valves include an end switch that must prove to ensure the valve is fully open.
3. Some boilers include a water flow switch that also must prove before the boiler is allowed to fire.
4. If the boiler has electronic ignition for burner control, the inducer fan will energize once there is a call for heat, purging any products of combustion from the venting system.
5. A pressure differential switch must then be activated, proving that the inducer fan is operating and the vent passage is clear.
6. Once the pressure differential switch proves, the burner ignition source is activated—this may be a spark igniter or hot surface igniter.
7. When the burner is energized, the flame signal is proven through a flame sensor.

Figure 16-7. Inspect the vent piping and combustion air piping for proper installation before initially firing the boiler.

8. The burner will remain lit until the boiler reaches its hot water temperature set point or until an individual zone is satisfied—at which time the zone valve will close.
9. The burner circuit and circulating pump are de-energized.
10. The inducer fan motor will continue to run on a post-purge cycle for approximately 10 to 30 seconds.

Once the boiler has successfully completed the proper sequence of operation, further set point adjustments may need to be made, including the *high-fire offset*, high and low temperature limits, and outside air reset adjustment.

16.4.1 Setting the Gas Pressure

Another important step during boiler startup is to set the proper gas pressure at the gas valve. The proper gas pressure setting is dependent on whether the fuel to be used is natural gas or LP. The gas pressures must be measured at both the inlet to the gas valve and at the manifold. Most gas valves will have service taps at both the inlet and manifold side of the valve, **Figure 16-8**. Follow these steps to correctly measure the gas pressures to the boiler:

1. Turn off the power to the boiler.
2. Turn off the valve or set it to pilot.
3. Using a *manometer*, connect to the inlet side of the gas valve at the pressure tap to measure incoming pressures.
4. Open the fuel supply valve and measure the inlet pressure. It should fall within these ranges:
   A. **Natural gas:** 5 in. WC minimum, 10.5 in. WC maximum
   B. **LP gas:** 11 in. WC minimum, 13 in. WC maximum
5. If the gas supply pressure is out of range, contact the gas utility or service supplier to determine the necessary steps to provide the proper gas pressure at the control valve.
6. If the gas supply pressure is within the specified range, proceed with measuring the manifold pressure.
7. Close the inlet gas supply valve and move the manometer to the outlet side of the gas valve. Be sure the plug for the inlet pressure tap is securely in place.
8. Open the inlet gas supply valve. Fire the boiler and observe the pressure reading. It should be within the following range:
   A. **Natural gas:** 3.5 in. WC minimum
   B. **LP gas:** 11 in. WC minimum
9. Be sure to check the installation instructions to ensure these readings are within the manufacturer's specifications.
10. If a pressure adjustment is needed, remove the pressure regulator adjustment cap on the gas valve. Adjust the pressure diaphragm screw clockwise to increase pressure and counterclockwise to decrease pressure.
11. Allow the boiler to operate for several minutes and observe the manifold pressure with the pressure regulator cap in place.

York International Corp.

Figure 16-8. Always set the proper gas pressure at the valve before initiating boiler start-up.

12. Before removing the manometer from the boiler, shut down the burner and disconnect the hose fitting from the pressure tap. Install the plug back into the pressure port.

TECH TIP

Two-Stage Gas Valves

Some boilers have two-stage gas valves that require setting both high- and low-fire manifold pressures. Follow the installation instructions for performing this procedure. Also, some high-efficiency boilers have special procedures for setting gas manifold pressures and for converting the boiler from natural gas to LP gas. Be sure to check the installation manual for these procedures as well.

16.5 Combustion Testing

An important part of boiler startup involves testing the products of combustion. In fact, some boiler manufacturers require this process as part of their warranty fulfillment. Regardless of whether the fuel being used is natural gas, LP, or fuel oil, a comprehensive *combustion analysis* taken with a digital analyzer will ensure that the boiler is operating at a safe level and performing at peak efficiency, **Figure 16-9**.

A combustion analysis should be performed as part of the boiler commissioning process for four primary reasons:

1. To verify that the boiler is operating safely
2. To calculate the boiler's combustion efficiency
3. To determine the proper emissions levels that the boiler should be producing
4. To comply with the manufacturer's warranty requirement

Bacharach, Inc.

Figure 16-9. An important part of boiler start-up is to perform a comprehensive combustion analysis.

Furthermore, an accurate combustion analysis will extend the longevity of the boiler as well as save the customer money and limit the amount of future unnecessary callbacks.

The combustion testing process starts with quality electronic equipment. Be sure that the analyzer being used has been accurately calibrated and that the sensors are fully functional. Most manufacturers of combustion testing equipment provide the proper process for taking accurate readings using their equipment. Be sure to familiarize yourself with the equipment before beginning the testing procedure. The two main categories of performing accurate combustion testing include taking critical measurements and performing accurate calculations.

The elements to be measured include:

- **Carbon monoxide.** Dangerous by-products produced by incomplete combustion
- *Stack temperature.* The temperature of the flue gases plus the combustion air temperature
- **Oxygen (O_2).** The amount of oxygen in the flue gases after combustion has occurred
- *Nitric oxide (NO).* A by-product of combustion—also referred to as nitrogen monoxide

The elements to be calculated include:

- **Boiler efficiency.** This is a calculation of the maximum heat available in the combustion process minus the stack losses.
- **Nitrogen oxides (NO_x).** This mixture includes various elements, including nitrogen dioxide (NO_2) and nitrogen trioxide (N_2O_3). Testing for NO_x is critical since some states and jurisdictions require the installation of low-NO_x boilers only.
- **Carbon dioxide (CO_2).** Carbon dioxide is considered a greenhouse gas and is a normal by-product of complete combustion.
- **Excess air.** This is air that passes through the combustion process without acting as an oxidizer. A certain amount of excess air is required to assure complete combustion. Excess air should be kept to a minimum as it can dilute flue gases and affect efficiency.
- **Air-free carbon monoxide (CO).** This is a calculated measurement of the undiluted carbon monoxide within the flue gases.

The proper levels of each of these elements varies depending on the targeted efficiency of the boiler and the type of fuel being used. Consult the boiler manufacturer and the maker of the combustion analyzer for the correct readings and proper adjustments to be made.

16.6 Testing Safety Devices

Another key element of a successful boiler commissioning process is to check the functionality of all safety devices. The devices to be tested include, but are not limited to, the following:

- **Low-water cutoff switch.** Drain the boiler while it is in operation to ensure the low-water cutoff switch de-energizes the burner circuit. This switch should reset, and the boiler should automatically fire after the proper water level has been restored. Some boilers offer a test switch that can be used to check the low-water cutoff switch, **Figure 16-10**.
- **High limit operation.** Most boilers are equipped with an automatic resetting high-limit switch that has a maximum set point of 200°F. To test this switch, allow the boiler water temperature to rise above the desired set point and observe that the burner circuit de-energizes. Allow the boiler to cool down, and ensure that this switch automatically resets.
- **Flow switch.** The flow switch should open if the circulating pump fails. To test it, disconnect power to the pump during normal boiler operation and observe that the burner circuit de-energizes. Ensure that the burner re-fires after power is restored to the pump and that flow is present in the system.
- **Pressure-relief valve.** The pressure-relief valve should open if the water pressure within the piping system exceeds the valve rating (normally 30 psi for residential applications), **Figure 16-11**. The pressure-relief valve can be

Goodheart-Willcox Publisher

Figure 16-10. This boiler offers a low-water cutoff switch with a test button to check its functionality.

tested by slowly opening the fast-fill handle on the pressure-reducing valve. Observe the pressure gauge during this test to ensure that the valve opens at the predetermined point. The valve should reset after the pressure in the system has been reduced. Remember to close the fast-fill handle upon completion.

Review the boiler installation guide to ensure that all safety devices have been tested properly before placing the boiler into continuous operation.

16.7 Freeze Protection

If the boiler is installed in an area that may experience below-freezing temperatures, adequate freeze protection is required. When this is the case, use only an inhibited *propylene glycol* solution designed for hydronic systems, **Figure 16-12**. Specially formulated freeze protection solutions prevent contamination of metallic system components. The proper concentration of glycol within the boiler water is typically 25% to 30%; however, follow the manufacturer's recommendation. Be sure to test the boiler water once a year for the proper glycol levels as recommended by the manufacturer of the glycol solution. In addition, the system's pumping requirements change when a freeze protection solution is required, so be sure that the circulator is sufficient for this operation.

SAFETY FIRST

Toxic Freeze Protection

If the hydronic system requires freeze protection, then propylene glycol is recommended. Do not use ethylene glycol because of its toxicity. Automotive antifreeze contains ethylene glycol and should not be used. Ingestion of this substance by children or pets could lead to death! Furthermore, there is the risk that it could be drained into a municipal water system—endangering the lives of others.

Goodheart-Willcox Publisher

Figure 16-11. The pressure-relief valve should be tested to ensure that it opens at its predetermined pressure set point.

Goodheart-Willcox Publisher

Figure 16-12. When freeze protection is necessary for the hydronic system, be sure that an inhibited propylene glycol solution is used.

Chapter Review

Summary

- Once the boiler has been installed, it is ready to be properly started up and commissioned.
- It is a good practice to flush the boiler and the hydronic system to remove any contaminants before performing a startup procedure.
- Before initially filling the boiler, a testing procedure should be performed to verify the water quality.
- Once the boiler water has met the water chemistry standards, the correct procedure for filling and purging air from the boiler needs to be performed.
- Before initially firing the boiler, a prestart check should be performed, which includes checking the gas piping, electrical system, and the venting system.
- The proper sequence of operation must be understood in order to effectively commission the hydronic system.
- An important step that must be included during boiler startup is to set the proper gas manifold pressure.
- A combustion analysis should be performed during the boiler startup to ensure that the proper levels of the products of combustion are maintained.
- Checking the functionality of all safety devices is a key element of a successful boiler commissioning process.
- Adequate freeze protection is required for hydronic systems that are operated in colder climates.

Know and Understand

1. Boilers can be flushed using water mixed with _____.
   A. sodium dioxide
   B. trisodium phosphate
   C. calcium carbonate
   D. hydrogen peroxide
2. Which of the following elements is typically *not* tested for when analyzing boiler fill water?
   A. Nitrates
   B. Total dissolved solids
   C. pH
   D. Water hardness
3. *True or False?* The time that it takes to purge air from the gas lines will typically be the same regardless of the size of the system.
4. Which device is used for measuring manifold gas pressures?
   A. Manometer
   B. Pitot tube
   C. Micrometer
   D. Anemometer
5. *True or False?* The manifold pressure for natural gas should be adjusted to a minimum pressure of 11 in. WC.
6. What type of air passes through the combustion process without acting as an oxidizer?
   A. Air-free carbon monoxide
   B. Combustion air
   C. Excess air
   D. Ventilation air

7. Which safety device can be tested by slowly opening the fast-fill handle on the pressure-reducing valve?
   A. Low-water cutoff switch
   B. Pressure-relief valve
   C. Flow switch
   D. High-limit temperature switch
8. Most high-limit temperature switches have a maximum set point of _____.
   A. 180°F
   B. 190°F
   C. 200°F
   D. 210°F
9. The calculated measurement of the undiluted carbon monoxide within the flue gases is known as _____.
   A. saturated carbon monoxide
   B. air-free carbon monoxide
   C. air-filled carbon monoxide
   D. unsaturated carbon monoxide
10. What type of solution should be used as a boiler freeze protector?
    A. Propylene glycol
    B. Methyl alcohol
    C. Epsom salt
    D. Ethyl alcohol

Apply and Analyze

1. Describe the steps that should be taken to ensure the boiler's proper water chemistry.
2. Explain the proper procedure for initially filling and purging the boiler before startup.
3. List the proper steps to be taken to ensure that the boiler was installed correctly and is fully functional before performing an initial startup.
4. List the steps of the boiler's proper sequence of operation that the technician must understand before firing the boiler for the first time.
5. Why are there different gas pressure manifold settings for natural gas and LP?
6. Describe the procedure for purging excess air from the gas lines.
7. Explain why an important part of boiler startup includes the testing of the products of combustion.
8. List and describe the elements to be measured and calculated during a comprehensive combustion analysis test.
9. What is the procedure for testing safety devices during a boiler startup?
10. What type of solution should be added to the boiler water for proper freeze protection? How much should be added?

Critical Thinking

1. What would happen if the air was not purged from the gas lines?
2. Describe some of the applications where boiler water freeze protection would be required.

17 Boiler Maintenance and Service

Chapter Outline
17.1 Boiler Maintenance
17.2 Boiler and Hydronic System Service

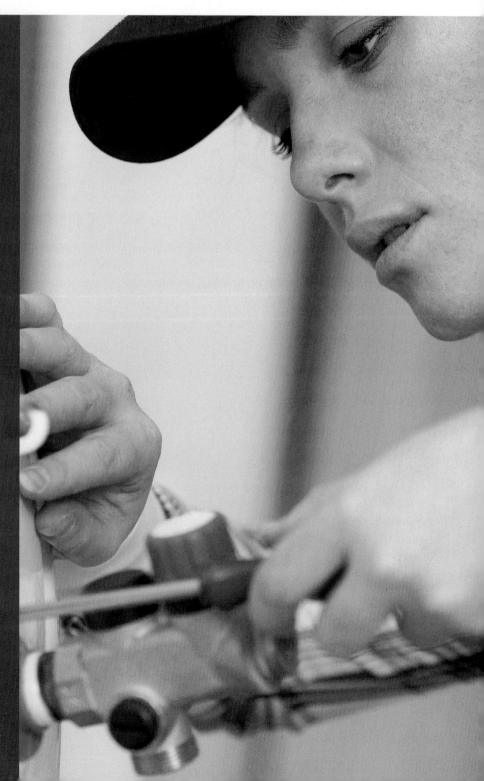

ALPA PROD/Shutterstock.com

Learning Objectives

After completing this chapter, you will be able to:
- List the steps taken during a preventive maintenance procedure.
- Describe the process for detecting leaks in the hydronic system.
- Outline the procedure for checking the venting and combustion systems during a maintenance procedure.
- Describe the process for cleaning and maintaining the boiler's burner and heat exchanger.
- Identify problems that can arise from improper system sizing or installation of the hydronic system.
- Explain the troubleshooting procedure for the boiler's burner circuit.
- Describe the steps taken for troubleshooting venting issues.
- List problems that can occur with the combustion air system.
- Outline the steps taken for effectively troubleshooting water circulation issues.
- Describe potential expansion tank problems.

Technical Terms

combustion leak detector
Giannoni-type heat exchanger
hopscotching
hydronic airlock
manifold gas pressure
microammeter
microamp
millivolt
multimeter
vent stack
vestibule

Once the boiler has been installed and is ready to be commissioned, the work begins in keeping it running reliably and efficiently. In order to achieve this goal, the technician must have a working knowledge of the proper service and repair of the boiler and hydronic heating system. This chapter focuses on preventive maintenance and service of the boiler and the complete hydronic system.

17.1 Boiler Maintenance

Just as with any other piece of HVAC equipment, maintenance is an important element in keeping the boiler and hydronic system running properly and efficiently. Routine boiler maintenance prevents many future problems and prolongs the life of the equipment. To effectively service and repair a boiler and its hydronic system, the technician should develop a plan to systematically inspect the equipment for any possible defects or problems and also be able to diagnose and repair issues as they become evident.

17.1.1 Visual Inspections

The first step in performing effective hydronic boiler maintenance is to look around. Visual inspections can reveal clues to more serious problems that might otherwise be overlooked. Begin by checking the boiler for any clear signs of rust or corrosion. These clues could reveal problems with combustion, water flow, or venting. Also survey the boiler room for combustible or corrosive products placed near the boiler. Good housekeeping practices can go a long way in extending the life of the boiler and its performance.

Next check for system water leaks. This should be performed both at the boiler and throughout the system's piping. Water leaks can cost money and will accelerate the rust and corrosion factor for both the boiler and associated piping components. In some areas, the piping may be covered with insulation. Peel this back in suspicious areas, then repair or replace it if any problems are found. Gas piping must also be checked for any leaks. Gas leaks not only cost money but are also a potential fire and explosion risk. Leak check all gas connections

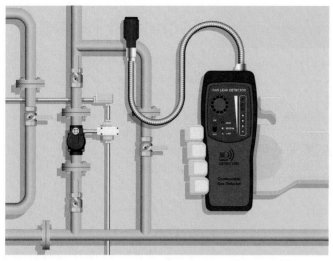

Figure 17-1. All gas connections and devices should be periodically leak checked using a quality combustion leak detector.

and devices periodically using soap bubbles or a quality *combustion leak detector*, **Figure 17-1**.

17.1.2 Checking the Water Side of the System

As previously mentioned, maintenance on the water side of the hydronic heating system begins with a visual inspection to determine if there are any leaks. If the boiler has a meter connected to the makeup water source, check this meter periodically for the amount of water usage. If the usage seems excessive, a leak is probably present within the piping system. Any leaks found should be repaired immediately; otherwise the boiler will continue to deliver fresh makeup water into the system. The continuous introduction of makeup water creates a situation that could reduce the life of the boiler because most domestic and municipal water contains minerals that can build up inside the heat exchanger, causing reduced heat transfer, overheating, and premature failure. Water leaks in piping located above ceilings can also cause property damage. Hydronic piping leaks can be caused by the following issues:

- A broken seal
- Corrosion caused by water contamination
- Damaged valves or circulating pump
- Faulty pipe connections

Most leaks can be repaired by simply draining the boiler and resoldering a defective joint. If a threaded fitting is leaking, check the conditions of the pipe threads and reapply pipe joint compound or replace the fitting. Leaking valves and circulating pumps may need replacement seals or bearings. If isolation valves are not present on each side of the valve or pump, the system needs to be drained before repairs can be made.

Leaks caused by corrosion are a bigger issue. If excessive corrosion is evident, first test the boiler water, **Figure 17-2**. If tests show contaminated boiler water, drain and flush the boiler before introducing newly treated water. Remember that most boiler corrosion is the result of too much oxygen present in the boiler water or a pH imbalance.

Other water-side preventive maintenance checks include the following:

- Verify that all system components are correctly installed and operational.
- Check the cold-fill pressure on the system. Verify that it is correct (usually a minimum of 12 psi).
- Watch the system pressure as the boiler heats up during testing to ensure that the pressure does not rise too high. An excessive pressure rise indicates a problem with the expansion tank sizing or possibly with the pressure-reducing valve.
- Inspect any automatic air vents and air separators. Remove each air vent cap and depress the valve to flush out the vent. Replace any caps and make sure the vents do not leak. Replace any leaking vents.
- Check the expansion tank. Inspect the tank for any leaks, especially around the fittings, and confirm that the air pressure within the tank is at its rated level, **Figure 17-3**.

Figure 17-2. Boiler leaks caused by excessive corrosion can result from the boiler water not being thoroughly tested.

- Check the circulating pump. Some pumps have oil ports that must be filled on a regular basis. Check with the pump manufacturer to determine this interval and any other maintenance requirements.
- If the boiler is a high-efficiency condensing model, remember to check the condensate trap. This trap must be filled with water at all times during boiler operation to avoid flue gases from escaping through the condensate drain line. Failure to keep the trap full of water could result in severe personal injury or even death.
- During routine maintenance, flush the condensate trap with fresh water to remove any sediment that has built up over time.

17.1.3 Checking the Combustion Side of the System

Maintenance on the boiler's combustion system includes checking the following:
- Venting system
- Combustion air system
- Burner components
- Heat exchanger

Goodheart-Willcox Publisher

Figure 17-3. The expansion tank should be checked for proper pressure.

17.1.3.1 Venting and Combustion Air

The venting and combustion air system should be examined at least once per year. Begin with the combustion air source. Combustion air intake sources that have dampers or louvers should be cleaned and checked for proper operation and for any obstructions, **Figure 17-4**. Check the tightness of any blades and

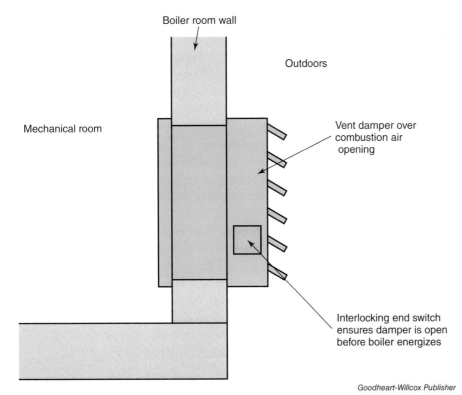

Goodheart-Willcox Publisher

Figure 17-4. Combustion air dampers and intakes should be checked for proper operation and for any obstructions.

Goodheart-Willcox Publisher
Figure 17-5. Check the vent stack for tightness and for any obstructions.

damper seals, and make sure the damper linkage moves freely. Check to see that the dampers open on a call for heat and that they open completely. If the combustion air source is made of PVC or PPE and has an outdoor screen, check to see that the screen is clean and corrosion free. Verify that the inlet pipe is connected and properly sealed.

The venting system should be visually checked for blockage, tightness, and deterioration, **Figure 17-5**. If the boiler uses a metal venting system, check to see that there are three screws securing every joint. Inspect the metal piping for any signs of corrosion or white powder. These may be caused by flue gases that are too cool before they are vented out of the building. Determine the cause of this issue and make the proper repairs. Also, repair any joints that show signs of leakage. Venting systems made of PVC or PPE piping need to be checked for tightness and to make sure that all joints are properly sealed and free of obstructions.

SAFETY FIRST

Preventing Carbon Monoxide Contamination
Failure to inspect the combustion air and vent system piping, as well as failure to make any required repairs, could result in severe personal injury or even death as a result of carbon monoxide poisoning!

17.1.3.2 Burner Maintenance

The boiler's burner assembly should be visually checked at startup after long shutdown periods or at least every six months. If the boiler has an atmospheric burner, observe the flame characteristics to determine if the flame is burning properly. A normal flame is blue with slight, orange tips. It should have a well-defined inner cone and no yellow streaks, **Figure 17-6**. The following flame characteristics can mean that there is a problem with the combustion process:

- **Yellow or lazy flame.** This can indicate a lack of combustion air caused by a lack of primary airflow as a result of a blockage or partial obstruction of airflow to the burners. It could also mean excessive gas pressure. These conditions must be corrected immediately.
- **Lifting flame.** This situation can be caused by overfiring of the burner or excessive primary air. Examine the venting system and ensure that the gas supply pressure is set correctly. Check to see that there is adequate combustion and ventilation air.

Some older boilers have adjustable air shutters on the burner tubes. Check to see that these are adjusted correctly to provide the correct flame shape and color. A combustion analyzer should be used to ensure that the boiler has adequate combustion air. On boilers with gas injection burners, there is no combustion air adjustment. On these types of burners, adequate combustion air must be supplied to the mechanical room to ensure proper burner operation.

Follow the manufacturer's maintenance guide for proper cleaning of the burner assembly. Most burners can be removed from the boiler and cleaned periodically using a vacuum or stiff bristle brush. Some burner assemblies can be cleaned by rinsing them with water. Allow the burner assembly to dry before reinstalling. Observe for any cracks

Braslavets Denys/Shutterstock.com
Figure 17-6. The burner flame should be blue with a well-defined cone and no yellow streaks.

or damage to the burner assembly and replace if necessary. Before reinstalling the burners, vacuum any dirt and debris from the combustion chamber. Once the burners have been properly cleaned and reinstalled, check the strength of the flame signal on boilers that incorporate a flame rod and flame rectification system. This testing procedure involves the use of a *microammeter*. Disconnect the flame rod wire from the burner ignition module and connect the leads of the microammeter in series with the flame rod and ignition module. Fire the burner and observe the flame signal. At high fire, the flame signal shown on the meter's display should be between 2 and 10 microamps. A lower-than-normal flame signal may indicate a fouled or damaged electrode. Clean the flame rod using a mild abrasive cloth such as a scouring pad. If cleaning the electrode does not improve the flame signal, check the ground wire on the ignition module to ensure it is securely fastened and in good condition. Otherwise, replace the flame rod.

17.1.3.3 Cleaning the Heat Exchanger

Proper cleaning of the heat exchanger should be done on an annual basis. Typically, the best time is at the end of the heating season. This allows more time for a thorough inspection. Some boilers allow for removal of the heat exchanger. Other models include fixed heat exchangers that must be cleaned internally. The manufacturer's maintenance guide will provide the best procedure for heat exchanger removal and cleaning.

On models with a fixed heat exchanger that cannot be removed, begin by shutting down the boiler and allowing it to cool. Use a vacuum to remove any dirt and debris that may have accumulated around the combustion chamber. Then use a long-handled brush to clean the interior lining and around the openings of the heat exchanger. Some water tube heat exchangers may also require special brushes to thoroughly clean the interior linings.

Giannoni-type heat exchangers found in high-efficiency condensing boilers can be removed for cleaning. These types of heat exchangers can be cleaned using soap, water, and a stiff nylon or other nonmetallic scrub brush to loosen the dirt and debris encrusted on the surface of the coils, **Figure 17-7**. When finished cleaning, apply a coating of white vinegar to the surface of the heat exchanger before reinstalling it. This will reduce the buildup of lime scale.

As with any heat exchanger removed from the boiler for cleaning, check the condition of any gasketing materials that seal the heat exchanger in place. Always replace seals and gaskets that appear to be damaged.

Braslavets Denys/Shutterstock.com

Figure 17-7. Many types of heat exchangers can be cleaned and serviced using soap, water, and a nonmetallic scrub brush to loosen dirt and debris.

17.1.4 Checking the Safety Features

One of the most important parts of a routine maintenance schedule is to check the proper operation of all boiler safety devices. The most important devices that need to be checked include, but are not limited to, the following:
- Pressure-relief valve
- Low-water cutoff switch
- Flow switch

SAFETY FIRST

PRV Piping
Before manually operating the pressure-relief valve, ensure that the discharge piping is directed to a suitable drain to avoid a potential scald hazard. The discharge piping on the relief valve must be full-sized without any restriction and installed in such a way as to permit complete drainage of both the valve and line.

17.1.4.1 Checking the Relief Valve

Pressure-relief valves should be tested, at a minimum, on an annual basis. Begin by inspecting the relief valve for any damage. Be sure the lever is in place and functional. To test the valve, lift the lever to verify flow. Upon releasing the lever, the valve should spring back to a closed position. If the valve fails to seat properly after it closes or continually leaks water, replace it. Verify that any dripping or leaking water from the valve is not due to system overpressurization, which could be caused by a defective expansion tank that is waterlogged or undersized.

17.1.4.2 Testing the Low-Water Cutoff Switch

The low-water cutoff (LWCO) switch is perhaps the most important safety device on the boiler, second only to the pressure-relief valve. Most modern boilers use a probe-type switch that senses the conductivity of the water and shuts down the burner circuit when the water level gets too low. The low-water cutoff switch should be installed on the boiler at the minimum safe water level, as determined by the boiler manufacturer.

The low-water cutoff can be tested when the boiler is empty. Energize the boiler and set the room thermostat to call for heat. Confirm that the burner will not operate without water in the system. If there is an LED light on the switch, the red light should illuminate.

TECH TIP

Testing LWCO Switches
When testing some low-water cutoff switches, the burner may energize briefly (one second or less) and then shut off to verify operation. This situation is normal.

Another method for testing this switch is accomplished when the boiler is full. Under normal operation, the LED display on the LWCO should be green, or the red LED display should be off. With the burner operating normally, slowly drain the boiler until the LED display turns red and the boiler shuts down. It is recommended that a second technician be present during this testing to shut down the system immediately if the situation becomes unsafe. *Do not allow the boiler water level to drop too low.* Shut down the system immediately if there is concern that the switch is not working properly. Once the switch has been properly tested for safe operation, check the threaded connections for any leaks. Tighten if necessary.

TECH TIP

Another Method for Testing LWCO Switches
A third method for testing the switch is to simply drain the boiler without firing the burner and check for continuity across the switch wiring. It should open when the water level falls below the preset level. This method, however, does not guarantee that the system shuts down under actual firing conditions.

17.1.4.3 Checking the Flow Switch

A flow switch is used to verify flow through the boiler before it is allowed to fire. The flow switch must be installed at the outlet of the boiler and wired in series with the 24-volt circuit that powers the gas valve or burner ignition module.

Check with the boiler manufacturer to determine the minimum water flow rate required to energize the burner circuit.

Checking the proper operation of the flow switch is similar to testing the low-water cutoff switch. Begin by setting the thermostat to call for heat. Observe that the circulating pump and burner circuit energize and that the boiler is firing properly. Disconnect power to the circulating pump. The burner circuit should immediately shut down. Another test that can be performed is to remove one wire from the terminal that connects the flow switch to the burner circuit. Again, the burner should shut down as soon as the wire is removed—even though the pump remains energized.

If for any reason the burner does not shut down, first remove the flow switch and check for any obstruction within the supply water piping. Also check to see that the paddle on the flow switch moves freely. Refer to Chapter 8, *Boiler Control and Safety Devices*, which discusses different types of flow switches.

CODE NOTE

Meeting Code Requirements

A water flow switch will meet most code requirements for a low-water cutoff device on hydronic boilers that require forced water circulation for proper operation. When in doubt, always check with local authorities having jurisdiction.

17.2 Boiler and Hydronic System Service

The ability to troubleshoot a hydronic system is one of the most important skills that the service technician can attain. Effective troubleshooting expertise can take a long time to achieve, but with the proper tools and a clear understanding of how the system operates, a technician can learn quickly how to diagnose service problems and successfully make the proper repairs.

The first step in learning how to troubleshoot a hydronic system is to understand how it is intended to operate. It has been said that you cannot know what is wrong with the system if you do not know what is right in the first place. Begin by spending some time reviewing the correct operation of the boiler and its peripheral components. Then study the piping arrangement and how individual zones are being controlled. By first becoming familiar with the overall system operation, the technician can diagnose the problem quickly and effectively.

GREEN TIP

Avoid Temporary Fixes

When a technician is in haste to complete a service call, the temptation is sometimes to fix the problem with whatever means available. This creates a situation known as a "temporary fix," which can lead to bigger problems. An improper or temporary fix cannot only cost the customer additional money from an avoidable callback, but can also have a greater impact on the environment.

Consider the amount of additional energy consumed because the system is not functioning to its maximum efficiency, or the amount of additional water used because a leaky boiler is not repaired properly.

By following proper service and repair procedures, technicians can ensure the boiler performs more efficiently, save the customer money, and reduce the environmental impact of the system. Remember, time spent performing improper service repairs—and the resulting callbacks—is time not spent repairing other systems in need.

17.2.1 Separating the Two Systems

There are many different methods the technician can use to approach troubleshooting. One method is to divide the system into two categories:

1. Issues dealing with the boiler
2. Issues dealing with the individual heating zones

As mentioned in earlier chapters, the overall hydronic system can be viewed as two separate systems divided between the boiler and the heating zones, **Figure 17-8**. The boiler is controlled by the aquastat or temperature sensing device that is intended to maintain a hot water temperature set point regardless of what is happening with individual zones. On the other hand, problems can occur with one or more zones, even when the boiler is working perfectly fine.

System problems with the boiler can occur because of the following issues:
- Gas and electrical problems
- Problems with water flow
- Control issues
- Problems with safety devices

Problems with individual zones can occur because of the following:
- Pump or zone valve issues
- Problems with the thermostat
- Air elimination issues
- Problems with the heat emitter

This section systematically covers troubleshooting according to the following categories:
- Sizing and installation issues
- Troubleshooting electrical issues
- Troubleshooting burner and venting issues
- Troubleshooting water circulation issues

17.2.2 Sizing and Installation Issues

In some cases, there may be nothing wrong with the boiler or any of the peripheral devices. When this is the case, the issue is typically due to incorrect system sizing or problems with system installation.

A
Goodheart-Willcox Publisher

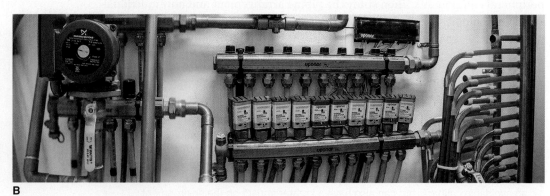
B
Lost_in_the_Midwest/Shutterstock.com

Figure 17-8. The first step in effective troubleshooting is to divide the hydronic system into two separate categories: the boiler and the individual heating zones.

TROUBLESHOOTING

Diagnosing Sizing Issues

To diagnose sizing problems, the technician should first ask the customer whether the system has ever worked properly. If there is not a definitive answer to this question, then further investigation is needed. Begin by reviewing the documentation of the project, including any available plans and specifications. Determine whether a comprehensive load calculation was performed, **Figure 17-9**. Observe the building envelope to see if it is well insulated. Look for such things as adequate sealing around doors and windows. If the condition of the structure is subpar, compare these findings to the Btu rating of the boiler and determine if there is a discrepancy between the two factors. If the boiler and hydronic system is undersized, it will not be able to keep up with demand under extreme weather conditions. Conversely, an oversized system will cycle frequently and cause temperature swings, which will lead to uncomfortable conditions for the building occupants.

Unfortunately, problems with incorrect system sizing are often expensive and sometimes impractical to remedy. Be sure you have well-documented evidence of these issues before addressing any possible solutions with the customer.

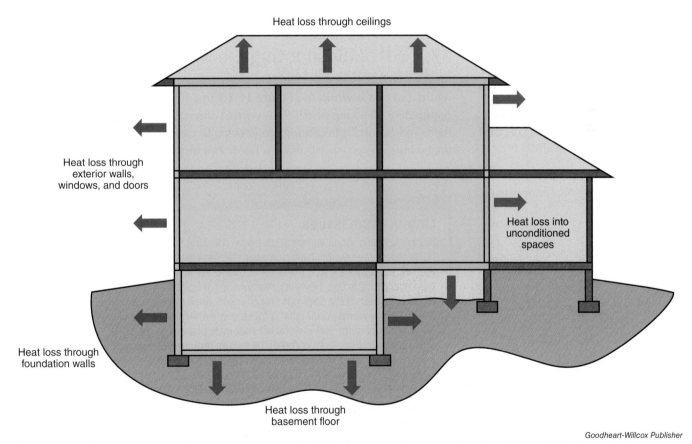

Goodheart-Willcox Publisher

Figure 17-9. Problems can arise with the performance of the hydronic system if a comprehensive load calculation is not first performed on the structure.

> **TROUBLESHOOTING**
>
> ### Diagnosing Installation Issues
>
> Issues with system installation can also be complex and sometimes difficult to diagnose. Begin by reviewing the manufacturer's installation and maintenance instructions. If none are available on-site, most installation manuals can be downloaded from the Internet. Compare the instruction manual with the actual installation. Problems that can occur as a result of incorrect installation practices can be linked to the following:
>
> - Incorrect vent sizing
> - Lack of proper combustion air
> - Incorrect hydronic piping or pump sizing
> - Undersized gas piping
> - Inadequate gas pressure at the burner manifold
> - Incorrect wiring practices
> - Undersized heat emitters
> - Poor thermostat location
> - Improper system startup

Unlike sizing issues, problems with the system installation can usually be easily remedied and at a much lower cost. Take time to thoroughly examine the overall system to determine whether the problem is a result of improper sizing or installation before proceeding.

17.2.3 Troubleshooting Electrical Issues

Most electrical issues can be diagnosed using a volt-ohm meter, sometimes referred to as a ***multimeter***, **Figure 17-10**. In addition, a microammeter may be required for checking whether the correct current is flowing through the burner flame rod circuit. By becoming proficient with electrical troubleshooting meters, the technician can solve problems faster and more efficiently.

Nazrul Iznan/Shutterstock.com

Figure 17-10. Most electrical issues can be diagnosed using a multimeter.

> **TROUBLESHOOTING**
>
> ### Line-Voltage Issues
>
> Begin troubleshooting electrical issues by checking for main power at the boiler disconnect switch. This may seem like an obvious observation, but many technicians assume that if the breaker seems set and the switch is turned on, then everything is fine. However, many hours of troubleshooting can be wasted if the service technician does not first check to see if the breaker is tripped, a fuse is blown, or a wiring connection is loose somewhere. Do not assume anything! Always first confirm that the *correct* voltage is present to the boiler, **Figure 17-11**. Most residential and light commercial boilers require 120 volts at the disconnect switch, which allows for a plus or minus differential of 10%. If the voltage to the boiler is out of this range, it may require the services of a licensed electrician to determine whether there is a problem with the electrical service to the building.
>
> The next step in electrical troubleshooting is to check to see that there is correct power to the circulating pump and burner circuit. Begin by confirming that the pump is operational. If the pump does not energize when there is a call for heat, check the pump contacts and relay located on the aquastat relay or the integrated circuit board. Again, check for any loose connections through the pump control circuit and also at the pump itself. If the pump runs for a short period of time and shuts down, the problem may be with the motor windings. Use an ammeter to measure the amp draw through the pump circuit when it is running and compare this reading with the required pump amperage rating. If the amp reading is above its rated level, the pump should be replaced.

Andrey_Popov/Shutterstock.com

Figure 17-11. Begin troubleshooting electrical circuits by confirming that the correct voltage is present at the boiler.

TROUBLESHOOTING

Low-Voltage Issues

Most boiler control circuits, including the burner circuit, are rated for 24 volts. If the burner circuit does not energize when there is a call for heat, begin troubleshooting at the control transformer. Check for 120 volts on the line side of the transformer and 24 volts on the load side. If the transformer does not produce 24 volts when it is energized, it is usually defective. Next follow the 24-volt control circuit to the room thermostat and confirm that the heating contacts are closed and voltage is passing through the contacts when there is a call for heat. One method for confirming that there is continuity through the burner circuit is to temporarily place a jumper wire across the contacts at the thermostat to see if the burner circuit energizes. This may determine whether there is a problem with the thermostat. Next check for power at the zone valve (if present). There should be either 24 volts or 120 volts to power up the valve motor, depending on its electrical rating. If voltage is present but the valve does not open, check to see if the valve is sticking. This can be a problem at the beginning of the heating season when the valve has not been active for an extended period of time. Some zone valves include end switches that prove the valve is open before the boiler is allowed to energize. Most zone valves include a manual opening lever that will allow for testing of the end switch, even when there is no call for heat. If the end switch does not close automatically once the motor has fully opened the valve, there may be a problem with the valve linkage or the switch itself.

Once it has been established that the zone circuit is working correctly, the next step is to troubleshoot the burner control circuit. This circuit may simply energize the gas valve or may have a more sophisticated integrated controller with a built-in connection board. Regardless of the type of burner controls, most circuits have a series of safety switches that must first close before the boiler is allowed to energize. These safety switches include, but are not limited to, the following:

- Low-water cutoff
- Flow switch
- High-temperature limit
- Pressure differential switch

Review Chapter 8, *Boiler Control and Safety Devices*, to ensure that you have a comprehensive understanding of how these devices operate. In addition, Chapter 4, *Gas Burners and Ignition Systems*, outlines the correct sequence of operation for various types of burner controls. Also, study the schematic wiring diagram included with the boiler to determine where all safety and control devices are located, and in what order within the circuit.

TROUBLESHOOTING

The Hopscotch Method of Troubleshooting

One preferred method of electrical troubleshooting is referred to as *hopscotching*. This technique allows the technician to easily determine the status of a safety switch using a voltmeter. To perform this method, attach one probe of the voltmeter to the common wire within the circuit. Attach the other probe before and after the switch to see if the safety is closed. If voltage is read before the switch and not after, the switch is open. Refer to the wiring diagram and move along the circuit, checking each safety switch in order.

If it has been confirmed that all safety controls are working properly, there should be 24 volts to the gas valve or to the ignition circuitry. Once this has been verified, it is now time to troubleshoot the flame circuit. Determine which type of ignition circuitry is being utilized and also review Chapter 4, *Gas Burners and Ignition Systems*, for a thorough understanding of the various types of burner controls and circuits available. Most burner circuits are controlled by either a standing pilot or by an intermittent ignition system.

TROUBLESHOOTING

Burner Circuits: Standing Pilots

If the boiler uses a standing pilot for flame safeguard control, begin by determining whether the pilot flame does not light or lights but the burner does not ignite. If the pilot does not light when depressing the gas valve pilot ignition button, or if it will not remain lit when the button is released, the issue is typically with the thermocouple, **Figure 17-12**. Test the thermocouple using a DC voltmeter and a lighter.

1. Remove the thermocouple and connect one lead of the DC voltmeter to the end of the thermocouple.
2. Connect the other lead anywhere on the side of the thermocouple's copper tubing.
3. Hold a flame to the end of the thermocouple's bimetal bulb and read the voltage output.
4. Within a short period of time, the thermocouple should produce at least 15 **millivolts** DC of power—typically with a maximum output of 30 millivolts DC.
5. If there is no voltage produced, or the signal is low, replace the thermocouple.

If the pilot remains lit, but the gas valve does not open, perform the previous test to determine if the thermocouple is producing enough voltage to hold the gas valve open. If it has been determined that the thermocouple is operable, test for 24 volts AC at the gas valve when the thermostat is calling for heat. If 24 volts is present but the burner does not ignite with the pilot lit, the gas valve is probably defective and should be replaced.

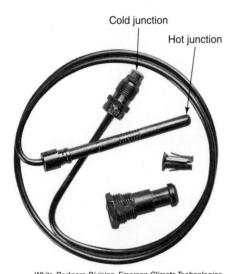

White-Rodgers Division, Emerson Climate Technologies

Figure 17-12. If the boiler has a standing pilot for burner control, the thermocouple may need to be checked.

TROUBLESHOOTING

Burner Circuits: Intermittent Ignition Systems

If the boiler uses an ignition module or an integrated circuit board for energizing the burner, the first step in troubleshooting is to verify that there is 24 volts AC to the board or module when there is a call for heat. If no voltage is present, go back to the transformer and troubleshoot through the safety circuits. Most electronic ignition controls utilize either spark ignition or a hot surface igniter to light the burner. The ignition modules and integrated circuit boards operate with *inputs* and *outputs*, **Figure 17-13**. When a 24-volt input energizes the burner circuit, the board or module signals an output to the spark ignition or hot surface to operate. Simultaneously, the pilot valve or gas valve opens through an output, and ignition takes place. Once the burner has ignited, a flame rod or the hot surface igniter sends a signal back to the burner circuit to hold the gas valve open.

TROUBLESHOOTING

Spark Ignition Systems

If a boiler with a spark ignition system fails to light, it may be due to a lack of spark, a lack of pilot gas, or both. Begin troubleshooting by verifying there is a spark. This is a visual observation. If no spark is present, first determine any of the following:

- The ignition transformer is defective.
- There is an incorrect gap between the spark igniter and the ground rod, **Figure 17-14**.
- The porcelain insulator on the spark igniter is cracked.

If a spark is present but the pilot fails to light, check for voltage at the pilot valve. If no voltage is present, either the pilot valve or the ignition module may be defective.

Fenwal Controls

Figure 17-13. Some boiler burner controls operate from an integrated control board such as the one shown here.

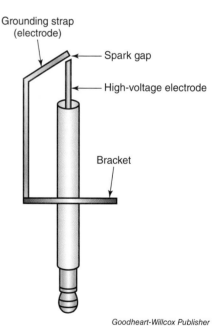

Goodheart-Willcox Publisher

Figure 17-14. Check for the correct gap between the grounding strap and high-voltage electrode when troubleshooting a spark ignition system.

TECH TIP

Using High-Voltage Test Equipment
Do not use a conventional voltmeter to troubleshoot the ignition transformer. This device can produce up to 10,000 volts and can permanently damage a conventional voltmeter. Special high-voltage test equipment must be used to check ignition transformers.

TROUBLESHOOTING

Hot Surface Ignition Systems
The sequence of operation for a hot surface ignition system is similar to that of spark ignition systems. Upon a call for heat, the igniter is energized and lights the pilot flame or the main burner. A flame rod is then used to verify a signal back to the ignition module that a flame is present. If the igniter fails to glow when energized, perform the following test:

1. Disconnect the leads to the igniter coil and check for voltage from the ignition module to the igniter.
2. If voltage is not present, determine whether the ignition module is defective.
3. If voltage is present, examine the igniter coil for any cracks or breakage. If there are any signs of damage, replace the igniter.
4. If the igniter coil looks acceptable, measure the ohms resistance through the coil—there should be between 30 and 100 ohms, **Figure 17-15**.
5. If the ohm reading is either zero or infinity, replace the igniter.

Sealed Unit Parts Co., Inc.

Figure 17-15. Check for the proper voltage and ohms resistance reading when troubleshooting a glow coil.

TROUBLESHOOTING

Checking the Flame Rod Signal
If the burner lights but fails to remain lit, check the flame rod signal using the following procedure:

1. Remove the wire that connects the flame rod to the ignition module, **Figure 17-16**.
2. Using a microammeter, connect one lead to the flame rod and the other lead to the flame module (connect in series with the flame rod).
3. Ignite the burner and read the flame signal—it should read between 2 and 10 *microamps*.
4. If there is not a significant signal, replace the flame rod.
5. If the flame rod is producing an adequate signal, clean the flame rod using a mild abrasive cloth. At times, dust and dirt can accumulate on the flame rod, impeding the signal.

TECH TIP

Manufacturer's Troubleshooting Matrix
Some boiler manufacturers' installation manuals include a troubleshooting matrix for diagnosing the burner's electronic ignition circuit. To save time, always consult this matrix when troubleshooting an electronic control system.

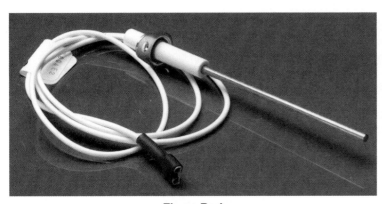
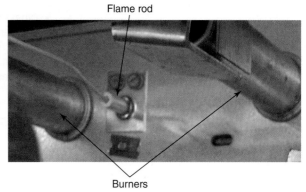

Flame Rod — *Emerson Climate Technologies*
Flame Rod Installed — *York International Corp.*

Figure 17-16. A flame rod is used to supervise the burner flame and will need to be checked using a microammeter.

17.2.4 Troubleshooting Burner and Venting Issues

Issues with the boiler's burner and its controls can be caused by problems with venting or a lack of combustion air. The venting system is used to carry the burner's products of combustion out of the building. Boilers need adequate combustion air in order for the burner to work effectively. Problems with the burner circuit can also arise, which can be linked to a mechanical issue. The following section covers both of these topics.

Whether the boiler is a standing pilot natural draft model, an 80% efficient model with a combustion inducer blower, or a high-efficiency model that has a direct vent system, the most obvious cause of venting problems is an obstruction in the vent itself. Obstructions can be caused by a number of issues and can result in ignition failures due to the pressure switch not closing—or worse, a dangerous situation in which the products of combustion are allowed to flow backward into the boiler room.

TROUBLESHOOTING

Checking Venting Problems

On natural draft boilers, begin by examining the venting system to determine whether there is an obstruction present. Also check the draft hood to ensure there is a clear path for the room air to be entrained into the ***vent stack***, **Figure 17-17**. In addition to obstructions caused by such things as nests, leaves, and vermin or by excessive snow loads, the venting itself may be corroded and collapsing onto itself. If this is the case, the venting system needs to be replaced.

Other problems due to venting issues include:

- **Faulty pressure switch:** Pressure switches are installed on boilers that have inducer fans to sense a negative pressure created by the boiler's draft. Upon sensing the proper draft, the switch will close and allow the boiler to fire. These sometimes fail and will need to be properly diagnosed. Use an ohmmeter to determine if the pressure switch closes when manually drawing air through the tubes connected to the pressure switch. If the switch does not close when performing this test, replace the pressure switch. In addition to checking the pressure switch itself, examine the hoses connected to the pressure switch for cracks or breakage—these conditions can also prevent the switch from closing.

(continued)

Figure 17-17. The draft hood is used on natural draft boilers to entrain room air into the venting system. Obstructions in the draft hood can lead to problems with proper venting.

Goodheart-Willcox Publisher

- **Clogged condensate drain:** High-efficiency boilers have condensate drains that can cause burner problems if they become clogged. This situation can also prevent the ignition pressure switch from closing, preventing flame ignition. Causes of a clogged condensate drain include dirt, debris, or frozen condensate. Because the ignition pressure switch senses the accumulation of condensate in the boiler's condensate trap, the burner will not fire until the drain has been cleared and the condensate is allowed to flow freely.
- **Improper exhaust gas recirculation:** On condensing boilers, improperly installed intake and exhaust vents on the outside of the building can create short-circuiting of the flue gases. This situation allows the exhausted flue gases to recirculate into the combustion air intake vent and is usually caused by installing the vent and intake piping too close together. When the exhausted gases are allowed to be drawn back into the combustion air intake, the boiler's burner does not have enough oxygen for proper combustion. This situation can be resolved by the use of a concentric vent kit or by properly installing both vent and intake piping at the proper distance apart. Most installation guides show the proper distances for vent terminations outside of the building.

TROUBLESHOOTING

Replacing the Flue Liner
Some medium-efficiency boilers may employ the use of a flue liner run through an existing chimney in order to meet code requirements. Depending on the age of the boiler and flue liner, the liner itself can deteriorate and collapse. Check for proper drafting at vent connections or at the draft hood by using a flame while the boiler is firing. If the boiler is not drafting properly, consider replacing the flue liner.

SAFETY FIRST

Flame Rollout Switch
All boilers, regardless of age or efficiency, should have a flame rollout switch located near the burner inlet inside the boiler's *vestibule*. When the burner vent becomes obstructed, the flame is starved for oxygen and will seek combustion air from wherever it can—including inside the boiler room! When this situation occurs, the burner flames will roll out into the boiler's vestibule, which can cause the control wiring to burn and other unsafe conditions. If the rollout switch becomes tripped, the technician must first determine the cause and make the proper repairs before resetting the switch and refiring the boiler.

SAFETY FIRST

Replacing the Pressure Switch
When replacing a pressure switch, be sure to check for the proper switch rating. This is expressed in inches water column (in. WC) and should be displayed on the switch's rating tag. Installing an incorrectly rated switch can result in the switch closing before the proper pressure is reached, or not closing at all.

CODE NOTE

Proper Vent Slope

Regardless of the type or efficiency of the boiler, proper sloping of the venting is essential. Make sure that the minimum upward slope of the vent from the boiler is 1/4″ of rise for every 1″ of horizontal run. Exhaust venting that is sagging or improperly sloping can collect condensate and restrict airflow, resulting in a tripped pressure switch. In addition, be sure that the vent pipe is supported every 5″ of horizontal run.

TROUBLESHOOTING

Checking Problems with Combustion Air

In addition to venting issues, problems with proper burner ignition can be linked to inadequate combustion air. In some mechanical rooms, the boiler may be large enough that combustion air needs to be supplied by mechanical means. The source may be from intake fans or from an opening that uses a mechanical damper for control. When this is the case, problems can arise from a faulty damper motor, improper damper linkage, a defective intake fan motor, or an intake blocked by debris or snow, **Figure 17-18**. All of these items would need to be addressed if it is determined that the boiler is not receiving enough combustion air.

In some cases, the combustion air intake was never sized properly. This situation can also lead to problems with burner performance and can lead to an unsafe situation within the building. Review Chapter 14, *Boiler System Design Considerations*, which discusses the proper procedure for determining whether the room in which the boiler is housed should be classified as a confined or unconfined space. In addition, use the sizing procedure found in Chapter 14 for properly sizing combustion air openings.

Burner issues can also arise because the ***manifold gas pressure*** was never checked or adjusted at the time of installation and startup. Both overfiring and under firing can lead to combustion issues, poor burner performance, and premature boiler failure. Check the manifold gas pressure using a manometer, and if necessary, adjust it to the following settings:

- For natural gas, the manifold pressure should be 3.5 in. WC.
- For LP, the manifold pressure should be 11 in. WC.

Also remember to check for the proper inlet gas pressure using a manometer, especially for boilers using LP gas.

Finally, an annual inspection and cleaning of the burner assembly will ensure that the boiler performs at the optimum combustion efficiency. Follow the boiler manufacturer's maintenance guide for the proper cleaning procedure. Also include a combustion analysis as part of an annual maintenance program.

Van Topics/Shutterstock.com

Figure 17-18. Outside air intakes need to be checked to ensure they are not blocked when troubleshooting problems with combustion air.

17.2.5 Troubleshooting Water Circulation Issues

Water circulation issues can be linked to either mechanical problems or air within the system. Any of these problems can create system noise and lead to poor space heating. Mechanical problems can usually be linked to issues with circulating pumps. Air within the system can be caused by a variety of problems. This section looks at the probable causes of poor water circulating issues.

TROUBLESHOOTING

Circulating Pump Issues

Begin by first determining that there is power to the circulating pump. Use a voltmeter to confirm that the pump is receiving 120 volts. Check to see that the aquastat relay or boiler control board is sending a signal to the circulating pump whenever there is a call for heat. Now confirm that the circulating pump is working properly, **Figure 17-19**. This can be done by measuring the water temperature into and out of the pump when the burner is firing. If the outlet temperature is not close to the inlet temperature, there may be a problem with the pump's impeller or linkage. A broken linkage needs to be replaced. Otherwise, de-energize the pump, valve off the water into and out of the pump and disassemble the pump to inspect the impeller. If there are signs of damage or if the impeller is seized up, it needs to be replaced. In some cases, it may be faster and more economical to replace the entire circulating pump. Other problems with the circulating pump can be a defective motor or a leaking bearing assembly. Depending on its age, both of these issues are usually remedied by simply replacing the circulating pump.

The next issue that can cause problems with water circulation is a ***hydronic airlock***. Oxygen-rich water that is introduced into the system as makeup water can cause a buildup in the amount of air trapped within the system. This excess

Dimitri Kalinovsky/Shutterstock.com

Figure 17-19. Check for proper operation of the circulating pump by measuring the inlet and outlet temperatures—they should be similar.

air can find its way to risers on radiators and to other heat emitters. Most excess air will be automatically eliminated by the system's air separator and air vent. However, when these components cannot remove enough excess air or if they become defective, problems can arise.

TROUBLESHOOTING

Problems with Air in the System

If the heat emitters have manual air vents, begin by bleeding as much air as possible from the heat emitter at each zone. When this step does not solve the problem, the system should be purged. Purging involves forcing makeup water through the system to remove any excess air from the hydronic system. To do this, the hydronic piping system needs to include a purge or drain valve and an isolation shut-off valve as shown in **Figure 16-3** in Chapter 16, *Boiler Startup*.

The following steps outline the proper procedure for purging the hydronic system:

1. Begin by shutting down the boiler and allowing it to cool. Avoid introducing cold water into a hot system, as this can cause the boiler to crack.
2. Connect a hose to the purge or drain valve and terminate the other end into a drain or bucket.
3. Open the shutoff valve to the makeup water connection and engage the fast-fill handle on the pressure-reducing valve to increase the inlet water pressure into the system.
4. Close the isolation valve near the boiler and open the purge or drain valve. This will divert water through the drain hose. Air should now be forced out of the system.
5. When a solid stream of water is coming out of the drain hose and no air is present, disable the fast-fill handle on the pressure-reducing valve and close the shutoff valve to the makeup water connection.
6. Close the purge or drain valve and open the isolation valve near the boiler.
7. Check the water pressure on the indicator dial and adjust the system water pressure to its proper level. This may involve adding or removing water from the system.
8. Energize the boiler's burner circuit and allow the system to heat up. Check for any leaks around the system.
9. When the system has reached its operating temperature, recheck the system water pressure for its correct level and determine whether each zone is operating properly.

If the hydronic system has a primary-secondary loop, or if it incorporates zones controlled by either circulating pumps or zone valves, the aforementioned procedure can be used along with the following modifications:

- **Primary-secondary loop:** When purging a piping system that has primary and secondary loops, begin by isolating the secondary loops so that only the primary loop will be purged. After the primary loop is complete, purge the secondary loops individually by moving the hose to the drain valve on each respective secondary loop. Purge each individual loop until they are clear of bubbles. Finish by completing Steps 6 through 9 above, **Figure 17-20**.

(continued)

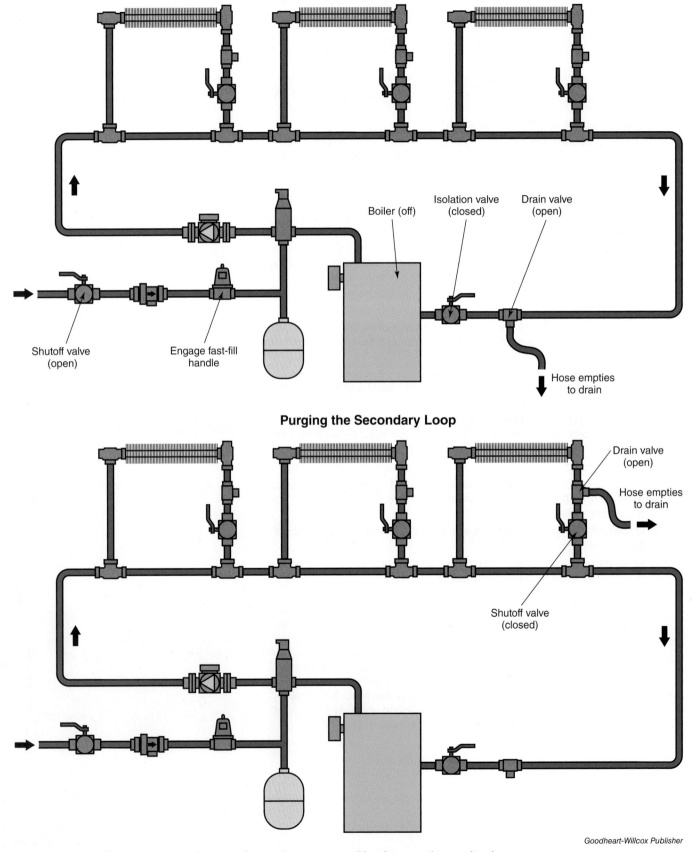

Figure 17-20. This illustration shows the setup for purging a system with primary and secondary loops.

- **Zoned system using circulating pumps:** Close the shutoff valves on the boiler return line and on all system zones. Attach the drain hose to each zone drain valve individually and flush each zone in sequence. When each zone is clear, move the drain hose to the boiler's main return line, and ensure that the entire system is clear of any bubbles. Finish by completing Steps 6 through 9, **Figure 17-21**.
- **Zoned system using zone valves:** Follow the steps used for zones with circulating pumps except manually open the individual zone valve for each zone being purged. Repeat the process for each zone, and finish by purging the entire system. Complete the process by following Steps 6 through 9, **Figure 17-22**.
- **Systems using glycol:** When glycol is used in the hydronic system, use a technique that reclaims the glycol solution during the purging process. This usually involves capturing the solution in a bucket or pail. Use a transfer pump to recirculate the glycol from the bucket back into the boiler supply line. Allow the solution to recirculate through the system until all air bubbles are removed. Make sure that the ends of each hose are fully submerged in the bucket to prevent any air from making it back into the system. When the system is air free, close all valves and start up the system. Check for any leaks, **Figure 17-23**.

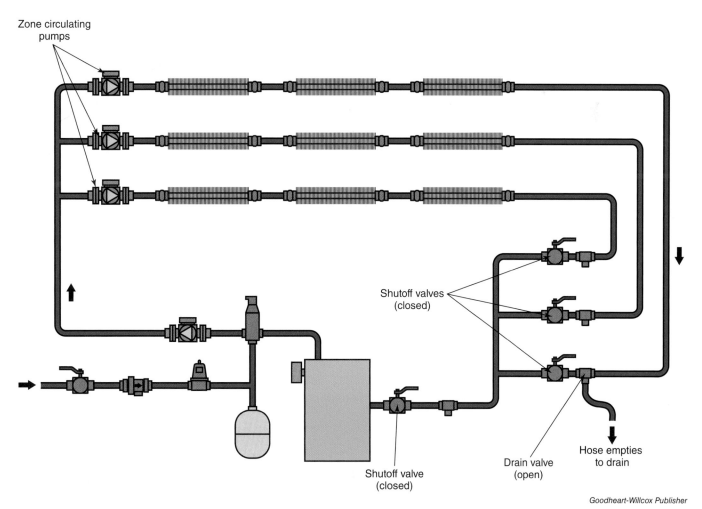

Figure 17-21. This illustration shows the setup for purging a system with zoned circulating pumps.

370 Hydronic Heating: Systems and Applications

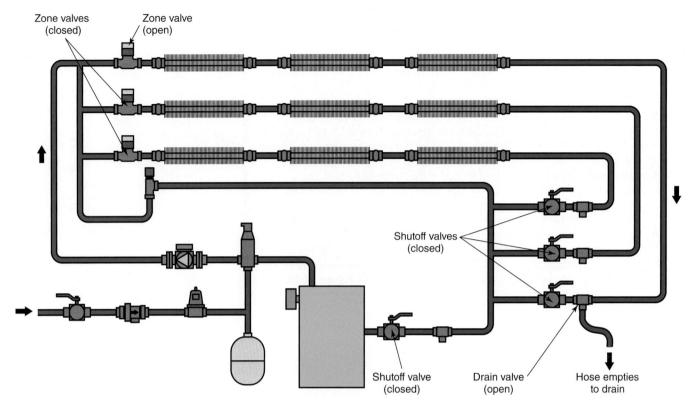

Figure 17-22. This illustration shows the setup for purging a system with zone valves.

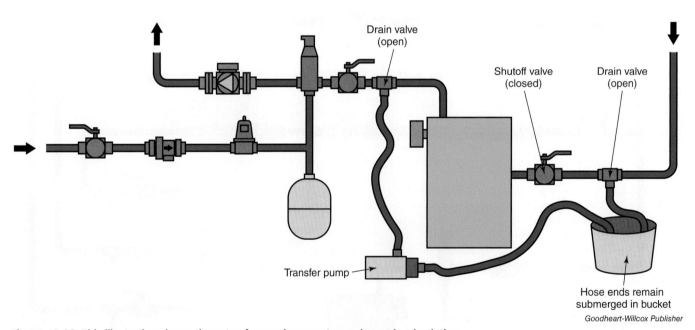

Figure 17-23. This illustration shows the setup for purging a system using a glycol solution.

TROUBLESHOOTING

Expansion Tank Problems

The expansion tank is a critical element in the hydronic system, as it allows a place for the water to expand when heated. However, the rubber bladder within the expansion tank can rupture or tear, leading to the tank becoming waterlogged. When this situation occurs, the pressure within the hydronic system rises while the boiler is firing to the point that the pressure-relief valve opens, releasing water until the pressure falls back to its normal level. If this situation is allowed to continue, water that is released from the boiler will need to be replenished by fresh, oxygen-rich makeup water. Over time, air and oxygen from the makeup water can cause problems with airlocks and eventually lead to corrosion within the system. If the pressure-relief valve continues to periodically leak over time, consider replacing the expansion tank. One way to check the expansion tank for a possible internal rupture is to measure the temperature at the top and bottom of the tank when the water supply has been heated to its normal operating level. If the temperatures are consistent, the tank may be in trouble. Be sure to check the tank pressure as a part of routine maintenance. An underinflated tank can also lead to the problems mentioned previously.

CAREER CONNECTIONS

Maintenance Contracts

A maintenance contract or service plan is an agreement between the home or building owner and the company performing the service or services. These plans or agreements are designed to prevent large-scale service bills from occurring by performing planned maintenance to the boiler and hydronic system. By checking the equipment on a regular basis, the technician not only ensures that the system is operating at optimum efficiency but can spot potential problems and service them before they become even bigger problems. Service may be performed on a monthly, semiannual, or annual basis, depending on the size and complexity of the hydronic system.

Most maintenance contracts or service agreements include a list of the services provided for an agreed upon price. If there are additional repairs or if replacement parts are needed, these generally are billed to the customer at a separate rate. In addition, most maintenance contracts do not include emergency repairs, so if the boiler fails in the middle of the night, it will cost extra.

In addition to providing "insurance" to the boiler and heating system by making sure it is routinely serviced, maintenance contracts can be a great benefit to companies who are not large enough to justify employing an in-house maintenance technician. In addition, the specialist who is providing the services can educate the home or building owner and their staff on how their system operates and what to look for on their own. This can build a rapport and trust with the customer, which in turn can foster return business.

Chapter Review

Summary

- Boiler maintenance begins with a visual inspection of the water piping, gas piping, venting system, and electrical controls.
- Checking for and repairing leaks is another important part of a comprehensive maintenance program.
- Maintenance on the boiler's combustion system includes checking the venting system, combustion air sources, burner components, and the boiler heat exchanger.
- Burner maintenance includes checking the flame signal and having an understanding of the correct flame characteristics.
- One of the most important parts of a routine maintenance schedule is to check the proper operation of all boiler safety devices.
- One troubleshooting method is to divide the system into issues dealing with the boiler and issues dealing with the individual heating zones.
- In some cases, problems with the hydronic system can be linked to improper sizing and installation issues.
- Troubleshooting electrical issues can be divided into problems with line voltage and problems with low voltage.
- Troubleshooting burner issues includes problems with ignition systems and controls.
- Problems with the boiler's burner system can be linked to issues with the venting system and with combustion air.
- Troubleshooting water circulation issues commonly deals with the circulating pump and with problems with air in the system.
- Following the proper purging procedure can eliminate trapped air within the system.
- Issues with the expansion tank can lead to system waterlogging and problems with the system's pressure-relief valve.

Know and Understand

1. Leaks caused by boiler corrosion can result from _____.
   A. too much oxygen present in the boiler water
   B. a pH imbalance in the boiler water
   C. poor sealing of joints
   D. Both A and B are correct.
2. A burner that has a lazy or yellow flame can be caused by _____.
   A. excessive gas pressure
   B. lack of combustion air
   C. an oversized vent
   D. the wrong size gas piping
3. What device is used to test the strength of the flame signal?
   A. Microammeter
   B. Ohmmeter
   C. Millivolt meter
   D. Milliamp meter
4. A zone valve has the proper voltage when there is a call for heat but does not allow the boiler burner to energize. What could be the problem?
   A. The zone valve is undersized.
   B. The zone valve end switch is not making contact.
   C. The aquastat relay is defective.
   D. The transformer is not sending voltage to the zone valve.
5. *True or False?* The boiler flow switch can be tested by disabling the circulating pump.
6. Frequent boiler cycling and temperature swings that lead to uncomfortable conditions for the building occupants could be caused by _____.
   A. an oversized system
   B. an undersized system
   C. lack of combustion air
   D. Both A and B are correct.
7. A transformer that measures 120 volts on the primary side but does not produce 24 volts on the secondary side indicates _____.
   A. a problem with the primary voltage
   B. a mis-wired transformer
   C. a poor electrical connection
   D. a defective transformer
8. What device is used to prove the flame with a standing pilot burner system?
   A. Flame rod
   B. Thermocouple
   C. Hot surface igniter
   D. Glow coil
9. A pressure switch that does not close when there is a call for heat can be caused by _____.
   A. a plugged vent
   B. a clogged condensate line
   C. air in the system
   D. Both A and B are correct.
10. A waterlogged expansion tank can result in _____.
    A. the pressure-relief valve opening
    B. leaks on the floor of the boiler room
    C. burner lockout
    D. water hammer

Apply and Analyze

1. What steps should be taken to repair boiler leaks?
2. What issues can result if the boiler does not have sufficient combustion air?
3. Describe the procedure for performing burner and heat exchanger maintenance on the boiler.
4. Explain why it might be advantageous to divide the hydronic system into two separate categories when performing troubleshooting procedures.
5. Differentiate between troubleshooting line-voltage issues versus low-voltage issues.
6. Why are there different procedures for troubleshooting various types of burner ignition circuits?
7. Explain the difference between troubleshooting venting issues and troubleshooting combustion air issues.
8. Describe the procedure for troubleshooting circulating pump issues.
9. How does purging a system differ if it has a zoned system using valves or pumps?
10. What are some methods for diagnosing a system with a defective expansion tank?

Critical Thinking

1. What steps could the service technician take to troubleshoot a boiler if there is no installation and service guide to follow?
2. At what point should the service technician discuss with the owner replacing the boiler?

18 Domestic Hot Water Production

Chapter Outline

18.1 Types of Equipment Used for Domestic Hot Water Production
18.2 Sizing for Domestic Hot Water Production
18.3 Tankless Boilers
18.4 Indirect Water Heaters
18.5 Combi Boilers
18.6 Hybrid Water Heaters

nikkytok/Shutterstock.com

Learning Objectives

After completing this chapter, you will be able to:
- Explain how the generation of domestic hot water can be an added benefit to a hydronic system.
- List the different types of equipment that can be used for domestic hot water production.
- Discuss the design factors that go into sizing the devices used for hot water production.
- Explain how the recovery rate of a hot water storage tank affects its production.
- List the advantages of a tankless hot water system.
- Describe how a tankless boiler system operates.
- Describe how an indirect water heater functions.
- List the steps taken when sizing an indirect water heater.
- Explain why an anti-scald valve is important to an indirect water heater.
- Describe the steps taken to service and maintain an indirect water heater.
- Explain how a combi boiler can be used for both comfort heat and domestic hot water production.
- Describe the steps taken to service and maintain a combi boiler.
- Describe how a hybrid water heater uses ambient air for heating water.
- Explain how to service a hybrid water heater.

Technical Terms

anti-scald valve
auxiliary storage tank
combination boiler
heat pump
hybrid water heater
indirect water heater
on-demand water heater
point-of-use heater
potable hot water
recovery rate
standby heat loss
temperature rise
thermostatic mixing valve
whole-house heater

An added benefit of utilizing a modern boiler or hydronic heating system is the ability to generate domestic hot water. Today's boilers offer a number of different options and features for producing *potable hot water* for residential and light commercial applications. Hot water can be used for many domestic applications such as cleaning, sanitizing, bathing, and laundry. It can also be applied to many commercial applications such as car washes, apartment buildings, bakeries, and grocery stores. Many households and commercial establishments can benefit from the additional features that these types of boilers provide.

This chapter focuses on the types of equipment available to keep a home or building warm while also producing an abundance of clean, hot water.

18.1 Types of Equipment Used for Domestic Hot Water Production

Along with the production of hot water for comfort applications, the hydronic industry provides a number of different types of equipment to generate potable hot water for both domestic and light commercial applications. Boilers available for use with potable hot water production generally fall into four categories:
- *Tankless Boilers.* As the name implies, these hot water generators utilize a heat exchanger to heat the water without the use of a storage tank, **Figure 18-1**. The tankless boiler transfers heat from the gas burner to the cold water that enters the boiler by means of a heat exchanger. When a hot water tap is turned on, the incoming water passes through the heat exchanger, which then heats the water to the desired set point.

Weil-McLain

Figure 18-1. Tankless boilers utilize a heat exchanger to heat water without the use of a storage tank.

- ***Indirect Water Heaters.*** This type of hot water generating system utilizes the building's existing boiler to heat water, **Figure 18-2**. A heat exchanger located inside of a buffer or auxiliary tank is piped into the building's existing boiler, which heats the potable water supply based on the occupants' demand.
- ***Combination Boilers.*** A combination or "combi" boiler offers both space heating and domestic hot water production in a single package, **Figure 18-3**. Most combi boilers are equipped with a primary and secondary heat exchanger. This type of package can provide both comfort heat and domestic hot water production during the winter months and operate only when there is a call for domestic hot water during the summer months.
- ***Hybrid Water Heaters.*** A hybrid water heater is essentially a heat pump water heater. This type of system blends together the mechanics of both a traditional and electric water heater, **Figure 18-4**.

Although there may be some crossover between these different types of boilers in regard to their function and design, or whether they are simply used for potable hot water production or for comfort heating purposes, the end results are the same: they offer the consumer alternatives to conventional "tank" type water heaters.

18.2 Sizing for Domestic Hot Water Production

There are several factors to consider when deciding which type of hot water system to purchase.

If the system is to be used exclusively for potable hot water, the first step in choosing the right system is to size the equipment for domestic hot water production. If the system is to be used for both domestic hot water production and for comfort heating, then other factors need to be taken into consideration.

Freeyoumind/Shutterstock.com

Figure 18-2. An indirect water heater utilizes the building's existing boiler to heat water.

Goodheart-Willcox Publisher

Figure 18-3. A combi boiler can provide both space heating and domestic hot water production in a single package.

Rheem Manufacturing Company

Figure 18-4. A hybrid water heater utilizes the principles of a heat pump to generate hot water.

Let us begin our discussion by first examining some of the design factors that go into sizing the devices used for hot water production:
- Residential peak usage for domestic hot water is typically a two-hour period during the day when the heaviest hot water draw will occur, typically from 7:00 to 9:00 am. To calculate residential peak usage, add up the following factors:
- The average hot water temperature for domestic usage is 120°F.
- Twenty gallons of water per person are required for the first two people.
- Five gallons of water per person are required for each person after the first two.
- Ten gallons of water are required for each full bath over the first bath.
- Ten gallons of water are required for an automatic dishwasher.
- Twenty gallons of water are required for an automatic washing machine.

Another important factor when sizing water heating equipment is *temperature rise*. This factor takes into account the incoming water temperature and the actual desired hot water temperature. For instance, the average incoming water temperature for most domestic applications is 50°F. If the desired hot water temperature is 120°F, then the temperature rise would be 70°F (120 – 50). Also, the designer will need to determine whether or not a storage tank will be utilized. Tanks used for hot water production are rated by their capacity in gallons and also by their recovery rate. By definition, a water heater tank's *recovery rate* is equal to the amount of hot water in gallons that can be produced in one hour after the tank is completely drained, **Figure 18-5**. Because the burner on a water heater is energized once hot water consumption begins, the actual recovery rate may be a larger number than the tank's physical capacity. Tank manufacturers will post both the temperature rise and the "first-hour rating" in their equipment specifications. This second number is the listed recovery rate.

TIMEOUT FOR MATH

Recovery Rates: Example 1
A family of four has two full baths plus an automatic dishwasher and a washing machine. If the incoming water temperature is 50°F and the desired hot water temperature is 120°F, what required recovery rate would be needed for a tank-type water heater?

First determine the requirements in gallons per minute based on the above information:

Two persons at 20 gallons/person = 40 gallons
Two persons at 5 gallons/person = 10 gallons
Second full bath = 10 gallons
Automatic dishwasher = 10 gallons
Automatic clothes washer = 20 gallons
Total two-hour peak hot water usage = 90 gallons
Answer: A recovery rate of 45 gallons per hour at a 70°F temperature rise would be required for the first two hours of use.

Tankless or *on-demand water heaters* are rated by the maximum temperature rise possible at a particular flow rate—typically in gallons per minute (GPM). Therefore, to size these types of systems, the designer needs to determine the flow rate and the temperature rise required for the application. For instance, first list the number of hot water devices that will be used at any one time. Then total the flow rate in gallons per minute of these devices. (The equipment manufacturer lists the GPM flow rate of their appliance or faucet.) This number will be the desired rate for an on-demand type heater.

Recovery Rate for Gas Water Heaters (GPH, 75% Efficiency)					
	Temperature Rise (°F)				
Input Rating (Btu)	60	70	80	90	100
30,000	45.5	39.0	34.1	30.3	27.3
35,000	53.0	45.5	39.8	35.4	31.8
40,000	60.6	51.9	45.5	40.4	36.4
50,000	75.8	64.9	56.8	50.5	45.5
60,000	90.9	77.9	68.2	60.6	54.5
70,000	106.1	90.9	79.5	70.7	63.6
80,000	121.2	103.9	90.9	80.8	72.7
90,000	136.4	116.9	102.3	90.9	81.8
100,000	151.51	129.9	113.6	101.0	90.9

Source: US Department of Energy

Figure 18-5. A water heater's recovery rate is equal to the amount of hot water in gallons that can be produced in one hour after the tank is completely drained.

TIMEOUT FOR MATH

Recovery Rates: Example 2

A demand-type water heater will need to be sized to meet the following criteria for household devices that often need to run simultaneously:
    One hot water faucet with a flow rate of 0.75 GPM
    One showerhead with a flow rate of 2.5 GPM
    One dishwasher with a flow rate of 1.5 GPM

If the incoming water temperature is 40°F and the desired hot water temperature is 125°F, what would be the required rate for this demand-type water heater?

Answer: A minimum flow rate of 4.75 gallons per minute and a minimum temperature rise of 85°F would be required for this equipment.

18.3 Tankless Boilers

Whether used exclusively for heating domestic water or for comfort heating, tankless boilers and water heaters have distinct advantages over conventional tank-type heaters. Tankless heaters or *gas-fired instantaneous heaters* do not use a tank for storage purposes. They work as an on-demand heating device, which means they energize only when there is a call for heat. Because of this, tankless heating systems offer considerable energy savings. One way that a tankless system saves money is because it does not have standby heat loss. With a conventional hot water storage tank, the burner or heating element cycles on a continuous basis to maintain the desired water temperature. This occurs even when the water is not being used. This excess energy used to maintain a constant water temperature is called **standby heat loss**. With a tankless system, this energy loss is absent, thus reducing energy consumption. Furthermore, tankless systems often have up to a 20-year life span as opposed to a conventional water heater, which may last only 10 to 15 years. Other advantages of tankless systems include the fact that they never run out of hot water and they take up less space than conventional water heaters and boilers.

One disadvantage of a tankless system, however, is the initial equipment cost. An average tankless system can cost up to three times more than a conventional water heater. This may not be the case if the system is used exclusively for comfort heat. Another drawback with a tankless system is the limited number of devices that can be served simultaneously. Typically, tankless systems provide hot water at a rate of two to five gallons per minute. This may not be sufficient if there are too many devices calling for heat at the same time.

Tankless water heaters and boilers operate by activating the heating source when there is a call for hot water. An internal flow switch within the appliance detects water flow and energizes the heating circuit, **Figure 18-6**. A robust heat exchanger heats the incoming water to the desired temperature and delivers that water to all devices being served. Tankless heaters are classified as either point-of-use or whole-house heaters.

Point-of-use heaters are typically smaller heaters that service only one or two outlets. These units are usually located close to the devices that they serve, **Figure 18-7**. For convenience's sake, they may use electric heating elements instead of being gas fired. Some applications for point-of-use heaters include bathrooms, hot tubs, and laundries.

Whole-house heaters are larger units with greater capacity. As the name implies, these heaters are sized and designed to service the entire heating or hot water requirements.

A combination of whole-house units and point-of-use heaters can be utilized to satisfy larger commercial applications.

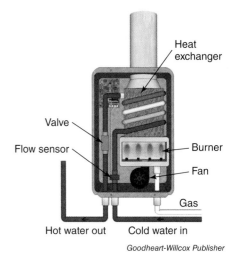

Goodheart-Willcox Publisher

Figure 18-6. This illustration shows how a tankless water heater operates.

Bosch Thermotechnology

Figure 18-7. Shown here is an example of a point-of-use water heater.

GREEN TIP

Saving Money with Tankless Water Heaters
According to HomeAdvisor.com, 15% of the energy loss experienced by most water heaters comes after the water has been heated and is sitting in the tank waiting to be used. A tankless water heater heats the water on demand, eliminating this form of wasted energy and reducing hot water energy costs by as much as 20%. In addition to this, tankless units tend to last five to 10 years or longer than traditional tanks—thereby saving money with less frequent replacements.

18.3.1 Installation and Control of Tankless Boilers

The installation of tankless boilers and water heaters is similar to that of a conventional boiler, with a few exceptions. Most tankless heaters are wall-hung, which means they need to be securely attached to the wall using the appropriate fasteners. Follow the same procedures for connecting the electrical wiring, gas piping, and water piping as with conventional boilers, **Figure 18-8**. Provide the appropriate safety devices as with a conventional boiler and test them for proper functionality. If the heater is to be installed where freezing may be an issue, follow the proper steps for freeze protection as outlined in the manufacturer's installation manual. Some manufacturers will specify the correct venting material for their product. In some cases, this may be polypropylene as opposed to PVC, CPVC, or ABS materials. Again, check with the installation guide for the proper material. Because most tankless heaters are condensing-type appliances, the piping configuration used is not as critical as it would be with a noncondensing appliance.

The controls for most tankless boilers and water heaters are self-contained microprocessor controllers, **Figure 18-9**. These types of controllers are menu-driven and easily set up and configured for either hot water generation for domestic usage or for comfort heating. If the unit is to be used specifically for comfort heating, an outdoor reset controller is typically included as standard equipment.

18.3.2 Service and Maintenance of Tankless Boilers

Service and maintenance on tankless boilers and water heaters requires an annual inspection that includes the following tasks:

1. Cleaning of the control compartments, burners, and circulating air passageways using pressurized air to remove dust and dirt.
2. Inspection of the venting system and combustion air intakes for any blockages or damage.
3. Visual inspection of the burner to ensure that the flame evenly spreads over the entire burner surface when in operation. Check to see that the flame is burning with a clear blue, stable flame.
4. In areas with hard water, the heat exchanger should be flushed periodically with a descaling solution to prevent the buildup of lime scale, **Figure 18-10**. Consult the manufacturer's maintenance guide for proper procedures.

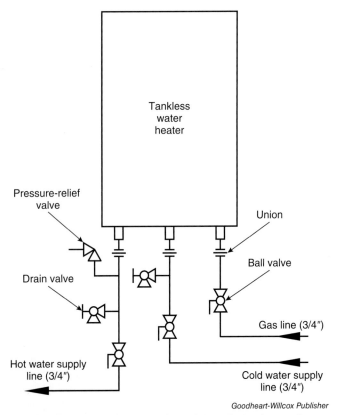

Figure 18-8. Follow the manufacturer's installation instructions for proper piping when installing a tankless water heater.

Figure 18-9. The controls for most tankless boilers and water heaters are self-contained, microprocessor-type controllers.

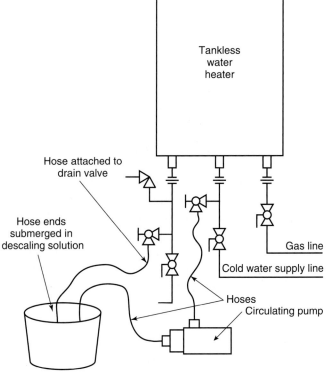

Figure 18-10. The heat exchanger on a tankless boiler needs to be flushed periodically with a vinegar solution if there is evidence of limescale buildup.

18.4 Indirect Water Heaters

Indirect water heaters utilize the existing hydronic heating system to generate domestic hot water. Sometimes called an integrated or combination water and space heating system, indirect water heaters use an *auxiliary storage tank* with virtually no controls to provide an abundant amount of hot water to homes and commercial buildings. This configuration could be a very efficient choice for buildings that utilize an existing boiler for comfort heating needs. The energy that is stored by the hot water tank allows the boiler to cycle on and off less often, which can save energy. If used with a high-efficiency boiler and a well-insulated storage tank, an indirect water heater may be the least expensive means of providing domestic hot water to a home or building.

The auxiliary tank used in an indirect water heating system does not produce its own heat but rather relies on the existing boiler connected to the hydronic system, **Figure 18-11**. Water in the auxiliary tank is heated by a coiled-type heat exchanger. A closed-loop piping system connects the boiler to the coiled heat

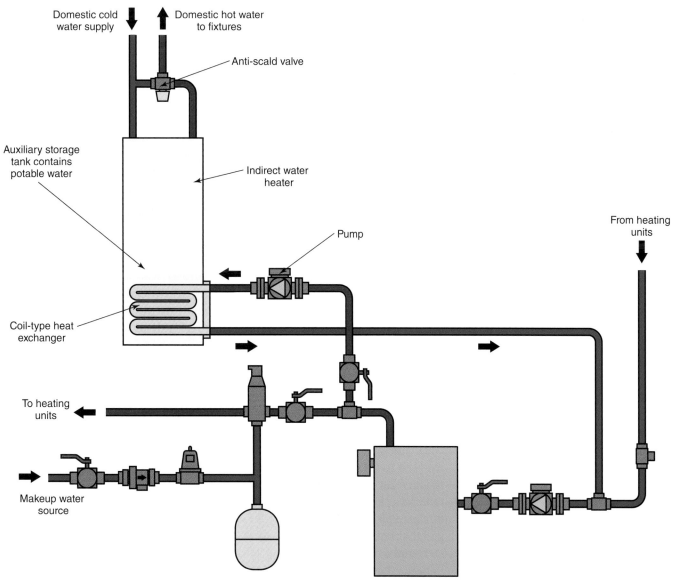

Figure 18-11. Indirect water heaters use a closed-loop heat exchanger connected to an existing boiler.

exchanger. This heat exchanger contains water or fluid that is heated by the building's boiler and circulates by means of a circulating pump—thus the concept of "indirect" heating, **Figure 18-12**. The boiler's water or fluid never directly mixes with the potable water found in the storage tank; it simply provides the heating source for the domestic hot water usage. Indirect systems can be used with boilers that utilize a number of heating fuels such as gas, oil, propane, and even solar energy.

Goodheart-Willcox Publisher

Figure 18-12. Water in the auxiliary tank is heated by a coiled heat exchanger.

GREEN TIP

The Green Advantage of Indirect Water Heaters
Because indirect water heating systems are installed in conjunction with an existing boiler or heat pump system, they allow you to heat domestic water while heating your home. If your current comfort heating system is compatible with heating water, energy costs can be greatly reduced because your system is performing both functions together. If you currently have a high-efficiency boiler or heat pump, this combination can be the most energy efficient heating system available.

18.4.1 Installation and Control of Indirect Water Heaters

Before installing an indirect water heater, important considerations must be made regarding proper sizing. The auxiliary tank must be sized according to hot water domestic usage. The average residence with one shower should require a 40-gallon tank. Smaller tanks should only be considered for residences with minimal hot water demands. In most cases, residences with multiple bathrooms or families of four or larger will require larger storage tanks. Factors such as high-flow showerheads, hot tubs, and the use of more than one appliance at the same time will increase hot water demand dramatically. If any of these factors are present, increase the tank size accordingly.

Boiler sizing is also important, considering the fact that in certain situations the boiler must provide both comfort heating and hot water generation simultaneously. In addition, the indirect water heater will need its own dedicated circulating pump that must also be sized correctly. Consult the manufacturer of the auxiliary storage tank for the proper flow ratings in GPM and the correct pressure drop at minimum flow when sizing the circulator.

Begin the installation by locating the water tank in an area where any potential leakage will not result in damage to areas adjacent to the tank or any floors located beneath the tank's location. When in doubt, install a suitable drain pan beneath the tank and connect a drain line between the pan and nearby drain. The tank should be installed as close as is practical to the boiler for easy access and service. Check any local codes for tank placement requirements.

Locate the tank close to the centralized piping system. It should not be exposed to freezing temperatures. Be sure that the controls, drain, and piping to and from the tank are easily accessible. There should be clearance around the tank for easy serviceability. Connect the piping into and out of the tank so that it is parallel to the boiler piping. Provide a circulating pump between the boiler and the tank.

Install a mixing or tempering valve, otherwise known as an ***anti-scald valve***, between the inlet and outlet piping of the domestic hot water. An anti-scald valve works by mixing cold water with the outgoing hot water to assure that the hot water temperature reaching the building's fixtures is at a safe level, **Figure 18-13**. The maximum hot water temperature will usually be maintained by the aquastat located near the tank. However, in some cases hot water usage patterns can cause the supply water temperature to rise significantly above the desired control setting. This anti-scald valve is typically classified as a ***thermostatic mixing valve***, which blends a controlled amount of cold water with hot water leaving the tank automatically to provide a more constant temperature to all building fixtures, **Figure 18-14**.

Scald Protection Point-of-Use Thermostatic Mixing Valve, (c) Caleffi North America, Inc.

Figure 18-13. An anti-scald valve such as this one mixes cold water with the outgoing hot water to assure that the supply hot water temperature is at a safe level for human skin.

SAFETY FIRST

Scalding Temperatures

Typical hot water temperatures for domestic usage range between 100°F and 120°F. In the United States, most authorities consider hot water at or below 120°F to be safe from scalding. Temperatures hotter than this can run the risk of scalding exposed skin. For protection against excessive hot water temperatures, an anti-scald valve—also called a tempering or mixing valve—should be installed at the outlet of the water heater. An anti-scald valve mixes cold water with the outgoing hot water to assure that hot water reaching a building fixture is at a temperature low enough to be safe for human skin.

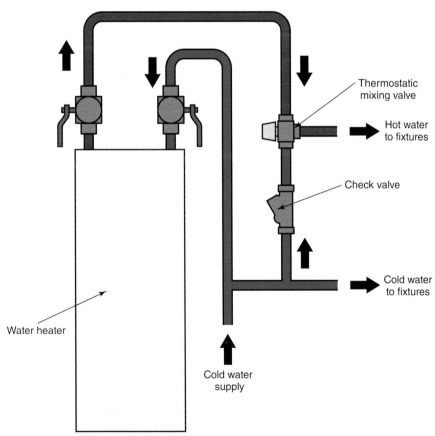

Goodheart-Willcox Publisher

Figure 18-14. Thermostatic mixing valves, such as the one shown in this system, can be used as anti-scald valves.

Goodheart-Willcox Publisher

Figure 18-15. An aquastat such as this one can be used as a temperature controller in an indirect hot water heating system.

SAFETY FIRST

Temperature/Pressure Relief Valve—Keep It Open!

Never plug or block the temperature/pressure (T&P) relief valve's outlet port or its discharge piping. Property damage or personal injury could result as the tank could become an explosive bomb!

Hot water controls for an indirect hot water heating system can be relatively simple. To control the water temperature, install an aquastat to the domestic tank—similar to a Honeywell L4006 control, **Figure 18-15**. This aquastat can control the circulating pump between the boiler and tank, or it can be used to start and stop the boiler to maintain the desired hot water supply temperature. Check with the boiler manufacturer's installation instructions to determine how to maintain parallel controls between the domestic hot water temperature and the hot water used for comfort heating purposes. Some modern boilers provide their own separate controls for controlling domestic hot water. Follow the manufacturer's instructions to ensure that the controls are installed and wired correctly.

18.4.2 Service and Maintenance of Indirect Water Heaters

Most indirect water heaters are simple devices that require very little maintenance. However, there are several items that should be checked on an annual basis or as needed to ensure that the supply of hot water is maintained:

1. Check to ensure that the supply boiler and domestic piping are free of leaks.
2. Make sure that the supply boiler is maintained correctly in accordance with the manufacturer's maintenance instructions.
3. If any water treatment is required for the boiler supply water to maintain its proper chemistry, check to ensure that this treatment is properly maintained.
4. Verify the supply boiler maintains its proper system pressure—typically around 12 psi for residential applications and 15 psi to 30 psi for commercial usage.
5. Manually open the temperature/pressure valve located on the domestic tank at least once per year to release some hot water. After water has flowed for several seconds, allow it to snap closed. If water continues to expel from the valve after closing, drain the system and replace the valve. If there is periodic weeping of water from the valve, it may mean that the pressure in the tank is set too high.

18.5 Combi Boilers

Combination boilers are often referred to as *combi boilers*. These boilers can be used to provide both space heating and domestic hot water. When combined into one package, combi boilers offer a space-saving option to conventional heating and hot water boiler applications. During the heating season, a combi boiler operates as needed to provide comfort heating to the building's structure and also provide domestic hot water for showers, sinks, and laundry facilities. During the summer months, the combi boiler only operates when there is a call for domestic hot water, **Figure 18-16**. Combi boilers offer very high operating efficiency because they operate using a low internal water volume and only when there is a need for heating or domestic hot water. Conventional hot water heating systems require a storage tank that must heat a large volume of water—whether or not it is needed. With a combi boiler, the customer is only paying to heat the water that is needed, rather than the entire tank. In addition, the initial cost of installing a combi boiler is generally quite competitive with conventional alternatives.

Combi boilers operate using two hot water heat exchangers. The first, or primary, heat exchanger is used for comfort heating. When there is a call for heat, the water in the closed heating loop passes through the heat exchanger, where the water is heated at a rapid rate. Some combi boilers may provide a small tank for storing heating water, and some may use only the provided heat exchanger. This heated water is piped and pumped through the hydronic heating system to the terminal units and back to the boiler—similar to a conventional boiler.

The secondary heat exchanger found in a combi boiler is used for heating domestic water, **Figure 18-17**. This system operates on an "on-demand" basis. In other words, the boiler only fires when a faucet is opened and the cold water supply passes through the secondary heat exchanger, thus eliminating the need for a hot water storage tank. A flow switch located in the boiler energizes the burner circuit whenever a faucet is opened. The boiler's efficiency is achieved by extracting a larger amount of heat from the flue gases that pass through the heat exchanger.

Advantages of a combi boiler include their space-saving design, endless amounts of hot water, and competitive equipment pricing versus a conventional heating and hot water system. However, typically only one hot water tap can be used at one time with a combi boiler, and if there is a mechanical failure,

neotemlpars/Shutterstock

Figure 18-16. This image shows how a combi boiler can be piped for both comfort heating and hot water domestic usage.

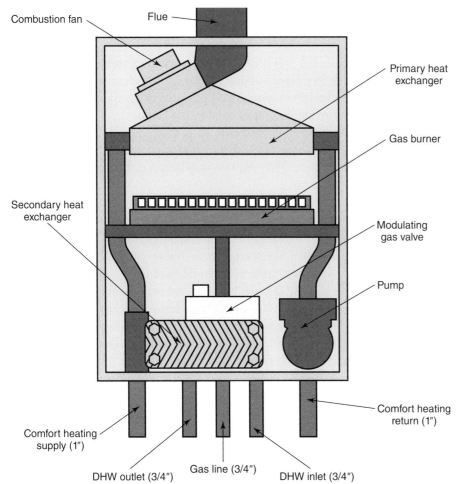

Goodheart-Willcox Publisher

Figure 18-17. This combi boiler uses two heat exchangers—one for comfort heating and one for domestic water heating.

the customer loses both comfort heating and domestic hot water usage. Combi boilers can be sized to meet the requirements of most residential and light commercial heating and hot water applications.

18.5.1 Installation and Control of Combi Boilers

The installation of a combi boiler begins by adhering to the required clearances laid out in the manufacturer's installation instructions. Once an appropriate location is confirmed, attach the boiler to the wall using the appropriate fasteners. Locate the boiler near the new or existing hot and cold water supply piping. Most combi boilers provide a 120-volt cord that can be plugged into a standard electrical receptacle. The boiler must be electrically grounded in accordance with the National Electrical Code or the Canadian Electrical Code. Do not rely on the gas or water piping as a means of grounding the boiler.

Gas piping must be sized according to the boiler size and capacity. Proper gas pipe sizing and material must conform with the National Fuel Gas Code, NFPA 54, and in Canada, the Natural Gas and Propane Installation Code, CAN/CSA B149.1.

Use the proper isolation and pressure-relief valves when installing the water piping. Several different piping configurations may be utilized for comfort heating. Follow the proper piping instructions or refer to Chapter 7, *Hydronic Piping Systems*, for the desired piping arrangement.

The boiler should be installed where there is no threat of freezing. Otherwise, provide the proper freeze protection according to the manufacturer's installation instructions.

Because combi boilers are classified as high-efficiency condensing boilers, the installer must adhere to proper venting practices, including the use of correct venting materials. In addition, ensure that the correct combustion air requirements are met, and provide for proper condensate disposal.

Combi boilers are often equipped with solid-state digital controls for both comfort heating and domestic hot water production. Detailed instructions for setup and configuring the desired parameters of these controls are included with the boiler installation guides.

18.5.2 Service and Maintenance of Combi Boilers

Annual service and maintenance of combi boilers should be performed by a qualified service technician. Service on these boilers usually involves proper cleaning, inspection, and prevention of the buildup of scale.

Use pressurized air to clean and remove dust and debris from the main burner after properly disassembling the appliance. Clean the heat exchanger and combustion air fan blades. Do not use spray cleaners, wet cloths, or volatile substances for cleaning, as they can be ignition hazards and break down the factory finish.

Annually inspect the venting system, including the air intake and exhaust, for potential blockage or damage. Keep combustion air fans free of dust and dirt by cleaning annually. Wipe down the temperature controllers, but do not use solvents on them.

Water quality can affect the heat exchanger. Hard water, which contains lime or calcium, can cause a buildup of scale. This is evident by the white, flaky material that can coat the inside of the heat exchanger, **Figure 18-18**. Scale can reduce the level of heat transfer inside the boiler and lower the efficiency of the heating system. The inside of the heat exchanger needs to be accessed in order to tell if limescale is present. If the buildup of lime scale is evident, flush the heat exchanger according to the manufacturer's guidelines with an acid solution and test the boiler water to see if treatment or conditioning is required.

PARADORN KOTAN/Shutterstock

Figure 18-18. Scale can build up inside the boiler heat exchanger if the supplied water is high in lime or calcium.

TROUBLESHOOTING

Descaling a Heat Exchanger

How do you know if a heat exchanger is "limed up" if you cannot see it?

One way to tell is if the boiler becomes unusually noisy. This condition can be caused by water becoming trapped in places where there is a buildup of scale. This water will overheat and begin to evaporate, which creates bubbles and steam. As the bubbles move through the system, they create noise. Scale buildup will cause the boiler to work harder to maintain the proper heating level, reducing the efficiency and costing more money.

One way to get rid of unwanted scale is to perform a power flush. This process involves a mixture of chemicals and water flushed through the heat exchanger at a high pressure. As this solution moves through the heat exchanger, it will also expel any rust or debris within the system—making the boiler cleaner and more efficient.

18.6 Hybrid Water Heaters

Traditional and hybrid water heaters both use a storage tank—but rather than using conventional fossil fuels to heat the water, hybrid water heaters incorporate the use of a *heat pump* to heat the domestic water supply. Heat pumps transfer heat from one area to another. In this case, the hybrid water heater uses heat from the surrounding air where the heat pump water heater is located and transfers this heat into the water by means of the refrigeration cycle.

Heat pumps have been used for years. Their operation is similar to that of a refrigerator or air conditioner. The heat pump system consists of a compressor, an evaporator, a condenser, and a metering device, **Figure 18-19**. A fan-powered evaporator coil filled with refrigerant absorbs heat from the air surrounding the system. This refrigerant travels through the compressor, where its pressure and temperature are increased. From the compressor, the refrigerant enters the condenser coil located within the boiler tank, where its heat is given up to the water located in the tank. Once the refrigerant has given up the heat that has been absorbed from the ambient air, it passes through a metering device, which drops the temperature and pressure so that it can recirculate through the evaporator, where it absorbs more heat. This process repeats itself over and over until enough heat has been transferred to adequately heat the tank water.

Hybrid water heaters offer a more efficient means of heating water compared to conventional methods and without the burning of fossil fuels. Typically, a hybrid water heater includes electric heating elements to supplement the need for heat and also provide a source of emergency heat should the heat pump system go down, **Figure 18-20**.

Because hybrid water heaters are typically installed in locations where the ambient air is between 40°F and 90°F, they are better suited for warmer climates such as Florida. They are not as efficient when installed in colder climates.

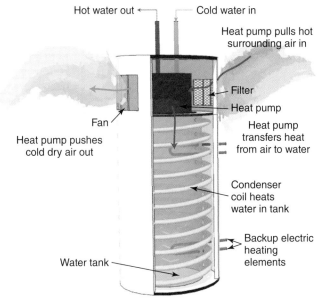

Source: US Department of Energy

Figure 18-19. This is a detailed illustration showing how a hybrid heat pump water heater operates.

bane.m/Shutterstock

Figure 18-20. Auxiliary electric heating elements are typically used in conjunction with a hybrid water heater.

18.6.1 Installation and Control of Hybrid Water Heaters

Since the installation location of the hybrid water heater will determine the efficiency of the appliance, avoid areas that may experience cold or freezing temperatures. The heater should be installed as near as practical to the area where there is the greatest heating demand. Long, uninsulated water lines can waste heat and energy.

The volume of the space where the tank is installed will have an effect on its performance. Because the heater needs an adequate volume of supply air from which to extract heat, it is suggested that the mechanical room have a total volume of at least 700 cubic feet. Enclosed spaces such as closets that are smaller than 700 cubic feet require that a louvered door be installed at the entrance. Otherwise, additional air can be ducted from an adjacent room or outside.

> **TECH TIP**
>
> **Ambient Air Requirements for Hybrid Water Heaters**
> Here is one rule of thumb to ensure that there are proper volumes of ambient air surrounding a hybrid water heater: if the air temperature surrounding the water heater drops more than 15°F during the heating cycle, the amount of ambient air circulation is insufficient for efficient operation. When this condition is the case, an auxiliary source of ambient air will be required.

Installation of hybrid water heaters is similar to that of conventional water heaters with regard to hot and cold water supply lines, as well as the temperature/pressure relief valve and its discharge line. What is different is that hybrid water heaters require a condensate line that needs to be piped to a nearby drain. Condensate develops from the frosting of the evaporator coil when it absorbs heat from air that has a temperature lower than 45°F.

The electrical connections may be 120 volts or 230 volts, depending on the size of the tank. If the tank has electric heating elements for supplemental heating, these require a 230-volt connection. All wiring must conform to local codes as well as the National Electrical Code NFPA 70 or the Canadian Electrical Code.

If the boiler is to be ducted to the outdoors or an adjacent space, the ductwork must conform to local HVAC codes and should not be connected to existing ductwork. This ductwork can be either rigid or flexible, but there may be a length limitation depending on the size of the boiler. Check with the manufacturer's installation instructions for proper sizing and installation of the heater ductwork.

Just as with combi boilers, most hybrid water heaters are supplied with microprocessor controls that can be programmed for operating temperature set points, as well as for occupancy scheduling and alarm signaling, **Figure 18-21**. Some are even equipped with Wi-Fi controls. The boiler

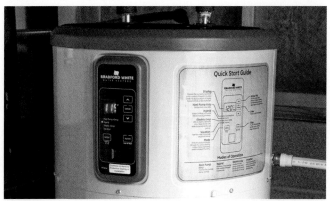

Source: US Department of Energy

Figure 18-21. Many hybrid water heaters incorporate electronic controllers that can be used to regulate hot water temperature.

manufacturer will provide all of the configuration instructions and product support for ease of operating the hybrid water heater.

18.6.2 Service and Maintenance of Hybrid Water Heaters

Routine service and preventive maintenance on a hybrid water heater is similar to that of a conventional water heater. The tank should be drained periodically to remove any sediment from the bottom. This is especially true for systems that use water high in calcium and magnesium. Most hybrid water heaters include an air filter for the evaporator coil. These filters should be cleaned or replaced on a regular basis, depending on the amount of dust and particulate found in the ambient air where the tank is located.

If the heater is to remain idle for an extended period of time, the power supply and water supply to the appliance should be turned off. Furthermore, the tank and piping should be drained if there is a chance they will be subjected to freezing conditions. After long periods of shutdown, the heater's operation should be checked by a qualified service technician. Make sure that the tank is completely filled before placing the unit back in operation. Most tanks are equipped with an anode rod designed to prolong the water heater's life against calcium buildup. This rod can be replaced by a qualified technician. Consult the manufacturer as to when the anode rod should be replaced.

CAREER CONNECTIONS

Plumbers, Pipefitters, and Steamfitters

Plumbers, pipefitters, and steamfitters assemble, install, and repair pipelines and piping systems that carry water, steam, air, and other liquids and gases. These tradespeople also install heating and cooling equipment as well as mechanical control systems and sprinkler systems. Plumbers and pipefitters are also responsible for resolving problems such as leaks and cracks in piping systems, as well as maintaining, cleaning, and repairing standard piping systems.

Pipefitting and plumbing careers are suitable for individuals who are passionate about practical work. A tradesperson in this field should know how to work in a fast-paced, demanding environment and have the ability to master new skill sets quickly. Because frequent heavy lifting and ladder climbing are required, a plumber or pipefitter needs to possess both patience and physical strength.

Employers are looking for plumbing and pipefitting candidates who have at least a high school diploma or equivalent. Most candidates have successfully attended a technical school as well as a training program, such as those offered by community colleges. Good customer communication skills and problem-solving abilities are very important.

Careers in plumbing and pipefitting usually begin at the apprenticeship level. Once the apprenticeship program has been successfully completed, the candidate can qualify for journeyman status certification or even licensing.

The United States Department of Labor statistics on plumbers, pipefitters, and steamfitters show national annual salary estimates as of May 2019 starting at $32,690 up to $97,170. The average annual salary is $55,160. The best way to increase salaries is to continue education and training through associations, manufacturers, and experience.

Chapter Review

Summary

- An added benefit of utilizing a boiler or hydronic heating system is the ability to generate domestic hot water.
- There are a number of different types of equipment to generate potable hot water for both domestic and light commercial applications.
- There are a number of design factors that go into sizing devices used for hot water production.
- Tanks used for producing domestic hot water are rated by their capacity in gallons and also by their recovery rate.
- Tankless water heaters are rated by the maximum temperature rise possible at a particular gallon per minute flow rate.
- Tankless heaters work as on-demand heating devices, which means they are energized only when there is a call for heat.
- One drawback of a tankless system is the limited number of devices that can be served simultaneously.
- The installation of tankless boilers and water heaters is similar to that of a conventional boiler.
- The controls for most tankless boilers and water heaters are self-contained, microprocessor-type controllers.
- Indirect water heaters use an auxiliary storage tank to provide hot water to homes and buildings.
- The auxiliary tank used in an indirect water heating system does not produce its own heat but relies on the existing boiler connected to the hydronic system.
- The auxiliary tank used with an indirect water heater must be sized according to hot water domestic usage.
- An anti-scald valve blends a controlled amount of cold water with hot water to provide a more constant temperature to the building.
- Most indirect water heaters require very little maintenance.
- Combi boilers can be used to provide both space heating and domestic hot water generation.
- Combi boilers operate using two hot water heat exchangers.
- Combi boilers are often equipped with solid-state digital controls for both comfort heating and domestic hot water production.
- Hybrid water heaters incorporate a heat pump to heat the domestic water supply.
- The installation location of the hybrid water heater will determine the efficiency of the appliance.
- Routine service and maintenance on a hybrid water heater is similar to that of a conventional water heater.

Know and Understand

1. Which type of water heating system incorporates an auxiliary tank with enclosed heat exchanger?
   A. Tankless boiler
   B. Indirect water heater
   C. Hybrid water heater
   D. Combi boiler
2. Which type of water heating system offers both space heating and domestic hot water production in a single package?
   A. Tankless boiler
   B. Indirect water heater
   C. Hybrid water heater
   D. Combi boiler
3. Which type of water heating system incorporates the mechanics of a heat pump system?
   A. Tankless boiler
   B. Indirect water heater
   C. Hybrid water heater
   D. Combi boiler
4. What is the typical peak two-hour period for residential domestic hot water usage?
   A. 7:00 to 9:00 pm
   B. 1:00 to 3:00 pm
   C. 7:00 to 9:00 am
   D. 10:00 am to 12:00 pm
5. *True or False?* Temperature rise takes into account the incoming water temperature and the actual desired hot water temperature.
6. _____ is the amount of hot water in gallons that can be produced in one hour after the tank is completely drained.
   A. Maximum output
   B. Recovery rate
   C. Temperature rise
   D. Standby heat loss
7. What activates the tankless water heater's heating source when there is a call for hot water?
   A. An airflow switch
   B. An internal flow switch
   C. An internal temperature sensor
   D. An internal pressure switch
8. How is the hot water temperature controlled with an indirect water heater?
   A. By using an aquastat to control the boiler or heating source
   B. By cycling the boiler or heating source whenever hot water is being used
   C. An internal flow switch within the tank controls the boiler cycling
   D. By a microprocessor controller
9. Typically, tankless water heaters provide hot water at a rate of _____ gallons per minute.
   A. 1 to 3
   B. 2 to 5
   C. 5 to 10
   D. 12 to 15
10. *True or False?* Because of their high efficiency, hybrid water heaters are well suited for use in colder climates.

Apply and Analyze

1. Explain how a tankless boiler generates hot water without the use of a storage tank.
2. Describe the procedure for sizing hot water systems based on domestic usage.
3. Explain why standby heat losses can affect the efficiency of conventional hot water storage tanks.
4. Describe where using a point-of-use water heater would be more practical than a whole-house water heater.
5. What type of controls are used for tankless boilers?
6. Describe how an indirect water heater uses a conventional boiler to generate hot water.
7. What is an anti-scald valve used for and how is it controlled?
8. How is the hot water temperature controlled with an indirect water heater?
9. Explain how a combi boiler can be used for both comfort heating and domestic hot water production.
10. Describe how a hybrid water heater operates.
11. How is supplemental heating incorporated when using a hybrid water heater?
12. What steps should be taken to maintain a hybrid water heater?

Critical Thinking

1. Explain how an auxiliary storage tank might be installed and controlled in conjunction with a combi boiler.
2. Would using fossil fuels as an auxiliary heating source be a good choice with a hybrid water heater? Why or why not?

19 Solar Thermal Storage

Chapter Outline
19.1 Passive Storage Systems
19.2 Active Storage Systems
19.3 Solar Collectors
19.4 Application Selection for Solar Thermal Storage
19.5 Thermal Storage System Installation
19.6 System Piping
19.7 Control Strategies for Solar Thermal Systems
19.8 Filling and Starting up the Systems
19.9 Additional Applications for Solar Thermal Storage

Learning Objectives

After completing this chapter, you will be able to:
- Explain the difference between an active and passive solar heating system.
- Describe how an integral collector storage unit operates.
- List the advantages and disadvantages of an integral collector storage unit.
- Describe how a thermosiphon system operates.
- Compare an integral collector storage unit and a thermosiphon system.
- Explain how a thermostatic mixing valve prevents scalding conditions.
- List the different configurations available for an active solar heating system.
- Identify the differences between open- and closed-loop solar heating systems.
- Explain why an antifreeze solution is used with a closed-loop storage system.
- Describe how a drainback system operates.
- Compare pressurized and unpressurized systems.
- Identify the differences between a flat-plate and an evacuated tube solar collector.
- List the applications for solar thermal storage.
- Describe how the sun's solar angles figure into the installation of a collector.
- Explain how a solar pathfinder is used to determine solar collector panel positioning.
- List the proper steps for roof mounting a solar collector panel.
- Outline how the solar collector panel is piped into the storage tank.
- List the piping components used when installing the solar thermal system.
- Describe the control strategies used for solar thermal systems.
- List the proper steps for the filling and start-up of a solar thermal system.

Technical Terms

active solar thermal storage system
closed-loop solar thermal storage system
declination angle
drainback system
evacuated tube collector
flat-plate collector
flush-and-fill cart
freeze protection valve
glazing
glycolic acid
insolation
integral collector storage (ICS) unit
negative temperature coefficient (NTC)
open-loop system
passive solar heating system
photovoltaic module
pressurized solar thermal storage system
slope angle
solar altitude angle
solar array
solar azimuth
solar collector
solar pathfinder
strap-on temperature device
thermistor
thermosiphon solar heating system
thermosiphoning
unpressurized system

Solar thermal storage is achieved by capturing the sun's energy in a solar collector, which allows the sun's radiation to pass through a selective material without allowing it to reradiate back out. The collector contains one of several different types of piping through which water is circulated. A number of different solar thermal storage configurations can be utilized. The most appropriate type depends on a number of factors, including geographic location, the amount of hot water that needs to be generated, and the specific heating application. This chapter discusses each type of these systems in detail.

GREEN TIP

Going Green with Solar Hot Water

Apart from home heating and air conditioning, the water heater is the largest energy drain on a home's budget. This makes an excellent area to focus on when going green.

Solar water heating is the greenest energy option available. It uses clean solar energy, which means no greenhouse gases are used to heat domestic water. And after an initial payback of four to eight years, the cost of hot water heating is as close to free of charge as it gets for the next 15 to 20 years!

Figure 19-1. A house utilizing passive solar heating typically includes south-facing windows.

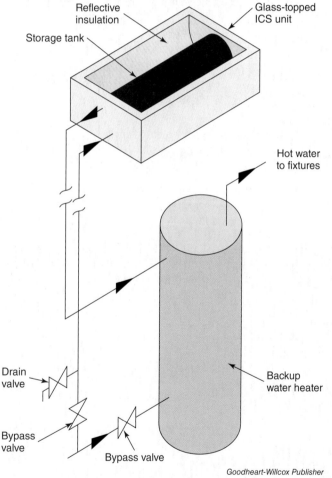

Figure 19-2. An integral collector storage device (ICS) consists of a solar collector and storage tank.

19.1 Passive Storage Systems

A *passive solar heating system* harvests energy from the sun without the use of electrical or mechanical devices. In its simplest term, this type of solar heating system can be compared to a home with south-facing rooms and windows that has an open-space floor plan to optimize thermal mass, **Figure 19-1**. This application is usually implemented during the home's design and construction phase, incorporating such facets as the orientation of the home in relation to the sun and construction materials that will maximize solar gains.

19.1.1 Integral Collector Storage Units (ICS)

A passive solar thermal storage system consists of a solar collector and some type of solar heating water storage device, such as an *integral collector storage (ICS) unit*. With an ICS unit, a 30- to 40-gallon storage tank is painted matte black and mounted in an insulated collector box that has *glazing* (glass) on the side exposed to the sun, **Figure 19-2**. The collector box is usually lined with aluminum foil–faced foam insulation that is sloped and curved toward the tank. The best installation design for this type of system is to use a single tank in the box and mount it horizontally, with the tank facing from east to west. The curved reflector surfaces are angled to reflect the sunlight onto the tank. When the sun shines through the glass, its radiation is absorbed by the tank and in turn the water is heated. With this type of system, the water itself is the solar collector. In most ICS units, the hot water is drawn from an outlet at the top of the tank where it is the hottest and cold water passes out through the bottom of the collector. It is then piped indoors to either the main water heater, to a backup water heater, or to a solar thermal storage tank.

One advantage of this type of system is its relatively simple design, which keeps installation costs low. In addition, ICS units do not require a great amount of maintenance because there are no moving parts. Furthermore, because no electrical devices are used to transfer water, there are essentially no operational costs. There are, however, several disadvantages to this type of passive thermal storage system. In colder climates, the water in the tank will cool down overnight—to as much as 30°F to 40°F below the supply water temperature. Therefore, hot water is only available during the daytime hours when there is a sufficient amount of direct sunlight on the tank. The optimal usage period for domestic hot water is typically between the hours of 12 p.m. and 8 p.m. Furthermore, certain areas of the United States should avoid using this type of system, especially where the nighttime temperature is below 60°F from late fall to early spring. This is due to the extreme overnight heat loss. In these areas, the ICS should be used on a seasonal basis or not at all. Another cautionary measure that needs to be taken into consideration is the tank weight. An empty tank itself can weigh up to 250 pounds. If it is filled with 30 to 40 gallons of water, the total weight can easily exceed 500 pounds. Therefore, consideration needs to be taken for proper rigging and for correctly mounted roof supports. Finally, it is essential that the proper type of glass be used. The glazing on the exterior of the box should be low-iron, tempered glass—never plastic, Teflon™, acrylic, or fiberglass.

19.1.2 Thermosiphon Systems

Some solar thermal storage units incorporate a sloped solar collector installed in conjunction with an insulated tank. The tank may be mounted either indoors or outdoors. This type of system is referred to as a ***thermosiphon solar hot water heating system***, **Figure 19-3**. Just as with ICS units, thermosiphon systems do not incorporate pumps or other controls. They utilize the principle of natural convection, wherein hot water rises and cold water falls. When the solar collector is heated by the sun's radiation, the water in the collector expands. Because the heated water has a lower density than the cooler water located at the bottom of the storage tank, a gentle circulation effect takes place. The warmer water rises to the top of the collector and into the tank where it is stored. This is why the tank is always located above the collector. Because the cooler water is heavier, it descends to the bottom of the collector. This process is known as ***thermosiphoning*** and takes place without the need of a circulating pump. Thermosiphoning occurs as long as the water in the collector is warmer than the water in the storage tank. As the sun's radiation diminishes, the water in the collector cools to a temperature below that of the storage tank, and the water's natural flow ceases. The water from the storage tank is piped directly into the water heater located inside the home. Certain precautions should be taken with this type of system because extreme water temperatures are possible during prolonged periods of sunny weather. To prevent scalding conditions, a thermostatic mixing valve, or tempering valve, can be installed on the outlet of the water heater. This valve automatically mixes a portion of the cold supply water with the hot water to maintain the desired water temperature delivered to the home, **Figure 19-4**.

Figure 19-3. A thermosiphon system uses natural convection to circulate water.

SAFETY FIRST

Maximum Hot Water Temperature

Most experts recommend that the temperature of domestic hot water not exceed 125°F. The International Plumbing Code states that the maximum hot water temperature for a shower or bathtub shall be 120°F. Water temperatures exceeding this set point can pose a serious risk of scalding burns, particularly to children. However, in order to sanitize dishes, a hot water temperature of 140°F is required. Some are concerned that lowering the water temperature will result in soap not working properly in dishwashers or washing machines. Actually, most soaps and detergents are designed to work best at temperatures between 120° and 125°F.

Thermosiphon systems are relatively inexpensive, and as with the ICS system, since there are no electrical or mechanical controls, they are quite reliable and pose few maintenance problems. Unlike the ICS system, thermosiphon systems incorporate an insulated tank and therefore maintain their water temperature

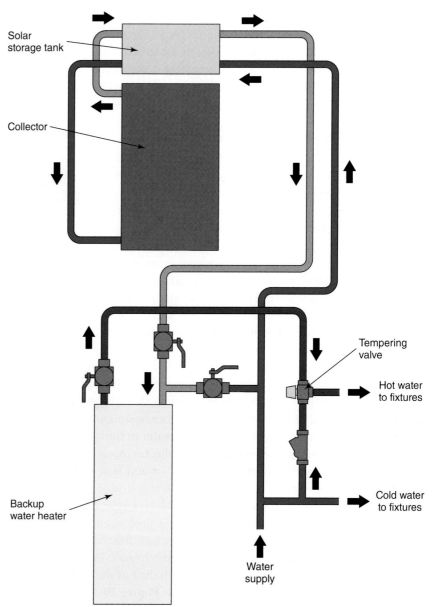

Figure 19-4. An example of a piping diagram for a thermosiphon system.

many hours after the sun goes down. However, like the ICS system, thermosiphon systems rely on the outdoor air temperature in order to function properly. This means that they will stop working if the outdoor air temperature falls below freezing. For this reason, thermosiphon systems are best suited for warmer climates or for seasonal usage only. In addition to being temperature sensitive, like the ICS system, storage tanks are quite heavy and usually require special roof reinforcement.

19.2 Active Storage Systems

What separates an *active solar thermal storage system* from a passive one is the incorporation of a mechanical means of fluid transfer. For instance, an active system that is used primarily for space heating uses a fan to circulate air through the system. This type of configuration may use an inline fan ducted to the solar collector, or simply use a conventional ceiling fan to circulate the room air, **Figure 19-5**.

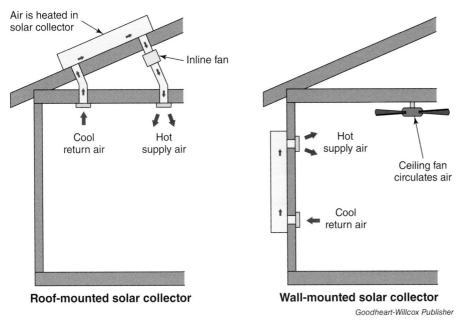

Figure 19-5. An inline fan ducted to a solar collector can act as an active solar system, or a ceiling fan circulating room air can be used.

The majority of active systems, however, incorporate an energy storage system to provide heat when the sun is not shining. These types of systems are typically liquid-based and use either water or an antifreeze solution circulated through a hydronic collector by means of a pump. They can be used for space heating applications, for heating of domestic water, or for a combination of both. Active solar thermal storage systems can utilize a number of different configurations including:
- Open-loop systems
- Closed-loop systems
- Drainback systems
- Pressurized and unpressurized systems

19.2.1 Open-Loop Systems

An *open-loop system* is one in which the hot water is piped from the solar collector directly into a storage tank or a domestic water heater. The system is called an open loop because the piping from the collectors to the storage tank or water heater is open to either the home's well water supply or the city water system. With this type of system, a pump is used to circulate the water from the solar collector to the storage tank or water heater. In some applications, a primary storage tank may be used in conjunction with a backup water heater, **Figure 19-6**. The water flow through the collector is controlled by measuring the differential temperature between the water in the storage tank and at the outlet of the collector. When the water temperature at the outlet of the collector is warmer than in the tank, a controller energizes the pump, which circulates the water through the storage tanks and back to the solar collector.

The open-loop thermal storage system is very efficient, simple to install, and very reliable. The circulating pump can be as small as 10 watts. If the pump runs off DC voltage, it can be powered directly by a *photovoltaic module*. A photovoltaic solar module absorbs sunlight as a source of energy to generate direct-current electricity.

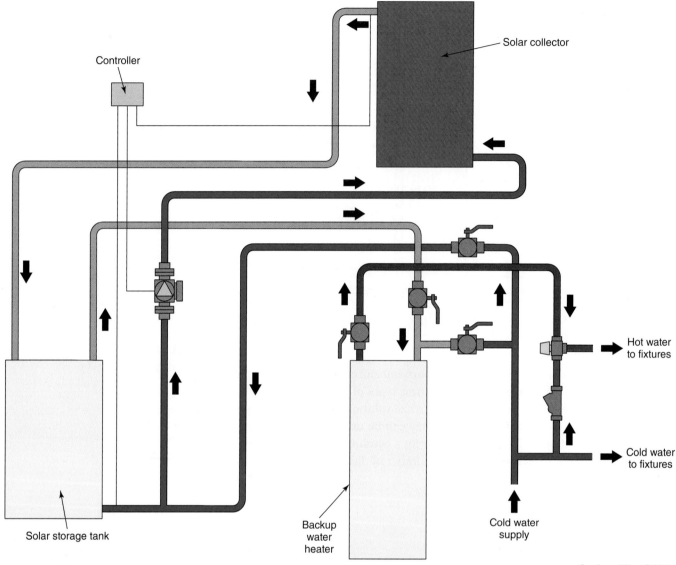

Figure 19-6. An example of an open-loop thermal solar system using two storage tanks.

There are, however, several drawbacks to an open-type system. Because they utilize the domestic water supply, open-loop systems cannot circulate an antifreeze solution through them and therefore are subject to freezing. If this type of system is used in areas where freezing temperatures are experienced, they must be drained whenever the outside air temperature is expected to fall below 35°F. In addition to freezing conditions, open solar thermal systems are subject to water quality. Water that is acidic or contains a high level of rust or dissolved solids (hardness) should be avoided. Hard or acidic water will corrode the copper piping or cause scale buildup and lead to premature failure of the collector.

19.2.2 Closed-Loop Systems

A *closed-loop solar thermal storage system* is similar to an open-loop system, except it circulates a heat transfer fluid through the solar collector and through a closed heat exchanger inside of the storage tank, **Figure 19-7**. The heat transfer fluid used in a closed-loop system usually consists of a glycol-water mixture. This type of mixture is necessary in any climate where there is a potential for freezing temperatures. The glycol used in these systems is usually food-grade propylene

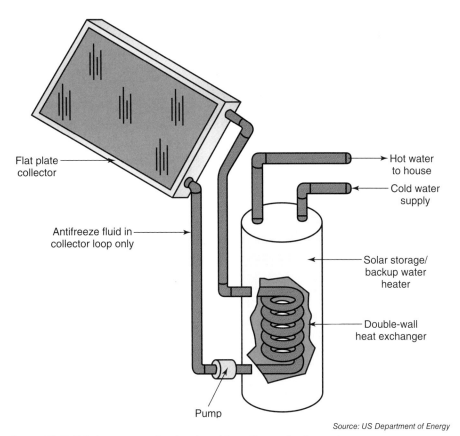

Figure 19-7. This is an example of a closed-loop thermal solar system.

glycol. The glycol acts as antifreeze for the fluid and is nontoxic in the event that it should ever come in contact with the domestic water supply. A single night of subfreezing temperatures can severely damage a solar collector that does not utilize some type of antifreeze. This situation can occur even in traditionally warmer locations such as Arizona, Texas, and Florida. This is why all closed-loop thermal storage systems used in the United States and Canada should employ some method of freeze protection, **Figure 19-8**.

Conversely, solar thermal storage systems that use glycol should never be designed to operate over 195°F on a continuous basis. Doing so will cause the glycol solution to degrade and quickly turn into *glycolic acid*. This will cause copper piping to corrode and eventually fail. In addition, glycol-filled systems should not be allowed to remain idle for long periods. If the solution is not circulated on a regular basis, it can become stagnant and form sludge and organic acids—especially during hot, sunny weather. This situation will also cause corrosion in the piping and collector, as well as substantially reduce the rate of heat transfer. Problems with freezing and glycol stagnation can be prevented by incorporating a drainback system into the thermal storage system.

19.2.3 Drainback Systems

Drainback systems offer a viable, alternative method of freeze protection for solar thermal storage systems and can be safely installed anywhere in the United States. This method works by simply draining all of the water or glycol solution from the solar collector and any exposed piping into a drainback tank whenever the system is not collecting solar energy. The drainback tank is located inside the building, where it is protected from outdoor elements, **Figure 19-9**. The drainback system relies on gravity, along with properly pitched piping, to quickly drain

DID YOU KNOW?

Propylene Glycol

Propylene glycol is a common additive in food beverages. It is biodegradable and will not concentrate in common water systems. Furthermore, its effects on aquatic organisms have shown to be practically nontoxic.

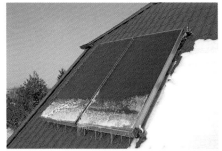

Figure 19-8. When using solar collectors for thermal storage in cold climates, freeze protection must be included.

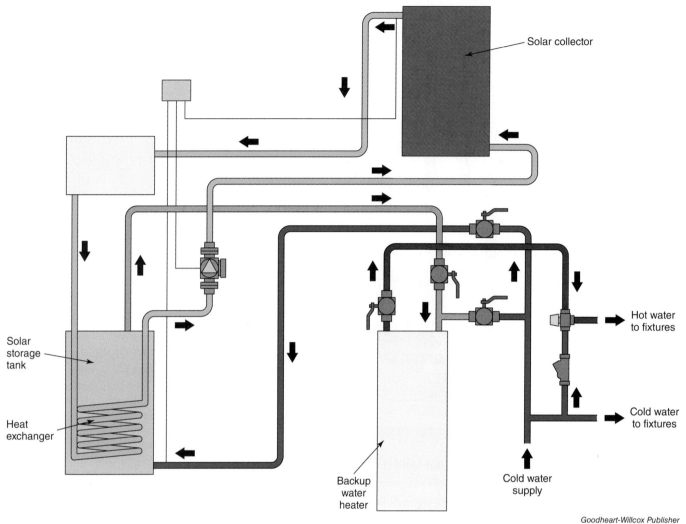

Figure 19-9. This is an example of a solar thermal storage system using the drainback method.

Figure 19-10. Automatic air vents should not be used on a closed-loop solar collector.

the water or glycol solution when the system's circulating pump shuts off. When the system is initially filled with water, the drainback tank is filled only to a predetermined level. When the circulating pump is de-energized and the system is drained back, all of the piping above the tank's water level, as well as the piping in the solar collector, is displaced with air. This prevents any damage to the exposed piping when the outside air temperature drops below freezing. All of the piping and other components below the drain tank's fill level are filled with liquid. When the system's circulating pump is energized, liquid is pumped up into the collector. By doing so, air is forced ahead of the liquid, displacing it and eventually returning to the drainback tank. This process causes a slight drop in the drainback tank's liquid level due to the fact that the liquid replaces the air in the solar collector and system piping. Because air is vital to its proper functionality, it is recommended that automatic air vents not be used with drainback systems, **Figure 19-10.**

TECH TIP

Air Vents in Solar Thermal Storage Systems
An air vent is a device that is installed at the highest point on the piping loop, usually at the outlet of the collector. When laying out the piping arrangement for a solar thermal storage system, it is important to know which type of system needs an air vent. The most common type of system that would incorporate an air vent is a closed-loop pressurized system. On this type of system, it is important to eliminate air that may become trapped in the piping or in the collector itself. Trapped air can cause a number of problems, including corrosion, undesirable noise, and poor or no heat transfer due to a lack of circulated heat transfer fluid.

19.2.4 Pressurized and Unpressurized Systems

A *pressurized solar thermal storage system* is one in which the circulation loop is closed to atmospheric pressure. This type of system is characteristic of the closed-loop and drainback-type systems. An *unpressurized system* is one in which the loop is open to the atmosphere, such as with an open-loop system. With any of the previously mentioned loop configurations, adhere to proper piping practices. These practices include the use of an expansion tank, pressure-regulating valve, water-regulating valve, high-temperature limit, pressure-relief valve, and the proper placement of the circulating pump.

19.3 Solar Collectors

Solar collectors act as the key component of the thermal storage system. They gather the sun's radiant energy and transform the radiation into heat. This energy is then transferred to the system's heat transfer fluid, which consists of either water or glycol. Nearly all liquid-based solar thermal storage systems use one of two types of solar collectors:
- Flat-plate collectors
- Evacuated tube collectors

Most residential and commercial applications for solar thermal storage that require fluid temperatures below 200°F will use flat-plate collectors. Those applications requiring liquid temperatures higher than 200°F usually use evacuated tube collectors.

19.3.1 Flat-Plate Collectors

Flat-plate collectors are the most common type of solar collector used with thermal storage systems. A typical *flat-plate collector* consists of an insulated metal or aluminum box covered with tempered, low-iron glass called *glazing*. This glazing can withstand high thermal stress as well as hazards such as hailstones. It utilizes a low iron oxide content to minimize the absorption of solar radiation as the sun's rays pass through it, **Figure 19-11**. The main component of the flat-plate collector is the absorber plate. This plate is usually an assembly of copper sheeting with copper tubing fastened to it. The top of the absorber plate is typically coated with a dark-colored paint or with a special coating that readily

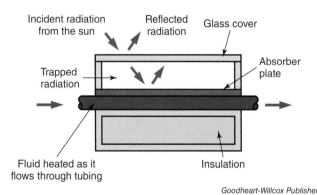

Figure 19-11. This illustration shows a detailed view of a flat-plate solar collector.

Goodheart-Willcox Publisher

absorbs radiation as the sun passes through the glazing and strikes its surface. As the absorber plate heats up, the sun's energy is transferred to the fluid that is circulating through the copper tubing connected to the plate. This takes place because the fluid is cooler than the absorber plate. As the fluid absorbs heat, it is pumped through the collector to the storage tank or heat exchanger. In areas where there is an average amount of solar energy available, flat-plate collectors are usually sized for approximately one square foot per gallon of hot water for single-day use, **Figure 19-12**.

19.3.2 Evacuated Tube Collectors

An ***evacuated tube collector*** consists of a row of glass tubes, each having concentric inner and outer walls. The air between these walls has been removed, forming a vacuum. This vacuum essentially eliminates any convective and conductive heat transfer between the two walls, forming the best possible thermal insulation for a solar collector. The result is exceptional performance, even at low ambient temperatures. Inside the glass, there is a copper tube that absorbs heat. This heat is carried up the tube to the heat pipe condenser located outside the end of the tube, **Figure 19-13**. Each heat pipe condenser is then connected to a common copper header located at the top of the collector, **Figure 19-14**. The header is enclosed within an insulated aluminum manifold, where the heated fluid flows from the collector to a storage tank or heat exchanger, **Figure 19-15**. As shown in **Figure 19-16**, solar energy is absorbed by the inner heat pipe and carried to the condenser, where it warms the fluid flowing through the manifold. The roof- or wall-mounted evacuated tube collectors are collectively known as a ***solar array***, **Figure 19-17**. Because the evacuated tube collector absorbs the sun's radiation, it can be effective even on a cloudy day.

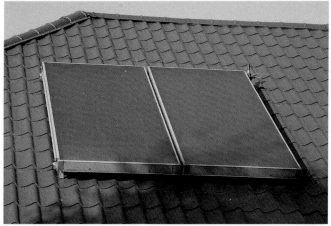

Aleksander Bolbot/Shutterstock.com

Figure 19-12. An example of roof-mounted, flat-plate collectors used for the production of hot water.

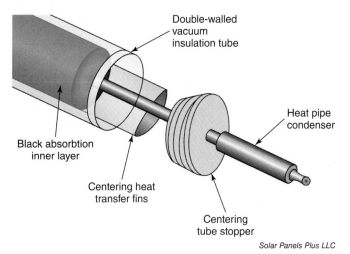

Solar Panels Plus LLC

Figure 19-13. A view of the inside of an evacuated solar collection tube.

TECH TIP

Evacuated Tube Collectors and Snowfall

Because they use radiation rather than convection for heat transfer, evacuated tube collectors will not heat up like flat-plate collectors. Thus, in the winter they will not melt large quantities of snow that fall on them at any one time. This can lead to an undesirable buildup that can block the volume of heat they can produce. Use a long-handled broom or roof rake to carefully remove unwanted snow. It may be difficult to clear the snow from the glass tubes without breaking them, so be careful.

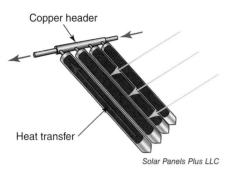

Figure 19-14. This illustration shows how evacuated tubes are connected to a copper header.

Figure 19-15. An aluminum manifold housing the main copper header on an evacuated solar collector.

Making a decision on which type of solar thermal storage system to invest in will depend on a number of factors, including geographic location, the amount of hot water needed, and the type of heating fuel available for auxiliary use. When making a large investment in solar energy, it is best to consult with local distributors and other customers who have purchased similar systems. This will ensure that the best purchase is made based on sound decision making and comparative analysis.

19.4 Application Selection for Solar Thermal Storage

Once a solar thermal storage configuration has been chosen, the next step is to properly install the equipment based on application. There are several excellent uses for solar thermal storage systems, suited for both residential and commercial applications.

It is important to remember that the application of solar thermal energy involves three steps: First, it must be collected. Next, it must be stored. And then it must be distributed. This is with the understanding that the medium being used to transfer this solar energy is water or a water/antifreeze solution. Collection of this energy takes place at the outdoor solar collector. The storage of solar energy involves the use of some type of tank or vessel. The distribution of the sun's energy can be accomplished through the building's existing plumbing system—either through the hot water system or central heating system—or it may be distributed through a dedicated solar thermal system, which includes a storage tank.

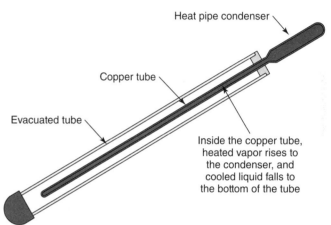

Figure 19-16. This illustration shows how heat transfer is achieved through the evacuated tube collector.

Figure 19-17. A group of wall-mounted solar evacuation tubes is known as an array.

Solar thermal energy can be applied to individual heating needs or a combination of applications. For residential applications, these uses include swimming pools, hot tubs and spas, space heating, and of course domestic hot water. For commercial applications, solar thermal energy can be applied to food service areas such as restaurants and bakeries, commercial car washes, and for the hot water needs of multiunit buildings such as apartments and condominiums. See **Figure 19-18**.

19.5 Thermal Storage System Installation

Before the solar equipment can be installed, the proper location and placement of the collector must be determined. In order to accurately determine the placement, a site survey is necessary. Before an accurate site survey can be accomplished, one must have an understanding of the relationship between the earth and the angle of the sun's rays.

Solar Thermal Applications—Residential

Swimming pool heating

Artazum/Shutterstock.com

Hot tub heating

RAFDC/Shutterstock.com

Solar Thermal Applications—Commercial

Car wash water heating

chuyuss/Shutterstock.com

Domestic hot water heating in multifamily housing

Sundry Photography/Shutterstock.com

Figure 19-18. Examples of residential and commercial uses of solar thermal energy.

CODE NOTE

Investigate Local Building Codes

Before installing a solar water heating system, investigate local building codes, zoning ordinances, and neighborhood association covenants regarding rules and regulations. Not every community initially welcomes solar panel installations. Nonetheless, the home or building owner must comply with existing building and permit procedures before installing a system. Most zoning requirements and building codes pertaining to renewable energy improvements are local issues.

Common problems that home and building owners encounter with building codes pertaining to solar panels include:

- Exceeding roof loads
- Obstructing side yards
- Erecting unlawful protrusions on the roof
- Systems located too close to streets or lot boundaries

Always find out the required rules and regulations from the local authority having jurisdiction before proceeding with an installation.

19.5.1 Solar Angles

The earth revolves around the sun on an axis that passes through the north and south poles. This axis is tilted at an angle of 23.44° with respect to the earth's orbital plane, which is also known as the *declination angle*, **Figure 19-19**. The tilt of the earth is the reason why there are four seasons in a year and accounts for the changes in the amount of daylight as the earth makes it annual orbit around the sun. The intensity of the sun's radiation is also significantly affected by the tilt of the earth. This effect is observed as the path of the sun changes across the sky. The sun's exact position in the sky can be determined by the measurement of two distinct angles. The first angle is known as the *solar altitude angle*, which is measured from the horizontal surface of the earth to the center of the sun. The second angle is the *solar azimuth*. This angle is measured starting from true north, which is considered 0°, and travels in a clockwise direction

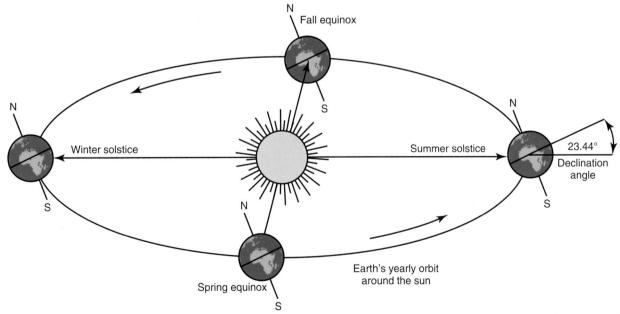

Goodheart-Willcox Publisher

Figure 19-19. The earth revolves around the sun on an axis angle of 23.44°. This is known as the declination angle.

until it intersects with the position of the sun, **Figure 19-20**. These two angles are in constant variation as the sun travels across the sky and also vary according to different latitudes and longitudes. The calculation of these angles plays an important role in determining where the solar collector should be mounted. Fortunately, these two angles have been precisely measured and can be calculated for any time and location anywhere on the face of the earth. Search online for the University of Oregon's Solar Radiation Monitoring Laboratory website, Sun Path Chart Program, which can generate a solar path diagram for any location and time.

19.5.2 Solar Collector Panel Positioning

Proper positioning of the solar collector will greatly improve the performance and longevity of the solar thermal storage system. One of the biggest issues with regard to site selection is the amount of shading that may interfere with the collector. Shading from obstacles such as trees, hills, nearby buildings, or other objects must be kept to a minimum. One device used to determine the effects of shading is a *solar pathfinder*. This device is placed at the location where the amount of shading is to be evaluated. After the pathfinder has been placed in the proper orientation, its clear hemispherical dome projects the reflections of nearby objects onto a special chart, which indicates when the location being evaluated is in the shade, **Figure 19-21**. Preferably, the solar collector should be unshaded for as long as the sun is shining. However, as a rule, no portion of the collector should be blocked by shading between the hours of 9:00 am and 3:00 pm for every day of the year.

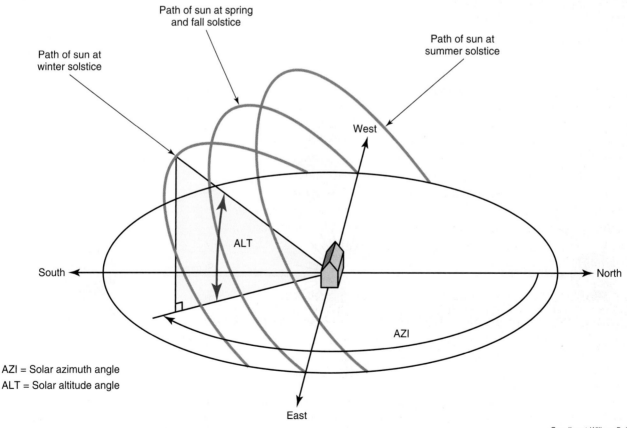

Figure 19-20. This illustration shows the azimuth and altitude angles of the sun in relationship to the earth.

The angle at which the solar collector is mounted is also of importance. Collectors mounted facing due south will receive the optimum amount of *insolation* (solar radiation). In the northern hemisphere, due south is equivalent to the azimuth angle being 180°. In some instances, because of the orientation of a given building, this exact angle may not be achievable. Fortunately, the total annual solar energy captured by the collector is not completely subject to the azimuth angle. Variations of up to 30° east or west of true south will only reduce the amount of annual solar energy collected by about 2.5%.

The other important factor when determining site selection and mounting is the *slope angle* of the solar collector. The ideal angle of slope depends on the latitude of the geographic location as well as the intended application of the system. For instance, collectors used for domestic water heating should be sloped at an angle equal to the local latitude. However, variations of ±10° on this angle will not greatly affect the performance of total solar energy collected. If the system is being used for a space heating application, the collector should be at a steeper angle to take advantage of the sun's angle during fall, winter, and spring, **Figure 19-22**. For this application, use the local latitude and add 10° to 20° to the slope. Using this steeper slope will actually reduce the amount of solar collection that takes place during the summertime, but it will prevent the overheating of the larger solar arrays often used for space heating applications. Though this practice may appear counterproductive, these larger arrays should provide most of the solar energy that will be needed during warmer weather. For dual applications of both space and domestic water heating, a slope angle of the local latitude plus 10° to 20° is also acceptable. For instance, a location at 44° north latitude may have a slope angle of approximately 60°. Regardless of the application, all solar collectors should be mounted with a tilt angle of at least 15° so that there is enough slope to ensure that normal rainfall will wash off any dirt, pollen, or other substances to help prevent soiling of the exterior glass casing.

Solar Pathfinder

Figure 19-21. A solar pathfinder is used to determine the effects of shading on solar panels.

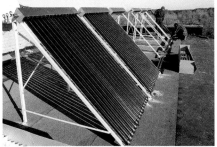

Serg_Kr/Shutterstock.com

Figure 19-22. If the solar collector is being used for a space heating application, it should be at a steeper angle to take advantage of the sun's angle during fall, winter, and spring.

19.5.3 Mounting the Solar Panel

Most solar thermal collectors are roof mounted. In order to achieve a successful installation, several factors need to be considered to ensure that the collector or array of collectors will maintain structural integrity with the roof and its components. In addition, the installer must take into consideration whether the collectors will be subject to extreme weather conditions such as tornados or hurricanes. In situations such as these, the wind forces acting on an elevated or raised collector create suction on the front of the panel and an uplifting force from behind. Always anticipate the worst possible weather scenario for the area where the collectors are to be installed and plan accordingly by following the manufacturer's instructions for proper roof mounting practices based on local weather extremities.

CODE NOTE

Installation of Solar Thermal Storage Systems

In the United States, installers are to follow Section 14 of the International Mechanical Code (IMC, 2018). This section covers the design, construction, installation, and repair of solar thermal storage systems.

In Canada, Standard CAN/CSA-F383-08 (R2013) specifies the requirements for the installation of packaged hot water systems that meet the requirements of the CSA F379 Series.

408 Hydronic Heating: Systems and Applications

Hill120/Shutterstock.com

Figure 19-23. In most cases, solar collectors can be flush mounted on the roof.

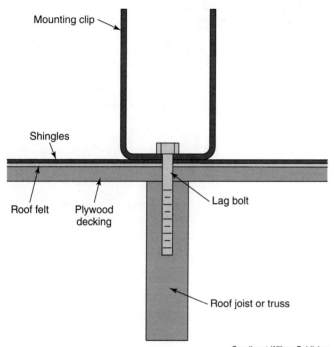

Goodheart-Willcox Publisher

Figure 19-24. Detail showing lag bolt mounting of roof bracket.

Bjorn Wylezich/Shutterstock.com

Figure 19-25. Solar collectors can weigh over 100 pounds, requiring more than one individual to lift them into place.

The solar collector manufacturer should provide the proper framework, mounting brackets, and fasteners to achieve a strong and stress-free installation. Before beginning the installation, verify that the roof structure is in a suitable condition for mounting the collectors. An inspection should be performed that ensures the roof shingles are in satisfactory condition and the roof deck and truss support systems are structurally adequate to support the weight of the solar collectors. In most cases, there is no problem with mounting a standard collector on a conventional roof, **Figure 19-23**.

The first step in mounting the solar collector or array of collectors is to locate the roof trusses or rafters to which the framework brackets or clips will be mounted. This can be done by locating the trusses or rafters in the attic, then carefully measuring the distance between them on the roof. After identifying the location of the trusses, the collector's mounting clips are attached to them using lag bolts. Typically, a sealant such as silicone caulk is applied to the underside of the mounting clip to prevent water from seeping through the roof penetration, **Figure 19-24**.

Stainless steel fasteners are the primary choice of most installers because they are less susceptible to rust and corrosion. Once the mounting clips are fastened to the roof, the next step is to install any framework that may be required. In some cases, the collector is attached directly to the mounting clips. This will depend on the pitch of the roof. When all framework is securely fastened, the collector is raised onto the roof and attached to the roof clips or to the framework. Note that flat-plate collectors can range in weight from 100 to 150 pounds, thereby requiring more than one individual to assist in mounting them in place, **Figure 19-25**.

Once the collectors are in place, the next step is to install the solar collector piping and make the proper roof penetrations where the piping will be run through. This process involves running the supply and return fluid piping through the proper roof flashing, as well as sealing this flashing to ensure that the project will be leak free, **Figure 19-26**. If there is any doubt about performing this operation, consult a local roofing contractor for advice or to assist in this step. In addition, remember that there usually will be temperature sensor wiring run to the roof. This sensor measures the temperature of the water leaving the solar collector and is used to control the circulating pump. For systems that use this wiring, a special fitting should be included in the flashing cap to protect the wire routing and sensor from damage.

CODE NOTE

Following Electrical Codes

In the United States, the National Electrical Code (NEC) includes many specifications regarding the installation of outdoor electrical circuits and equipment. In Canada, the Canadian Electrical Code, CE code, or CSA C22.1 is the standard published by the Canadian Standards Association pertaining to the installation of outdoor wiring. When wiring outdoors, the primary safety concerns involve shielding against moisture and corrosion, preventing physical damage, and proper grounding practices. Always refer to these code books and follow all safety procedures to ensure a safe and legal installation.

Copyright Goodheart-Willcox Co., Inc.

19.6 System Piping

Once the solar collector or array has been successfully mounted, the piping is then installed below the roof to connect the collector to the storage tank. Depending on the type of system chosen (active or passive, open or closed loop), this piping arrangement can vary considerably. Consult Chapter 7, *Hydronic Piping Systems*, to determine the type of piping arrangement best suited to the particular application, **Figure 19-27**.

Regardless of the solar thermal storage application, there are piping best practices that should be adhered to and accessories that should be used, especially with a closed-loop, pressurized system. The typical accessories used in a closed-loop system are outlined in the following sections.

19.6.1 Expansion Tanks

An expansion tank is a small, pressurized vessel used to divert the expansion of the water or water/antifreeze mix when it becomes heated. When water becomes heated, it expands. If

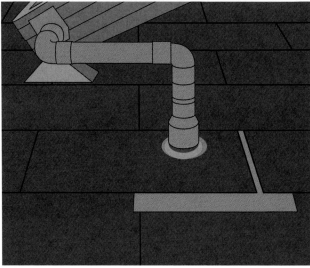

Goodheart-Willcox Publisher

Figure 19-26. Piping must be routed through the appropriate roof flashing and sealed properly to ensure that it will be leak-free.

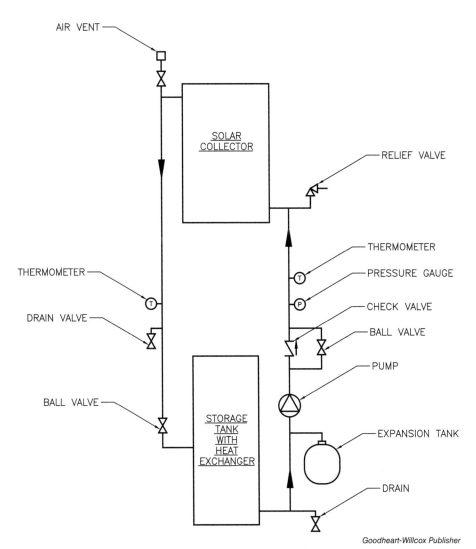

Goodheart-Willcox Publisher

Figure 19-27. This illustration shows a typical piping configuration for a solar thermal storage unit.

this expansion is not diverted or relieved in some fashion, it could rupture the piping. The expansion tank contains a rubber bladder inside of it that separates the air from the water. When the water is heated, the rubber bladder allows the water to expand. When the water cools, the bladder contracts, while pressure in the air chamber maintains a positive pressure in the closed loop, **Figure 19-28**. This bladder must be compatible with glycol. The air side of the tank contains compressed air, the pressure of which can be regulated by means of a fitting—similar to filling an automotive tire with air. Typically, these are precharged with nitrogen at 12 psi to 15 psi. The expansion tank should be installed downstream of the indoor tank or heat exchanger and upstream of the circulating pump, **Figure 19-29**.

19.6.2 Air Vents and Air Separators

It should be a standard practice to install an air vent and air separator on any closed-loop system. Water contains a certain percentage of dissolved oxygen, and when confined to a sealed, pressurized system, this oxygen breaks free from the water and creates trapped air. If left unchecked, air in a sealed system can cause corrosion and air locks, which will prevent the water from circulating through the system.

As the name implies, an air separator is used to separate air from the water as it flows through the system. The air separator incorporates either a mesh screen or deflectors that cause the air to collide and separate from the water. As more air bubbles are separated, they get larger, break loose, and travel up into the air vent, **Figure 19-30**.

Automatic, float-type air vents are mounted on top of the air separator and contain a float inside of their chamber that allows air to escape from the piping while preventing water from leaking out. As air collects inside of the chamber this float drops, allowing air to be vented out. Once the air is vented out, water refills the float's chamber—which allows water to refill the chamber, closing off the valve. The air vent should be installed vertically at the highest point in the system, typically on the return water line, **Figure 19-31**.

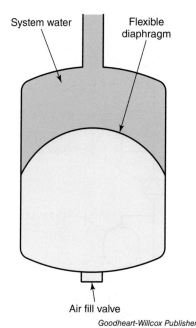

Goodheart-Willcox Publisher

Figure 19-28. A cutaway view of a common expansion tank.

Goodheart-Willcox Publisher

Figure 19-29. An example of an installed expansion tank.

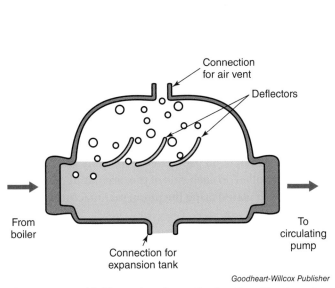

Figure 19-30. This illustration shows the functionality of an air separator.

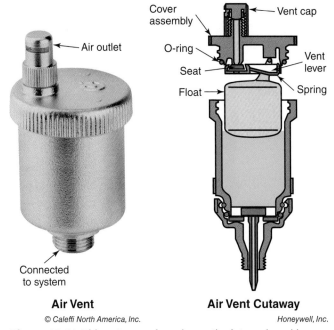

Figure 19-31. This cutaway view shows the internal workings of an automatic air vent.

19.6.3 Pressure-Relief Valves

Pressure-relief valves, sometimes referred to as safety relief valves, are used to prevent the pressure from becoming too great in the solar thermal storage system. When the predetermined pressure is reached, the valve opens, releasing the fluid to the atmosphere and preventing the system pressure from reaching unsafe levels, which could damage the collector and related equipment. Pressure-relief valves are typically set to open between 30 psi and 75 psi, depending on the configuration of the system. They should be installed near the bottom header of the collector, **Figure 19-32**.

19.6.4 Freeze Protection Valves

Freeze protection, or freeze prevention, valves are used in geographic areas where frozen pipes can be an issue. They are mounted near the outlet of the roof-mounted solar collector to protect the collector tubes from freezing during cold weather. The *freeze protection valve* should be installed so that it not only protects the collector from freezing but also protects the piping between the collector and the valve. In an indirect thermosiphon system, the freeze protection valve can be used to protect both supply and return piping. Freeze protection valves contain a wax-like material that fills a small enclosure. When this material reaches the freezing point, it changes volume, inducing a small flow of water through the relief port, **Figure 19-33**.

Figure 19-32. A pressure-relief valve is used to prevent the pressure from becoming too great in the solar thermal storage system.

19.6.5 Check Valves

A check valve permits fluid to flow in one direction only. It also reduces the amount of heat loss in the system at night by preventing a convectional flow of heat from the warm storage tank to the cool roof-mounted solar collectors. This is especially true if the system is using glycol as antifreeze, due to the fact that glycol will siphon faster than water when it is cold. The check valve can be installed on either the supply or return line near the indoor storage tank. It

ICECAL® - Anti-freeze safety device, © *Caleffi S.p.A.*

Figure 19-33. Freeze protection valves are used to prevent the collector tubes from freezing during cold weather.

should be mounted on a vertical line where the greatest amount of fall would occur. It is recommended that a spring-type check valve be used versus swing-type, which does not seat well enough to prevent thermosiphoning. The valve itself can only be opened by the force of the water or antifreeze overcoming the closing tension of the spring, **Figure 19-34**. When the circulating pump is de-energized, the spring automatically closes the valve, preventing the backflow of water into the storage tank.

19.6.6 Pressure and Temperature Gauges

Pressure gauges and thermometers are used to monitor the system's vital signs and should be installed where they can be easily read—usually at eye level. Typically, a pressure gauge with a range of 0–60 psi is used, especially with a glycol system. If pressure gauges are installed on each side of the circulating pump, the flow through the system can be calculated using the pump manufacturer's pump curves based on pressure drop.

A thermometer should be installed on the return side of the solar collector, before the storage tank or heat exchanger. The range of the thermometer should be 0–250°F, **Figure 19-35**. A second thermometer should be installed on the outlet side of the storage tank or heat exchanger. These two temperatures allow for easy monitoring of the system's efficiency. With constant flow through the system, and under full sun conditions, there should be a differential temperature between 5°F and 20°F across the storage tank or heat exchanger. Two additional thermometers can be installed on the inlet and return sides of the outdoor solar collector. The differential between these two temperatures should be less than 20°F.

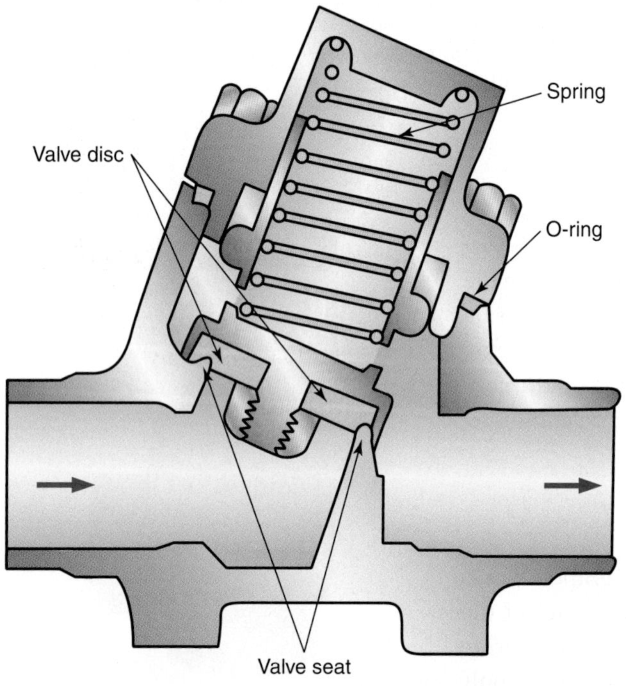

19.7 Control Strategies for Solar Thermal Systems

Once the solar collector and indoor storage tank have been installed and all piping has been completed, the next step is to install the system controls. Controls are an integral part of the total solar thermal energy system, ensuring that the system runs at peak efficiency. No matter what type of system is being controlled, the principles of control logic are similar. The following concepts are used in control systems for solar thermal energy:

- **Controlled medium.** The substance being controlled in the system. In this case, the controlled medium is the water/antifreeze solution that flows through the loop.
- **Controlled medium temperature.** The actual temperature of the substance being controlled. In this case, it is the temperature of the water/antifreeze solution that flows through the loop.
- **Controlled device.** The device that regulates the flow through the system. In this case, it is the circulating pump.
- **Temperature setpoint.** The desired temperature of the controlled medium.
- **Inputs.** Typically, these are temperature sensors connected to the controller.
- **Outputs.** The signals from the controller used to energize/de-energize the circulating pump.

Cross Section of Check Valve

Danfoss; Superior Refrigeration Products

Figure 19-34. This cross-section illustration shows the internal workings of a spring-type check valve.

- **Controller.** The device that receives signals from the temperature sensors, compares them to the set point value, and sends the appropriate output signal to the controlled device.
- **Control loop.** The arrangement of the input device, controller, and output device within the system.

One of the most common control strategies for active solar thermal energy systems with an enclosed glycol loop is to monitor differential temperatures. This type of system strategy monitors two temperature sensors as a means of controlling the respective circulating pumps. One sensor is located at the outlet of the solar collector. The other sensor monitors the temperature of the water in the storage tank, **Figure 19-36**. The controller is constantly monitoring the difference in temperature between these two sensors and their respective set points. When the temperature of the sensor located at the collector outlet exceeds its set point, the controller energizes the circulating pump for the glycol loop. Likewise, when the temperature of the storage tank exceeds its set point, the potable water loop circulating pump is energized. The normal differential set point on the controller is typically between 5°F and 10°F. As long as the temperature at the outlet of the collector or at the storage tank is 5–10 degrees higher than its set point, the respective pump will run continually. This would be the normal scenario during a sunny day. When these temperatures are equal to or below set point, the pump will shut off. Advanced controllers can also operate the circulating pump at variable speeds that are proportional to the differential temperature between the collector and storage tank. As the temperature differential rises, the pump speed increases. This strategy is used to reduce the electrical consumption of the pump under partial sun conditions, **Figure 19-37**.

Goodheart-Willcox Publisher

Figure 19-35. Combination pressure and temperature gauges are vital for monitoring the solar thermal storage system's operation.

Photo courtesy of Watts

Figure 19-36. This type of controller monitors two temperature sensors as a means of controlling the circulating pump.

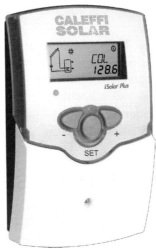

iSolar Plus, © Caleffi North America, Inc.

Figure 19-37. This solar thermal controller can operate the circulating pump at variable speeds that are proportional to the differential temperature between the collector and storage tank.

Taco Comfort Solutions

Figure 19-38. An example of a circulating pump that runs on DC power and is energized by photovoltaic solar cells.

As an additional feature, there are circulating pumps available today that run on DC power (direct current) and are energized by photovoltaic solar (PV) cells. The strategy behind this concept is to reduce the operating cost of the circulating pump and take advantage of the sun's rays to modulate the pump speed. As the sun's energy level increases, the circulating pump automatically increases its speed, allowing for greater thermal transfer through the collector, **Figure 19-38**. Another strategy is to use both an AC-powered (alternating current) and DC-powered pump, so that if power is lost to the AC pump, the DC pump can maintain flow through the system. When incorporating DC pumps with PV modules, be sure to match the proper pump with its appropriate solar panel. Also, be sure that the DC pump has enough capacity to circulate fluid through the solar loop. Not all DC pumps have the same rated output as AC pumps. Remember that antifreeze is more difficult to pump than straight water due to its higher viscosity—something to keep in mind when sizing the circulator.

Today's modern digital controllers incorporate specialized temperature sensors called thermistors for controlling solar thermal systems. By definition, a ***thermistor*** is a resistor made of a semiconductor material in which the electrical resistance varies with a change in temperature. With a ***negative temperature coefficient (NTC)*** type thermistor, the ohms resistance decreases as the temperature of the thermistor increases. Thermistors are utilized in modern control sequences because of their accuracy and reliability. They can be installed by two different methods. One way is to attach the sensor directly to the piping to measure the water temperature. This is commonly known as a ***strap-on temperature device***, and the installer should make sure the sensor is making good contact with the pipe, **Figure 19-39**. The other method is to install a recessed well into the piping and insert the sensor into the well. This method can be more accurate than the strap-on sensor and is less susceptible to damage and neglect, **Figure 19-40**.

Other features that may be included with today's modern digital controllers include:

- Controlling multiple stage units
- LCD displays on the controller
- Demand limit strategies
- Time clock scheduling
- Outdoor temperature reset function
- Night setback function
- Data logging

Goodheart-Willcox Publisher

Figure 19-39. An example of a strap-on temperature sensor.

19.8 Filling and Starting up the System

When the temperature controls have been installed and all piping is complete, it is time to fill the system and perform a proper start-up procedure. However, before the system is filled, it must first be properly leak tested. Note that the expansion tank and air vent should be installed after the system has been leak tested and cleaned to prevent damage to these devices. Temporarily cap off the fittings where these components are to be installed.

19.8.1 Cleaning and Flushing the System

First, fill the system with air until the pressure is between 50 psi and 60 psi—or at least three times the normal operating pressure. Note that leaks will be easier to detect with air than with water. Once the system is pressurized, use a solution of mild soap and water to swab every joint and fitting for potential leaks. Dishwashing soap is acceptable, but a better suggestion is to use a commercial leak-detecting fluid, which is available at any HVAC supply house. Look for any bubbling at each joint that may indicate a potential leak. When this situation occurs, tighten any joints that may be suspected of leaking and reinspect by reswabbing with the bubble solution. Another method of leak detection is to record the test pressure on the gauge and wait at least 30 minutes to see if it falls. If the pressure holds, the system should be leak-free. If the pressure falls, use the bubble-test method of leak detection and repair any leaks. If possible, keep the system pressurized overnight before filling it to ensure that it is leak-free.

Once the system has been verified as leak-free, the next step is to clean the lines. Use a mixture of one cup of trisodium phosphate (TSP) per one gallon of water as a cleaning solution. It will typically require several gallons of solution to ensure the system gets cleaned properly. A positive displacement pump works best for pumping the solution into the system. Once filled, run the circulating pump for 30 minutes to clean out any flux, pipe thread compound, or other substances from the lines, and then drain the cleaning solution and flush the system with clean water.

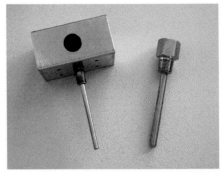

Goodheart-Willcox Publisher

Figure 19-40. Well-type temperature sensors can be more accurate than strap-on sensors and are less susceptible to damage and neglect.

19.8.2 Charging the System

At this point, the expansion tank and air vent should be installed. However, before installing the expansion tank, be sure to charge it with air to the correct pressure—typically 12 psi to 15 psi. It will not harm the system if the tank exceeds 15 psi, as long as this pressure does not exceed the manufacturer's recommendations for the solar collector or system components.

Before filling the system with water or glycol solution, cover the solar collector to prevent the solution from being heated too quickly while filling. Next, open the air vent located near the solar collector. This will force the system's air to the highest point of the loop. The pressure-relief valve may also be opened for the initial fill. Close it once fluid begins to flow out of the valve. Once again, use a positive displacement pump to charge the system. Introduce the fluid solution at the lowest point of the system to force the air upward to be displaced into the atmosphere. A *flush-and-fill cart* can come in handy at this point, **Figure 19-41**. Once the system has been filled, energize the circulating pump to force any air out of the piping. The air vent should remain open until all air bubbles have been displaced from the system. Record the operating system pressure and monitor this pressure for several days to ensure it stays constant. If during this period a pressure drop of less than 10 psi is observed, then the system is more than likely air free.

The temperature controls can now take over the monitoring and control of the system. Over the course of the first several weeks of operation, periodically check the system pressures and temperatures to ensure they are in range. As mentioned earlier, the temperature through the heat exchanger or storage tank should maintain a differential between 10°F and 25°F. This will indicate that the heat exchanger is functioning properly.

HYDROFLUSH™ Pump Cart, © Caleffi North America, Inc.

Figure 19-41. A flush-and-fill cart can fill the solar thermal system.

Additional maintenance points to monitor or observe are as follows:
- Take note if the pump is running at the "wrong time," such as at night.
- Wash the solar collector annually with water and a soft brush.
- Spray the collector with water during prolonged periods of no rain.
- Install labeling on all piping showing the direction of flow.
- Periodically check all fittings for possible leaks.
- Oil the circulating pump motor on a semiannual basis if required by the manufacturer.
- Unless there is a problem with stagnation, the glycol solution should not need to be replaced for 10 years.

19.9 Additional Applications for Solar Thermal Storage

At this point, most of the information presented has dealt with utilizing the solar thermal storage system for domestic hot water applications. However, there are several other applications that are just as suited for solar thermal energy systems.

19.9.1 Swimming Pools

The intent of any pool heater is to extend the length of the swimming season by beginning earlier in the spring and extending this time into the fall. What makes solar pool heaters so attractive is the relatively low cost of operation compared to conventional pool heaters. Solar pool heating systems use unglazed, low-temperature collectors typically made from polypropylene. The intent of using this type of collector is for it to operate just slightly above the temperature of the surrounding air. By doing so, the collector is capable of heating a large volume of water by only a few degrees. Remember that heating a swimming pool with a solar collector is a marathon, not a sprint. Also, raising the temperature of the water only a few degrees above the ambient air will prevent heat stratification and maintain a more consistent pool water temperature. By implementing a pool cover or blanket when the pool is not in use, water temperatures of 18°F to 25°F above ambient can be achieved, **Figure 19-42**.

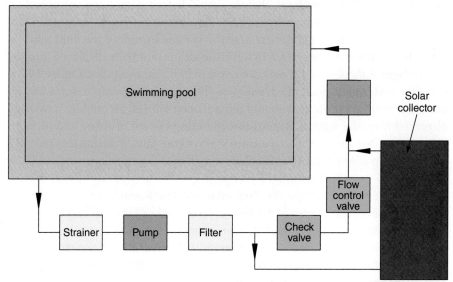

Goodheart-Willcox Publisher

Figure 19-42. This illustration shows a typical piping arrangement for a solar pool heating system.

The pool's solar collector is typically sized to be approximately 50% to 100% of the surface area of the pool. A rule of thumb is that a 3°F temperature rise can be achieved for each 20% of the pool surface area that is added to the solar collector. For instance, an 8°F temperature rise can be achieved for a pool area that is 512 ft² with a collector size of 256 ft². However, the temperature rise will increase to 15°F if the collector size is increased to 512 ft² for the same size pool.

Pool solar collectors are best installed on south-facing roofs, although any location that is unshaded during the middle six hours of the day will work—even for a freestanding collector. In fact, 80% of the solar radiation gained from the collector will typically occur during a four-hour period. For a south-facing solar collector, this would be between the hours of 10:00 am and 2:00 pm. Remember that shading from trees or nearby buildings will greatly reduce the capacity of any solar pool collector.

Another added benefit of using solar energy to heat a swimming pool is that the pool's existing circulating pump can be used to circulate water through the collector as well as through the pool filtration system. Just as with domestic water heating applications, the solar pool heater incorporates two temperature sensors for controlling the system. One sensor is located in the PVC piping downstream of the circulating pump. It measures the actual temperature of the water leaving the pool. The other sensor is located in the middle of the collector array.

Solar pool heating systems typically use a diverter valve. When the temperature controller senses that the collector temperature is 5°F to 8°F warmer than the pool water sensor, it energizes the diverter valve to direct the flow of water to the collector and then back to the pool. When the controller senses that the differential temperature between collector and pool is too low, it de-energizes the diverter valve, which blocks the flow of water to the collector, **Figure 19-43**. The pool water sensor also acts as a high limit that stops the flow to the collector should the pool water temperature exceed its set point.

Solar pool heating systems generally cost between $7 and $12 per square foot of pool area, depending on system design and collector type. This can provide a return on investment of between 18 months and 7 years as compared to conventional heating methods such as gas-fired or electric pool heaters, which can have an annual operating cost of between $2500 and $3000. Clearly, solar pool heating is one of the most cost-effective uses of solar energy available.

3-way Diverting, © Caleffi North America, Inc.

Figure 19-43. A diverter valve can be used to control water temperature when heating swimming pools.

19.9.2 Hot Tubs and Spas

Solar heating of hot tubs and spas is similar to pool heating. However, in order to achieve increased temperatures of 100°F and higher, a dedicated solar collector may be necessary. Also, because it is difficult to achieve these higher temperatures during the evening hours, a backup source of heat is usually required. One strategy for incorporating a hot tub along with an existing swimming pool is to use the pool's collector to increase the tub or spa temperature along with the pool's water temperature, and then finish increasing the spa's temperature up to its set point using an auxiliary heating source. As with swimming pools, the use of a spa cover is essential for saving money on energy costs, **Figure 19-44**.

19.9.3 Space Heating

The two main methods of utilizing solar energy for space heating are classified as radiant heating and forced-air

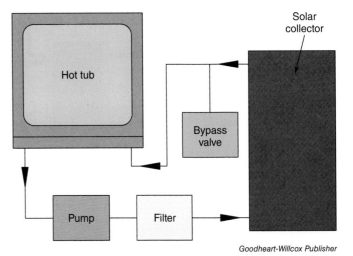

Goodheart-Willcox Publisher

Figure 19-44. An example of a typical piping layout for a solar hot tub.

systems. Both types of systems are dependent on whether the structure is classified as new construction or is being retrofitted for space heating needs. Generally, it is a better idea to incorporate a solar thermal space heating system into a well-insulated, tightly constructed building. Therefore, this type of system bodes well for new construction. Regardless of which type of system is preferred, utilizing solar thermal energy systems for space heating will usually satisfy 50%–80% of the annual space heating loads.

19.9.3.1 Radiant Solar Heating

Radiant heating is accomplished by circulating hot water through a network of tubing—usually embedded in the floor, or through terminal heating devices such as convectors or radiators. Once the terminal device is heated, it radiates warmth throughout the conditioned space. The defining characteristic of solar radiant heating is the low water temperature at which it operates. Whereas conventional radiant heating systems operate at water temperatures of up to 180°F, solar radiant systems typically function at a water temperature of no greater than 120°F—the reason being lower design temperatures allow for greater solar energy yields. These lower temperatures equate to a reduction in the size of the solar collector. Radiant solar heating is typically delivered to the space either through heated floor slabs or through radiant panels, **Figure 19-45**.

Heated floor slabs incorporate plastic or cross-linked polyethylene (PEX) tubing that is spaced close together and embedded into the subflooring of the structure. This design is well suited for use with floor coverings that have a low R-value, such as tile, vinyl, or hardwood surfaces. The tubing is typically spaced 6″ apart to maintain a room temperature of 70°F when the water in the circuit can maintain at least 88°F.

Spacing of the tubing can be increased up to 12″ apart if the supply water temperature can maintain a range between 95°F–98°F. Tube sizing for radiant

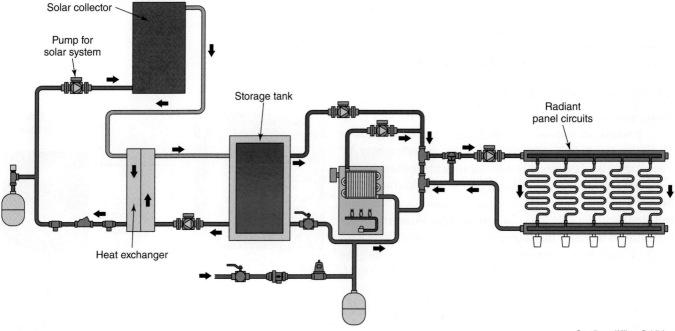

Figure 19-45. This piping arrangement is used for radiant floor heating.

floor heating is typically 1/2″ diameter. Separate zones can be controlled independently of each other by means of a space thermostat and controlling zone valve. The main water supply and return lines are connected to a manifold station, where individual zones can be controlled. Each circuit contains a zone valve, which opens when an individual thermostat calls for heat.

Radiant panels can also be integrated into walls and ceilings to offer an alternative to floor heating. Just as with solar floor heating systems, radiant panels operate at relatively low supply water temperatures. To achieve an adequate heating output, these panels require a high surface area relative to the rate of heat to be delivered to the space. They also favor a relatively low internal resistance between the tubing and the surface area that is releasing heat into the space. Radiant panels are typically covered with drywall or a similar wall covering, which makes them indistinguishable from a standard interior wall or ceiling. Because they typically have a low thermal mass, radiant wall panels respond quickly to changes in internal load conditions or zone setback schedules. With a supply water temperature of 110°F, the typical radiant panel will deliver approximately 32 Btu/hr/ft² in order to maintain a space temperature of 70°F. A heating load analysis on the conditioned space will determine the required Btu/hr generation to heat the space. Once this is known, the correctly sized panel can be incorporated. Most manufacturers can provide output ratings for the radiant panels they produce, **Figure 19-46**.

The amount of solar collector area needed to heat with radiation depends on many factors. These include the amount of solar energy available, the solar collector's efficiency, local geographic climate, and the heating requirements of the conditioned space. Heating requirements are based on insulation levels, the tightness of the house, and the lifestyle of the occupants. Generally, the area of solar collector needed is approximately equal to 10% to 30% of the building's floor area in square feet.

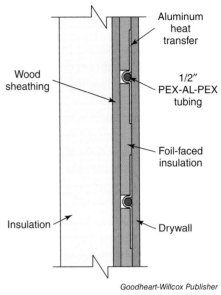

Goodheart-Willcox Publisher

Figure 19-46. This illustration shows a layout for radiant wall heating.

CAREER CONNECTIONS

Careers in Solar Thermal Power

The solar thermal power industry employs a wide range of occupations within a number of major industry segments. These include research and development, engineering, sales, manufacturing, construction, and maintenance.

Careers in research and development include physicists, chemists, and research scientists. Median annual salaries for these occupations range from $68,000 to over $100,000 per year.

Engineers design and test solar equipment to ensure its performance and reliability. Occupations such as these receive an annual salary of between $50,000 and $90,000.

Sales consultants in the solar industry are responsible for site evaluations, product assessment, and even project management. Consultants can receive up to $15,000 per month as full-time commissioned sales representatives.

Manufacturing careers range from machine tool operators to welding, soldering, and brazing workers, to industrial process managers. Most production workers are trained on the job and gain valuable expertise with experience. Skilled trade positions may attend formal training programs or apprenticeships. Median annual wages for these positions can range from $27,000 to $77,000.

Construction and maintenance of solar thermal storage systems include the skilled trades such as electricians, plumbers, pipefitters, HVAC technicians, and even roofers. Skilled trades workers in these areas generally receive between $30,000 to $50,000 per year as median annual wages.

Chapter Review

Summary

- Solar thermal storage is achieved by capturing the sun's energy in a solar collector.
- There are a number of different solar thermal storage configurations that can be utilized.
- A passive solar heating system harvests energy from the sun without the use of electrical or mechanical devices.
- In an integral collector storage system, the collector and tank are combined into one single unit. Integral collector storage systems have a simple design and do not require any mechanical or electrical devices.
- Thermosiphon solar hot water heating systems utilize the principle of natural convection.
- Active solar storage systems incorporate the use of an energy storage system that will provide heat when the sun is not shining.
- In an open-loop system, the hot water is piped from the solar collector into a storage tank, which is connected to the municipal water system.
- A closed-loop storage system circulates a heat transfer fluid through the solar collector and through a closed heat exchanger inside of a storage tank.
- The heat transfer fluid used in a closed-loop system usually consists of a glycol-water mixture.
- A drainback system works by draining all the water from the solar collector into a tank whenever the system is not collecting solar energy. The drainback system relies on gravity to quickly drain the water when the system's circulating pump shuts off.
- A pressurized storage system is one in which the circulation loop is closed to atmospheric pressure.
- An unpressurized system is one in which the loop is open to the atmosphere, such as with an open-loop system.
- Solar collectors are categorized as either flat-plate or evacuated tube.
- A flat-plate collector consists of an insulated metal or aluminum box covered with tempered, low-iron glass.
- An evacuated tube collector consists of a row of glass tubes from which the air has been removed, forming a vacuum.
- In order to accurately determine the placement of the solar collector panels, a site survey is necessary.
- Installation of the solar collectors is dependent on the positioning of the panels based on the sun's angles and the amount of shading.
- A solar pathfinder is used to determine the amount of shading when mounting the solar panels.
- It is important to secure roof-mounted panels correctly because of potential severe weather conditions.
- Once the solar collector has been mounted, the piping is connected to the storage tank.
- The typical piping accessories used in a closed-loop system include an expansion tank, an air separator, an air vent, a pressure-relief valve, a freeze protection valve, and a check valve.
- Controls are an integral part of the solar thermal system and are vital to ensuring that the system runs at peak efficiency.
- Temperature sensors are located at the collector and storage tank to control circulating pumps.
- Before filling and starting up the system, leak testing and proper system flushing should be performed.
- Maintenance procedures on the solar thermal system include general cleaning, periodic leak testing, and lubrication of circulating pumps.
- Applications for solar thermal storage include swimming pools, spas, and space heating.

Know and Understand

1. A(n) _____ harvests energy from the sun without the use of electrical or mechanical devices.
   A. passive solar heating system
   B. active solar heating system
   C. neutral solar heating system
   D. drainback system
2. Which type of solar unit incorporates a sloped solar collector installed in conjunction with an insulated tank?
   A. Thermo-drainback system
   B. Thermosiphon system
   C. Integrated drainback system
   D. Active solar thermal storage system
3. *True or False?* A closed-loop system is one in which the hot water is piped from the solar collector directly into a storage tank or a domestic water heater.
4. Open-type systems cannot circulate an antifreeze solution through them because _____.
   A. there is no threat of freezing
   B. they require a special type of circulating pump
   C. they utilize the domestic water supply
   D. they are more susceptible to corrosion
5. A drainback system is used to _____.
   A. improve the efficiency of open-type solar collectors
   B. prevent any damage to exposed piping when the outside air temperature drops below freezing
   C. allow antifreeze to freely flow back to the solar collector
   D. quickly drain the system in the event of a leak
6. Most residential and commercial applications for solar thermal storage that require fluid temperatures below 200°F will use which type of solar collector?
   A. Flat-plate collector
   B. Evacuated tube collector
   C. Neither A nor B.
   D. Both A and B.
7. Which type of solar collector is more effective on a cloudy day?
   A. Flat-plate collector
   B. Evacuated tube collector
   C. Photovoltaic solar module
   D. None of the above.
8. *True or False?* A device that is used to determine the effects of shading is called a solar pathfinder.
9. The sun's exact position in the sky can be determined by the measurement of the _____.
   A. solar altitude angle
   B. solar azimuth
   C. Both A and B are correct.
   D. Neither A nor B are correct.
10. What is the normal differential temperature set point for controlling the circulating pumps on a solar storage system?
    A. Between 5°F and 10°F
    B. Between 10°F and 20°F
    C. Between 20°F and 30°F
    D. Between 25° and 30°F

Apply and Analyze

1. Explain the difference between a passive and active solar heating system.
2. How is an integral collector storage unit similar to a thermosiphoning system? How do they differ?
3. What is the difference between an open-loop and a closed-loop solar thermal storage system?
4. Why does a closed-loop solar thermal storage system require the use of an antifreeze solution?
5. What does the drainback system rely on to drain the water or glycol solution from the collector when the system's circulating pump shuts off?
6. Explain how a flat-plate solar collector operates.
7. Describe the difference in functionality between a flat-plate and evacuated tube collector.
8. What factors determine the best location for a roof-mounted solar collector?
9. Explain what a solar pathfinder is used for.
10. Describe the steps taken to secure a solar array onto a roof.
11. What safety and control components are installed along with the piping between the storage tank and solar collector?
12. What is the functionality of each of these safety and control components?
13. Describe the control strategies used for solar thermal storage systems.
14. What are the proper steps for flushing, filling, and starting up the solar thermal storage system?

Critical Thinking

1. What steps should be taken to supplement an existing hydronic system with a solar thermal storage system?
2. Which areas of the world would best benefit by using a thermosiphoning system? Why would these areas benefit the most from this type of system?

20 Outdoor Wood Boilers

Chapter Outline
20.1 Sizing the Boiler
20.2 Installation of Outdoor Boilers
20.3 Additional Applications
20.4 Boiler Maintenance

Learning Objectives

After completing this chapter, you will be able to:
- Describe how an outdoor boiler can be a viable source of comfort heating.
- List the advantages of an outdoor boiler versus an indoor biomass appliance.
- Explain the process for sizing an outdoor boiler.
- Identify the type of heat exchanger used on an indoor air handling unit in conjunction with an outdoor boiler.
- Outline how heat is transferred from the outdoor boiler into the building.
- Describe the process for setting the outdoor boiler in place.
- Explain why insulated PEX piping is the preferred choice for the outdoor boiler's underground piping.
- Detail the process for installing the underground heating pipes for an outdoor boiler.
- Describe the types of fittings used for connecting the hot water piping to the indoor heat exchanger.
- Identify the types of controls used both indoors and outdoors on a biomass boiler.
- Explain the process for wiring the indoor and outdoor controls together on an outdoor boiler.
- Describe other types of biomass that can be used in conjunction with an outdoor boiler.
- List the various applications for an outdoor boiler.
- Differentiate the maintenance on an outdoor boiler from that of a conventional boiler.

Technical Terms

biomass heating
creosote
dual fuel boiler
firebox
frost line
heat load calculation
indoor air pollution
indoor heat exchanger
line-voltage thermostat
outdoor boiler
propylene glycol
storage hopper

With a growing awareness of environmental stewardship sweeping throughout the world, it only makes sense to consider renewable resources as a source of heating fuel. One viable alternative method for heating homes and businesses is to use a biomass fuel in conjunction with an outdoor boiler. This is especially true in rural areas, where forests are abundant. Renewable biomass heating resources that can be utilized for an outdoor boiler include wood, corn, and cherry pits—although wood is the predominant choice. This chapter covers the proper sizing, installation, piping, and control wiring for outdoor wood boilers.

20.1 Sizing the Boiler

Wood has been used as a reliable heating fuel for centuries. While burning wood is typically associated with indoor fireplaces, there are also wood-burning stoves and furnaces that can be used inside buildings as a source of comfort heating. However, because of the potential risk of house or building fires, the trend in burning wood today is to move the appliance outdoors for safety—namely in the form of an outdoor boiler. An **outdoor boiler** is an unpressurized appliance that heats water at a distance from the building to be heated by burning biomass fuel. There are several other advantages to using an outdoor boiler when compared to an indoor wood-burning appliance, **Figure 20-1**. These advantages include the following:
- Because the wood is stored and burned outdoors, there are no problems with dirt, soot, or insects inside the home or building.

Goodheart-Willcox Publisher

Figure 20-1. Outdoor wood boilers such as this have increased in popularity over the past several years.

- There is less possibility of *indoor air pollution* caused by a wood burning appliance located inside the building.
- Typically, there is no increase in the building owner's property insurance by having the unit outdoors as opposed to indoors.

When considering what size outdoor wood boiler to purchase, the first step is to perform a *heat load calculation* on the building, **Figure 20-2**. A detailed description of this task can be found in Chapter 13, *Building Heating Loads and Print Reading*. Once heat loss for the structure is calculated, consult a boiler manufacturer or dealer to determine what sizes are available as well as the efficiency of the chosen model. Traditionally, outdoor wood- or corn-fired boilers have maintained a heating efficiency of around 50%. This factor has improved significantly over the past several years as the demand for *biomass heating* has increased. Today, there are outdoor wood boilers with efficiencies certified by the EPA of close to 80%.

Once the proper boiler has been selected, the next step is to size an *indoor heat exchanger* that will be installed in conjunction with the building's forced-air heating system. This system may be the building's furnace or an air handling unit. An air handler typically consists of a cabinet with a circulating fan or blower and a heating device such as an electric coil. Typically, a water-to-air heat exchanger is used inside the building and is installed within the supply air ductwork, **Figure 20-3**. If the boiler manufacturer or wholesaler does not furnish an indoor heat exchanger, it is best to consult an HVAC wholesaler or heating coil manufacturer to assist in sizing this item. These suppliers can best determine the proper dimensions of the coil to fit the existing ductwork of the forced-air system and ensure that the proper heating output is achieved.

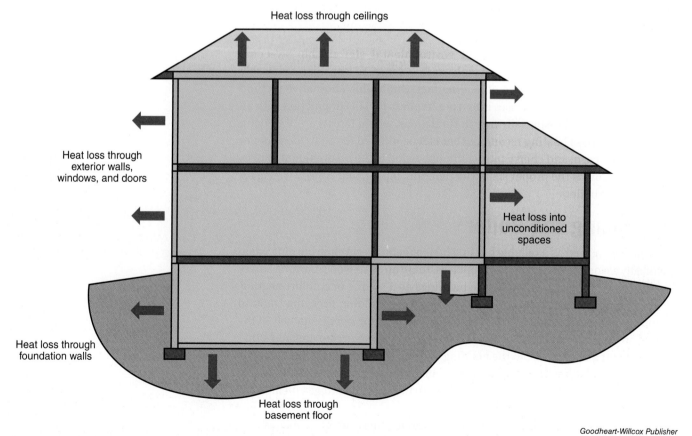

Goodheart-Willcox Publisher

Figure 20-2. The first step in sizing an outdoor wood boiler is to perform a heat load calculation on the building.

TECH TIP

Choosing the Right-Size Boiler
Be careful not to choose an outdoor wood boiler that is oversized. Problems will occur during mild weather when there is minimal demand for heat, causing the wood to simply smolder inside the combustion chamber without really burning. Also, choose a boiler with a large access door. This will make it easier to load wood into the boiler.

sspopov/Shutterstock.com

Figure 20-3. This image shows an example of a water-to-air heat exchanger.

20.2 Installation of Outdoor Boilers

Outdoor boilers are often located between 50′ and 100′ from the home or building. This is to reduce the risk of fire to the structure and, more importantly, to improve the venting of the products of combustion. A boiler located too close to the structure can allow smoke and soot to enter the structure, causing problems with indoor air quality. However, in some instances it is convenient to have a storage shed nearby to store the wood, **Figure 20-4**.

Outdoor boilers use water as a medium for heat transfer. The boiler heats water that is transferred through underground piping to the indoors and through a water-to-air heat exchanger. This heat exchanger is mounted in the supply plenum of the forced-air furnace, **Figure 20-5**. It is important to understand that, unlike a typical residential hot water boiler, the outdoor boiler's configuration is not a pressurized system. The water circulated through the boiler is vented to the atmosphere. This is due to safety reasons—specifically to prevent an explosion in the event that the outdoor boiler runs out of water. A pressurized hot water system requires additional safety features and falls under a separate set of building code regulations. Another important point to emphasize is that an outdoor boiler typically circulates a mix of water and ***propylene glycol*** to guard against freezing.

Goodheart-Willcox Publisher

Figure 20-4. This image shows an outdoor wood boiler, loaded and ready to fire up. The adjacent building is used for storing wood.

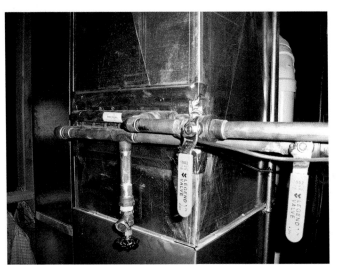

Goodheart-Willcox Publisher

Figure 20-5. This image shows an example of a heating coil mounted in the supply ductwork of a residential furnace.

> **GREEN TIP**
>
> **Clean, Green Wood Heat from an Outdoor Boiler**
> The difference between an efficient and an inefficient outdoor boiler can mean the difference between burning 15 cords of wood per year versus 30 cords. Fortunately, today's outdoor boilers meet the standards that are required to make them truly efficient. In fact, new combustion technology means that outdoor boiler systems now burn clean enough to be considered truly green.
> Using a process called gasification, EPA certified models are now more than 90% cleaner than traditional outdoor boilers. Besides cleaner air, a higher efficiency boiler also means more heat from less wood.

20.2.1 Setting the Boiler

Outdoor boilers can weigh as much as 3000 lb when filled with water. Therefore, it is important to ensure that the boiler is mounted on a solid base. Typically, a 4″ concrete pad is poured to support the boiler. Most manufacturers provide a template with their product showing the required dimensions of the concrete pad, **Figure 20-6**. The amount of concrete may vary, but typically half of a cubic yard will suffice. Remember that the pad should extend beyond the front of the boiler to allow the operator sufficient space for loading the firebox. The *firebox* is the chamber inside the boiler where the fuel is loaded and where combustion occurs. An alternative to pouring concrete for a base is the use of concrete blocks. An ample number of blocks should be used at each corner of the boiler, or they should be set around the complete perimeter of the boiler. Be sure that these blocks rest on a firm foundation in the ground to prevent shifting of the boiler once the blocks settle.

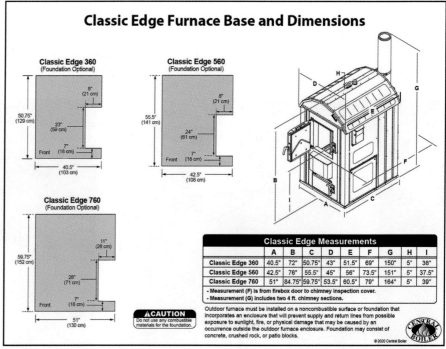

Central Boiler, Inc.

Figure 20-6. The foundation plans for an outdoor wood boiler are available from the manufacturer and will indicate the proper dimensions for the concrete pad on which the boiler sits.

20.2.2 Piping Installation

The first step in running the underground piping between the boiler and building is to dig a trench below the *frost line*, the maximum depth that frost will penetrate the ground during the coldest period of the heating season. This will prevent excessive heat loss and prevent water from freezing in the pipes if the system does not use antifreeze. This depth varies depending on geographic location. **Figure 20-7** shows typical frost line depths throughout the United States. Digging the trench requires a trenching machine or small backhoe, which can be rented. Otherwise, contract with a local excavator, landscaper, or septic tank installer.

The piping of choice for underground installation is PEX, which stands for cross-linked polyethylene. It is flexible and can withstand high and low temperature fluctuations. PEX is also easy to install and highly resistant to chemicals found in the plumbing environment. The smooth interior of PEX will not corrode and is also very resistant to freezing and breakage, **Figure 20-8**. Although it is not mandatory to install PEX lines below the frost line, doing so will prevent a large amount of heat loss through the system. Two lines of 1″ to 1 1/4″ PEX need to be installed for supply and return water. These lines should be encased in high-density urethane insulation and wrapped in a high-density polyethylene jacket to prevent any underground moisture or frost from coming into contact with the hot water lines, **Figure 20-9**. If this should happen, substantial heat loss will result, which will significantly affect the amount of hot water available to heat the building. Another suggestion is to run the hot water piping through oversized PVC pipe. Remember that the PEX piping will expand and contract

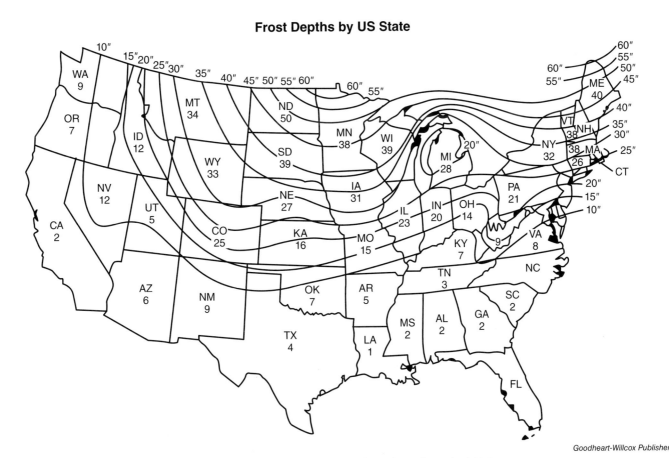

Goodheart-Willcox Publisher

Figure 20-7. This map indicates the frost line depths for various regions throughout the United States.

Figure 20-8. PEX is the preferred underground piping for outdoor wood boilers.

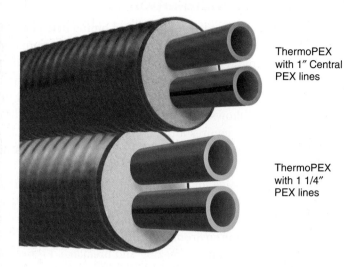

Figure 20-9. Underground piping lines should be encased in high-density urethane insulation and wrapped in a high-density polyethylene jacket to prevent any moisture or frost from coming into contact with them.

when heated and cooled, and the PVC pipe will allow for this without affecting the pipe's performance. At the very least, use an oversized sleeve through the foundation wall where the piping enters the building to allow for expansion and contraction.

TECH TIP

Installing an Outdoor Boiler in the Winter

An outdoor boiler can be installed during the winter when the ground is frozen. Simply lay the insulated water lines on the ground on top of a bed of straw, and cover the lines with more straw to help insulate them. Some heat loss will occur, but it should be minimal. Then when spring comes, the water lines can be buried.

Connect the PEX piping to the indoor heat exchanger using special compression fittings or push fittings provided by the piping supplier, **Figure 20-10**. These fittings may require a special tool, which can also be provided by the PEX supplier. At the outdoor boiler, the piping will be connected to a circulating pump. On larger outdoor boilers, or if the boiler is a significant distance from the building, it is not unusual to use two circulating pumps—one on the supply line and one on the return.

20.2.3 Wiring and Controls

When installing the piping in the outdoor trench, include the electrical wiring that runs to the boiler. This wiring is typically 12-gauge, 3-conductor, 110-volt cabling that should be run in its own 3/4″ PVC electrical conduit underground. One suggestion when installing both the electrical conduit and hot water piping is to include a nylon cord through both the electrical and hot water PVC sleeves in case an additional circuit is needed in the future. This cord will act as a pulling cable for future wiring if necessary. Next, install a new 15-amp circuit

Figure 20-10. Push fittings such as the one shown here can be used to join PEX piping to the indoor heat exchanger.

breaker in the building's breaker panel, and connect the boiler wiring from the indoor breaker to the outdoor hot water controller. This circuit will also power the circulation pumps. Include a light fixture and convenience outlet as part of the outdoor wiring circuit. The light fixture will provide convenience lighting during the evening hours, and the outlet will offer power for any electrical tools that may be needed during installation or repairs.

The controls for the outdoor boiler are relatively simple. They are divided between controls for the outdoor equipment and the indoor equipment. Most outdoor boilers are controlled by a ***line-voltage thermostat*** that senses the hot water temperature, **Figure 20-11**. When this temperature falls below its set point, the contacts on the thermostat close, which in turn opens a combustion damper on the outdoor unit. This damper allows combustion air to enter the fire chamber and increase the size of the flame, thus raising the water temperature. When the thermostat reaches its set point, it closes the damper and the flame is reduced to a smolder. Regardless of the heating demand, there is always a fire burning in the outdoor unit.

Indoors, a room thermostat senses the space temperature and energizes the furnace circulating fan on a call for heat. The fan blows air across the hot water coil, warming up the space to its temperature set point. When the space temperature set point is satisfied, the fan is de-energized.

The last control scenario deals with the circulating pump. There are two schools of thought with regard to controlling the circulator. One thought is that the pump should be energized continuously. This is true if there is no antifreeze solution in the water—otherwise the piping would freeze. If the system does contain antifreeze, the circulating pump could be cycled on and off by either the outdoor thermostat, which controls the boiler water temperature, or whenever the indoor space thermostat calls for heat. Cycling the pump with the boiler's outdoor hot water thermostat is usually a better option. By doing this, there will always be enough hot water present in the indoor coil when there is a call for heat by the indoor thermostat. The indoor fan motor could also be wired in series with a thermostat strapped to the return water line near the indoor heat exchanger coil. This thermostat will energize the fan only when the coil contains hot water, which will prevent cold air from blowing through the building when there is a call for heat, **Figure 20-12**.

Goodheart-Willcox Publisher

Figure 20-11. This image shows the location of the thermostat controls for the outdoor wood boiler.

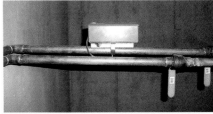

Goodheart-Willcox Publisher

Figure 20-12. This is a strap-on thermostat. It is located on the return water piping of the indoor heat exchanger and used to control the system circulating pump.

GREEN TIP

Dual Temperature Controllers
Some modern boilers come equipped with a device used to reset the hot water temperature set point based on the outdoor air temperature. These dual temperature controllers sense both the outdoor air temperature and the hot water supply temperature. Here is how they work: as the outdoor air temperature increases, the boiler's hot water temperature set point automatically decreases. Lowering the hot water set point during milder weather saves money and reduces wear on the boiler. There is still some debate as to whether to use this type of temperature controller on an outdoor boiler. Some say it is not worth the additional cost because a wood-fired boiler is more efficient when it is operating at a higher temperature.

Figure 20-13. In some cases, the outdoor boiler may use other types of solid fuels, such as corn or wood pellets.

20.3 Additional Applications

Some outdoor boilers use corn or wood pellets as a source of heating fuel, **Figure 20-13**. When this is the case, a ***storage hopper*** will be required. This hopper may be integral to the boiler or externally mounted, **Figure 20-14**. It should be of significant size so that it does not need to be filled too often. In addition, the boiler will require an auger device to automatically feed corn or wood pellets into the combustion chamber. Some modern corn and pellet boilers include electronic controls that automatically ignite the fuel source at each heating cycle. These controls can also incorporate self-modulating adjustments for low-, medium-, and high-fire settings, depending on the heating demand output, **Figure 20-15**. Modern outdoor boilers can also include ***dual fuel*** capability. In addition to biomass, these units can switch to oil, propane, or natural gas whenever necessary—such as in an emergency situation.

Indoors, the boiler can be utilized for other heating applications as well. These include:

- *Heating of domestic hot water using a "sidearm" heat exchanger:* This device connects to the piping on the water heater as shown in **Figure 20-16** and **Figure 20-17**. The domestic hot water heat exchanger supplements the existing water heater controls to generate an economical source of hot water.
- *Radiant floor heating or baseboard heat instead of forced air:* Radiant floor heating has gained immense popularity over the past several years. An outdoor boiler can utilize an existing radiant floor heating system and the owner can save money on heating bills. The layout shown in **Figure 20-18** displays how an outdoor boiler can be incorporated into both radiant floor heating along with domestic water heating.
- *Connecting to an existing boiler,* **Figure 20-19.** An outdoor boiler can be piped in parallel with an existing indoor hot water boiler where a conventional hydronic system is being used for comfort heating. This is a great way of

Figure 20-14. This is an example of a storage hopper used with outdoor boilers to house corn and wood pellets.

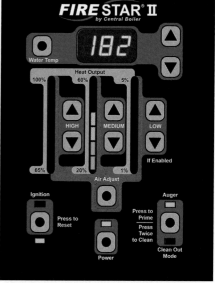

Figure 20-15. Electronic controllers such as this can be used on corn or wood pellet boilers to control firing rates.

reducing the cost of heating bills, regardless of which type of heating fuel is used with the indoor boiler. Other applications for outdoor boilers include:

- *Swimming pool and hot tub applications,* **Figure 20-20.** Outdoor boilers are an excellent source of heat for swimming pools and hot tubs. Regardless of which type of fuel is currently used for these heating applications, an outdoor boiler can easily be piped in parallel with the existing heating system to offer energy savings year around.
- *Greenhouse heating applications:* Greenhouses require a large amount of energy to keep plants warm during cold days. Outdoor boilers can be used to reduce energy bills for this application when installed in conjunction with indoor fan coil units or radiant floor heating.
- *Heating of multiple buildings:* If there is an abundant supply of biomass fuel available, an outdoor boiler can easily be sized to heat multiple buildings.

20.4 Boiler Maintenance

The maintenance of an outdoor boiler is similar to that of an indoor wood-burning stove. This maintenance includes inspection of the firebox for wear and inspection of the chimney for **creosote** buildup. Creosote is a black or brown residue that adheres to the inner walls of the venting system. It can appear as a crust or flaky buildup and can be sticky like tar. Creosote is highly combustible and can reduce the

Goodheart-Willcox Publisher

Figure 20-16. This sidearm heat exchanger is used to supplement the generation of domestic hot water.

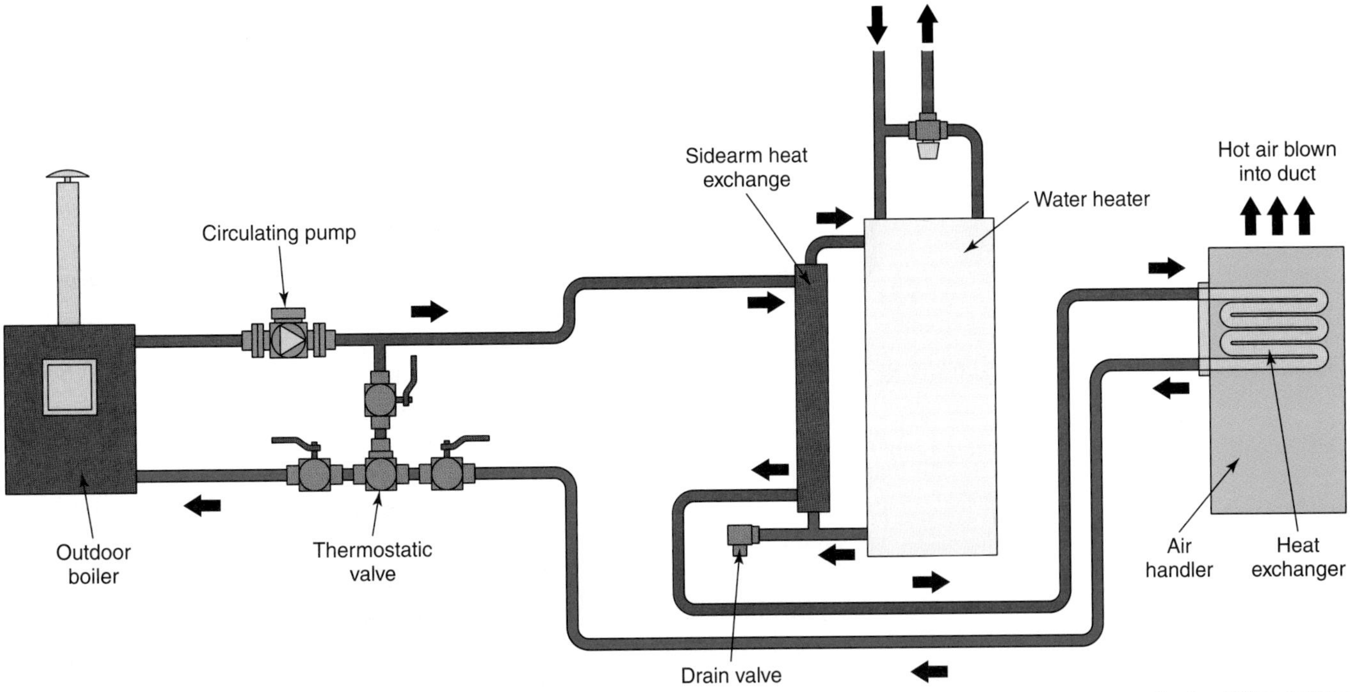

Goodheart-Willcox Publisher

Figure 20-17. This illustration shows how a sidearm hot water heat exchanger can be piped in conjunction with the air handler's space heating heat exchanger.

432 Hydronic Heating: Systems and Applications

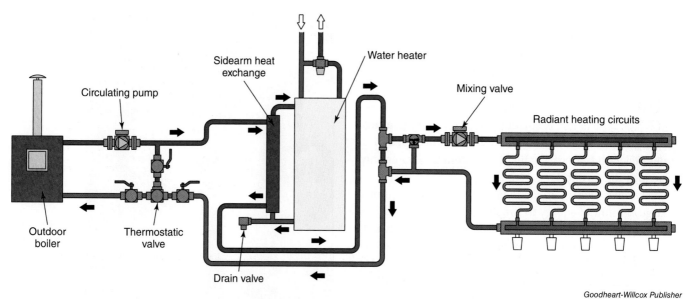

Figure 20-18. Radiant floor heating is another viable application for outdoor boilers. This illustration shows an example of piping the boiler in conjunction with a sidearm heat exchanger.

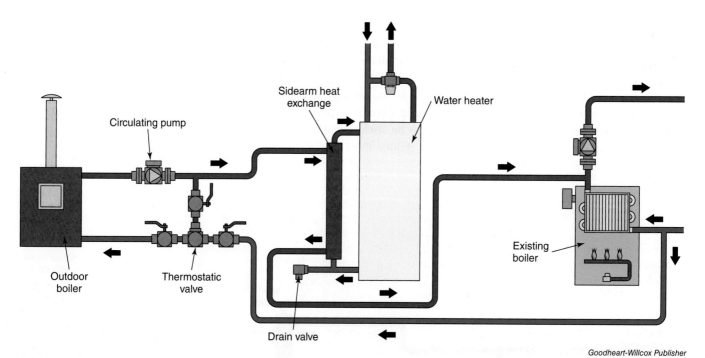

Figure 20-19. This illustration shows how an existing boiler can be piped to utilize an outdoor boiler as a supplemental heating source.

efficiency of the outdoor boiler. If it is allowed to build up in sufficient quantities, and the flue temperature is high enough, the result could be a chimney fire. The conditions that encourage the buildup of creosote include a restriction in the air supply, the burning of damp wood, and cooler than normal flue temperatures. Prevent creosote buildup by burning only hard, dry wood in the outdoor boiler. In addition, ensure that controls are in good working order and properly calibrated. An overall seasonal cleaning is always recommended to ensure the equipment is performing to its maximum potential.

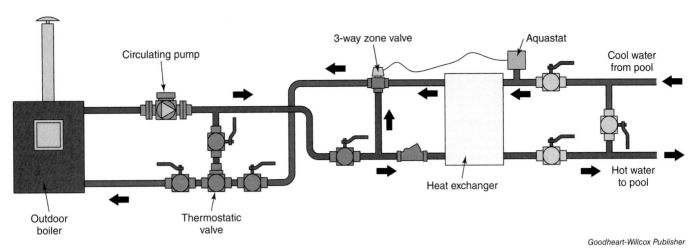

Figure 20-20. Outdoor boilers can also heat swimming pools and hot tubs. This illustration is using an auxiliary heat exchanger to temper the water that is supplied to a hot tub or swimming pool.

Goodheart-Willcox Publisher

SAFETY FIRST

Removing Creosote from an Outdoor Boiler

Part of a routine maintenance program on an outdoor boiler should include the cleaning and removing of creosote from the firebox and chimney flue.

This procedure can be done several ways:

1. If the creosote buildup is light and flaky, it can usually be removed from the flue by using an approved chimney brush. Thoroughly brush the inside liner of the chimney flue, then remove the residue from the firebox with a brush, shovel, or vacuum cleaner.
2. If the buildup of creosote is dense, use an approved creosote removal product. This type of product is available in granular form, as a brick, or wood block that is combusted inside the firebox to break down the creosote to an ash. Once it is broken down, remove the creosote by brushing the inside liner of the chimney flue, and remove the residue from the firebox.

Routine cleaning of the chimney flue and firebox may need to be done several times per year depending on the quality of the wood being burned and the length of the heating season. Always follow the boiler manufacturer's maintenance instructions for proper cleaning and always use an approved creosote removal product.

In addition to these maintenance items, the water supply level must be maintained to ensure proper system efficiency. The system's water level should be checked periodically to ensure it is at its proper height. Add water to the system when necessary. To do this, a permanent water line should be connected from the building's domestic water supply to the indoor water lines of the boiler as a means of adding makeup water. The water lines may also need to be intermittently purged of any air that becomes trapped in the lines. If the system includes water filters, these must be cleaned periodically as well.

One more maintenance task is to remove ash from the outdoor unit as needed—usually once or twice per month during the heating season. Some units include an ash auger that automatically empties ash buildup from the outdoor unit.

Chapter Review

Summary

- Outdoor boilers have become a viable way of using renewable energy to heat homes and buildings.
- There are several advantages to using an outdoor wood boiler compared to an indoor wood-burning appliance, including safety, the elimination of dirt, soot, and insects inside the home, less indoor air pollution, and lower property insurance rates.
- The first step in sizing an outdoor boiler is to perform a comprehensive heat load calculation on the building to be heated.
- Typically, a water-to-air heat exchanger is used for heating the inside of the building and connected to the outdoor boiler by underground piping.
- The outdoor boiler is located between 50′ and 100′ away from the building to be heated to prevent fire damage and provide for proper venting.
- Outdoor boilers are nonpressurized appliances that are vented directly to the atmosphere.
- A concrete pad is usually used to mount the boiler outdoors to provide support.
- The underground piping used with an outdoor boiler is typically PEX pipe. It is buried in a trench below the frost line, and it must be thoroughly insulated to prevent excessive heat loss. Outdoor boiler piping is typically connected to the indoor coil by push fittings.
- Hot water controls on the outdoor boiler include a thermostat that modulates a ventilation damper to maintain the hot water temperature set point.
- The indoor controls are used to cycle the circulating pump and the furnace circulating fan.
- If antifreeze is not used in the hot water system, the circulating pump should be energized continuously.
- Storage hoppers are generally used in conjunction with wood pellet or corn boilers and include feed augers for supplying the fuel to the outdoor boiler.
- Outdoor boilers can be used for various other applications, including heating domestic water, radiant floor heating, and supplementing existing indoor fossil fuel boilers.
- The maintenance on an outdoor boiler is similar to that of an indoor wood-burning stove.
- Inspection of the outdoor boiler venting system should be performed to determine if there is any buildup of creosote.
- The water level in the outdoor boiler should be monitored to maintain a safe level.
- One maintenance task is to remove ash from the outdoor unit as needed.

Know and Understand

1. What is an advantage of using an outdoor boiler as opposed to an indoor wood-burning appliance?
   A. There are no problems with dirt, soot, or insects inside the home or building.
   B. There is less possibility of indoor air pollution caused by a wood burning appliance located inside the building.
   C. There is typically an increase in the building owner's property liability insurance.
   D. Both A and B are correct.
2. Traditionally, outdoor wood or corn fired boilers have maintained a heating efficiency of around _____.
   A. 40%
   B. 50%
   C. 60%
   D. 70%
3. When an air handler is used indoors with an outdoor boiler as a means of heat transfer, what type of heat exchanger is typically used?
   A. Air-to-air
   B. Water-to-air
   C. Water-to-water
   D. Geothermal
4. What distance should the outdoor boiler be located from the home or building?
   A. 5′ to 10′
   B. 50′ to 100′
   C. 100′ to 200′
   D. 200′ to 400′
5. *True or False?* An outdoor boiler is classified as a *pressurized system.*
6. Why should the boiler's concrete pad extend beyond the front of the boiler?
   A. To provide for expansion from the weather
   B. Because it is a code requirement
   C. To allow the operator sufficient space for loading the firebox
   D. To prevent grass fires in the surrounding yard
7. *True or False?* The frost line is the minimum depth that frost will penetrate the ground during the coldest period of the heating season.
8. Why is PEX the piping of choice for underground installation with an outdoor boiler?
   A. It is easy to install.
   B. It is highly resistant to chemicals.
   C. It can withstand high and low temperature fluctuations.
   D. All of the above.
9. What device does the hot water thermostat control on the outdoor wood boiler?
   A. The indoor fan motor
   B. The hot water control valve
   C. The boiler's combustion damper
   D. The storage hopper's feed auger
10. Under what conditions would the hot water circulating pump be allowed to cycle on and off?
    A. When the air temperature is above freezing
    B. When there is glycol antifreeze in the system
    C. When there is no flame in the outdoor boiler
    D. When the piping is laid above ground in a bed of straw

Apply and Analyze

1. What are some of the advantages of using an outdoor boiler as opposed to using an indoor wood-burning appliance?
2. Explain why a heat load calculation on the building to be heated is necessary when sizing the outdoor boiler.
3. Describe the heat transfer device used indoors in conjunction with an outdoor boiler.
4. List the steps necessary for setting and piping the outdoor boiler.
5. Why is it necessary to insulate the underground piping between the outdoor boiler and the building to be heated?
6. Explain how the piping between the outdoor boiler and the building to be heated could be installed if the ground were frozen.
7. Describe the components used and their functionality to control both the outdoor boiler and the indoor air handler.
8. Explain when it would be necessary to allow the hot water circulating pump to run on a continuous basis.
9. List the types of biomass fuels that can be used with an outdoor boiler.
10. Explain how an outdoor boiler can be used to supplement the heating of the domestic water supply.
11. Describe the procedure for maintaining the outdoor biomass boiler.

Critical Thinking

1. Explain why the burning of biomass fuels in an outdoor boiler is not considered a contributor to air pollution.
2. What are some factors that would need to be considered when using an outdoor biomass boiler in an urban area?

Glossary

10-2 rule. When the boiler venting passes through a pitched roof, the vent shall be terminated at least 2′ above the highest point of the roof within a 10′ radius of the termination. (15)

A

above grade. That part of a building which is exposed above the ground. (13)

active solar thermal storage system. A solar heating system that incorporates a mechanical means of fluid transfer. (19)

adapter. Fittings that have both socket and threaded ends. Also called *connectors*. (6)

air change method. Technique for calculating infiltration by estimating the number of times the total volume of air contained within the structure is displaced every hour. (13)

air lock. Trapped air within a hydronic system that blocks the water from circulating. (6)

air pressure switch. A gas burner ignition system safety device that proves positive air pressure throughout the venting system before the ignition system is allowed to be energized. (4)

air scoop. An air separation device with a series of deflectors or baffles, which create turbulence as the water circulates through them, resulting in a drop in pressure and the removal of air from the water. (6)

air separator. A device that uses a mesh screen to separate air from the water as it flows through the hydronic system. (6)

air vent. Air removal device through which air is eliminated from the hydronic system. (6)

American National Standards Institute (ANSI). An organization that oversees the development of standards for products, services, processes, systems, and personnel in the United States. (2)

American Society for Testing and Materials (ASTM). An international standards organization that develops and publishes voluntary consensus technical standards for a wide range of materials, products, systems, and services. (5)

American Society of Heating, Refrigerating, and Air-Conditioning Engineers (ASHRAE). A global professional association whose mission is to advance HVAC system design and construction through the development of standards and sustainability within the industry. (1)

American Society of Mechanical Engineers (ASME). A US professional association that promotes the art, science, and practice of multidisciplinary engineering and allied sciences. (3)

Annual Fuel Utilization Efficiency (AFUE). A measurement of how efficiently a heating appliance can utilize the fuel being consumed. (1)

Annual Fuel Utilization Efficiency (AFUE). A thermal efficiency measurement of space-heating boilers. It is the percent of heat produced for every dollar of fuel consumed. (3)

anti-scald valve. A mixing or tempering valve, installed between the inlet and outlet piping of the domestic hot water, that mixes cold water with the outgoing hot water to assure that the hot water temperature reaching the building's fixtures is at a safe level for human skin. (18)

aquastat. A control device used to maintain the boiler hot water temperature by operating the circulator pump and burner circuit. (10)

aquastat relay. A common control device on earlier residential and light commercial boilers, the aquastat relay is a temperature-controlled switch designed to cycle the boiler's burner controls based on the hot water temperature. (8)

arc flash. An explosion or discharge that occurs when electrical current flows through an air gap between conductors within a breaker panel or main disconnect. (2)

architect's scale. Measuring device that consists of a ruler that has multiple scales to allow for easier measuring when estimating projects. (13)

ash content. The amount of noncombustible particulate found within fuel oil. (5)

ASHRAE Handbook of Fundamentals. A book published by the American Society of Heating, Refrigeration, and Air-Conditioning Engineers, which is used extensively by HVAC engineers and technicians and covers the basic principles and data used in the HVAC/R Industry. (6)

atmospheric burner. Burners that rely on gravity or the natural buoyancy of heated air to vent the products of combustion through the flue or chimney. (4)

atomization. The process of breaking fuel oil into tiny droplets to allow it to ignite within the combustion chamber. (5)

authority having jurisdiction. The organization, office, or individual responsible for approving layout drawings, equipment, installation, or a procedure. (15)

Note: The number in parentheses following each definition indicates the chapter in which the term can be found.

automatic gas shutoff valve. A 24-volt powered gas valve that opens when there is a call for heat from the space thermostat or the boiler's aquastat. (4)

auxiliary storage tank. A hot water storage tank, separate from the boiler, that does not produce its own heat in an indirect water heating system, but rather relies on the existing boiler connected to the hydronic system to heat a coiled heat exchanger in the tank. (18)

B

backflow preventer. A safety device used in the hydronic heating system to prevent the boiler water from flowing backward into a domestic water source. (8)

balancing valve. A type of flow control valve designed to regulate the proper flow in GPM through respective terminal units. (7, 9)

ball valve. A type of isolation valve that prevents the flow of water by means of a small sphere, or ball, within the body of the valve. (9)

base-mounted circulator. A circulator pump that is mounted on the floor and secured to a rigid base. (10)

below grade. That part of a building's foundation wall that is below the ground. (13)

below-floor suspended tubing. A radiant heating system in which heat is transferred from the bottom of the flooring, through the tubing to the floor above. (12)

biomass. A fuel source from the burning of organic matter, such as wood, corn, and other biological materials. (3)

biomass heating. Heating with fuel from organic matter, especially plant matter. (20)

blower door test. A method for determining heat loss due to infiltration by reducing the air pressure inside of the building and measuring the flow of outside air inward through unsealed cracks and openings around the building. (13)

boiler. A pressurized vessel that contains water or other fluid that is heated to a certain temperature by burning fossil fuels or other combustible materials. (1)

boiler drain valve. A device used to drain water from the boiler. (16)

buffer tank. A storage vessel for the heated water after it leaves the heat pump. (12)

building management system (BMS). A computer-based control system installed in buildings that controls and monitors the building's mechanical and electrical equipment such as ventilation, lighting, power systems, fire systems, and security systems. (15)

built-up wall. A wall that contains the various building components used to construct the wall. (13)

C

cabinet unit heater. Fan coil units that are flush-mounted inside of a wall or mounted against a wall. (11)

cad cell relay. A light-sensitive semiconductor, made up of cadmium sulfide, which responds to light by changing its ohms resistance. It serves as a safety device by confirming ignition. (5)

Canadian Standards Association (CSA). A global organization dedicated to the safety and sustainability of standards development and in testing, inspection, and certification. (2)

capillary action. Capillary action occurs in soldering when the solder is drawn toward the source of heat—in this case the torch. (15)

carbon dioxide reading. Data that is expressed as a percentage of flue gas volume from a combustion analysis to determine the efficiency of the oil burner. (5)

carbon monoxide (CO). An odorless, colorless gas that is lethal to humans, inhibiting the blood's ability to carry oxygen throughout the body. It is a by-product of incomplete combustion in the boiler's burner equipment. (2)

carbon monoxide reading. Data that is expressed in parts per million (ppm) of flue gases from a combustion analysis to determine the amount of unburned fuel from combustion. (5)

carbon residue. The amount of carbon left in a sample of oil after it is converted from a liquid to a vapor by boiling the oil in an oxygen-free atmosphere. (5)

cascade boiler system. A group of boilers piped together and controlled by a single building management system. (15)

Category B-vent. Double-walled vent piping that must be used when a vent needs to pass through a wall, ceiling, or roof. (14)

cavitation. The formation of bubbles or cavities in the hydronic system water, developed in areas of low pressure around an impeller. The imploding or collapsing of these bubbles triggers intense shock waves inside the circulating pump, causing significant damage to the impeller and/or pump housing. (6)

centrifugal pump. A type of circulating pump that creates differential pressure between the inlet and discharge of the pump through centrifugal force. This forces water through the hydronic system. (10)

check valve. A type of flow control valve that allows water to flow in one direction and automatically prevents backflow in the opposite direction through the hot water piping. May be either swing type or spring-loaded type. Also called a *one-way valve*. (9)

close-coupled pump. A direct coupled circulating pump in which the motor shaft and impeller are connected together so that the motor bearings must absorb the entire torsion load. (10)

closed-loop solar thermal storage system. A solar heating system that circulates a heat transfer fluid, usually a glycol-water mixture, through the solar collector and through a closed heat exchanger inside of the storage tank to prevent freezing. It is not open to the domestic water supply. (19)

closely spaced tees. The adjacent branch tees located on the boiler supply header of a primary-secondary piping system to create hydraulic separation. (7)

coefficient of performance. The ratio of the heating output of a geothermal heat pump compared to the input of energy necessary to operate the unit. (12)

combination boiler. Boilers that can be used to provide both space heating and domestic hot water. Frequently referred to as *combi boilers*. (18)

combination valve. Gas valve found on modern residential and light commercial boilers that incorporates all of the features usually found separate on older gas trains: solenoid valve, built-in regulator, pilot supply line and controls, ignition controls, safety shutoff, and manual on-off control. Also known as *redundant gas valve*. (4)

combustible floor base kit. A durable base made of flame-resistant material used to mount the boiler on a floor made up of flammable material. (15)

combustion. The rapid expansion of gases resulting in oxidation of fuel and oxygen, producing heat and light. (4)

combustion air. Air that is supplied to combustion appliances to be used in the combustion of fuels and the process of venting combustion gases. (4)

combustion analysis. The testing of the products of combustion to ensure that the boiler is operating safely and at peak efficiency. (5, 16)

combustion leak detector. A device used to detect possible leaks in a boiler gas line. (17)

comfort zone. Sets the boundaries in which an average person maintains optimum comfort based on the temperature and humidity of the conditioned space. A comfort zone can be deduced using a psychometric chart. (1)

concentric vent kit. A venting assembly that allows both the intake for combustion air and the exhaust vent of the boiler to pass through a standard roof or sidewall. This is an alternative to the standard two-pipe intake/vent used in the basic boiler installation instructions. (15)

condensate. The moisture created as a by-product during the combustion process in high-efficiency boilers. (15)

condensate removal pump. A special pump used to remove the moisture created as a by-product of combustion in high-efficiency boilers. (15)

condensate trap. A plumbing trap located in the boiler's condensate line used to prevent the products of combustion from entering into the building. (16)

condensing boiler. A high-efficiency boiler that contains a secondary heat exchanger to extract more heat from the burned fuel and in which combustion gases condense into liquid, releasing latent heat into the boiler's water. (3)

conduction. A type of heat transfer when there is physical contact between two types of materials. (1)

confined space. A non-inhabited area with restricted entry and exit, typically closed off from an outside source of ventilation, such as tanks, storage bins, pits, tunnels, and pipelines. It cannot support the proper amount of combustion air for areas, such as the boiler. (2, 14)

control device. Hydronic heating devices that increase system efficiency, reduce energy consumption, and make the controlled space more comfortable for the building occupants. (8)

convection. A type of heat transfer where heat moves through a fluid source, such as air or water. (1)

cord. A unit of measurement for wood in the United States and Canada, equal to 128 cubic feet of volume, or approximately 4′ × 4′ × 8′. (3)

coupling. A bearing assembly. (10)

crack method. A method used to measure heat loss through infiltration by finding the air leakage through cracks around doors and windows. (13)

cracking. The distillation process of refining fuel oil from crude oil. (5)

creosote. A highly combustible, black or brown residue that adheres to the inner walls of the venting system. It can appear as a crust or flaky buildup and can be sticky like tar. (20)

cross-linked polyethylene (PEX) tubing. A durable, flexible, corrosion-resistant piping material that is popular for hydronic applications such as in-floor radiant heating because it can withstand repeated heating and cooling cycles. (6)

current relay. A type of flow switch that proves flow through the hydronic circuit by detecting the current running through electrical leads of the circulating pump, closing the relay contacts and allowing the boiler's burner circuit to energize. Also called *current sensing switch*. (8)

D

deadheading. A situation that occurs when the pump is forced to run when there is no supply water and the pump impeller continues to rotate the same volume of water in the pump casing without allowing this water to pass through. This can result in the pump overheating and possible damage. (7)

decibel (dB). A unit used to measure the intensity of a sound. (2)

declination angle. The 23.44° tilt of the axis at which the earth revolves around the sun with respect to the earth's orbital plane. (19)

delta-P circulator. A circulator that utilizes a variable speed motor to modulate the pump speed based on differential pressure. (10)

delta-T circulator. A circulator that utilizes a variable speed motor to modulate the pump speed based on differential temperature. (10)

design temperature difference. The difference between the desired indoor temperature and the outdoor design air temperature. (13)

detail drawing. Blueprints or construction drawings that show a specific part of a piece of equipment at a larger scale. (13)

dew point. The temperature below which water vapor begins to condense. (3)

diaphragm valve. A gas valve that uses differential pressure to hold shut the gas valve when it is de-energized and to push open the valve when there is a call for heat. (4)

dielectric union. A special fitting with a threaded steel female fitting on one end and a female copper or brass sweat fitting on the other end, used to connect Schedule 40 black pipe and copper tubing to prevent galvanic corrosion. (6)

differential pressure bypass valve. A valve used to control the excess fluid flow in the piping system by acting as a bypass while ensuring adequate flow to the remaining circuits. It prevents multiple loop systems from becoming overpressurized. (7, 9)

differential pressure switch. A safety device flow switch that measures the difference in pressure generated across the pump inlet and outlet to prove flow. (8)

diffusion. The movement of particles from an area of higher concentration to lower concentration. (12)

dilution air. The air that combines with the flue gases and lowers the concentration of the emissions. (5)

direct spark ignition system. A gas boiler ignition system that ignites the main burner directly from a spark generated from a transformer located within the ignition module. (4)

direct vent boiler. High-efficiency condensing boilers that receive all of their combustion air from the outdoors through a dedicated sealed opening. Also known as a *sealed combustion boiler*. (14)

distillation quality. The ability of the oil to become vaporized. (5)

diverter tee. A tee containing a special cone inside that acts as a venturi to restrict the flow of water through the tee outlet. This creates a pressure drop through the main circuit, which allows a certain amount of fixed water flow through the branch terminal unit. Also called a *monoflo* or *venturi* tee. (7)

diverting valve. A temperature control valve used to divert water to separate control devices, including domestic hot water heating. (9)

draft. The amount of vacuum or suction that exists inside the heating system, measured in inches of water column pressure (in. WC). (5)

draft hood. Part of the venting system that entrains room air into the venting system, mixing it with the products of combustion. (14)

draft test. Combustion analysis test performed to ensure that the products of combustion pass through the boiler at the proper rate. (5)

drainback system. A method of freeze protection for solar thermal storage systems that works by draining all of the water or glycol solution from the solar collector and any exposed piping into a drainback tank whenever the system is not collecting solar energy. (19)

drain-waste-vent (DWV) piping. PVC piping used for condensate drains and for venting. (6)

drip leg. A short gas pipe extending from a tee at the beginning of a gas train that serves as a trap and prevents moisture and debris from entering the gas valve. (4)

dry radiant hydronic system. Radiant hydronic systems that incorporate tubing without a poured substrate, such as above-floor prefabricated panels and below-floor staple-up or suspended tube systems. (12)

dry-base boiler. A boiler in which the area under the combustion chamber is dry and the boiler water is contained in a space above the combustion chamber. (3)

dual fuel boiler. Outdoor boilers that can switch from using biomass to another fuel such oil, propane, or natural gas whenever necessary. (20)

E

electrodes. Stainless steel devices located near the end of the nozzle orifice that use a spark to ignite the vaporized fuel oil in the combustion chamber. (5)

electronically commutated motor. A motor that uses microprocessor electronic controls to vary its speed. (10)

elevation drawing. A scaled view that looks directly at a vertical surface from floor to ceiling. (13)

end suction circulator. A type of centrifugal pump that has a casing with the suction coming in one end and the discharge coming out of the top. (10)

enthalpy. The total heat (both sensible and latent) content of a substance. (1)

equivalent length. The equivalent length of a pipe fitting is the length of pipe of the same size as the fitting that would provide the same resistance as the fitting. (7)

escutcheon plate. A decorative plate used to conceal the opening of a floor or wall where piping has penetrated. (11)

evacuated tube collector. A type of solar collector that gathers the sun's radiant energy and turns the radiation into heat using a row of glass tubes from which the air has been removed, forming a vacuum. This vacuum essentially eliminates any convective and conductive heat transfer between the two walls, forming the best possible thermal insulation for a solar collector. The result is exceptional performance, even at low ambient temperatures. (19)

excess air. Secondary and dilution air needed for proper combustion and venting. (5)

expanded foam board (EPS). A lightweight, rigid, closed-cell type of insulation. (12)

expansion tank. Tank required in a hydronic system to allow for the expansion of water as it is heated. (6)

extruded foam board (XPS). A rigid insulation that is formed with a polystyrene polymer. (12)

F

fan coil unit. A convector that uses an integral fan to transfer heat throughout the conditioned space, either with or without the use of connected ductwork. (11)

fast-fill lever. The lever located on the pressure-reducing valve used to permit rapid filling of the hydronic system. (16)

feet of head. The measurement that relates the amount of energy in a fluid (water) to the height of an equivalent static column of that same fluid or water. This is the resistance—determined by the number of fittings, valves, terminal units, and overall length of piping—that the pump has to overcome to achieve proper system flow. Also called *head pressure* or *pump pressure*. (7, 10)

feet per second (FPS). Measurement used to express the velocity of a fluid's flow through a pipe. (7)

finned-tube baseboard heater. A type of heat emitter that uses convection through fins fitted over the supply tubing to increase the amount of heat transfer between the fluid in the tubing and the surrounding air. (11)

firebox. The chamber inside the outdoor boiler where fuel is loaded and where combustion occurs. Also known as a *combustion chamber*. (20)

fire-tube boiler. A boiler that contain tubes surrounded by water, in which combustion gases from the burner pass through the inside of the tubes, heating the water. (3)

flame proving device. A safety device found on gas burning appliances that monitors the flame for safe, continuous operation. (4)

flame rectification system. A gas boiler safety system that detects the electrical current in the pilot or main burner flame and keeps the main gas valve open as long as a flame is present. If there is an interruption in the flame signal, the ignition module recognizes this and immediately shuts down the main burner circuit. (4)

flame rod. A gas boiler ignition system device that is immersed in the pilot or main burner flame when ignited and sends a signal back to the ignition module, proving the flame is lit and allowing the main gas valve to open. (4)

flame rollout switch. A gas burner ignition system safety device found inside the boiler vestibule near the gas burners that will trip due to an excessive increase in temperature in the event of a flame rollout. (4)

flash point. The lowest temperature at which fuel oil vapors will ignite above a pool of liquid for a short period of time when exposed to a flame. (5)

flat-plate collector. A type of solar collector that consists of an insulated metal box covered with tempered, low-iron glass and an absorber plate inside. The absorber plate transfers the sun's energy to fluid circulating through copper tubing connected to the plate. (19)

flexible coupling. A coupling that consists of metal bars connected to stainless steel springs or two separate brass collars connected to a single, continuous helical spring. They can connect a standard frame motor to a base-mounted pump to provide a certain amount of flexibility. They act as sacrificial connectors, being designed to break in order to protect the motor bearings and pump shaft. (10)

floor plan. A plan view that shows the layout of equipment and piping when looking down from above, shown as they appear in an imaginary horizontal section five feet above the floor level. (13)

flow coefficient (Cv). The factor used to size globe valves. The C_v is the volume of water flow in gallons per minute (GPM) that passes through the valve at a pressure drop of 1 psi at 68°F. (9)

flow control valve. Valve used to modulate, restrict, or prevent water from flowing in a certain direction. Flow control valves include check, globe, pressure-reducing, and balancing valves. (9)

flow switch. A safety device that ensures proper water flow through the boiler before the burner is activated. (8)

flue. A duct, pipe, or opening in the boiler's chimney for conveying exhaust gases from the fire side of the heat exchanger to the outdoors. (3)

flue liner. Vent liners made of corrugated stainless steel or aluminum and connected directly to the boiler to prevent damage from flue gas condensation. (14)

flush-and-fill cart. A portable, preassembled cart used to fill and flush solar, geothermal, and hydronic heating systems. (19)

FM gas manifold. Type of gas train typically found on larger commercial and industrial boilers that is compliant with Factory Mutual Insurance requirements including: additional manual shutoff valves, low-gas pressure switches, and two electric safety shutoff valves. (4)

fossil fuel. Plant and animal remains transformed into oil, natural gas, and coal from heat and pressure within the earth. (5)

free area. The sum of the areas of all the spaces between the bars or fins of a grille opening measured in square inches. (14)

freeze protection valve. Safety device in a solar thermal storage system that protects both the collector tubes and the piping between the collector and the valve from freezing during cold weather. (19)

friction loss. Pressure losses resulting from the roughness of the piping material and from the turbulence of the fluid as it flows through the pipe. (7)

frost line. The maximum depth that frost will penetrate the ground during the coldest period of the heating season. (20)

fuel line filter. Traps that remove particulates and impurities from the fuel oil that could damage the system before the oil reaches the pump. (5)

fuel oil pump. Positive displacement-type pumps that deliver oil from the storage tank to the combustion chamber and regulate the pressure at which the oil is delivered into the burner for ignition. (5)

fuel oil. A petroleum product refined from crude oil that is used for heating. Also called *heating oil*. (5)

G

gallons per minute (GPM). Measurement used to express the rate of water flow through a pipe. (7)

galvanic corrosion. A process that occurs when copper tubing is joined directly to Schedule 40 black pipe and the system's water is slightly conductive, causing iron or steel components to prematurely corrode and eventually fail. Also known as *bimetallic corrosion*. (6)

gas ignition module. The system that controls safe and reliable burner ignition in modern boiler combustion systems through a series of operations and lockout requirements. (4)

gas manifold. A pipe on a gas boiler that distributes fuel into multiple burners through small sockets called *spuds*. (4)

gas orifice. The opening on the burner spud that distributes fuel into a boiler's burner. (14)

gas train. A series of components that safely feeds fuel into the burner of a gas boiler. (4)

gate valve. A type of isolation valve that uses a disc that moves up and down when the valve is turned open and closed to allow or stop water flow. (9)

geothermal heat pump. A heating source that uses a refrigeration cycle to transfer heat absorbed from the earth into the hydronic system. (12)

Giannoni-type heat exchanger. A serpentine tube-in-tube–type stainless steel heat exchanger invented by Rocco Giannoni and used to transfer large volumes of heat within a small space. (17)

glazing. A tempered low-iron glass that is used with solar thermal heating systems that can withstand high thermal stress and hazards. It utilizes a low iron oxide content to minimize the absorption of solar radiation as the sun's rays pass through it. (19)

globe valve. A type of flow control valve named for its spherically shaped body and designed for modulating the flow of water. (9)

glow coil. A burner ignition device made of silicon carbine in a hot surface ignition system that illuminates when it is energized. (4)

glycolic acid. A chemical that forms when a glycol solution degrades due to excessive heat; it causes copper piping to corrode and eventually fail. (19)

ground fault circuit interrupter (GFCI). Receptacle or breaker that will detect the faintest short to ground and immediately break the circuit, interrupting the flow of power to the connected tool. (2)

grounding prong. The third prong on a three-prong plug designed to fit into the grounded part of the receptacle. Its purpose is to provide an alternative path for electricity to flow in the event that the tool or appliance develops a shorted circuit. (2)

gun-type burner. A gun-type burner atomizes fuel oil by forcing the oil through a nozzle and spraying it into a gun-like airflow atomic nozzle. The liquid forms microscopic particles or globules, which are well mixed and partly evaporated before being ignited in the combustion chamber. (5)

H

handwheel. The handle on a gate valve that requires several rotations to fully open or close the valve. (9)

heat. Energy added to a substance that causes the substance to rise in temperature. (1)

heat emitter. Device in the hydronic heating system that transfers heat from the water located in the hydronic piping into the space to be heated, either by convection or radiation. Also called the *terminal device*. (11)

heat exchanger. A device that transfers heat from one fluid to another without allowing the two fluids to make physical contact. (3)

heat load calculation. Figure that determines a structure's heat loss based on the geographic location, the type of building materials used, and how well the building is resistant to the infiltration of unwanted and uncontrolled outdoor air. (13, 20)

heat motor actuator. A motor that uses a bimetal strip that warps to open a zone valve when electrical current is applied. (9)

heat pump. A system that transfers heat from one area to another. (18)

heat transfer. The movement of thermal energy from one substance to another, creating a difference in temperature. (1)

heating load. The heating requirement necessary to maintain a desired temperature in a building. (3)

high temperature limit. A gas burner ignition system safety device that trips in a variety of high-temperature situations including excessive fuel gas pressure, lack of water flow through the boiler's heat exchanger, or a blocked exhaust vent or flue. (4)

high-fire offset. The device on a temperature controller used to set the number of degrees below set point that the high-fire stage shuts down. At that point, the boiler will continue to operate at the low-fire stage until the temperature set point is reached. (16)

high-mass boiler. A boiler that takes longer to heat up than other types of boilers but tends to hold its heat longer due to its heavy weight and large water capacity. (3)

high-mass radiant system. A radiant system with much higher capacity for energy storage; for example, embedding radiant heating tubes into concrete. (12)

high-mass terminal unit. Terminal devices made of heavier materials that can retain heat for longer periods of time, even after the heating source has cycled off. (11)

high-pressure boiler. A high-pressure boiler has an operating water temperature of above 350°F and an operating pressure above 300 psi. (3)

hopscotching. An electrical troubleshooting technique that works by starting at a point where voltage exists and moving one lead at a time from point to point toward the nonfunctioning load until voltage is lost. (17)

hose bib. A small faucet-like device connected to the boiler as a means of draining water from it. (15)

hot surface ignition system. An ignition system in which the burner is ignited directly using a silicon carbide glow coil. (4)

hot water circulating pump. Device used to efficiently transport hot water through the hydronic system. (10)

housekeeping pad. A concrete pad on which the boiler is placed, used to keep debris and water away from the boiler. (15)

hybrid water heater. A heat pump water heater that uses heat from the surrounding air where the boiler is located and transfers this heat into the water by means of the refrigeration cycle. (18)

hydraulic separation. The state of the primary and secondary loop pumps in a primary-secondary piping system operating independently of each other and acting as if the other pump does not exist. (7)

hydrocarbon. Any one of a family of combustible, efficient, clean-burning, natural gases such as methane, ethane, propane, and butane. (3) Liquid made up of hydrogen and carbon in chemical composition, from which fuel oil is made. (5)

hydronic airlock. A buildup of excess air in a hydronic circulating line that blocks the flow of water through the system. (17)

hydronic heating system. A system in which conditioned water is circulated to an occupied space to heat the area. (1)

hydronic towel warmer. Radiators that provide warmth to the surrounding space heater and also warm towels and bathrobes for personal use. (11)

I

ideal heating curve. A graphical representation of the optimum comfort level in degrees Fahrenheit from floor to ceiling for an average person at rest. (12)

ignition point. The temperature at which fuel oil will ignite and continue to burn as a vapor as it rises from a pool of liquid. (5)

ignition transformer. Device on an oil burner that generates the spark used by the electrodes to ignite the fuel oil in the combustion chamber. (5)

impeller. A circular disc in a circulating pump that uses turning vanes to create pressure through centrifugal force to move water through the hydronic system. (10)

inches water column. A unit of measurement used to determine very low levels of pressure. (15)

indirect water heater. A hot water generating system that utilizes the building's existing boiler to heat water using a heat exchanger located inside an auxiliary tank, which heats the potable water supply based on the occupants' demand. (18)

indoor air pollution. Toxic airborne chemicals within a building that can result from the burning of fuel and cause respiratory issues. (20)

indoor heat exchanger. Device used to transfer heat from the water heated from an outdoor boiler to a building's forced-air heating system. (20)

induced draft boiler. A boiler that includes an inducer fan to vent the products of combustion by drawing or pulling them through the heat exchanger. (14)

infiltration. The transfer of heat that occurs in a building as a result of uncontrolled cold air entering and exiting the structure. Also called *convection*. (13)

inline circulator. A type of centrifugal pump that has the inlet and discharge ports in a straight line in relation to the hot water piping. (10)

insolation. Solar radiation. (19)

Institute of Boiler and Radiator Manufacturers. A nationally recognized certification organization for pressurized vessels and hydronic devices. (11)

integral collector storage (ICS) unit. A passive solar heating system in which the collector and tank are combined into one single unit and the water itself is the solar collector. (19)

integrated ignition control module. The sophisticated system on newer gas boilers that controls ignition with more advanced electronics for greater control and functionality, including self-diagnostics. (4)

intermittent pilot ignition system. A gas boiler ignition system with an ignition control module that only lights the pilot light when there is a call for heat and a fast response time to close the main gas valve after a pilot failure. (4)

interpolate. A method of estimating new data points within the range of two different known data points. (11)

IRI gas manifold. Type of gas train typically found on larger commercial and industrial boilers that is compliant with Industrial Risk Insurers requirements including: additional manual shutoff valves, high- and low-gas pressure switches, two electric safety shutoff valves, and a normally open gas bleed valve between the safety shutoff valves. (4)

isolation valve. Valve capable of isolating specific areas or components from the rest of the hydronic heating system for service or repair. Isolation valves include gate and ball valves. (9)

K

kilowatt (kW). A measurement of electrical power equal to 1000 watts, or approximately 3410 Btu of energy. (3)

L

laminar flow. Water flowing in more of a straight line through a piping system. (7)

latent heat. The amount of heat required to change the state of matter without a change in temperature. (1)

line-voltage thermostat. A device used with an outdoor boiler to open and close the combustion damper in response to the hot water temperature. (20)

liquefied petroleum (LP). A gas derived from crude oil that can contain a mix of propane and butane and that will condense into a liquid state when pressurized. (3)

load calculation. A mathematical design tool used to determine the heat loss of a building based on the design outdoor temperature for a geographic location. (7)

lockout/tagout. A safety procedure requiring that electrical power to any piece of machinery must first be shut off, locked out at the disconnect switch, and tagged to identify the service technician before work begins on the equipment. (2)

longest length method. A method for sizing gas piping that applies the maximum operating conditions by setting the length of pipe to size any given part of the system to the maximum value. (14)

LonWorks. A networking platform specifically created to address the needs of control applications. (15)

low water cutoff. A boiler safety device that detects when the fluid level in the boiler drops below the minimum safe operating level specified by the boiler manufacturer. If this occurs, the low water cutoff will cut off the burner and shut down the boiler. (2)

low-level CO detector. A carbon monoxide detector, recommended in all boiler rooms, that will alarm when levels rise above 6 parts per million (ppm). (2)

low-loss header. A device that provides hydraulic separation between the boiler circuit and the heating circuit by regulating the flow rate and pressure. (7)

low-mass boiler. A boiler with a lightweight design that contains a minimal volume of water, such as copper-tube boilers. (3)

low-mass radiant system. A radiant system with a very limited capacity to store heat energy, such as finned tube baseboard radiation. (12)

low-mass terminal unit. Terminal devices made of lightweight materials such as aluminum that do not have the ability to retain heat for long periods of time and therefore need to function constantly when there is a call for heat. (11)

low-NO$_x$ boilers. A boiler designed with controlled mixing of air and fuel at each burner in order to reduce harmful nitrogen oxide emissions. (3)

low-pressure boiler. A low-pressure boiler has an operating pressure of up to 160 psi and an operating water temperature of up to 250°F. (3)

low-water cutoff (LWCO). A safety device that de-energizes the boiler's burner circuit if the water level within the boiler falls below a predetermined point. (8)

M

manifold gas pressure. The pressure that the gas valve delivers to the burners to supply the fuel to the boiler. (17)

manifold station. The central distribution point where the supply and return ends of each individual loop in a radiant system are joined together and piped back to the boiler or heating device. (12)

manometer. An instrument for measuring gas pressure acting on a column of fluid in which a difference in the pressures acting in the two arms of the tube causes the liquid to reach different heights in the two arms. (16)

Manual J. A manual published by the Air Conditioning Contractors of America used for determining the amount of heating and cooling that a residence requires to keep its occupants warm in the heating months and cool and dry in the cooling months. (13)

Manual N. A manual published by the Air Conditioning Contractors of America designed to instruct contractors and others on how to meet the heating and air conditioning requirements for commercial buildings. (13)

medium-pressure boiler. A medium-pressure boiler has an operating pressure of 160 to 300 psi and an operating water temperature of between 250°F and 350°F. (3)

microammeter. An instrument for measuring electric current in microamperes. (17)

microamp. A unit of current measured as a millionth of an ampere. (17)

millivolt. A unit of electricity measured as one-thousandth of a volt. (17)

mixing valve. A specialized valve that blends two sources of water to create a desired outlet water temperature. (9)

mod/con boiler. A combination modulating and condensing boiler. (3)

ModBus. A data communications protocol used with programmable logic controllers. (15)

modulating boiler. A boiler capable of changing its heat output by varying or "modulating" the input of fuel and air used for combustion in order to match a building's heat load. (3)

modulating gas valve. A gas valve found on modern boilers that modulates from fully open down to 20% of its maximum operating capacity for improved efficiency and advanced comfort from the heating system. (4)

multimeter. A handheld tester used to measure electrical voltage, current (amperage), resistance, and other values. Also known as a *volt-ohm meter*. (17)

N

national pipe thread taper (NPT). A standard for the tapered threads used in the joining of pipe fittings that pull together tightly, making a fluid-tight seal. (6)

near-boiler piping. A collection of devices located near or on the boiler. (15)

negative draft. The state in which the flue stack must have a negative differential pressure in relation to the boiler room in order for proper venting to take place. (14)

negative temperature coefficient (NTC). A type of thermistor in which the ohms resistance decreases as the temperature of the thermistor increases. (19)

net stack temperature. An indicator of proper combustion efficiency, this measurement is the ambient air temperature around the boiler subtracted from the measured flue gas temperature. (5)

neutralizing filter. A device located on the condensate line of condensing boilers, used to neutralize the acid contained within the condensate. (15)

nitric oxide. A colorless gas with the formula NO. It is one of the principal oxides of nitrogen. (16)

nonintegrated ignition module. A less sophisticated ignition module found on older boilers that controls the spark or hot surface ignitor, operates the gas valve opened and closed, and monitors the burner using a flame rod and lockout feature. (4)

normally closed valve. A valve that is always closed unless power is applied to the solenoid coil. (4)

nozzle. A device made of stainless steel or brass that atomizes the fuel oil. (5)

O

Occupational Safety and Health Administration (OSHA). An organization run by the US Department of Labor whose purpose is to ensure safe and healthful working conditions for working men and women by setting and enforcing standards and by providing training, outreach, education, and assistance. (2)

oil burner. The device that controls the combustion of the fuel oil within the boiler's heat exchanger. (5)

oil burner motor. Motor that provides power to both the blower and fuel pump simultaneously within the oil burner assembly. (5)

oil deaerator. A device that automatically removes air from the fuel oil system. (5)

on-demand water heater. Tankless water heater, which energizes only when there is a call for heat. Also known as a *gas-fired instantaneous heater* or *tankless boiler*. (18)

one-pipe delivery system. A fuel oil delivery system that includes a single supply line between the oil tank and fuel oil burner. (5)

one-pipe system. A piping arrangement in which all of the main water supply passes through only one pipe between the boiler's supply and return flow inlets and individual terminal units are piped in parallel to the main primary loop as branch circuits. A portion of the water flowing through the main circuit is diverted through each terminal unit by supply and return tees. (7)

open-loop system. A solar thermal storage system in which the hot water is piped from the solar collector directly into a storage tank or a domestic water heater. The piping from the collector to the storage tank or water heater is open to either the home's well water supply or the city water system. (19)

order of operations. The proper sequence for completing a mathematical calculation: parentheses, exponents, multiplication and division (left to right), addition and subtraction (left to right). (1)

orifice. The predrilled hole on a threaded spud of a gas manifold that meters the flow of gas at a predetermined rate into the burner. (4)

outdoor boiler. An unpressurized appliance that burns biomass fuel to heat water at a distance from the building to be heated. (20)

outdoor reset controller. A control device that aims to reduce operational costs without sacrificing comfort by automatically adjusting the boiler's hot water temperature set point based on the outdoor air temperature. (8)

oxygen barrier. A material paired with tubing in radiant heating systems that prevents oxygen molecules from diffusing into the water through the tubing walls, which can corrode any ferrous-based materials. (12)

oxygen reading. Data that is used from a combustion analysis to determine the proper amounts of fuel and air mixtures to the burner. (5)

P

paddle switch. A flow switch safety device that proves flow using a blade suspended within the water piping. (8)

parallel pumping. A circulating pump configuration in which two pumps are installed parallel to each other to provide additional flow capacity with a minimal increase in head pressure. (10)

passive solar heating system. A solar heating system that harvests energy from the sun without the use of electrical or mechanical devices. (19)

personal protective equipment (PPE). Safety devices worn to protect the eyes, ears, head, face, hands, feet, and respiratory system. (2)

PE-RT tubing. Tubing made up of a polyethylene resin that has been designed to operate at raised temperatures. It is constructed with five layers of materials that give it significant strength and chemical resistance. (12)

PEX (cross-linked polyethylene tubing). A durable, flexible, corrosion-resistant piping material that is popular for hydronic applications—such as in-floor radiant heating—because it can withstand repeated heating and cooling cycles. (12)

PEX-AL-PEX. Tubing that has an internal layer of aluminum sandwiched between two layers of conventional PEX tubing. The aluminum layer adds a coil memory feature that allows it to retain its shape when the tubing is bent. (12)

pH. A scale used to specify the acidity or basicity of an aqueous solution. (16)

photovoltaic module. A solar module that absorbs sunlight as a source of energy to generate direct current (DC) electricity. (19)

pilot gas line. A gas line that tees off the main supply line or gas valve to supply gas to the pilot assembly. It may have its own manual shutoff valve and sometimes a separate pilot pressure regulator on older boilers. (4)

pilot generator. See *thermopile*. (4)

pilot valve. A small, automatic valve located on the burner pilot line that limits the flow of gas to the pilot assembly. (4)

pipe friction chart. A chart used to calculate head loss in a hydronic system based on the size of the pipe and the amount of flow in GPM. (7)

piping. A component that carries heated water or fluid away from the boiler to the terminal device in a conditioned space. It is also used to return water back to the boiler. (1)

point of no pressure change. The point where the expansion tank is connected to the hydronic heating system. This is the point in the system where the pressure always remains constant, regardless of what the system is doing. (10)

point-of-use heater. Smaller tankless water heater that services only one or two outlets, typically located close to the device. (18)

polyvinyl chloride (PVC) piping. Piping material used for condensate drains and, in some cases, venting on high-efficiency boilers. (6)

potable hot water. Hot water supply that is safe to drink. (18)

pour point. The lowest temperature at which fuel oil can be stored and handled. (5)

power burner. A type of gas burner found on commercial boilers that incorporates a blower to force both the primary and secondary air and the fuel into the burner tube. (4)

prefabricated radiant floor panel. Radiant heating system with heat transfer panels designed to fit over an existing wooden subfloor or over slab-on-grade concrete where the existing floor cannot support the additional weight of a thin-slab installation or where the finished flooring requires a large number of fasteners for completion. (12)

pressure drop. The difference in pressure between two points in the hydronic system. Pressure drop is caused by frictional forces that create a resistance to flow. (9)

pressure regulator. A gas train component used to reduce the inlet gas pressure to a level specified by the boiler manufacturer. (4)

pressure relief valve. A safety device that protects the boiler and hydronic system piping by opening at or below the maximum allowable working pressure to release pressure in the event that the boiler's burner circuit fails to de-energize. (8)

pressure-reducing valve. A type of flow control valve that delivers the correct amount of makeup water to the hydronic heating system at the correct pressure. Also called a *water-regulating valve* or a *feed water valve*. (9)

pressure/temperature gauge. Tool that provides readings for supply temperature and pressure to indicate whether the boiler is operating normally. (6)

pressurized solar thermal storage system. A solar heating system in which the circulation loop is closed to atmospheric pressure, characteristic of closed-loop and drainback systems. (19)

primary air. Air that mixes with the gas at the inlet of a burner. (4)

primary control unit. A device on oil-fired burners that controls the boiler combustion process and ensures a proper safety shutdown. (5)

primary-secondary piping. A piping arrangement in which the first, or primary, circuit simply circulates water throughout the boiler and provides a main supply of water for the individual terminal circuits. The secondary circuits are used to feed water through their individual terminal units, and each secondary circuit acts independently of the others. (7)

propylene glycol. A colorless, odorless, slightly syrupy liquid that belongs to the same chemical group as alcohol and is used as an antifreeze in hydronic systems. (16, 20)

psychrometric chart. A means for determining proper indoor temperature and humidity conditions based on the properties of air and moisture. (1)

pump affinity law. Rules used in hydronic design to indicate the influence of the pump performance based on volume capacity, feet of head, and power consumption. (7)

pump motor. The device used to drive the circulating pump by rotating the pump's impeller. (10)

pump performance curve. A graphical representation of a pump's performance based on testing conducted by the manufacturer. (7, 10)

purging valve. A specialty valve that assists in air removal, specifically during the initial filling of the system. It is also used to remove water in the system when it needs to be flushed out. (9, 16)

R

radiant ceiling panel. A low-mass radiant system installed in a ceiling. (12)

radiant heating system. A hydronic heating system in which thermal radiation is used to transfer heat from the piping to *objects* in the space to be heated, rather than the *air*. (12)

radiant panel. A terminal unit that transfers heat through radiation and delivers heat directly from its surface to the objects that are in the line of sight. (11)

radiant wall panel. A low-mass radiant system installed in a wall or ceiling. (12)

radiation. A type of heat transfer that occurs through light waves. (1)

radiator. Terminal devices that transfer heat primarily by means of radiation. (11)

reciprocal. The inverse of a given value, which can be found by dividing 1 by the value. (13)

recovery rate. The amount of hot water in gallons that can be produced in one hour after the tank is completely drained. (18)

redundant gas valve. Combination gas valves that typically have dual shutoff seats for extra safety protection. (4)

relative humidity. The amount of water vapor in the air as a percentage of the maximum amount of water vapor that the air could hold at the same temperature. (1)

R-value. The rating system used to grade insulation products or the material's insulating properties, which is determined by the amount of mass found in a building's structural components or their thermal resistance to heat transfer. (13)

S

Safety Data Sheet (SDS). A report on a particular chemical, including its properties, potential hazard to human health and the environment, and the protective measures required when handling, storing, and transporting the chemical. (2)

safety device. Hydronic heating devices that protect the boiler, hot water piping, mechanical room, and building occupants from dangers that can occur when controlling a pressurized vessel for comfort heating and hot water generation. (8)

Saybolt Seconds Universal (SSU). A measurement of fuel oil's viscosity that describes the time elapsed for an oil to flow through a calibrated orifice at a defined temperature, usually 100°F. The total number of seconds determines the viscosity rating. (5)

scale. Correlating a larger measurement to a smaller measurement so that large structures can be accurately depicted in construction drawings. (13)

Schedule 40 piping. The most common type of steel piping used for water and gas in boiler installations with NPT as a standard for the tapered threads used in the joining of pipe fittings. (6)

Scotch Marine boiler. The most common type of fire-tube boiler, notably used for marine service on ships because of its compact size. (3)

secondary air. Air that mixes with the ignited fuel when it leaves the burner, improving combustion efficiency by burning the fuel more completely and entraining the combustion gases through the boiler's vent or flue. (4)

self-contained breathing apparatus (SCBA). A supplemental oxygen device often used in confined spaces or when working with hazardous vapors. (2)

sensible heat formula. A formula commonly used throughout the HVAC industry to compare the heat carrying capability of air and water. (1)

sensible heat. The amount of heat that can be measured by temperature—typically in degrees Fahrenheit or degrees Celsius—with no change of matter state. (1)

series loop arrangement. The simplest piping system that typically consists of one boiler, one pump, and one zone. (7)

series pumping. A circulating pump configuration in which two pumps are installed one after the other to add head pressure without additional flow capacity. (10)

set point. The target value at which a controlling device attempts to maintain the desired temperature. (1)

side shield. Personal protective equipment that attaches to prescription glasses to prevent debris from entering around the outside edges of the glasses and injuring the eye. (2)

slab-on-grade construction. A structural engineering practice whereby a concrete slab that is to serve as the foundation for the structure is formed from a mold set into the ground. (13)

slab-on-grade radiant piping. A radiant system in which a layer of concrete is poured over an existing grade, with the radiant tubing imbedded into the concrete. (12)

slope angle. The angle at which a solar collector is mounted based on the latitude of the geographic location as well as the intended application of the system. (19)

snow and ice melt system. A radiant system that automatically clears snow and ice from many types of outdoor surfaces. (12)

socket fusion. A method of joining HDPE tubing and fittings by heating and fusing them together. (12)

solar altitude angle. The measurement from the horizontal surface of the earth to the center of the sun. (19)

solar array. The roof- or wall-mounted evacuated tube collectors as a collective group. (19)

solar azimuth. The angle measured starting from true north, which is considered 0°, and traveling in a clockwise direction until it intersects with the position of the sun. (19)

solar collector. The key component of a solar thermal storage system that gathers the sun's radiant energy and transforms the radiation into heat to be transferred to the system's heat transfer fluid. (19)

solar pathfinder. A device used to determine the effects of shading when positioning solar panels. (19)

solar thermal storage. A renewable energy heating source that stores and uses energy from the sun to heat the water. (12)

solenoid valve. A valve that is normally closed until power is applied to its solenoid coil, which acts as an electromagnet, lifting the valve and allowing gas to pass through. (4)

solid-state circuitry. Electronic control modules that control hot water temperatures and often offer additional features such as multipoint temperature monitoring, night setback control, outdoor air reset control, and control of domestic hot water. (8)

spark igniter. A series of thermocouples joined together designed to supervise the pilot light and hold the pilot valve open when lit. It is capable of generating a higher dc voltage output than a thermocouple. Also known as *thermopile*. (4)

specific heat. The amount of heat required to raise the temperature of a material by 1°F. (1)

split coupling. A solid steel device that connects the pump to the shaft of a standard frame motor with no flexibility. (10)

spring-loaded check valve. A type of flow control valve that works by allowing a small spring inside the valve to close against the valve disc if the water is not flowing in the appropriate direction. It can be installed in both horizontal and vertical piping. (9)

spud. A small socket on a gas manifold that distributes fuel into a burner. (4)

stack relay. A heat-sensing device that de-energizes the burner if it fails to detect heat within the flue piping or stack. Also known as a *stack switch*. (5)

stack temperature. The temperature of the flue gases when the boiler is operating. (16)

standby heat loss. The excess energy used to maintain a constant water temperature when hot water is not in use. (18)

standby losses. Water that migrates through the secondary circuit and its terminal unit when the zone is not calling for heat. (7)

standing pilot. An early boiler ignition system in which a continuously burning pilot flame is used as a heat source to ignite the burner whenever there is a call for heat. (4)

static pressure. The pressure generated by the weight of the water within the system. (10)

static pressure. The pressure generated by the weight of the water within the system. (7)

stoichiometric combustion. The ideal combustion process in which fuel is burned completely. Also known as *perfect combustion*. (4)

storage hopper. An integrated or externally mounted bin for holding wood pellets or corn to be used as fuel for an outdoor boiler. (20)

strap-on temperature device. A sensor that straps on directly to the piping to measure the water temperature. (19)

stratification. The stack effect, or layering of warm air from floor to ceiling of a structure. (1)

sustainable design. Discipline that involves the incorporation of materials and processes that reduce energy costs, improve productivity, and decrease the amount of environmental waste to reduce the amount of negative impact on the environment while improving the health and comfort of building occupants. (13)

swing check valve. A type of flow control valve that uses a swinging disc attached to the top of the valve by a hinge to allow or prevent water flow through the valve, thus preventing water from flowing in the wrong direction. It should be installed in a horizontal position. (9)

system pressure. Pressure produced by the amount of force generated by the hot water circulating pump. (7)

T

temperature difference. The measure of the relative amount of energy between two substances. (1)

temperature droop. The extreme temperature fluctuation between the actual space temperature and design set point. (1)

temperature rise. A factor that takes into account the incoming water temperature and the actual desired hot water temperature. (18)

terminal device. A hydronic heating component that emits or transfers heat into the air to maintain a specific set point temperature within the space. It is located within the conditioned space. Also called a *heat emitter*. (1, 11)

thermal density. A measurement of a material's ability to conduct heat. (13)

thermal mass. An object's capacity to absorb and store heat energy. (7)

thermal memory. A quality of tubing that allows it to be heated and returned back to its original shape if it becomes kinked. (12)

thermal resistance. The ability of a material to resist heat transfer through conduction. (13)

thermal transmittance. The rate of transfer of heat through a material. (13)

thermistor. A resistor made of a semiconductor material in which the electrical resistance varies with a change in temperature. (19)

thermocouple. A device that generates a small voltage when the pilot flame is lit, which sends a signal back to the gas valve to keep the pilot valve held open. (4)

thermopile. A series of thermocouples joined together designed to supervise the pilot light and hold the pilot valve open when lit. It is capable of generating a higher dc voltage output than a thermocouple. Also known as *pilot generator*. (4)

thermosiphon. A passive solar thermal storage unit that incorporates a sloped solar collector installed in conjunction with an insulated tank. Thermosiphons utilize the principle of natural convection. (19)

thermosiphoning. The process by which circulation takes place in a thermosiphon solar heating system without the need for a circulating pump. The warmer water rises to the top of the collector and into the tank where it is stored, and the cooler water descends to the bottom of the collector. (19)

thermostat. A component that senses the temperature of a heating or cooling system and performs actions so that the system's temperature is maintained near a desired set point. (1)

thermostatic mixing valve (TMV). A type of mixing valve with a thermostatic element that reacts to changes in water temperature at the outlet of the valve. The element adjusts the mixing ratio of inlet water to maintain a desired outlet temperature. (9, 18)

thermostatic radiator valve (TRV). A self-contained, spring-loaded valve mounted directly onto terminal units to control space temperature automatically by modulating the flow of hot water into the terminal unit. (9)

thin-slab radiant flooring system. A radiant heating system that incorporates either a formulated concrete or poured gypsum underlayment material over radiant tubing over an existing floor. (12)

three-piece circulator. Designed for residential and light commercial applications, this type of circulating pump consists of the pump body, an isolated motor that allows for easy service or replacement, and a coupling. (10)

Toolbox Talk. A weekly safety meeting typically performed during construction projects to keep safe working practices at the forefront of everyone's daily responsibilities. (2)

total dissolved solids (TDS). A measure of the dissolved combined content of all inorganic and organic substances present in water in molecular, ionized, or micro-granular suspended form. (16)

transmission losses. The thermal transfer of heat energy through a building's envelope as a result of heat conduction naturally occurring between the indoors and outdoors through walls, windows, doors, ceilings, and the basement floor. (13)

trisodium phosphate. An inorganic compound used as a cleaning agent to initially flush a hydronic system. (16)

turbulent flow. The movement of water through a pipe in a chaotic or swirling motion. (7)

turndown ratio. The ratio of maximum heat output capacity to the lowest firing point level of heat output at which the boiler will operate efficiently or controllably. (4)

turndown ratio. The rate at which a modulating boiler can throttle its output. (3)

two-pipe delivery system. A fuel-oil delivery system in which a second line is routed from the fuel oil burner back to the storage tank and any excess oil that is not used for combustion is returned back to the storage tank. (5)

two-pipe direct return. A piping arrangement in which one pipe supplies water from the boiler to the terminal units, which are piped in a parallel arrangement, and one pipe returns water to the boiler. (7)

two-pipe reverse return. A piping arrangement in which the first terminal unit on the system closest to the boiler is the first to receive inlet water but the last unit to return water back to the boiler. (7)

two-stage gas valve. A gas valve that provides a dual firing rate: a first stage, or low-fire stage, which maintains a lower gas manifold pressure, and a second stage, or high-fire stage, that provides a higher gas pressure to the burner. (4)

U

unconfined space. A space that will support the proper amount of combustion air for the boiler that has at least 50 cubic feet of open area for every 1000 Btu of input. (14)

under-cabinet fan coil unit. Space-saving heat emitter designed to be installed under cabinets or fit inconspicuously inside a wall or floor. It incorporates a centrifugal-type fan or blower that gently blows heated air into the space. Also called *kick-space heater* or *toe-kick heater*. (11)

unit heater. Small, compact heating devices that consist of a heating coil and propeller fan housed in a sheet metal cabinet, good for heating an unconditioned space such as a garage or shop. (11)

unit ventilator. Terminal device consisting of a heating coil, fan or blower, dampers, and temperature controls that is typically floor-mounted and provides a good heating source for individual spaces and offers the ability to incorporate outdoor air for ventilation or cooling. They are traditionally used in classrooms. (11)

unpressurized system. A solar thermal storage system in which the loop is open to the atmosphere, such as with an open-loop system. (19)

U-value. The measurement of thermal transmittance defined by the quantity of heat in Btu per hour that will flow through 1 square foot of material in 1 hour at a temperature difference between the indoors and outdoors of 1 degree. (13)

V

vapor barrier. Polyethylene sheeting that prevents the migration of water vapor from the soil into a concrete slab. (12)

variable frequency drive (VFD). A type of adjustable-speed device used in electro-mechanical drive systems to control AC motor speeds and torque by varying motor input frequency and voltage. (10)

vent stack. The vertical portion of the boiler vent or chimney that extends through the roof. (17)

venturi. An hourglass shape near the opening of an atmospheric burner that creates a narrow constriction, causing the air and gas to mix and accelerate as they pass along through the burner, promoting good combustion. (4)

vestibule. The enclosed portion of the front of the boiler where the gas valve and burner are located. (17)

viscosity. The measurement of the oil's resistance to flow and a measurement of the thickness of the oil under normal temperatures. (5)

volute. The circulating pump's housing that contains the impeller and captures the velocity of the hot water as it enters the outermost diameter of the impeller, converting it into pressure. (10)

W

water hammer. A condition in which trapped air causes undesirable noise within the hydronic system. (6)

water hardness. The amount of dissolved calcium and magnesium in water. (16)

waterlogged. A condition in compression tanks in which water takes the place of the air cushion and the tank completely fills with water. (6)

water-to-water heat pump. A heat pump that uses water for both the source of heat and to distribute the heat to the conditioned space. (12)

water-tube boiler. A boiler in which water is contained inside tubes within the heat exchanger and the combustion gases from the burner pass around the outside of these tubes. (3)

wet radiant hydronic system. Radiant hydronic systems that include a poured substrate such as slab on grade and concrete or gypsum thin slab. (12)

wet rotor circulator. Small- to medium-sized circulating pump that combines the motor, pump shaft, and impeller into one assembly contained in a single housing. The motor is cooled and lubricated by the water enclosed in the hydronic heating system. (10)

wet-base boiler. A boiler in which the water is heated throughout—both above and below the combustion chamber—surrounding it. (3)

whole-house heater. Larger, tankless heaters sized and designed to service the entire heating or hot water requirements of a building. (18)

Z

zone isolation valve. A specific type of valve used to control the flow of water from the main water supply header to an individual zone branch in a hydronic heating system. (16)

zone valve. Two-way or three-way control valves used to control water flow to individual zones. (9)

Index

A

abbreviations on blueprints, 289, 291
above-floor prefabricated panel systems, radiant flooring, 251–252
above gap, electrode, 99
above-grade walls, 280
ACCA (Air Conditioning Contractors of America)
    Commercial Load Calculation, Manual N, 276
    definition, 17
    Residential Load Calculation, Manual J, 276
    R-value building materials tables, 274–275
accidents
    avoiding, 21
    boiler room, 35
    truck or van, 33
    See also safety
account managers, 127
acetylene gas, 33–34
active solar storage systems
    closed-loop systems, 398–399
    definition and functions, 396–397
    drainback systems, 399–400
    pressurized and unpressurized, 401
adapters, copper pipe fittings, 113
AFUE (Annual Fuel Utilization Efficiency)
    boiler fuel and efficiency consideration, 295–298
    copper-tube boilers, 49
    definition, 7
    high-efficiency boilers, 46
    lower-efficiency boilers, 45
    mid-efficiency boilers, 46
air
    compression tanks, 117–118
    dilution, 89
    draft importance, 105
    excess, 89
    fuel oil combustion, 89
    human body and temperature, 9
    indoor pollution, 424
    sensible heat formula, heat delivery, 13–15
    See also ventilation system
air bands, adjusting oil burner, 96
air change method, heat loss through infiltration, 283
Air Conditioning Contractors of America (ACCA)
    Commercial Load Calculation, Manual N, 276
    definition, 17
    Residential Load Calculation, Manual J, 276
    R-value building materials tables, 274–275
air dampers, inspecting, 351–352
air eliminating devices, 326–327
airflow proving switch, 79
air-free carbon monoxide, 344
air in water system, troubleshooting, 366–370
air pressure switch, ignition safety, 79
air purifying breathing protection devices, 25
air removal devices
    air separators and air scoops, 123–124
    air vents, 124–126
    automatic or float-type, 126
    baseboard tees, 125
    cavitation, 122–123
    checking the cap on, 126
    defining functions, 124
    hygroscopic, 125
    importance and function, 122
    manual, 124–125
    solar system piping, 410–411
    water hammer, noise, 122
    See also ventilation system
air requirements, boiler combustion and ventilation, 299–303
air scoops, 123–124
air separators
    air scoops, 123–124
    defining function, 123, 326
    placement, 124
    solar system piping, 410–411
Airtrol fitting
    definition, 118
    diagram, 119
    waterlogging, compression-type tanks, 118
allergen circulation, 11–12
aluminum heat transfer plates, 253
aluminum ladders, 32
American National Standards Institute (ANSI)
    ANSI Z223.1 gas piping installation, 327
    blueprints, 288
    Canada, boiler installation, 315
    definition, 32
American Society for Testing and Materials (ASTM)
    condensate piping compliance, 329
    definition, 86
    Standard D396, 87
American Society of Heating, Refrigeration, and Air-Conditioning Engineers (ASHRAE)
    definition, 9
    Handbook of Fundamentals, 120, 276
    R-values publications, 276
    standards, 17
    winter design temperatures, 273
American Society of Mechanical Engineers (ASME)
    Boiler and Pressure Vessel Code, 156, 315
    CSD-1 2009, 29, 315
    definition, 29
American Society of Safety Professionals (ASSP), 161
Americans with Disabilities Act (ADA), 264
Annual Fuel Utilization Efficiency (AFUE)
    boiler fuel and efficiency consideration, 295–298
    copper-tube boilers, 49
    definition, 7
    high-efficiency boilers, 46

lower-efficiency boilers, 45
mid-efficiency boilers, 46
anode rods, 46
ANSI (American National Standards Institute)
   ANSI Z223.1 gas piping installation, 327
   blueprints, 288
   Canada, boiler installation, 315
   definition, 32
anti-scald valve, 382–383
aquastat relay
   check set-in point, 206
   definition and components, 162
   differential temperature control setting, 163–164
   high-limit control setting, 163
   indirect water heaters, 384
   internal components, 163
   low-limit control setting, 162–163
   operation sequence, 164
   safety tip, 207
   sensing bulb, 162
   temperature control settings, 164
   troubleshooting noisy, 208
arc flash
   causes, 28
   definition, 27
   PPE, 28
architect's scale, 286
ash content, fuel oil, 88
ASHRAE (American Society of Heating and Refrigeration and Air-Conditioning Engineers)
   definition, 9
   Handbook of Fundamentals, 120, 276
   R-values publications, 276
   standards, 17
   winter design temperatures, 273
ASME (American Society of Mechanical Engineers)
   Boiler and Pressure Vessel Code, 156, 315
   CSD-1 2009, 29, 315
   definition, 29
ASSP (American Society of Safety Professionals), 161
ASTM (American Society for Testing and Materials)
   condensate piping compliance, 329
   definition, 86
   Standard D396, 87
atmospheric burners
   combustion blowers, 69
   diagram, 70
   primary air, 69
   secondary air, 69
   types and functions, 68
authority having jurisdiction, 315
automatic air vents, 126
automatic gas shutoff valve, 63
auxiliary storage tank, 381

B

back brace, 25
backflow preventers
   cleaning and leak repair, 161–162
   code requirements, 161
   defining functions, 160–161, 326
   definition, 7
   installation location, 162
   internal components, 161
   municipal swimming pools, 162
back injuries, lifting safety, 24
balancing valves
   definition and function, 179, 327
   two-pipe reverse return system, 145–146
ball valve
   definition and function, 174
   inner workings, 175
   lockout, 30
baseboard tees, 125
baseboard units. *See* finned-tube baseboard units
base-mounted circulators, 194–196
Bell & Gossett, 144
below-floor suspended tubing systems, radiant heating, 252–253
below-grade walls, 280
bidirectional triode thyristor, 104
bimetallic corrosion, 114
biofuels, 8
biomass
   definition, 8
   natural carbon cycle, 45
   types, 44–45
biomass heating, 424
blower door test, 283–284
blower types, 69
blueprints
   abbreviations and symbols, 287–291
   common hydronic drawing abbreviations, 291
   detail drawings, 287, 289
   elevation drawings, 287–288
   floor plans, 286–287
   history and development, 285
   isometric drawing, example, 286
   piping symbols, 290
   reading, 284–285
   scale and architect's scale, 286
   types, 284–285
body protection, 23–25. *See also* personal protective equipment (PPE)
boiler components, installing
   balancing valves, 327
   boiler safety, 326
   boiler trim, 325
   check valves, 327
   control devices, 326–327
   hose bibb valves, purge and drain, 327
   makeup water components, 326
   near boiler piping, 324–325
boiler control devices
   air separation and eliminating devices, 326–327
   circulating pump, 326
   expansion tank, 326
   pressure and temperature gauges, 327
boiler drain valves, 338
boiler installation
   authority having jurisdiction, 315
   boiler location, 316–317
   cascade boiler, 332–333
   clearance considerations, 317–318
   codes, 314–315
   combustible floor base kit, 318
   concentric vent kit, 320
   condensate disposal, 328–329
   condensing boiler, 319–320
   field wiring, 329–323
   garages, code, 318
   gas pipe, 327–329
   housekeeping pad, 316
   hydronic piping, 320–327
   natural draft and induced draft, 318–319
   new construction, 314
   pitched roof ventilation, diagram, 319
   safety, 314
   warranties, 315
boiler maintenance
   combustion systems, 351–353
   outdoor wood boiler, 431–433
   safety features check, 353–355
   visual inspections, 349–351
   water leaks and water-side maintenance, 350–351
boiler operator license, 43

Index

boiler room safety, 35–37
    CO exposure, 36–37
    explosions, 36
    low water cutoff, 35–36
boilers
    biomass fuel for, 44–45
    burner ignition controls, 78–80. *See also* boiler safety components
    cast-iron construction, 46
    condensing, 49–51, 241–242
    control technologies, 7, 326–327
    conventional, radiant heating system, 240–241
    copper-tube, low-mass, 48–49
    definition, 4, 42
    domestic hot water (DHW) production, 375–376
    dry firing, 157
    efficiency, construction materials, 45–49
    electric, 44, 53–54
    examples, 5
    flushing and cleaners, 337
    forced convection, 13
    fuel oil operated, 43
    fuel systems, 7, 8
    gas-fired burners, 68–70
    ignition systems, 70–78
    liquefied petroleum (LP) operated, 43–44
    lower-efficiency, 45
    low-NO_X, 54
    low-pressure, 41–42
    medium-pressure, 42
    mid-efficiency, 45–46
    mod/con, 52
    modulating, 51–53
    natural gas operated, 43
    outdoor wood, 8
    pressure-reducing valves, 177
    pressure-relief valves, 7
    series loop piping arrangement, 137–139
    solenoid gas valves, 65
    steel and stainless steel, 46–48
    temperature droop, 5–6
    water return temperature, 137–139
boiler safety components
    burner ignition controls, 78–80
    flow switch, checking, 354–355
    inspecting, 344–345, 353–354
    low-water cutoff (LWCO), 326
    pressure-relief valve, 326, 353
    testing low-water cutoff (LWCO), 354
    water flow switch, 326

boiler selection
    application-based, 298
    combustion and ventilation air requirements, 299–303
    fuel and efficiency considerations, 295–298
boiler startup
    combustion testing, 343–344
    filling and purging the system, 338–340
    flushing process, 337
    freeze protection, 345
    ignite burner, operation sequence, 341–343
    prestart checklist, 340–341
    safety devices, testing, 344–345
    water quality, 337–338
boiler trim, 325
boiler venting
    categories, 304
    code requirements, 304
    condensing boilers, 305–306
    importance of proper, 303–304
    induced draft boilers, 305
    materials, 306
    natural draft boilers, 304–305
breaker panel, electrical safety, 27–28
Btu (British thermal unit), 14–15
Btu rating
    biomass fuels, 45
    commercial and industrial gas trains, 63–64
    cord wood, 45
    electric boiler, 44
    finned-tube baseboard heaters, 217
    FM gas manifolds, 64
    fuel oil, 43
    liquefied petroleum, 43–44
    low-pressure boiler, 42
    medium-pressure boiler, 42
    modulating boilers, 52–53
    modulating gas valves, 67–68
    natural gas, 43
    number 2 fuel oil, 86
    two-stage gas valves, 67
buffer tank, 242–243
building management system (BMS), wiring, 332
built-in diagnostics, integrated boiler control, 167
burner circuits
    intermittent ignition systems, 361
    standing pilot, 360
burner ignition
    boiler startup, 341–343
    safety devices, 78–80

burner maintenance, 352–353
    lifting flame, 352
    microammeter, testing with, 353
    yellow or lazy flame, 352
Burn Wise (EPA emissions data), 45
bushings, iron and steel pipe fittings, 112
butane, boiler fuel, 43–44
Butz, Albert, 6
Butz Thermo-Electric Regulator Company of Minneapolis, 6
bypass valve, 147–148

C

cabinet unit heaters, 226
cad cell relay
    definition and function, 102–103
    operation sequence diagram, 104
Canadian Gas Association (CGA) Standard B149 Installation Code for Gas Burning Appliances and Equipment, 299, 304
Canadian Standards Association (CSA)
    codes, 32, 161, 246
    CSA B149.1 gas piping code, 327
    field wiring, 329
    solar panels, 408
caps, iron and steel pipe fittings, 112
carbon dioxide (CO_2) readings, 107, 344
carbon monoxide (CO)
    air-free, 344
    combustion analysis, 343
    contamination, preventing, 352
    definition, 36
    exposure effects, 36–37
    poisoning prevention, 37
    reading, combustion analysis, 107, 343
carbon residue, fuel oil, 88
careers
    hydronic system designer, 151
    maintenance contracts, 371
    plumbers, pipefitters, and steamfitters, 389
    professional organizations, 17
    sales representative, 55
    skills, 127
    solar thermal power, 419
    test and balance (TAB) technician, 209
    troubleshooting systems, 169
    See also education; workplace skills
Carrier, Willis, 10

cascade boiler system, 332–333
cassette fan coil unit, 226
cast-iron boiler
 definition, 5
 dry-base, 46, 47
 example, 46
 outdoor reset control, 166
 types and uses, 46
 wet-base, 46, 47
cavitation, water pressure, 122–123
ceiling panels, radiant heating, 253–254
Centers for Disease Control and Prevention (CDC), 36
centrifugal pump, 192
certification and training (NATE), 291
check valves
 definition and function, 175, 327
 emergency heat requirements, 176
 solar system piping, 411–412
 spring-loaded, 176, 412
 swing check valve, 175–176
 tilting disc check valve, 175–176
chemical hazards, Safety Data Sheets (SDS), 33–34
chlorinated polyvinyl chloride (CPVC) piping, 5
chlorine, boiler water quality, 338
circuit board, gas ignition, 77
circulating pump
 base-mounted, 195–196
 components and operation, 192–193
 DC, photovoltaic (PV) solar cells, 414
 definition and function, 326
 Delta-P (ΔP) circulators, 205
 Delta-T (ΔT) circulators, 205
 diagram, 148
 hot water, 192
 hydronic loop, 199
 impeller, 192
 incorrect placement, 198
 inline circulators, 193–195
 installation and placement, 196–199
 location, 196
 performance, 199–202
 point of no pressure change, 196
 positive displacement pumps, 193
 pump curves, 199–201. See also circulating pump performance curve
 selecting, 204
 sizing, calculating, 203
 static pressure, 199
 swimming pool solar system, 417
 troubleshooting, 206–209, 366
 types of, 193–196
 variable-speed circulators, 204–206
 zoned system, purging, 369
circulating pump performance
 heating loop pressure, 198
 hydronic circulators, 193
 positive displacement pumps, 193
 pump curves, 199–201. See also circulating pump performance curve
 series and parallel pumping, 201–202
circulating pump performance curve
 components of, 200–201
 defining functions, 134
 developing, 200
 energy conversion, 199
 establishing, diagram, 200
 manufacturer performance curve, 137
 performance components, graph, 201
 piping system design and, 199
 sample performance and motor current table, 201
clamps, 31
closed-loop solar thermal storage system, 398–399
CO (carbon monoxide)
 air-free, 344
 combustion analysis, 343
 contamination, preventing, 352
 definition, 36
 exposure effects, 36–37
 poisoning prevention, 37
 reading, combustion analysis, 107, 343
coal-fired boiler, 43
codes
 backflow prevention, 161
 boiler installation, 314–315
 bushings, 113
 combustion air venting installation, 302
 combustion and ventilation air requirements, 299
 condensate piping compliance, 328
 electrical, mounting solar panels, 408
 field wiring requirements, 329
 garages, boiler installation, 318
 gas piping installation, 327
 lockout/tagout, 30
 low-water cutoff requirements, 158
 manually operated remote shutdown switch, 29
 outdoor airflow for ventilation, 231
 pipe reducers, 113
 PP piping, 51
 pressure-relief valves and discharge tubes, 156
 PVC piping, 51, 116
 PPV piping, 116
 radiant piping materials, 246
 reduced bushings, 328
 solar thermal storage, 405
 water flow switch, 355
 venting requirements, 304
 vent slope, 365
coefficient of performance, 242
cold contacts, stack relay, 102
combi boiler. See combination boilers
combination boilers
 advantages, 385–386
 definition, 376
 descaling heat exchanger, 386
 heat exchangers, 385
 installation and control, 386
 service and maintenance, 386
 space-saving, efficiency, 384
 troubleshooting, descaling, 387
combination gas valve, 66
combustible floor base kit, 318
combustion
 blower types, 69
 boiler, 13, 36–37
 boiler clearances, 317–318
 boiler selection and air requirements, 299–303
 by-products of perfect, 60, 89
 cast-iron boilers, 46
 components, 60
 condensing boiler, 50
 copper-tube, low-mass boiler, 49
 definition and components, 59
 fire-tube boilers, 47
 fuel oil, 88–90
 gas train, 60–63
 high-efficiency and sealed, 302
 incomplete, 89
 mid-efficiency boilers, 45
 modulating boilers, 51
 natural draft, 68–69
 power of, 3
 process, fuel oil, 88–90
 stoichiometric, 59–60
 test, 106
 water-tube boiler, 48

combustion air
    damper, 332
    gas-fired burners, 68
    troubleshooting, 365
combustion analysis
    calculating elements, 344
    carbon dioxide (CO_2) reading, 107
    carbon monoxide (CO) reading, 107, 343
    draft test, 106
    net stack temperature, 106
    nitric oxide, 343
    oxygen reading, 107, 343
    performing, 343
    sealing hole, hex-head countersunk plug, 107
    stack temperature, 343
    testing boiler startup, 343
    testing by-products of combustion, 105–106
    testing kit, 106–107
combustion components, maintenance checks
    burner maintenance, 352–353
    heat exchanger, cleaning, 353
    venting and combustion air, 351–352
combustion leak detector, 350
comfort factor, hydronic heating, 11–12
comfort zone, 10. *See also* human comfort
commercial gas train components, 63–64
Commercial Load Calculation, Manual N (ACCA), 276
comparing fuel costs, 296–297
compression tool, 325
compression-type expansion tanks, 117–118
concentric vent kit, 320
condensate, 328, 364
condensate disposal, 328–329
condensate pump, 328–329
condensate system
    diagram, 341
    inspect and fill, 340
condensate trap, 340
condensing boiler
    acidic flue gases, vent material, 51
    definition and uses, 5, 50
    outside and inside views, 50
    radiant heating, 241–242
    ventilation termination requirements, 319–320

    venting requirements, 305–306
    water temperature, 49–50
conduction
    body warmth, 9
    heat loss through, 272–280
    heat transfer through, 13, 14
    See also heat loss through conduction
conductivity sensor, low-water cutoff, 157–158
confined space
    boiler combustion, 300–303
    definition, 25, 35
    lack of oxygen, 25
    safety, 35
    See also space
connectors, copper pipe fittings, 113, 114
contacts, stack relay, 102
continuing education, 17
continuous pilot light, 45
continuous retry with 100% shutoff, 77
control devices
    aquastat relay, 162–164
    backflow preventer, 160–162
    boiler, 326–327
    cascade boilers, 333
    combination boilers, 386
    defining functions and components, 155
    electronic controls, 166–169
    fan coil unit (FCU), 227–228
    flow switch, 159–160
    hybrid water heaters, 388
    indirect water heaters, 384
    field wiring, 329–332
    outdoor reset controller, 165–166
    outdoor wood boiler, 429
    radiant heating, 261–263
    solar system, 413
    solar thermal systems, 412–414
    tankless water boilers, 379–380
    two-stage modulating controller, 167
    unit heaters, 230
    See also digital controls
control loop, solar system, 413
Controls and Safety Devices (CSD), 29, 315
convection
    body warmth, 9
    circulation, 191
    heat transfer through, 13, 14
convective currents principle, 3
conventional boilers, radiant heating, 240–241

copper heat exchangers, cast-iron boilers, 46
copper press fittings, 220
copper-tube piping, 5, 8
    example, 49
    joining procedure, 322–324
    low-mass boiler, 48–49
    pipe fittings, 113, 114
    tube sizes, system design, 135
cord, wood, 45
corrosion
    cast-iron boilers, 46
    pipe fittings, 114
    reducing, anode rods, 46
couplings
    definition, three-piece circulator, 193
    iron and steel pipe fittings, 112
cracking, 86
crack method, heat loss through infiltration, 280
creosote buildup, removal, 431–433
cross-linked polyethylene (PEX) piping
    definition and uses, 5, 113–114
    fittings, 113–115
    outdoor wood boiler, 427–428
    radiant heating systems and, 244–245
    See also PEX tubing
CSA (Canadian Standards Association)
    codes, 32, 161, 246
    CSA B149.1 gas piping code, 327
    field wiring, 329
    solar panels, 408
current relay, 159–160
customer service representatives, 127

D

damper
    barometric type, 105
    combustion air damper relay, 332
    definition, 6
    inspecting, 351–352
Damper-Flapper, 6
DC (direct current power), 414
deadheading pumps, 145, 183
decibels (dB), 23
declination angle, solar collection, 405
Delta-P (ΔP) circulators, 205
Delta-T (ΔT) circulators, 205
design considerations
    domestic hot water (DHW) production, 376–378
    heat load calculations, 272

heat loss through conduction, 272–280
hydronic piping, 320–321
one-pipe system with multiple zones, 143
piping arrangements, 136–139
piping system, 134–136, 199
primary-secondary piping system, 150
radiant heating systems, 254–262
series loop piping arrangement, 139
snow and ice melt, radiant heating, 265–266
temperature difference, 272
two-pipe direct return system, 144–145
two-pipe reverse return system, 146–147
winter temperatures, 273
detail drawings, 287, 289
dew point temperature, 49–50
diaphragm-type expansion tank
advantages, 119
defining components and function, 118–119
recharging procedure, 120
diaphragm valve
definition and functions, 65–66
pilot valve, 65
dielectric union, 114–115
diesel fuel, 86
differential pressure bypass valve (DPBV)
definition and function, 147–148, 185–186
installation procedure, 186
differential pressure switches, 159–160
digital controls
energy saving, 228, 229
hydronic heat systems, 7
solar systems, 414
digital pressure gauge, 62
dilution air, 89
direct-coupled pumps, 195
direct spark ignition system, 74–75
direct vent boilers, 302
discharge tubes, 156
distillation quality, fuel oil, 86
diverter tees
defining functions, 140
diagram, 140
hot water heating, competition, 144
piping diagram using, 141
piping system using, 141–142
sizing, 142

diverting valves, 182–183
domestic hot water (DHW)
combination boilers, 376, 384–387
electronic control devices, 166, 167
hybrid water heaters, 376, 387–389
indirect water heaters, 376, 381–384
on-demand water heaters, 377
outdoor wood boiler, 431–432
potable hot water, 375
production equipment, 375–376
recovery rate, tank-type water heater, 377
sizing for, design factors, 376–378
tankless boilers, 375, 378–380
temperature rise, 377
DPBV (differential pressure bypass valve)
definition and function, 147–148, 185–186
installation procedure, 186
draft, fuel oil boiler venting, 105
draft hood, 304, 364
draft test, combustion analysis, 106
drainback systems, solar thermal storage, 399–400
drain valve, 327
drain-waste vent (DWV) piping, 115
drip leg, 60–61
dry-base boiler, 46–47
dry-bulb temperature, 10
dry-chemical fire extinguisher, 26
dry-firing, 157
dry radiant hydronic systems, 246
dual fuel capacity, 430
dual temperature controls, 429
ductwork
flue, 44
forced-air systems, 11–12, 14
piping, 12
dust allergies, 11–12
dust mask, 25

E

earplugs, 23
education
certification and training (NATE), 291
continuing, 17
HVAC careers, 127
hydronic system designer, 151
plumbers, pipefitters, and steamfitters, 389
See also careers; workplace skills

efficiency
boiler and fuel selection, 295–298
boiler construction materials, 45–49
boiler types, 42
coefficient of performance, geothermal heat pump, 242
copper-tube boiler, 49
devices and technology, 7
electric boilers, 53
employee, 80
geothermal energy, 243
high-efficiency boilers and sealed combustion, 302
hybrid water heater, 387
hydronic heating, 12
occupancy scheduling, 228
outdoor wood boiler, 426
pump selection, 200
steady-state, condensing boiler, 49
See also green practices
elbows, iron and steel pipe fittings, 112
electrical fire, 26
electrical issues, troubleshooting, 358–363
electrical safety, 26–29
aquastats, 207
arc flash, 27–28
breaker or service disconnect, 27
fire extinguishers, 27
importance, 26
injuries from electricity, 26
lockout/tagout, 29–30
tools for, 28
electric boiler
definition, 44
disadvantages, 53–54
efficiency, 53
environmental impact, 54
types and sizes, 53
electric or electrical zone valves, 261
electrode gap, 99–100
electrodes
fuel oil burner, 99–101
triac or triode, AC current, 104
electronically commutated motors (ECM), 204–205
electronic controls
defining features and benefits, 166
diagnosing defective control board, 168–169
fully modulating control module, 167–168
high-efficiency two-stage control module, 167

installation and service, 167
integrated boiler control, built-in diagnostics, 167
solid-state circuitry, 166
solid-state modulating control board, 168
wireless, benefits, 166
electronic ignition, 45
module, grounding, 75
elevation drawings, 287–288
emergency heating, thermostatic radiator valve (TRV), 185
emergency stop button, 29
employee efficiency, 80
end suction pumps, 195–196
energy conversion, hot water heating and, 199
enthalpy (Btu/lb), 10, 15
environmental impact of electric boilers, 54. *See also* green practices
Environmental Protection Agency (EPA)
Burn Wise, emissions data, 45
underground storage tanks, 90
equivalent length of pipe, piping system design, 134
ethanol, fuel, 8
evacuated tube solar collectors, 402–403
excess air, 89
expanded foam board (EPS), 248
expansion compensator
example, 221
installing, 220
expansion tanks
calculating pressure, 121–122
checking pressure, 350–351
compression-type, 117–118
defining functions, 116–117, 326
diaphragm-type, 118–120
pressure measurements, 121
sizing, 120–121
solar system piping, 409–410
troubleshooting, 371
extension cord, 28
extruded foam board (XPS), 248
eye protection, 22

F

face protection, 22
face shields, 22
Factory Mutual (FM), 64
fan coil unit (FCU)
cabinet unit heaters, 226
cassette, 226
configurations, 226
defining function, 225
digital controller, 228, 229
examples, 226
modulating hot water valve, 228–229
occupancy scheduling, 228
sizing, 226–27
variable fan speed control, 227
fast-fill lever
filling and purging the system, 338
pressure-reducing valve, 178
feed water valve, 177–178
feet of head
diagram, 200
piping system design, 134
pressure and, 199
water pressure, 133
feet per second (FPS), 132
female pipe thread (FPT), 124
fiberglass ladders, 32
field wiring, boiler installation
cascade boiler system, 333
codes, 329
control components, circuit, 331
line-voltage connections, 329–330
low-voltage connections, 330–331
terminal strip, diagram, 330
finned-tube baseboard units
available space, 221
components, diagram, 216
definition and types, 215–216
expansion compensators, 220, 221
function, 216
heat load, 217–218
installation procedure, 219–222
one-row or two-row, 219
sizing procedure, 218–219
steel-finned, 217
fire extinguishers
choosing correct, 26
environmentally friendly, 26
fire types and correct, 27
types, 26–27
fire safety, 26
fire-tube heat exchanger, 46–48
first aid, electric shock, 26
fittings
copper press fittings, 220
copper tube sizes, 135
pipe fittings, 111–116
press-type, 324
soldering, 220
spacing tees, 149, 150, 151
flame-proving devices, 70
flame rectification circuit, 80

flame rectification pilot and probe, 80
flame rod
cleaning, 76
definition, 73
examples, 74
flame rectification system, 79
measuring signal strength, 79
testing, 353
troubleshooting signal, 362–363
flame rollout switch, 79, 364
flashover, 27
flash point, fuel oil, 86
flat-plate solar collector, 401–402
flexible couplings
defining functions, 194–195
example, 195
floating volts, 75
float-type air vents, 126
flooring systems, radiant heating, 8. *See also* radiant heating systems
floor-mounted boiler, 52
floor or slab sensors, 262
floor plans, 286–287
floor-standing, high efficiency boiler, 42
flow coefficient (C_V), 177
flow control valves
balancing valves, 179
check valves, 175–176
definition and function, 174
globe valves, 176–177
pressure-reducing valves, 177–178
flowmeters, 262
flow rate
gallons per minute (GPM), 200
piping system design, 135
flow switches
current relay, 159–160
defining functions, 159
differential pressure switches, 159–160
installation locations, 159–160
puddle switches, 159
testing, 344
wiring, 332
flow table, 132
flue, 44
flue liner
definition and use, 305–306
replacing, 364
fluid flow, physics, 131–132
fluid velocity, 132
flush-and-fill cart, 415
flushing
boiler, 337, 387
solar system, 415

flux paste, 323
FM Approvals, 64
FM gas manifolds, 64
FM Global, 64
footwear, safety, 23–24, 27
force, torque and, 95
forced-air heating
    allergens, 11–12
    Btu capacity, 14
    humidity levels, 12
forced air heating curve, 239
forced convection heating, 13
forced-draft combustion blower, 69
fossil fuel, 85
four-way mixing valve, 181–182
FPT (female pipe thread), 124
freeze protection
    boiler, definition and function, 345
    outdoor wood boiler, 429
    solar thermal storage, 398–400
    toxicity, safety, 345
    See also glycol systems
freeze protection valve, solar thermal storage, 411, 412
front gap, fuel oil electrodes, 100
frost line, 427
fuel
    AFUE efficiency, 7
    boiler selection and efficiency considerations, 295–298
    boiler systems, 4, 7–8, 43–45
    comparing costs, 296–297
    dual capacity, 100
    ratio, combustion, 59–60
    types, 4
fuel line filters, 94
fuel oil
    ash content, 88
    burner components, 94–104
    carbon residue, 88
    characteristics, 85
    combustion, 88–89
    composition, 86
    definition, 43
    delivery. *See* fuel oil piping systems
    distillation quality, 86
    flash point, 86
    heating values, 86
    ignition point, 86
    IMC and NFPA rules, 92
    number 2, red oil, 86
    pour point, 87–88
    storing, 43
    temperature and viscosities, 87
    underground storage tanks, 90
    viscosity, 86–87
    water and sediment content, 88
fuel oil boiler components
    burner nozzle, 97–99
    cad cell relay, 102
    definition, 94
    draft, importance, 105
    electrodes, 99–101
    ignition transformer, 100–101
    oil burner motor, 94–95
    primary control unit, 101–102
    solid state igniter, 101
    stack relay, 102–103
    testing high-voltage ignition, 101
    troubleshooting fuel pump, 97
fuel oil piping systems
    clearing clogged lines, 94
    converting one-pipe system to two-pipe system, 93
    fuel line filters, 94
    installing and servicing, 92
    oil deaerators, 92
    one-pipe, 90–91
    two-pipe, 90–91
fuel oil pump
    definition and function, 96–97
    single- or two-stage, 97
fully modulating control module, 167–168

G

gallons per minute (GPM), 131–132, 203
galvanic corrosion, 114
galvanized piping, 8
gas control valves
    combination gas valve, 66
    definition and functions, 64–65
    diaphragm gas valves, 65–66
    modulating gas valves, 67–68
    solenoid gas valves, 65
    two-stage gas valves, 67
gases, natural, 43
gas-fired burners
    atmospheric burners, types, 68–69
    definition and components, 68
    gas ignition modules, 76
    power burners, 69
gas-fired instantaneous heaters, 378. *See also* tankless boiler
gasification technology, 45, 426
gas ignition circuit board, 77
gas ignition modules
    circuit board, 77
    combination gas valves, 66
    continuous retry with 100% shutoff, 77
    definition and types, 76
    integrated ignition control modules, 76
    nonintegrated ignition modules, 76
    non-100% shutoff, 77
    100% shutoff, 76–77
    wiring terminals, abbreviations, 77
gas leaks, inspecting for, 349–350
gas manifold, 68
gas orifices (spuds), 306–307
gas piping
    appliance size in Btu, 308
    converting natural gas boiler to propane, 306–307
    pipe capacity, 309
    sizing, longest length method, 307–309
gas piping installation
    layout diagram, 328
    materials, joining and cutting, 327–328
    sealants, 328
    supports and requirements, 327
gas pressure
    measuring and setting, 342–343
    setting inlet, 67
gas pressure regulator
    definition, 60
    diagram, 62
    example, 61
    safety, positioning, 63
gas train, commercial and industrial components, 63–64
    Btu ratings, 63–64
    FM gas manifolds, 64
    IRI gas manifolds, 64
gas train, residential and light commercial components, 60–63
    automatic shutoff valve, 63
    definition, 60
    drip leg, 60
    early-model components, 60–61
    manual shut off valves, 60–61
    modern components, 60–61
    older style, example, 61
    pilot gas line, 63
gas valve safety, 66
gate valves
    definition, 173
    function, 174
    handwheel, 173
    inner workings, 174
    lockout device, 30

Copyright Goodheart-Willcox Co., Inc.

geothermal ground loops, 246
geothermal heat pumps
    buffer tank, 242–243
    coefficient of performance, 242
    defining functions, 242
    water-to-water geothermal heat pump, 242
GFCI (ground fault interrupter) receptacle, 28, 29
Giannoni-type heat exchanger, 353
glazing
    flat-plate solar collector, 401–402
    integral collector storage (ICS) unit, 394–395
globe valves
    definition and function, 176
    flow coefficient (C_V), 177
    installation and sizing requirements, 177
    pressure drop, 177
gloves, 24
glow coil, 362
glycolic acid, 399
glycol systems
    purging, 369–370
    solar thermal storage, 398–399
    See also propylene glycol
goggles, 22
greenhouse applications, outdoor wood boiler, 431
green practices
    dual temperature controls, 429
    electric boilers and environment, 54
    fire extinguisher, environmentally friendly, 26
    geothermal energy, clean and efficient, 243
    green building organizations, 271
    high-efficiency boiler and sealed combustion, 302
    indirect water heater, advantage, 382
    natural carbon cycle, 45
    occupancy schedules, energy saving, 228
    on-demand ignition, 71
    outdoor wood boiler, efficiency, 426
    pressure/temperature gauge accuracy, 116
    solar hot water, 393
    standing pilot ignition, 71
    tankless water heaters, save money, 379
    temporary repairs, avoiding, 355
    underground fuel storage, EPA, 90
    using modulation valves and variable speed pumps, 187
    waste heat, sidewalks and snow melt, 265
    wireless controls save energy and money, 166
ground fault circuit interrupter (GFCI) receptacle, 28, 29
grounding
    boiler, 329
    electronic ignition module, 75
    gap, checking for correct, 36
    power tools, 28, 29, 31–32
grounding adapters, 32
grounding clip, 32
gun-type oil burner, 95, 100

H

hand tool safety, 31
handwheel, gate valve, 173
hard hats, 23–24
hard water, 49
headphones, 23
head protection, 23–24
health
    CO exposure effects, 37
    hydronic heating, 11, 12
    insulation, 12
hearing loss, 23
hearing protection, 23
heat
    air and water delivery, 14
    definition, 4
    early sources, 4
    human comfort, 8–9
    source, combustion, 59
    See also temperature
heat content Btu/lb (kJ/kg), 16
heat emitters, 4–5. *See also* terminal devices
heat exchanger
    cast-iron boilers, 46
    cleaning, 353
    closed-loop, indirect water heater, 381–382
    condensing boiler, secondary, 50, 51
    copper-tube, low-mass boiler, 48–49
    definition, 44
    descaling, 387
    fire-tube, 46, 47–48
    gas-fired burners, 68
    Giannoni-type, 353
    lower-efficiency boilers, 45
    mid-efficiency boiler, steel and copper, 45
    outdoor, wood boiler, 424
    primary and secondary, combination boilers, 385
    sidearm, outdoor wood boiler, 430–431
    stainless steel, high-efficiency boilers, 46
    tankless boiler, 380
    water-to-air, example, 425
    water-tube boilers, 48
    welded tubes, 46, 47
heating effect factor, 219
heating oil, 85. *See also* fuel oil
heating values, fuel oil, 86
heat load
    calculating, 272
    finned-tube baseboard unit, 217–218
    fire-tube boilers, 48
    modulating boilers, 52
    outdoor wood boilers, calculating, 424
    radiant heating system design, 254–255
heat loss
    calculating, 276
    insulation, 12
    standby, 378
heat loss through conduction
    analyzing components, 273
    calculating, 278–279
    constants for heat transmission (U- and R-values), chart, 277
    design temperature difference, 272
    heating design temperatures for select cities, chart, 274
    heat loss transfer formula, 276–278, 280
    heat transmission multipliers, 278
    materials R-values, 273
    R-value, 272
    slab-on-grade construction, 280
    soil temperature, 280
    transmission losses, 272
    winter design temperatures, 273
heat loss through infiltration or convection
    accounting for other infiltration sources, 283
    air change method, 283
    blower door test, 283–284
    causes, 280

crack method, 280
definition, 280
heat motor actuators, 183
heat pump
    geothermal, 242–243
    hybrid water heaters, 387
heat stratification, 11
heat transfer
    conduction, 13
    convection, 13
    definition, 4
    human body, 8–9
    methods, 12–14
    physics, 3–4, 12–17
    radiation, 13
    through piping, 13–17
    water, 12
heat transmission, constants for (U- and R-values) chart, 277
Herschel, John, 285
hex-head countersunk plug, seal combustion test hole, 107
high-density polyethylene polymer (HDPE), 244, 246
high-efficiency two-stage control module, 167
high-limit control setting, aquastat relay, 163
high-limit operation, testing, 344
high-mass boiler, 46
high-mass radiant heating, 239–240
high-mass radiator panels, 222–223
high-pressure boilers, 42
high temperature limit safety switch, 78–79
history
    blueprints, 285
    radiant heating, 239
    thermostats, 6
hollow cone spray pattern, 98–99
HomeAdvisor.com, 379
Honeywell, Mark, 6
Honeywell Heating Specialty Company, 6
hopscotch method, troubleshooting, 360
horsepower, 200
hose bibb valve, 327
hot contacts, stack relay, 102
hot surface ignition (HSI) systems
    cleaning, 76
    defining functions, 75
    diagnosing, 76
    troubleshooting, 362

hot tubs, solar heating, 417
hot water boilers, 42
hot water circulating pump, 192
hot water generation, solar, 8, 393
hot water heating, diverter tees, 144
hot-water piping, 5
hot water production. *See* domestic hot water (DHW)
hot water supply sensor, 331
housekeeping pad, boiler, 316
human comfort
    comfort zone, 10
    core body temperature heating, 8–10
    heat transfer process, 8–9
    preferences and standards, 9–10
    radiant heating and, 238–239
humidity levels
    reducing, 11–12
    standards, 9
HVAC professional organizations, 10, 17
HVAC technician, responsibilities, 8
HVAC wholesalers, 127
hybrid water heaters
    definition, 376
    heat pumps, 387
    installation and control, 388–389
    operation and efficiency, 387
    service and maintenance, 389
hydraulic separation, 149
hydrocarbons, 43, 86
hydronic airlock, 366
hydronic circulators, 193
hydronic heating systems
    basic configurations, 4–8
    benefits, 11–12
    components, 7–8
    definition, 4
    example of basic, 5
    materials, 5
    modern configurations, 7–8
    safety devices, 7
hydronic piping
    boiler components, 324–327
    configuration determination, 321
    designing, 320–321
    joining, 321–324
    MAAP gas, 324
    press-type fittings, use, 324
    soldering copper tubing, 322–324
    use and application, 321
hydronic towel warmer, 232–233
hygroscopic air vents, 125

I

IBR (Institute of Boiler and Radiator Manufacturers), 217
ideal heating curve, 238–239
IFGC (International Fuel Gas Code)
    boiler installation, 314–315
    codes, 113, 307, 309
    gas piping installation, 327
ignition
    burner, boiler startup, 341–343
    continuous pilot light, 45
    electrodes, fuel oil burner, 99
    electronic, 45
    fuel oil, 89–90
    hot surface ignition, troubleshooting, 362
    spark, troubleshooting, 361
ignition point, fuel oil, 86
ignition systems
    defining types and functions, 70
    direct spark ignition systems, 74–75
    flame-proving devices, 70
    gas ignition modules, 76–77
    hot surface ignition (HSI) systems, 75–76
    intermittent pilot ignition systems, 73–74
    sequence of operations, 78
    standing pilot ignition systems, 70–73
ignition transformer, fuel oil burner
    defining functions and components, 100–103
impeller, circulating pump, 192
impeller trim, 200
incomplete combustion, 89
indirect water heaters
    anti-scald valve, 382–383
    auxiliary storage tank, 381
    closed-loop heat exchanger, 381–382
    definition, 376
    green advantage, 382
    installation and control, 382–384
    safety tip, pressure-relief valve, 384
    scalding temperatures, safety, 383
    service and maintenance, 384
    sizing, 382
    thermostatic mixing valve, 383
indoor air pollution, 424
indoor heat exchanger, 424
indoor heating design, 9

induced draft boiler
    ventilation termination, 318–319
    venting requirements, 305
induced draft combustion blower, 69
industrial gas train components, 63–64
Industrial Risk Insurers, 64
infrared waves, 9
inlet gas pressure, setting, 67
inline circulators
    coupling, 193
    definition, 193
    examples, 194
    flexible couplings, 194–195
    split couplings, 193–194
    three-piece, 193
    wet rotor, 195
inline fan, air circulation, 397
input rating, 42
inshot single port burner, 68
inside sales representatives, 127
inspection
    accident prevention, 22
    boiler, leaks, 338
    condensate system, 340
    gas leaks, 340
    hand tools, 31
    power tools, 31
    thermopile, 72
    valves, parallel pumping, 202
    venting system, 318
    visual, boiler, 349–350
insulation, heat loss, 12
integral collector storage (ICS) unit, 394–395
integrated ignition control modules, 76
intermittent pilot ignition systems
    advantages, 73
    defining functions and components, 73
    sequence of operation, 74
Internal Revenue Service (IRS), 86
International Fuel Gas Code (IFGC)
    boiler installation, 314–315
    codes, 113, 307, 309
    gas piping installation, 327
International Mechanical Code (IMC)
    boiler installation, 315
    code, 246
    fuel oil and piping rules, 92
International Plumbing Code (IPC), 161, 246
International Residential Code (IRC), 246
interpolate, 219
in.WC (inches water column), 62

IRI gas manifolds, 64
iron and steel pipe fittings, 111–113
isolation valves
    ball valves, 174
    defining functions, 173, 326
    gate valves, 173–174
isometric drawings, 285–286

K
kerosene
    definition, 86
    pour point, 87–88
kick space heaters, 231–232
kilopascal (kPa), 116
kilowatts (kW), 44

L
ladder safety, 31–33
laminar flow pattern, water, 132
latent heat, 15
Leadership in Energy and Environment Design (LEED), 17, 271
leaks
    backflow preventers, 161–162
    combustion, 350
    gas, 340, 349–350
    radiant tubing, 249
    solar thermal storage system, 414–415
    water, 350–351
LEED (Leadership in Energy and Environment Design), 17, 271
lift, hot water circulation, 199
lifting procedure, 24
light commercial gas train components, 60–63
light oils, 86
Linde Group, 324
line-voltage connections, troubleshooting, 358
    wiring, 329–330
line-voltage thermostat, 429
liquid petroleum (LP)
    boiler fuel, 43–44
    definition, 8
load calculation
    piping system design, 134
    software, 284
local sensing, 74
lockout/tagout safety procedure, 29–30
longest length method, gas pipe sizing, 307–309
LonWorks, building management system, 332

lower-efficiency boilers, 46
low-level CO detector, 37
low-limit control setting, aquastat relay, 163
low-loss header, 149, 150
low-mass boiler, copper-tube, 48–49
low-mass radiant heating, 239–240
low-mass radiator panels, 222
low-NO_X boilers, 54
low-pressure boiler, 41–42
low-voltage connections
    troubleshooting, 359–360
    wiring, 330–331
low-water cutoff (LWCO) switch
    code requirements, 158
    conductivity sensor, 157–158
    defining functions, 157, 326
    definition, 35–36
    float type, 158
    radiant floor heating, 158
    switch location, image, 159
    testing, 344, 354
low-water cut-outs, 7
LP gas conversion, 341

M
MAAP gas, 324
maintenance contracts, 371
makeup water
    backflow preventer, 326
    definition, 177
    isolation valves, 326
    keeping on or turning off, 179
    pressure-reducing valve, 326
manifold gas pressure, 365
manifold station, 249
manometer, 179, 342
manual air vents, 124–125
Manual J, 276
manually operated remote shutdown switch, 29
Manual N, 276
manual reset button, 79
manual reset function, direct spark ignition, 75
manuals, power tool, 31
manual shutoff valve, 60–61, 66
math
    estimating water volume, 121
    expansion tank pressure, 121–122
    fluid flow, volume and rate, 131–132
    fuel oil burner nozzle, 99
    gallons per minute, 132

heat load, 217–218
heat loss, 278–279
order of operations, 16
pressure requirements, 133
pump size, 203
recovery rate, 377
sizing expansion tanks, 120–121
sizing fan coil units, 226–227
space and combustion, 299
total loop length, 259
measurement units
   cord of wood, 45
   fluid flow, 131–132
   gallons per minute (GPM), 131–132, 200, 203, 239
   parts per million (ppm), CO, 37, 107
   pressure measurements, 121
   torque and force, 95
   velocity (FPS), 132
   water pressure, 116
medium-efficiency industrial boiler, 42
metals, joining dissimilar, 115
microammeter, flame testing, 353
microamps, 362
microprocessor-based integrated controller, 77, 380
microprocessor controllers, radiant heating, 262–263
mid-efficiency boilers, 45–46
millivolts, 360
Minneapolis Heat Regulator Company, 6
mixing valves
   definition and functions, 180
   diverting valves, 182–183
   four-way, 181–182
   preventing boiler condensation with, 182
   thermostatic mixing valves (TMV), 180–181
   thermostatic radiator valves, 184–185
   three-way, 180
   zone valves, 183
ModBus, building management system, 332
mod/con boilers, 52
modulating boilers
   definition and uses, 51–53
   heat load, 52
   temperature controls, 52–53
modulating control module, 167–168
modulating gas valves
   definition and functions, 67
   turndown ratio, 67–68
modulating hot water valve, 228
modulating motor, 68
modulating valves, environmentally friendly, 187
Monoflo tee, 140, 144
multi-lock hasp, 30
multimeter, 358

N

NATE (North American Technical Excellence) exams, 291
National Board of Boiler and Pressure Vessel Inspectors, 22, 35
National Electric Code (NEC), 329, 330, 408
National Fire Protection Association (NFPA), 92, 327
National Fuel Gas Code ANSI 7223, 299, 304, 318
National Oceanic Atmospheric Administration (NOAA), 280
national pipe thread (NPT) taper, 111–112
natural carbon cycle, 45
natural draft boilers
   venting requirements, 304–305
   ventilation termination requirements, 318–319
natural draft combustion, 68–69
natural gas
   boiler fuel, 8
   definition, 43
   distributing and storing, 44
near boiler piping, 324–325
NEC (National Electric Code), 329, 330, 408
negative draft, 304
negative pressure, 12
negative temperature coefficient (NTC), 414
net positive suction head (NPSH), 200
net stack temperature, 106–107
neutralizing filter, 328
NFPA (National Fire Protection Association), 92, 327
nitric oxide, combustion analysis, 343
nitrogen, 59–60
nitrogen oxides (NO$_X$), 54, 344
noise exposure, maximum permissible, 23
nonintegrated ignition module, 76
non-100% shutoff, gas ignition module, 77
normally closed valve, 65

North American Technical Excellence (NATE) certifications and training, 291
nozzle, fuel oil burner, 97–99
number 1 fuel oil, 87–88
number 2 fuel oil, 86

O

obstructions, inspecting vents for, 22
occupancy scheduling, efficiency, 228
Occupational Safety and Health Administration (OSHA), 22–23, 30, 37. *See also* Safety Data Sheets
odorants, 43
oil, boiler fuel, 8
oil burner components
   air bands, adjusting, 96
   burner blower, 95
   burner motor, 94–95
   cad cell relay, 102–104
   electrodes, 99–101
   fuel oil pump, 96–97
   motor, 96
   nozzle, 97–99
   primary control unit, 101–102, 104
   split-capacitor motor (PSC), 95
   stack relay, 102
oil burner motor
   definition, 94
   split-phase, 95
oil deaerators, 92
oil-fired gun burner, 95
on-demand hot water heaters, 377–378
on-demand ignition, 71
100% shutoff, gas ignition modules, 76–77
one-pipe delivery system
   converting to two-pipe system, 93
   definition, 90
   diagram, 140
one pipe loop with diverter tees
   definition, 213
   diagram, 215
one-pipe system with multiple zones
   defining functions and characteristics, 139–140
   design considerations, 143
   diverter tee piping recommendations, 141–143
   terminal units piped, diagram, 140
   thermostat radiator valve, diagram, 143

open-loop system, active solar storage system, 397–398
operating pressure
    electric boiler, 44
    high-pressure boilers, 42
    liquefied petroleum, 43–44
    low-pressure boilers, 41–42
    medium-pressure boilers, 42
operations sequence
    aquastat relay, 164
    cad cell relay, 104
    ignition systems, 78
    intermittent pilot ignition systems, 74
    stack relay, fuel oil burner, 103
order of operations, math, 16
OSHA (Occupational Safety and Health Administration), 22–23, 30, 37
    See also Safety Data Sheets
outdoor reset control, 53
    benefits, 166
    cast-iron boilers, 166
    components, image, 166
    defining operational parameters, 165
    radiant heating, 262
    reset ratio, 165
outdoor temperature sensor, 332
outdoor winter design temperatures, 276
outdoor wood boiler
    ash removal, 433
    biomass heating, 424
    choosing correct size, 425
    connecting to existing, 430–431
    creosote buildup, 431–433
    definition, 8, 423
    dual fuel capacity, 430
    efficiency, 423–424
    example, 45
    foundation, 426
    heat load calculation, 424
    indoor air pollution, 424
    indoor heat exchanger, 424
    installing, 425
    multiple buildings, heating, 431–432
    piping installation, 427–429
    radiant floor or baseboard heat, 430, 432
    sidearm heat exchanger, domestic hot water production, 430
    sizing, 423–424
    storage hopper, 430
    swimming pool and hot tub, 431, 433
    winter installation, 428
    wiring and controls, 428–429
oxidation
    anode rods, 46
    nitrogen oxides (NO_X), 54
oxyacetylene torch, 33, 34
oxygen
    combustion, 59, 107, 343
    definition, 33–34
oxygen barrier, PEX tubing, 245
oxygen regulator, 33

P

parallel pipe arrangement, 213–214
parallel pumping, valve inspection, 202
parts per million (ppm), 37, 107
passive solar heating system
    definition and function, 394
    hot water temperature, safety, 395
    integral collector storage units (ICS), 394–395
    thermosiphon systems, 395–396
PEMDAS, 16
perfect combustion, 59–60, 89
personal protective equipment (PPE)
    body protection, 23–25
    eye and face protection, 22
    hand tools, 31
    head protection, 23
    hearing protection, 23
    power tools, 31
    respiratory, 25
perspective drawings, 285
perspiration, 9
PE-RT tubing, 245, 246
petroleum, formation, 85
PEX-AL-PEX tubing, 245
PEX tubing
    color, 245
    definition and uses, 244
    oxygen barrier, 245
    PEX-AL-PEX type, 245
    types of, 244
    See also cross-linked polyethylene (PEX) piping
pH levels, boiler water, 338
photovoltaic module, 397
physics of heat transfer, 3–4, 12–17
physics of water
    fluid flow, 131–132
    fluid velocity, 132
    pump pressure, 134
    static pressure, 134
    system head, 133
    system pressures, 133
    understanding, need for, 131
pilot assembly, 73
pilot flame, 70–71
pilot gas line, 63
pilot generator, 71–72
pilot light, 60, 72
pilot lines, clearing, 70
pilot supply line, 63
pilot valve, 65
pipefitter, career, 389
pipe fittings
    bushings, 113
    copper tube, 113
    defining functions, 111
    iron and steel, 111–113
    joining dissimilar metals, 115
    PEX, 113–114, 115
    pipe reducers, 113
    PVC, 115
    specialty types, 114–115
pipe friction chart, 132, 136
piping
    condensate piping compliance, 329
    definition, 5
    ductwork or, 12
    gas, 306–309
    gas piping installation, 327–329
    heat transfer through, 13–17
    hot-water, 5
    hydronic, boiler installation, 320–327
    joining copper, procedure, 322–324
    near boiler, 324–325
    outdoor wood boiler, insulated, 427–428
    PP, 51, 116
    PVC, 51, 116
    radiant floor heating, 418
    radiant heat systems, 244
    socket fusion, join with, 245, 246
    solar hot tub layout, 417
    solar swimming pool layout, 416
    solar system, 409–412
    symbols on blueprints, 290
    types, 8
piping arrangements
    factors determining, 136
    one-pipe system with multiple zones, 139–143
    series loop arrangement, 136–137
piping system
    balancing valves, 179
    choosing terminal units, 213–215
    designing, 134–136

flow rate table, 135
layout and design considerations, 134
one-pipe loop with diverter tees, 213, 215
parallel or two-pipe loop, 213, 214
plot pump system curve, 136
primary-secondary piping system, 147–150
series loop, 213, 214
split loop, 213, 215
two-pipe direct return system, 144–145
two-pipe reverse return system, 145–147
piping system curve, plot, 136
plot plans, 285
plugs, iron and steel pipe fittings, 112
plumber, career, 389
point of no pressure change, 196
point-of-use heater, 379
polypropylene (PP) piping, 51, 116
polyvinyl chloride (PVC) piping, 115, 116
potable hot water, 375. *See also* domestic hot water (DHW)
potassium-based fire extinguisher, 26
pounds per square inch (PSI), 116, 121
pour point, fuel oil, 87–88
power burners
    definition and functions, 69
    example, 70
power tool safety, 28–29, 31–32
PP (polypropylene) piping, 51, 116
PVC (polyvinyl chloride) piping, 115, 116
prefabricated radiant floor panels, 251–252
press-type fittings, 324
pressure drop, globe valves, 177
pressure gauge
    definition and function, 327
    solar system piping, 412
pressure measurements, 121
pressure-reducing valve
    components, 178
    defining purpose and function, 177, 326
    examples and installation requirements, 178
    fast-fill lever, 178
    safety tip, 384
pressure regulator
    definition, 34, 61
    early model gas train, 61
    fuel oil pump, 97
    vent, 61

pressure-relief valve
    components and example, 156
    definition, 7, 155, 326
    discharge tubes, 156
    example, 345
    function of, 155–156
    inspecting, 353
    internal workings, 157
    rating plate, 157
    safety tip, cap removal, 157
    solar system piping, 411
    testing, 156, 344–345
    T&P valve, 157
pressure surges, water hammer, 122
pressure/temperature gauge, 116
pressurized gases, 32–33
pressurized solar thermal storage system, 401
pressurized tanks, 33
prestart checklist, 340–341
primary air, atmospheric burners, 69
primary control unit, oil-fired burner
    definition and function, 101–102
    excess fuel accumulation, 104
primary-secondary loop, purging, 367–368
primary-secondary piping system
    boilers and, 240–241
    defining characteristics, 147–148
    design considerations, 150
    diagram, 148, 242
    hydraulic separation, 149
    illustration, 321
    low-loss header, 149
    standby losses, reducing, 151
propane
    boiler fuel, 43–44
    converting natural gas to, 44
    converting natural gas boiler to, 306–307
    definition, 33
propylene glycol
    freeze protection, 345
    solar thermal storage, 398–399
    See also freeze protection; glycol systems
proving switch, 332
PRV piping, 353
psychrometric chart, 10
puddle switches, 159
pump affinity laws, piping system, 134
pump curves, 199–201
pump deadheading, 145
purchasing agents, 127
purging air from gas lines, 340

purging valve
    definition and function, 186, 327
    use, 339

R

radiant floor piping configurations, 418
    above-floor prefabricated panels, 251–252
    below-floor suspended tubing systems, 252–253
    concrete or gypsum thin-slab floors, 250–251
    defining types of, 246
    insulated floor panel system, 251
    preparing tubing for control joints, 249
    repairing radiant tubing leaks, 249
    slab-on-grade radiant piping, 246–247
radiant heating systems
    balancing radiant floor system, 254
    condensing boiler, 241–242
    conventional boilers, 240–241
    definition and uses, 8, 13
    designing, 254–262
    floor or slab sensors, 262
    geothermal heat pumps, 242–243
    high-mass, 239–240
    history, 239
    human comfort, 238–239
    ideal heating curve, 238–239
    low-mass, 239–240
    mechanical code for piping, 246
    PE-RT tubing, 245
    PEX tubing, 244–245
    piping materials, 244
    principles of use, 237
    snow and ice melts, 263–266
    solar thermal storage, 243
    wall and ceiling panels, 253–254
radiant heating systems controls
    electric or electronic zone valves, 261
    flowmeters, 262
    microprocessor controllers, 262–263
    outdoor reset controllers, 262
    space thermostats, 261
    temperature gauges, 262
radiant heating systems design
    adjusting loop lengths and tubing size, 260
    calculating specific parameters, 254–261

GPM flow, calculate, 259
head loss, chart to calculate, 260
outdoor wood boiler, 430, 432
R- value, select carpets and pads, 257
surface temperature chart, 256
temperature limitations, 256
total loop length, calculating, 259
water temperature for 4-inch concrete slab, graph, 258
radiant panels, radiator, 222
radiant solar heating, 418–419
radiant wall heating, layout, 419
radiation
   body warmth, 9
   heat transfer through, 13, 14
radiator
   benefits, 223
   configurations and colors, 225
   definition and types, 222
   installing, 223–224
   panel ratings, 223
   sizing, 223
   steel panel, 5, 6
   wall-mounted example, 224
reading blueprints, 283–284
reciprocal, 276
recovery rate, domestic hot water (DHW) production, 377
red fuel oil, 86
reduced bushings, 328
redundant gas valves, 66
refining fuel oil, 86
relative humidity
   reducing, 11–12
   standards for, 9
renewable resources, boiler fuel, 8
reset ratio, outdoor reset control, 165
residential gas train components, 60–63
Residential Load Calculation, Manual J (ACCA), 276
respirator, 25
respiratory protection, 25
return water temperature, 49
ribbon burners, atmospheric burner, 68
risers, drilling holes for, 219
rollout switch, examples, 79
R-value
   converting to U-values, 280
   definition, 272
   thermal density, 274
   thermal resistance, 275

S

safety
   accidents, 35
   boiler installation, 314
   boiler operator license, 43
   boiler room, 35–37
   boiler safety components, 326
   chemical hazards, 33–34
   clearing clogged fuel lines, 94
   CO poisoning prevention, 37
   confined spaces, 35
   creosote buildup, outdoor wood boiler, 433
   discharge piping, PRV, 353
   electrical, 26–29, 207
   fire, 26
   flame rollout switch, 364
   fuel accumulation, excess, 104
   gas pressure regulator, 63, 66
   gas valve, 66
   grounding prongs, 32
   high-voltage ignition, testing tools, 101
   hot water, scald danger, 383
   importance, 21
   ladders, 32
   lockout/tagout, 29–30
   LP gas conversion, 341
   maximum hot water temperature, 395
   natural gas boiler, converting, 307
   odorants, natural gas, 43
   personal protective equipment (PPE), 22–25
   potential issues, 22
   pressure switch, replacing, 364
   pressure-relief valve cap removal, 157
   pressurized gases, 32–33
   preventing carbon monoxide contamination, 352
   proper ventilation, 25
   purging air, boiler gas piping, 340
   spark ignition circuits, 74
   tank transport, 33
   testing boiler safety devices, 344–345
   tool and equipment. *See* tool and equipment safety
   toxic freeze protection, 345
   water, electricity and, 27
   wet surfaces, 27
Safety Data Sheets (SDS)
   definition, 33
   requesting from vendor, 35
   sections, 34
safety devices
   burner ignition controls, 78–80
   defining functions and components, 155
   hydronic systems, 7
   low-water cutoff, 157–178
   pressure-relief valves, 155–156
safety glasses, 22
safety harness, 24–25
safety meetings, 21
safety relief valves, 411
sales representative, career, 55
scale, 286
schedule 40 piping, 111
Schlemmer, Oliver, 144
Scotch Marine boiler, 47–48
sealed combustion boilers, 302
secondary air, atmospheric burner, 69
secondary heat exchanger, 50, 51
self-contained breathing apparatus (SCBA), 35
semisolid cone spray pattern, fuel oil burner nozzle, 98–99
sensible heat, 15
sensible heat formula, 7
   air, 14–15
   definition, 14
   heat loss through infiltration, 280
   modifying, 17
   use in troubleshooting, 7
   water, 15
sensible heat ratio, 10
sensing bulb, aquastat relay, 162
series and parallel pumping design
   configuration tips, 201
   definition, 201
   diagram example, 202
   energy generation and transfer, 199–201
series loop piping arrangement
   boiler return water temperature, 137–139
   defining characteristics, 136
   design considerations, 139
   diagram, 137, 214
   disadvantages, 136–137
   terminal device, 213
set point, 6
sidearm heat exchanger, 432
side shields, eye protection, 22
single port burner, 68
single-stage fuel pump, 97

slab-on-grade construction, 280
slab-on-grade radiant piping
    choice considerations, 247
    definition and uses, 246–247
    installation procedure, 247–249
    insulation types, 247–248
    manifold station, 249
    pattern examples, 248
    rolled fabric, 248
    vapor barrier, 247
    welded wire fabric, 248
slotted burner, 68
slow-opening valve, 74
snow and ice melt, radiant heating
    applications, 264–265
    definition and uses, 263
    designing, 265–266
    functions, 264
    installing, 266
    waste heat, 265
socket fusion, 245, 246
soft copper, 113
solar angles, solar thermal storage, 405–406
solar array, 402
solar azimuth angle, solar collection, 405
solar collector
    angles for, 405–406
    definition and functions, 401
    evacuated tube, 402–403
    flat-plate collector, 401–402
    illustration, 8
    panel positioning, 406–408
    snow and, 402
    solar altitude angle, 405
    solar azimuth angle, 405–406
    storage tank, 244
    swimming pool system, 417
    vacuum tube, example, 243
solar pathfinder, 406–407
solar swimming pool heating systems, 416–417
solar system piping
    air vents and air separators, 410–411
    check valves, 411–412
    expansion tanks, 409–410
    freeze protection valves, 411
    pressure and temperature gauges, 412
    pressure-relief valve, 411
solar thermal controller, 413–414
solar thermal power career, 419
solar thermal storage
    active storage systems, 396–401

air vents, 401
application selection, 403–404
building codes, 405
charging the system, 415–416
cleaning and flushing, 415
closed-loop systems, 398–399
code, installing, 407
control strategies and devices, 412–414
drainback systems, 399–400
filling and starting the system, 414–416
flush-and-fill cart, 416
freeze protection, 398–400
function and components, 243
green energy, solar hot water, 393
hot tubs and spas, 417
installation, site survey, 404
integral collector storage (ICS) units, 394–395
leak testing, 414–415
mounting, diagrams, 397
open-loop systems, 397–398
passive storage systems, 394–396
pressurized and unpressurized, 401
radiant solar heating, 419–419
space heating, 417–419
swimming pools, 416–417
system piping, 409–412
thermosiphon systems, 395–396
soldering copper tubing, 322–324
solenoid valve, 65
solid cone spray pattern, fuel oil burner nozzle, 98–99
solid-state circuitry, electronic controls, 166
solid-state modulating control board, 168
space
    boiler installation, clearances, 317–318
    comfort conditions, 9–10
    confined, boiler combustion, 300–303
    finned-tube baseboard units, 221
    radiators, 223
    unconfined, boiler combustion in, 299–300
space thermostats, 261
spark igniter, 73, 361
spark ignition circuits, safety, 74
spas, solar heating, 417
specialty check valve, 176
specific heat, 14

split-capacity motor (PSC), 95
split-case pumps, 195–196
split couplings, 193–194
split loop pipe configuration, 213, 215
split-phase motor, 95
split-series loop system diagram, 139
spotters, 35
spray patterns, fuel oil burner nozzle, 98–99
spring-loaded check valves, 176
spuds, gas-fired burners, 68
stack relay, fuel oil burner, 102–103
stack temperature, combustion analysis, 343
stamped burner, atmospheric burner, 68
standby heat loss, 378
standby losses, reducing, 151
standing pilot ignition systems
    definition, 70
    example, 71
    pilot generator, 71–72
    troubleshooting, 360
static pressure
    definition and use, 199
    water, 133
steady-state boiler efficiency, 49
steam, heat source, 15
steamfitter, career, 389
steel and stainless steel boilers
    corrosion reduction, 46
    fire-tube heat exchangers, 46–48
    water-tube, 48
    welded heat exchange, 46
steel pipe fittings, 111–113
steel-toed boots, 23, 24
stoichiometric combustion, 59–60
storage hopper, 430
strap-on-temperature device, 414, 429
stratification, heat, 11
supplemental oxygen device, 35
swimming pools
    solar heating, 416–417
    outdoor wood boiler, 431, 433
swing check valve, 175–176
system head, water, 133
system pressures, 133

T

Taco Comfort Solutions, 144
tankless boiler
    advantages and disadvantages, 378–379
    definition, 53, 375, 378

domestic hot water production, 375
efficiency, 379
heat exchanger, 380
installation and controls, 379–380
operation, illustration, 379
point-of-use, 379
service and maintenance, 380
standby heat loss, 378
whole-house heater, 379

tankless heater, 378. *See also* tankless boiler

tankless water heater (on-demand water heaters), 377–378

technology
boiler control, 7
efficiency, 7
gasification, 45
load calculation software, 284
modern hydronic system, 7
software, snow and ice melt radiant heat systems, 265
thermal targeting, 167
See also control devices; tech tips

tech tips
air separator placement, 124
air vent caps, 126
ASHRAE standards, 10
cast-iron boilers, 166
circulating pumps, 208, 340
combustion test, 107
concentric vent kit, 320
converting natural gas to propane, 44
converting R-values to U-values, 280
copper press fittings, 220
electrode gap, 100
electronic ignition module grounding, 75
finned-tube baseboard unit types, 219, 221
flame rod, 76
fuel oil burner nozzles, 99
gas pressure, 62
heat loss through infiltration, 283
heat transfer methods, 13
heat transmission multipliers, 278
heating affect factor, 219
high-voltage test equipment, using, 362
hot surface igniter, 76
hydronic circulators, 193
ignition module wiring terminals, 77
inlet gas pressure, 67
loop lengths and tubing size, 260

low-water cutoff switch, 354
makeup water, 179
manufacturer's troubleshooting matrix, 362
mixing valves preventing boiler condensation, 182
noisy aquastats and pump relays, 208
outdoor boiler installation, 428
outdoor reset control, 166
petroleum-based cleaners, 337
PEX color, 245
positive displacement pumps, 193
preparing tubing for control joints, 249
PSI, PSIG, and PSIA, 121
pump pressure, 134
radiant floor heating, 158, 251, 254, 256
risers, drilling holes for, 219
rolled fabric, 248
series pumping, 201
sizing diverter tees, 142
snowfall and evacuated tube collectors, 402
solar thermal storage, air vents, 401
soldering, 220
spider webs, clearing from pilot lines, 70
static pressure, 134
two-stage gas valves, 343
valve inspection, parallel pumping, 202
valves and emergency heat requirements, 176
variable speed pump, advantages, 186
venting system, examining, 318
welded wire fabric, 248
See also technology

tees
iron and steel pipe fitting, 112
Monoflo tee, 140, 144
spacing diagram, 150, 151
spacing, 149
Venturi Tee, 140, 144

temperature
air, body temperature, 9
dew point, condensing boilers, 49–50
freeze protection, 345, 398–400, 411
high-pressure boilers, 42
human core body, 8
low-pressure boilers, 41–42
maximum hot water, 395

medium-pressure boilers, 42
net stack, testing, 106–107
return water, series loop piping arrangement, 137–139
sensible heat formula, 14–17
swimming pools, solar heating, 417
viscosity and fuel oil, 87

temperature control
solar systems, 414
two-stage, modulating boiler, 52–53
valves, 180–182
See also control devices

temperature difference, 14
temperature droop, 5–6
temperature gauges
definition and function, 262, 327
solar system piping, 412

temperature sensors, 415
temperature setpoint, 412
terminal devices
definition, 4–5, 213
example, 5
fan coil units (FCU), 225–229
finned-tube baseboard units, 215–222
hydronic towel warmer, 232–233
issues with, 6
piping configuration, choosing, 213–215
radiators, 222–225
under-cabinet fan coil units, 231–232
unit heaters, 229–230
unit ventilators, 230–231

terminal strip, line-voltage connections, 330

terminal units
convection, 13
diverter tees on, 140
piping diagrams, 142
thermostatic radiator valve, 143

test and balance (TAB) technicians, 209
thermal density, 274
thermal memory, 244
thermal resistance, R-value, 276
thermal transmittance, U-value, 276
thermistor, solar systems, 414
thermocouple, 70–72
thermocouple adapter, 73
thermopile, 70–72
diagnosing defective, 73
inspecting, 72
thermosiphoning, 395–396, 411
thermostat
example, 332

wiring, 331
See also control devices
thermostatic mixing valves (TMVs)
    defining functions, 180–181
    illustration, 182
    indirect water heater, 383
thermostatic radiator valves (TRVs), 143
    circulators and, 205
    components of, 184–185
    definition and functions, 184
    emergency heating, 185
    settings, 184
    troubleshooting, 185
thermostat, 6. *See also* control devices
thin-slab radiant flooring system, 250–251
three-piece circulator, 193–194
three-way mixing valves, 180–182
three-way zone valves, 183
tilting disc check valve, 175–176
tinted goggles, 22, 23
Titanic, Scotch Marine boiler, 48
toe-space heaters, 231–232
tool and equipment safety
    hand tools, 31
    high-voltage ignition, testing, 101
    insulated handles, 28
    ladders, 32
    power tools, 31–32
torque, 95
total dissolved solids (TDS), 338
towel warmer, 13, 232–233
T&P valve, 157
transmission losses, 272
transporting pressurized gas, 33
triac terminal, 104
trisodium phosphate (TSP), 337, 415
troubleshooting
    air in water system, 366–370
    burner and venting issues, 363–365
    burner circuits, standing pilots, 360
    checking the thermopile, 72
    circulating pumps, 206–209
    clogged condensate drain, 364
    combustion air, 365
    control board, 168–169
    descaling heat exchanger, 387
    electrical issues, 358–363
    expansion tank issues, 371
    faulty pressure switch, 363
    flame rod signal, 362–363
    flue liner, replacing, 364
    fuel oil pumps, 97
    high-voltage testing equipment, using, 362
    hopscotch method, electrical, 360
    hot surface ignition systems, 362
    improper exhaust gas recirculation, 364
    installation issues, 358
    intermittent ignition systems, 361
    line-voltage issues, 358
    live circuits, 27
    low-voltage connections, 359–360
    manufacturer's troubleshooting matrix, 362
    repairing radiant tubing leaks, 249
    sensible heat formula, 7
    separating the two systems, 356
    sizing and installation issues, 356–338
    sizing issues, 357
    skills for, 169
    spark ignition system, 361
    specific, sensible, and latent heat, 15
    standing pilots, 360
    stuck thermostatic radiator valves (TRVs), 185
    study system and piping arrangement, 355
    system installation issues, 358
    system size issues, 357
    thermopile, diagnosing defective, 73
    venting problems, 363–364
    water circulation issues, 365–371
tubing cutters, 322–323
turbulent flow pattern, water, 132
turndown ratio
    definition, 52
    modulating gas valves, 67–68
two-pipe delivery system, fuel oil
    converting to one-pipe system, 93
    definition, 90
    illustration, 91
two-pipe direct return system
    deadheading, 145
    defining functions, 144
    design considerations, 144–145
    diagram, balancing valves, 146
    example, 145
    pressure drop, 144
two-pipe loop arrangement
    diagram, 214
    terminal device, 213
two-pipe reverse return system
    defining characteristics, 145–146
    design considerations, 146–147
    diagram, 147
    differential pressure bypass valve, 147–148
two-stage fuel pump, 97
two-stage gas valves
    definition and functions, 67
    example, 67
    setting inlet gas pressure, 67
    setting manifold pressures, 343
two-stage high efficiency control module, 167

U

UA (United Association of Journeymen and Apprentices of the Plumbing and Pipe Fitting Industry of the United States and Canada), 17
unconfined space, boiler combustion, 299–300
under-cabinet fan coil units, 231–232
unions, iron and steel pipe fittings, 112
United Association of Journeymen and Apprentices of the Plumbing and Pipe Fitting Industry of the United States and Canada (UA), 17
unit heaters
    advantages and disadvantages, 229–230
    applications, 229
    control strategies, 230
    defining function, example, 229
unit ventilators, 230–231
unpressurized solar thermal storage unit, 401
upshot single port burner, 68
US Bureau of Labor and Statistics, 151
US Department of Commerce, 86
US Department of Labor, 21, 389
US Green Building Council (USGBC), 17, 271
U-tube manometer, 62
U-value
    converting to R-values, 280
    thermal transmittance, 276

V

vacuum tube solar collector, 243
valves
    anti-scald, 382–383
    automatic gas shutoff, 63
    balancing, 145–146, 179, 327
    ball, 174–175

boiler, 327
boiler drain, 338
bypass, 147–148
check, 175–176
diaphragm, 65–66
differential pressure bypass valve (DPBV), 185–186
flow control, 174–179
globe, 177
hose bibb, 327
inspecting parallel pumping, 202
isolation valves, 173–174
mixing, 180–182
modulating gas, 67–68
modulating hot water, 228–229
normally closed, 65
pilot, 65
pressure-reducing, 177
pressure-relief, 7
purging, 186, 327, 329
purging valve, 186
redundant gas, 66
slow-opening, 74
solenoid, 65
specialty check, 176
spring-loaded check, 176
temperature control valves, 180–182
thermostatic mixing valves (TMVs), 180–181
thermostat radiator, 143
three-way mixing, 180–181
tilting disc check, 175–176
T and P, 187
two-stage gas, 67, 343
zone valves, 183, 339, 369–370
vapor barrier, radiant floor systems, 247
variable fan speed control, 227
variable frequency drives (VFDs), 205–206
variable speed circulators
 definition, 204
 Delta-T and Delta-P type, 205
 electronically commutated motors (ECMs), 204–205
 variable frequency drives (VFDs), 205–206
variable speed pump
 definition and advantages, 187
 environmentally friendly, 187
velocity, fluid, 132
ventilation system
 air removal devices, 124–125
 air requirements, 299–303
 boiler selection, 298

check, prestart checklist, 341
code, combustion, 302
concentric vent kit, 320
confined spaces and, 35
definition, 25
fuel oil boiler, 105
inspecting, boiler maintenance, 351–352
obstructions, inspecting, 22
outdoor airflow for, code, 231
polypropylene materials, 116
pressure regulator, 61
proper, 25
proper boiler, 303–306
PVC materials, 116
solar thermal storage, 401
troubleshooting, 363–364
unit ventilators, 230–231
vent slope, code, 365
ventilation termination
 condensing boilers, 319–320
 natural draft and induced draft, 318–319
vents. *See* ventilation system
vent slope, code, 365
vent stack, 363
venturi opening, atmospheric burner, 69
Venturi Tee, 140, 144
versatility, hydronic heating, 12
vestibule, 364
viscosity, fuel oil
 definition, 86–87
 temperature and, 87
vises, 31
visual inspections, boiler maintenance, 349–350
volt-ohm meter, 358
volute (pump housing)
 definition, 192–193
 example, 193

W

wall-hung boilers
 high-efficiency, 42
 modulating, 52
wall panels, radiant heating, 253–254
warranties, boiler, 315
waste heat, snow and ice melt, 265
water
 air delivery, 14
 copper-tube boiler, and hard, 49
 electric conductivity, 27
 fire safety and, 26
 flow patterns, 132

heat-carrying ability, 13–14
heat transfer, 12
physics of, 131–134. *See also* physics of water
piping hot, 8
potable hot water, 375. *See also* domestic hot water (DHW)
sensible heat formula for, 15–17
solar generation of hot, 8, 393
solar heating, 8
specific heat, 14
steam, 15
water and sediment content, fuel oil, 88
water circulation issues
 air in system, 366–370
 circulating pump, 366
water flow switch, 326
water hammer, pressure surge, 122
water hardness, 338
water leaks, inspecting boiler, 350–351
waterlogged compression-type expansion tank, 117–118
water pressure, cavitation, 122–123
water quality, boiler, 337–338
water temperature, condensing boilers, 49–50
water-to-water geothermal heat pump, 242
water treatment specifications, 49
water-tube boiler, 48
water volume, estimating, 121
welded heat exchanger tubes, 47–48
welded wire fabric (WWF), 248
wet-base boiler, 46, 47
wet radiant hydronic systems, 246
wet rotor circulator
 defining functions, 195
 example, 194
whole-house heater, 379
winter design temperatures, 273
wireless controls, energy and money savings, 166
wiring
 boiler in field, 330–331. *See also* field wiring, boiler installation
 building management system (BMS), 332
 flow switch, 332
 outdoor wood boiler, 428–429
 prestart checklist, inspecting, 341
 thermostat, 331
wiring terminal, abbreviations, 77
wood, fuel source, 8

wood-burning boilers, emissions, 45
wooden ladders, 32
workplace skills
    employee efficiency, 80
    HVAC wholesalers, 127
    hydronic system designer, 151
    North American Technical Excellence (NATE) certifications and training, 291
    plumbers, pipefitters, and steamfitters, 389
    TAB technicians, 209
    troubleshooting, 169
    See also careers; education
wrap-around safety glasses, 22
WWF (welded wire fabric), 248

Z

zone isolation valves, 339
zone valves
    configurations, 183
    definition, 183
    electric, 261
    heat motor actuators, 183
    purging, 369–370
    three-way, preventing pump deadheading, 183
zoning flexibility, 11–12